W9-ADK-353

Core Curriculum

Introductory Craft Skills

Trainee Guide
Fifth Edition

PEARSON

Boston   Columbus   Indianapolis   New York   San Francisco   Amsterdam
Cape Town Dubai London Madrid Milan Munich Paris Montreal Toronto Delhi
Mexico City Sao Paulo Sydney Hong Kong Seoul Singapore Taipei Tokyo

NCCER

President: Don Whyte
Director of Product Development: Daniele Dixon
Core Curriculum Project Manager: Patty Bird
Senior Manager of Production: Tim Davis
Quality Assurance Coordinator: Debie Hicks

Desktop Publishing Coordinator: James McKay
Permissions Specialist: Adrienne Payne
Production Specialist: Adrienne Payne
Editor: Tanner Yea

Writing and development services provided by Topaz Publications, Liverpool, NY

Lead Writer/Project Manager: Troy Staton
Desktop Publisher: Joanne Hart
Art Director: Alison Richmond

Permissions Editor: Tonia Burke, Andrea Labarge
Writers: Troy Staton, Carol Herbert, Thomas Burke,
 Darrell Wilkerson, Cliff Bennett

Pearson Education, Inc.

Director, Global Employability Solutions: Jonell Sanchez
Head of Associations: Andrew Taylor
Editorial Assistant: Kelsey Kissner
Program Manager: Alexandrina B. Wolf
Project Manager: Janet Portisch
Operations Supervisor: Deidra M. Skahill
Art Director: Diane Ernsberger
Digital Product Strategy Manager: Maria Anaya
Digital Studio Project Managers: Heather Darby,
 Tanika Henderson
Directors of Marketing: David Gesell, Margaret Waples
Field Marketer: Brian Hoehl

Composition: NCCER
Printer/Binder: RR Donnelley/Owensville
Cover Printer: Phoenix Color/Hagerstown
Text Fonts: Palatino and Univers

Credits and acknowledgments for content borrowed from other sources and reproduced, with permission, in this textbook appear at the end of each module.

2   16

Perfect bound: ISBN 10:   0-13-413098-7
ISBN 13:   978-0-13-413098-9
Case bound: ISBN 10:   0-13-413143-6
ISBN 13:   978-0-13-413143-6

PEARSON

Preface

To the Trainee

Welcome to the world of construction! Construction is one of the largest industries, offering excellent opportunities for high earnings, career advancement, and business ownership.

Work in construction offers a great variety of career opportunities. People with many different talents and educational backgrounds—skilled craftspersons, managers, supervisors, and superintendents—find job opportunities in construction and related fields. As you will learn throughout your training, many other industries depend upon the work you will do in construction. From houses and office buildings to factories, roads, and bridges—*everything* begins with construction.

New with *Core Curriculum: Introductory Craft Skills*

NCCER and Pearson are pleased to present the fifth edition of *Core Curriculum: Introductory Craft Skills.* This full-color textbook includes nine modules for building foundation skills in construction. NCCER has enhanced the *Core Curriculum* to appeal to an international market. There are new features to show how construction impacts countries around the world.

We are also excited to provide a revised "Basic Safety (Construction Site Safety Orientation)" module that now aligns to OSHA's 10-hour program. This means that instructors who are OSHA-500 certified are able to issue 10-hour OSHA cards to their students who successfully complete the module. Combined with an NCCER credential, the OSHA 10-hour card will show employers a credible and valuable training record. While aligning to the OSHA-based standards of the United States, this module enhances safety practices and discusses how these can change state-to-state and country-to-country. Also, the successful completion of this module will award a Construction Site Safety Orientation credential.

We keep math "real" for students in this edition of "Introduction to Construction Math" by emphasizing application over theory-related exercises. By keeping math "real," the language of math is much easier to understand. As a companion piece to this module, a workbook is also available for instructors to use to supplement classroom activities.

The "Introduction to Basic Rigging" module includes basic safety requirements for working around rigging and cranes, and rigging equipment identification. This module has been reduced in size and hours. It is an elective, and as such is not required for successful completion of the *Core Curriculum.*

"Basic Communication Skills" now includes content on nonverbal communication, and explains the importance of electronic messaging in the construction industry. "Introduction to Material Handling" now presents the basics of knot tying, as knots are critical with any material handling.

We invite you to visit the NCCER website at **www.nccer.org** for information on the latest product releases and training, as well as online versions of the *Cornerstone* magazine and Pearson's NCCER product catalog.

Your feedback is welcome. You may email your comments to **curriculum@nccer.org** or send general comments and inquiries to **info@nccer.org**.

NCCER Standardized Curricula

NCCER is a not-for-profit 501(c)(3) education foundation established in 1995 by the world's largest and most progressive construction companies and national construction associations. It was founded to address the severe workforce shortage facing the industry and to develop a standardized training process and curricula. Today, NCCER is supported by hundreds of leading construction and maintenance companies, manufacturers, and national associations. The NCCER Standardized Curricula was developed by NCCER in partnership with Pearson Education, Inc., the world's largest educational publisher.

Some features of the NCCER Standardized Curricula are as follows:

- An industry-proven record of success
- Curricula developed by the industry for the industry
- National standardization providing portability of learned job skills and educational credits
- Compliance with Office of Apprenticeship requirements for related classroom training (*CFR 29:29*)
- Well-illustrated, up-to-date, and practical information

NCCER also maintains a Registry that provides transcripts, certificates, and wallet cards to individuals who have successfully completed a level of training within a craft in the NCCER Standardized Curricula. *Training programs must be delivered by an NCCER Accredited Training Sponsor in order to receive these credentials.*

Special Features

In an effort to provide a comprehensive user-friendly training resource, we have incorporated many different features for your use. Whether you are a visual or hands-on learner, this book will provide you with the proper tools to get started in the construction industry.

Introduction

This page is found at the beginning of each module and lists the Objectives, Performance Tasks, and Trade Terms for that module. The Objectives list the skills and knowledge you will need in order to complete the module successfully. The Performance Tasks give you an opportunity to apply your knowledge to real-world tasks. The list of Trade Terms identifies important terms you will need to know by the end of the module.

Notes, Cautions, and Warnings

Safety features are set off from the main text in highlighted boxes and organized into three categories based on the potential danger of the issue being addressed. Notes simply provide additional information on the topic area. Cautions alert you of a danger that does not present potential injury but may cause damage to equipment. Warnings stress a potentially dangerous situation that may cause injury to you or a co-worker.

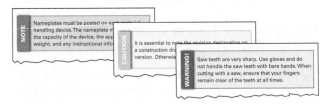

Color Illustrations and Photographs

Color illustrations and photographs are used throughout each module to provide vivid detail. These figures highlight important concepts from the text and provide clarity for complex instructions. Each figure is denoted in the text in *italic type* for easy reference.

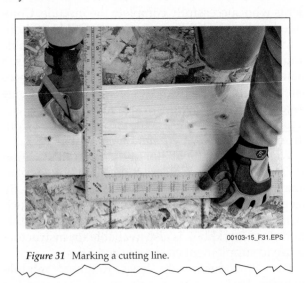

00103-15_F31.EPS

Figure 31 Marking a cutting line.

Special Features

Features present technical tips and professional practices from the construction industry. These features often include real-life scenarios similar to those you might encounter on the job site.

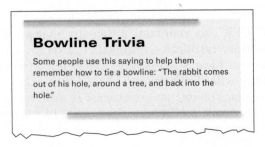

Bowline Trivia

Some people use this saying to help them remember how to tie a bowline: "The rabbit comes out of his hole, around a tree, and back into the hole."

Around the World

The Around the World features introduce trainees to a global construction perspective, emphasizing similarities and differences in standards, codes, and practices from country to country.

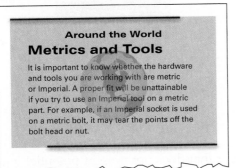

Going Green

Going Green looks at ways to preserve the environment, save energy, and make good choices regarding the health of the planet. Through the introduction of new construction practices and products, you will see how the "greening of the world" has already taken root.

Did You Know?

The Did You Know? features introduce historical tidbits or modern information about the construction industry. Interesting and sometimes surprising facts about construction are also presented.

Trade Terms

Each module presents a list of Trade Terms that are discussed within the text, defined in the Glossary at the end of the module, and reinforced with a Trade Terms Quiz. These terms are denoted in the text with **bold blue type** upon their first occurrence. To make searches for key information easier, a comprehensive Glossary of Trade Terms from all modules is found at the back of this book.

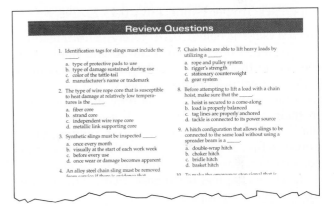

Review Questions

Review Questions are provided to reinforce the knowledge you have gained. This makes them a useful tool for measuring what you have learned.

Section Review

The Section Review features helpful additional resources and review questions related to the objectives in each section of the module.

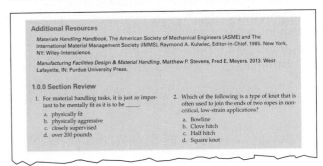

Enhance your training with these great supplemental *Core* companions. The following resources can be used alone or in combination with the *Core Curriculum*. Visit our online catalog at **www.nccer.org** or contact your Pearson Sales Representative to purchase any of these items to supplement your learning.

Applied Construction Math

Paperback Trainee Guide: ISBN 0-13-227298-9

Applied Construction Math: A Novel Approach features a story that students can relate to and math skills they never thought they could grasp. Its innovative style motivates students to follow the lessons by associating math with events they may encounter in their real lives. Students will see that learning math can be something as exciting as building a new house as they follow along with Mr. Whyte and his construction class as they build the perfect house.

Thirteen chapters teach basic math skills including the following topics:

- Division
- Decimals/Percentages
- Reading Measurements
- Calculating Area
- Powers of Ten
- Linear Measure, Angles, Volumes,
- Pressure and Slopes
- Solving for Unknowns
- Square Inches, Feet, and Yards
- Volume

Tools for Success: Critical Skills for the Construction Industry 3/E

Paperback Trainee Guide: ISBN 0-13-610649-8

The *Tools for Success* workbook includes classroom activities to help students navigate their way through intangible workplace issues such as conflict resolution, diversity, problem-solving, professionalism, and proper communications techniques.

Your Role in the Green Environment 3/E

Paperback Trainee Guide: ISBN 0-13-294863-X

Geared to entry-level craft workers or to anyone wishing to learn more about green building, this module provides fundamental instruction in the green environment, green construction practices, and green building rating systems.

NCCER is also a United States Green Building Council Education Partner, and, as such, is committed to enhancing the ongoing development of building industry professionals.

NCCERconnect
One Industry. One Training Program. One Online Solution.

NCCERconnect: An Interactive Online Course

Ideal for blended or distance education, NCCERconnect is a unique web-based supplement in the form of an electronic book that provides a range of visual, auditory, and interactive elements to enhance your training. It can be used in a variety of settings such as self-study, blended/distance education, or in the traditional classroom environment! It's the perfect way to review content from a class you may have missed or to practice at your own pace.

Features:

- **Online Lectures** – Each ebook module features a written summary of key content accompanied by an optional Audio Summary so if you need a refresher, this tool is always available.
- **Video Presentations** – Throughout, you'll find dynamic video presentations that demonstrate difficult skills and concepts! Of special note are the safe/unsafe scenarios shot on a live construction site testing your knowledge of the four 'high hazards' presented in the Basic Safety module.
- **Personalization Tools** – With the "highlighter" and "notes" options you can easily personalize your own NCCERconnect ebook to keep track of important information or create your own study guide.
- **Review Quizzes** – Short multiple-choice concept check quizzes at the end of each module section act as the ideal study tool and provide immediate feedback. Additionally, you'll find fill-in-the-blank trade terms quizzes, applied math questions, and comprehension questions at the end of each module.
- **Active Figures** – Interactive exercises bring key concepts to life, including animation in the Introduction to Construction Drawings module that will help you make the mental transition from a flat, 2-dimensional plan to a 3-dimensional finished structure.

Visit **www.nccerconnect.com** to view a demo.

NCCERconnect is available with:
Core Curriculum
Carpentry Levels 1-4
Construction Technology
Electrical Levels 1-4
Electronic Systems Technician Levels 1-4
HEO Levels 1-2
HVAC Levels 1-4
Plumbing Levels 1-4
Welding Levels 1-4
Your Role in the Green Environment

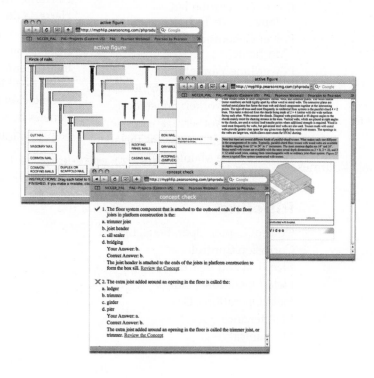

NCCER Standardized Curricula

NCCER's training programs comprise more than 80 construction, maintenance, pipeline, and utility areas and include skills assessments, safety training, and management education.

Boilermaking
Cabinetmaking
Carpentry
Concrete Finishing
Construction Craft Laborer
Construction Technology
Core Curriculum:
  Introductory Craft Skills
Drywall
Electrical
Electronic Systems Technician
Heating, Ventilating, and Air
  Conditioning
Heavy Equipment Operations
Highway/Heavy Construction
Hydroblasting
Industrial Coating and Lining
  Application Specialist
Industrial Maintenance Electrical
  and Instrumentation Technician
Industrial Maintenance
  Mechanic
Instrumentation
Insulating
Ironworking
Masonry
Millwright
Mobile Crane Operations
Painting
Painting, Industrial
Pipefitting
Pipelayer
Plumbing
Reinforcing Ironwork
Rigging
Scaffolding
Sheet Metal
Signal Person
Site Layout
Sprinkler Fitting
Tower Crane Operator
Welding

Maritime

Maritime Industry Fundamentals
Maritime Pipefitting
Maritime Structural Fitter

Green/Sustainable Construction

Building Auditor
Fundamentals of Weatherization
Introduction to Weatherization
Sustainable Construction
  Supervisor
Weatherization Crew Chief
Weatherization Technician
Your Role in the Green
  Environment

Energy

Alternative Energy
Introduction to the Power Industry
Introduction to Solar Photovoltaics
Introduction to Wind Energy
Power Industry Fundamentals
Power Generation Maintenance
  Electrician
Power Generation I&C
  Maintenance Technician
Power Generation Maintenance
  Mechanic
Power Line Worker
Power Line Worker: Distribution
Power Line Worker: Substation
Power Line Worker: Transmission
Solar Photovoltaic Systems Installer
Wind Turbine Maintenance
  Technician

Pipeline

Control Center Operations, Liquid
Corrosion Control
Electrical and Instrumentation
Field Operations, Liquid
Field Operations, Gas
Maintenance
Mechanical

Safety

Field Safety
Safety Orientation
Safety Technology

Supplemental Titles

Applied Construction Math
Tools for Success

Management

Fundamentals of Crew Leadership
Project Management
Project Supervision

Spanish Titles

Acabado de concreto: nivel uno
Aislamiento: nivel uno
Albañilería: nivel uno
Andamios
Carpintería:
  Formas para carpintería, nivel tres
Currículo básico: habilidades
  introductorias del oficio
Electricidad: nivel uno
Herrería: nivel uno
Herrería de refuerzo: nivel uno
Instalación de rociadores: nivel uno
Instalación de tuberías: nivel uno
Instrumentación: nivel uno, nivel
  dos, nivel tres, nivel cuatro
Orientación de seguridad
Paneles de yeso: nivel uno
Seguridad de campo

Acknowledgments

This curriculum was revised as a result of the farsightedness and leadership of the following sponsors:

ABC Merit Shop Training Program Inc. dba
  CTC of the Coastal Bend
ABC National
ABC South Texas Chapter
Central Cabarrus High School
Cianbro Corporation
Industrial Management & Training Institute
KBR Industrial Services

Lake Mechanical Contractors, Inc.
Maintenance & Construction Technology
  Alliance
River Valley Technical Center
The Shaw Group, Inc.
The Southern Company
Starcon
TIC - The Industrial Company

This curriculum would not exist were it not for the dedication and unselfish energy of those volunteers who served on the Authoring Team. A sincere thanks is extended to the following:

Tony Ayotte
Mark Bonda
Bob Fitzgerald
Todd Hartsell
Harold (Hal) Heintz

Erin M. Hunter
Sidney Mitchell
Jan Prakke
Brett Richardson
Fernando Sanchez

Michael Sandroussi
John Stronkowski
Chris Williams
Ralph Yelder Jr.

A final note: This book is the result of a collaborative effort involving the production, editorial, and development staff at Pearson Education, Inc. and NCCER. Thanks to all of the dedicated people involved in the many stages of this project.

NCCER Partners

American Fire Sprinkler Association
Associated Builders and Contractors, Inc.
Associated General Contractors of America
Association for Career and Technical Education
Association for Skilled and Technical Sciences
Construction Industry Institute
Construction Users Roundtable
Construction Workforce Development Center
Design Build Institute of America
GSSC – Gulf States Shipbuilders Consortium
ISN
Manufacturing Institute
Mason Contractors Association of America
Merit Contractors Association of Canada
NACE International
National Association of Minority Contractors
National Association of Women in Construction
National Insulation Association
National Technical Honor Society
National Utility Contractors Association
NAWIC Education Foundation
North American Crane Bureau
North American Technician Excellence
Pearson

Pearson Qualifications International
Prov
SkillsUSA®
Steel Erectors Association of America
U.S. Army Corps of Engineers
University of Florida, M. E. Rinker School of
  Building Construction
Women Construction Owners & Executives,
  USA

Contents

Module One

Basic Safety (Construction Site Safety Orientation)

This module complies with OSHA-10 training requirements. It explains the safety obligations of workers, supervisors, and managers to ensure a safe workplace. Discusses the causes and results of accidents and the impact of accident costs. Reviews the role of company policies and OSHA regulations. Introduces common job-site hazards and identifies proper protections. Defines safe work procedures, proper use of personal protective equipment, and how to safely work with hazardous chemicals. Identifies other potential construction hazards, including hazardous material exposures, welding and cutting hazards, and confined spaces. (Module ID 00101-15; 12.5 Hours)

Module Two

Introduction to Construction Math

Reviews basic mathematical functions such as adding, subtracting, dividing, and multiplying. Defines whole numbers, fractions, and decimals, and explains their applications to the construction trades. Explains how to use and read various length measurement tools, including standard and metric rulers and tape measures, and the architect's and engineer's scales. Explains decimal-fraction conversions and the metric system, using practical examples. Also reviews basic geometry as applied to common shapes and forms. (Module ID 00102-15; 10 Hours)

Module Three

Introduction to Hand Tools

Introduces trainees to hand tools that are widely used in the construction industry, such as hammers, saws, levels, pullers, and clamps. Explains the specific applications of each tool and shows how to use them properly. Also discusses important safety and maintenance issues related to hand tools. (Module ID 00103-15; 10 Hours)

Module Four

Introduction to Power Tools

Provides detailed descriptions of commonly used power tools, such as drills, saws, grinders, and sanders. Reviews applications of these tools, proper use, safety, and maintenance. Many illustrations show power tools used in on-the-job settings. (Module ID 00104-15; 10 Hours)

Module Five

Introduction to Construction Drawings

Familiarizes trainees with basic terms for construction drawings, components, and symbols. Explains the different types of drawings (civil, architectural, structural, mechanical, plumbing/piping, electrical, and fire protection) and instructs trainees on how to interpret and use drawing dimensions. A set of four oversized drawings is included. (Module ID 00105-15; 10 Hours)

Module Six

Introduction to Basic Rigging (Elective)

Provides basic information related to rigging and rigging hardware, such as slings, rigging hitches, and hoists. Emphasizes safe working habits in the vicinity of rigging operations. (Module ID 00106-15; 7.5 Hours)

Module Seven

Basic Communication Skills

Provides trainees with techniques for communicating effectively with co-workers and supervisors. Includes practical examples that emphasize the importance of verbal and written information and instructions on the job. Also discusses effective telephone and email communication skills. (Module ID 00107-15; 7.5 Hours)

Module Eight

Basic Employability Skills

Introduces trainees to critical thinking and problem-solving skills. Reviews effective relationship skills, effective self-presentation, and key workplace issues such as sexual harassment, stress, and substance abuse. Also presents information on computer systems and their industry applications. (Module ID 00108-15; 7.5 Hours)

Module Nine

Introduction to Material Handling

Recognizes hazards associated with material handling and explains proper techniques and procedures. Introduces material handling equipment, and identifies appropriate equipment for common job-site tasks. (Module ID 00109-15; 5 Hours)

Glossary

Index

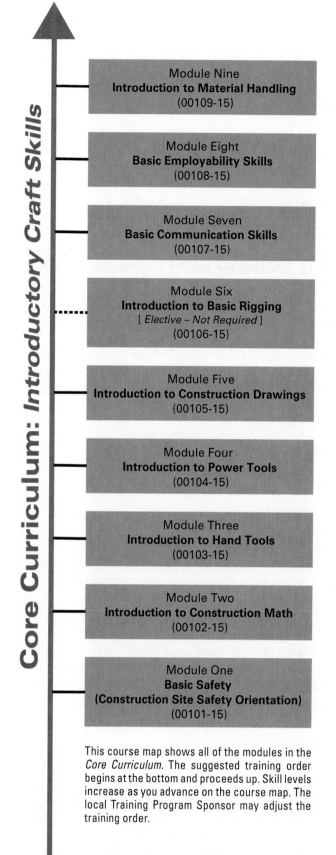

Core Curriculum: Introductory Craft Skills

Module Nine
Introduction to Material Handling
(00109-15)

Module Eight
Basic Employability Skills
(00108-15)

Module Seven
Basic Communication Skills
(00107-15)

Module Six
Introduction to Basic Rigging
[*Elective – Not Required*]
(00106-15)

Module Five
Introduction to Construction Drawings
(00105-15)

Module Four
Introduction to Power Tools
(00104-15)

Module Three
Introduction to Hand Tools
(00103-15)

Module Two
Introduction to Construction Math
(00102-15)

Module One
Basic Safety
(Construction Site Safety Orientation)
(00101-15)

This course map shows all of the modules in the *Core Curriculum*. The suggested training order begins at the bottom and proceeds up. Skill levels increase as you advance on the course map. The local Training Program Sponsor may adjust the training order.

00101-15

Basic Safety (Construction Site Safety Orientation)

OVERVIEW

Work at construction and industrial job sites can be hazardous. Most job-site incidents are caused by at-risk behavior, poor planning, lack of training, or failure to recognize the hazards. To help prevent incidents, every company must have a proactive safety program. Safety must be incorporated into all phases of the job and involve employees at every level, including management.

Module One

Trainees with successful module completions may be eligible for credentialing through the NCCER Registry. To learn more, go to **www.nccer.org** or contact us at **1.888.622.3720**. Our website has information on the latest product releases and training, as well as online versions of our *Cornerstone* magazine and Pearson's product catalog.

Your feedback is welcome. You may email your comments to **curriculum@nccer.org,** send general comments and inquiries to **info@nccer.org** , or fill in the User Update form at the back of this module.

This information is general in nature and intended for training purposes only. Actual performance of activities described in this manual requires compliance with all applicable operating, service, maintenance, and safety procedures under the direction of qualified personnel. References in this manual to patented or proprietary devices do not constitute a recommendation of their use.

Objectives

When you have completed this module, you will be able to do the following:

1. Describe the importance of safety, the causes of workplace incidents, and the process of hazard recognition and control.
   a. Define incidents and the significant costs associated with them.
   b. Identify the common causes of incidents and their related consequences.
   c. Describe the processes related to hazard recognition and control, including the Hazard Communication (HAZCOM) Standard and the provisions of a safety data sheet (SDS).
2. Describe the safe work requirements for elevated work, including fall protection guidelines.
   a. Identify and describe various fall hazards.
   b. Identify and describe equipment and methods used in fall prevention and fall arrest.
   c. Identify and describe the safe use of ladders and stairs.
   d. Identify and describe the safe use of scaffolds.
3. Identify and explain how to avoid struck-by hazards.
   a. Identify and explain how to avoid struck-by and caught-in-between hazards.
   b. Identify and explain how to avoid caught-in and caught-between hazards.
4. Identify common energy-related hazards and explain how to avoid them.
   a. Describe basic job-site electrical safety guidelines.
   b. Explain the importance of lockout/tagout and describe basic procedures.
5. Identify and describe the proper use of personal protective equipment (PPE).
   a. Identify and describe the basic use of PPE used to protect workers from bodily injury.
   b. Identify potential respiratory hazards and the basic respirators used to protect workers against those hazards.
6. Identify and describe other specific job-site safety hazards.
   a. Identify various exposure hazards commonly found on job sites.
   b. Identify hazards associated with environmental extremes.
   c. Identify hazards associated with hot work.
   d. Identify fire hazards and describe basic firefighting procedures.
   e. Identify confined spaces and describe the related safety considerations.

Performance Tasks

Under the supervision of your instructor, you should be able to do the following:

1. Properly set up and climb/descend an extension ladder, demonstrating proper three-point contact.
2. Inspect the following PPE items and determine if they are safe to use:

   - Eye protection
   - Hearing protection
   - Hard hat
   - Gloves
   - Fall arrest harnesses
   - Lanyards
   - Connecting devices
   - Approved footwear

3. Properly don, fit, and remove the following PPE items:

   - Eye protection
   - Hearing protection
   - Hard hat
   - Gloves
   - Fall arrest harness

4. Inspect a typical power cord and GFCI to ensure their serviceability.

Trade Terms

Accident
Arc welding
Brazing
Combustible
Competent person
Confined spaces
Cross-bracing
Excavation
Flammable
Flash burn
Flash point
Ground
Ground fault
Ground fault circuit interrupter
  (GFCI)
Guarded
Hand line

Hazard communication standard
  (HAZCOM)
Hydraulic
Incident
Lanyards
Lockout/tagout (LOTO)
Management system
Maximum intended load
Midrail
Occupational Safety and Health
  Administration (OSHA)
Permit-required confined space
Personal protective equipment
  (PPE)
Planked
Pneumatic
Proximity work

Qualified person
Respirator
Safety culture
Safety data sheet (SDS)
Scaffold
Shielding
Shoring
Signaler
Six-foot rule
Spoil
Toeboard
Top rail
Trench
Welding curtain
Wind sock

Industry Recognized Credentials

If you are training through an NCCER-accredited sponsor, you may be eligible for credentials from NCCER's Registry. The ID number for this module is 00101-15. Note that this module may have been used in other NCCER curricula and may apply to other level completions. Contact NCCER's Registry at 888.622.3720 or go to **www.nccer.org** for more information.

Note

The successful completion of this module will award a Construction Site Safety Orientation credential.

Contents

Topics to be presented in this module include:

Contents (continued)

Figures and Tables

Figures and Tables (continued)

1.0.0 SAFETY AND HAZARD RECOGNITION

Objectives

Describe the importance of safety, the causes of workplace incidents and accidents, and the process of hazard recognition and control.

a. Define incidents and accidents and the significant costs associated with them.
b. Identify the common causes of incidents and accidents and their related consequences.
c. Describe the processes related to hazard recognition and control, including the Hazard Communication (HAZCOM) Standard and the provisions of a Safety Data Sheet (SDS).

Trade Terms

Accident: According to the US Occupational Safety and Health Administration (OSHA), an unplanned event that results in personal injury and/or property damage.

Combustible: Capable of easily igniting and rapidly burning; used to describe a fuel with a flash point at, or above, 100°F (38°C).

Competent person: A person who is capable of identifying existing and predictable hazards in the surroundings or working conditions that are unsanitary, hazardous, or dangerous to employees, and who has authorization to take prompt corrective measures to eliminate them.

Confined space: A work area large enough for a person to work, but arranged in such a way that an employee must physically enter the space to perform work. A confined space has a limited or restricted means of entry and exit. It is not designed for continuous work. Tanks, vessels, silos, pits, vaults, and hoppers are examples of confined spaces. Also see *permit-required confined space*.

Flammable: Capable of easily igniting and rapidly burning; used to describe a fuel with a flash point below 100°F (38°C).

Ground fault: Incidental grounding of a conducting electrical wire.

Hazard Communication (HAZCOM) Standard: The Occupational Safety and Health Administration standard that requires contractors to educate employees about hazardous chemicals on the job site and how to work with them safely.

Hydraulic: Powered by fluid under pressure.

Incident: Per the US Occupational Safety and Health Administration (OSHA), an unplanned event that does not result in personal injury but may result in property damage or is worthy of recording.

Management system: The organization of a company's management, including reporting procedures, supervisory responsibility, and administration.

Occupational Safety and Health Administration (OSHA): An agency of the US Department of Labor. Also refers to the Occupational Safety and Health Act of 1970, a law that applies to more than 111 million workers and 7 million job sites in the United States.

Personal protective equipment (PPE): Equipment or clothing designed to prevent or reduce injuries.

Pneumatic: Powered by air pressure, such as a pneumatic tool.

Respirator: A device that provides clean, filtered air for breathing, no matter what is in the surrounding air.

Safety culture: The culture created when the whole company sees the value of a safe work environment.

Safety data sheet (SDS): A document that must accompany any hazardous substance. The SDS identifies the substance and gives the exposure limits, the physical and chemical characteristics, the kind of hazard it presents, precautions for safe handling and use, and specific control measures.

Trench: A narrow excavation made below the surface of the ground that is generally deeper than it is wide, with a maximum width of 15 feet (4.6m). Also see *excavation*.

When you take a job, you have an obligation to your employer, co-workers, family, and yourself to work safely. You also have an obligation to make sure anyone you work with is working safely. Your employer is likewise obliged to maintain a safe workplace for all employees. The ultimate responsibility for on-the-job safety, however, rests with you; safety is part of everyone's job. In this module, you will learn to ensure your safety, and that of the people you work with, by obeying the following rules:

- Follow safe work practices and procedures, both regulatory and corporate.
- Inspect safety equipment before use.
- Use safety equipment properly.

To take full advantage of the wide variety of training, job, and career opportunities the construction industry offers, you must first understand the importance of safety. Successful completion of this module will be your first step toward achieving this goal. Later modules offer more detailed explanations of safety procedures, along with opportunities to practice them.

On a typical job site, there are often many workers from many trades in one place. These workers are all performing different tasks and operations. As a result, the job site is constantly changing and hazards are continually emerging. These hazards can jeopardize your safety. Your employer should make every effort to plan safety into each job and to provide a safe and healthful job site. Ultimately, however, your safety is in your own hands.

Safety training is provided to make you aware that hazards exist all around you every day. The time you spend learning and practicing safety procedures can save your life and the lives of others.

Safety is a learned behavior and attitude. It is a way of working that must be incorporated into the company as a culture. A safety culture is created when all the workers at a job site or in an organization see the value of a safe work environment and support it through their actions. Creating and maintaining a safety culture is an ongoing process that includes a sound safety structure and attitude, and relates to organizations as well as individuals. Everyone in the company, from management to laborers, must be responsible for safety every day they come to work.

There are many benefits to having a safety culture. Companies with strong safety cultures usually have the following characteristics:

- Fewer at-risk behaviors
- Lower incident and accident rates
- Less turnover
- Lower absenteeism
- Higher productivity

A strong safety culture can also improve a company's safety record, which leads to winning more bids and keeping workers employed. Contractors with poor safety records are sometimes excluded from bidding, so good safety performance is essential. Factors that contribute to a strong safety culture include the following:

- Embracing safety as a core value
- Strong leadership
- Establishing and enforcing high standards of performance
- The commitment and involvement of all employees
- Effective communication and commonly understood and agreed-upon goals
- Using the workplace as a learning environment
- Encouraging workers to have a questioning attitude and empowering them to stop work when faced with potential hazards.
- Good organizational learning and responsiveness to change
- Providing timely response to safety issues and concerns
- Continually monitoring performance
- Positive reinforcement when proper safety practices are demonstrated by employees

Did You Know?

Safety First

Safety training is required for all activities. Never operate tools, machinery, or equipment without prior training. Always refer to the manufacturer's instructions.

Around the World

GOST

While OSHA serves to protect workers by setting safety standards in the United States, other systems are used internationally. One such set of technical standards used on a regional basis is known as GOST. GOST standards are more far-reaching than OSHA standards, as they cover a much broader range of topics than worker safety alone. The first set of GOST standards were published in 1968 as state standards for the former Soviet Union. After the Soviet Union was dismantled, GOST became a regional standard used by many previous members of the Soviet Union. Although countries may also have some standards of their own, countries such as Belarus, Moldova, Armenia, and Ukraine continue to use GOST standards as well. The standards are no longer administered by Russia, however. Today, the standards are administered by the Euro-Asian Council for Standardization, Metrology and Certification (EASC).

1.1.0 Incidents and Accidents

Incidents and accidents can occur at any job site. Both at-risk behavior and poor working conditions can cause these undesirable events. You can help prevent such events by using safe work habits, understanding what causes them, and learning how to prevent them.

The terms *incident* and *accident* are often used interchangeably. However, according to the US Occupational Safety and Health Administration (OSHA), an incident is an unplanned event that may or may not result in property damage. However, an incident is worthy of being documented so that steps can be taken to prevent it from recurring. When an incident occurs, no personal injury has occurred.

An accident is defined as an unplanned event that results in personal injury and/or property damage. Therefore, an event that results in property damage alone could be considered an incident or an accident. If personal injury or a fatality has occurred, the event is definitely an accident.

There are varying opinions on the use of these two terms, however. The US National Safety Council defines an incident as an unplanned, undesired event that adversely affects the completion of a task. In this definition, there is no mention of injury or property damage. Other safety organizations across the globe are likely to have their own definitions of these terms as well, or may use completely different terms.

The most important thing to understand is that both incidents and accidents are undesirable events that have a negative effect on both projects and workers. Do not be surprised when you hear the terms used interchangeably by other workers. The definitions provided by OSHA are used here to provide context for these terms as they are used throughout this module.

The lessons you will learn in this module will help you work safely. You will be able to spot and avoid hazardous conditions on the job site. By following safety procedures, you will help keep your workplace free from incidents and accidents, and protect yourself and others from injury or even death.

1.1.1 Incident and Accident Categories

Incidents and accidents are often categorized by their severity and impact, as follows:

- *Near-miss* – An unplanned event in which no one was injured and no damage to property occurred, but during which either could have happened. Near-miss incidents are warnings that should always be reported rather than overlooked or taken lightly.
- *Property damage* – An unplanned event that results in damage to tools, materials, or equipment, but no personal injuries.
- *Minor injuries* – Personnel may have received minor cuts, bruises, or strains, but the injured workers returned to full duty on their next regularly scheduled work shift.
- *Serious or disabling injuries* – Personnel received injuries that resulted in temporary or permanent disability. Included in this category would be lost-time incidents, restricted-duty or restricted-motion cases, and those that resulted in partial or total disability.
- *Fatalities* – Deaths resulting from unplanned events.

Studies have shown that for every serious or disabling injury, there were 10 injuries of a less serious nature and 30 property damage incidents. A further study showed that 600 near-miss incidents occurred for every serious or disabling injury.

There are four leading causes of death in construction work. These are often referred to as the "big four", the "fatal four", or the "focus four". They include falls; struck-by hazards; caught-in or caught-between hazards; and electrical hazards (*Figure 1*). Deaths from falls far exceed all other causes.

Here are explanations of the four leading hazard groups:

- Falls from elevation are incidents involving failure of, failure to provide, or failure to use appropriate fall protection.
- Struck-by accidents involve unsafe operation of equipment, machinery, and vehicles, as well as improper handling of materials, such as through unsafe rigging operations.
- Caught-in or caught-between accidents involve unsafe operation of equipment, machinery, and vehicles, as well as improper safety procedures at **trench** sites and in other **confined spaces**.
- Electrical shock accidents involve contact with overhead wires; use of defective tools; failure to disconnect power source before repairs; or improper **ground fault** protection.

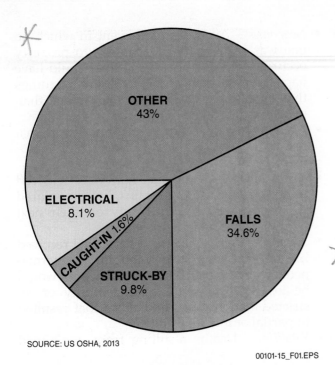

OTHER
43%

ELECTRICAL
8.1%

CAUGHT-IN 1.6%

STRUCK-BY
9.8%

FALLS
34.6%

SOURCE: US OSHA, 2013

00101-15_F01.EPS

Figure 1 The four high-hazard areas.

1.1.2 Costs

Incidents and accidents cost billions of dollars each year and cause much needless suffering. The National Safety Council estimates that the organized safety movement has saved more than 4.2 million lives since it began in 1913. This section examines why incidents and accidents happen and how you can help prevent them.

Accidents that result in injury or death can have a lasting effect not only on the victims, but on their families, co-workers, and employers. An injured worker who is disabled by an accident faces potentially huge medical bills. On top of those expenses, the worker's family faces loss of the income they rely on. It's convenient to think that insurance will take care of the costs, but it may not. Workers who are injured because they violated established safety rules may have their insurance claims denied and may be dismissed from the company because of the violation. A worker who is injured in a fall because he or she was not using fall protection could be refused

compensation. How these issues are handled varies dramatically among companies and countries.

Employers and co-workers can be affected because many contract awards are based, in part, on a company's safety record. Therefore, incidents and accidents can also result in the loss of future jobs, which affects the company's financial position. This can mean layoffs, hiring freezes, or inability to purchase new equipment or tools. In this way, these events affect not only injured employees and their families, but everyone on the job site.

1.2.0 Incident and Accident Causes

When an incident or accident occurs on the job site, it can often be attributed to one of the following causes:

- Failure to communicate
- At-risk work habits
- Alcohol or drug abuse
- Lack of skill
- Intentional acts
- Unsafe acts
- Rationalizing risks
- Unsafe conditions
- Housekeeping
- Management system failure

Each of these causes is discussed further in the sections that follow.

1.2.1 Failure to Communicate

Many incidents happen because of a lack of communication. For example, you may learn how to do things one way on one job, but what happens when you go to a new job site? You need to communicate with the people at the new job site to find out whether they do things the way you have learned to do them. If you do not communicate clearly, incidents can happen. Remember that different people, companies, and job sites do things in different ways.

Making assumptions about what other workers know and what they will do can cause incidents.

The Fatal Four

Out of 3,945 worker fatalities in US private industry during the 2012 calendar year, 775, or 19.6 percent, were in construction. The leading causes of worker deaths on construction sites were falls, followed by struck-by-object, electrocution, and caught-in/between. These fatal four were responsible for nearly three out of five (56 percent) construction worker deaths in 2012 reports, as reported by the US Bureau of Labor Statistics. Eliminating the fatal four would save the lives of 435 workers in America every year.

NCCER – *Core Curriculum* 00101-15

Half-Measures

Most workers who die from falls are wearing harnesses but failed to tie off properly. Always follow the manufacturer's instructions when wearing a harness. Know and follow your company's safety procedures when working on roofs, ladders, and other elevated locations and make sure you have an adequate anchor point at all times.

Don't assume, for example, that all workers understand what you are saying; some workers have limited language skills, especially outside of their native language. Also, don't use terms or jargon that other people may not understand.

> **CAUTION**
>
> Never assume anything. It never hurts to ask questions, but disaster can result if you don't ask. For example, do not assume that an electrical power source is turned off. First ask whether the power is turned off, then check it yourself to be completely safe.

All work sites have specific markings and signs to identify hazards and provide emergency information (*Figure 2*). Learn to recognize these types of signs:

- Informational
- Safety
- Caution
- Danger
- Temporary warnings

Informational markings or signs provide general information. These signs are blue. The following are considered informational signs:

- No Admittance
- No Trespassing
- For Employees Only

Toolbox Talks

Toolbox talks are one way to effectively keep all workers aware and informed of safety issues and guidelines. Toolbox talks are 5- to 10-minute meetings that review specific health and safety topics. These are very common at construction sites of all types.

INFORMATION SIGN

SAFETY SIGN

CAUTION SIGN

DANGER SIGN

00101-15_F02.EPS

Figure 2 Communication tags and signs.

Safety signs give general instructions and suggestions about safety measures. The background on these signs is white; most have a green panel with white letters. These signs tell you where to find such important areas as the following:

- First-aid stations
- Emergency eye-wash stations
- Evacuation routes
- Safety Data Sheet (SDS) stations
- Exits (usually have white letters on a red field)

Caution markings or signs tell you about potential hazards or warn against unsafe acts. When you see a caution sign, protect yourself against a possible hazard. Caution signs are yellow and have a black panel with yellow letters. They may give you the following information:

- Hearing and eye protection are required
- Respirators are required
- Smoking is not allowed

Danger markings or signs tell you that an immediate hazard exists and that you must take certain precautions to avoid an incident. Danger signs are red, black, and white. They may indicate the presence of the following:

- Defective equipment
- Flammable liquids and compressed gases
- Safety barriers and barricades
- Emergency stop button
- High voltage

Safety tags are temporary warnings of immediate and potential hazards. They are not designed to replace signs or to serve as permanent means of protection. Learn to recognize the standard incident and accident prevention signs and tags (*Table 1*).

1.2.2 At-Risk Work Habits

Some examples of at-risk work habits are procrastination, carelessness, and horseplay. Procrastination (putting things off) is a common cause of incidents. For example, delaying the repair, inspection, or cleaning of equipment and tools can cause incidents and even accidents. If you try to push machines and equipment beyond their operating capacities, you risk injuring yourself and your co-workers.

Machines, power tools, and even a pair of pliers can hurt you if you don't use them safely. It is your responsibility to be careful; tools and machines don't know the difference between wood or steel, and flesh and bone.

Work habits and work attitudes are closely related. If you resist taking orders, you may also resist listening to warnings. If you let yourself be easily distracted, you won't be able to concentrate. If you aren't concentrating, you could cause a significant problem.

Your safety is affected not only by how you do your work, but also by how you act on the job site. This is why most companies have strict policies for employee behavior. Horseplay and other inappropriate behavior are forbidden. Workers who engage in such behavior on the job site may be fired.

These strict policies are provided for the worker's protection. There are many hazards on construction sites. Each person's behavior—at work, on a break, or at lunch—must follow the principles of safety.

A person pulling a practical joke on a co-worker could consider it just having fun, but in fact, it could cause the co-worker serious, even fatal, injury. If you horse around on the job, play pranks, or don't concentrate on what you are doing, you are showing a poor work attitude that can lead to a serious incident.

Table 1 Tags and Signs

Basic Stock (background)	Safety Colors (ink)	Message(s)
White	Red panel with white or gray letters	Do Not Operate Do Not Start
White	Black square with a red oval and white letters	Danger Unsafe Do Not Use
Yellow	Black square with yellow letters	Caution
White	Black square with white letters	Out of Order Do Not Use
Yellow	Red/magenta (purple) panel with black letters and a radiation symbol	Radiation Hazard
White	Fluorescent orange square with black letters and a biohazard symbol	Biological Hazard

1.2.3 Alcohol and Drug Abuse

Alcohol and drug abuse costs the construction industry millions of dollars a year in incidents, accidents, lost time, and lost productivity. The true cost of alcohol and drug abuse is much more than just money, of course. Substance abuse can cost lives. Just as drunk driving kills thousands of people on our highways every year, alcohol and drug abuse kills on the construction site.

Using alcohol or drugs creates a risk of injury for everyone on a job site. Many states have laws that prevent workers from collecting insurance benefits if they are injured while under the influence of alcohol or illegal drugs.

Would you trust your life to a crane operator who was under the influence of drugs? Would you bet your life on the responses of a co-worker under the influence of alcohol or drugs? Alcohol and drug abuse have no place in the workplace. A person on a job site who is under the influence of alcohol or drugs is an incident or accident waiting to happen—possibly a fatal accident.

People who work while using alcohol or drugs are at risk of incident or injury; their co-workers are at risk as well. That's why most employers have a formal substance abuse policy that you should be aware of and follow. Avoid any substances that can affect your job performance; the life you save could be your own.

You do not have to be abusing drugs such as marijuana, cocaine, or heroin to create a job hazard. Many prescription and over-the-counter drugs, taken for legitimate reasons, can affect your ability to work safely. Amphetamines, barbiturates, and antihistamines are only a few of the prescription-controlled legal drugs that can affect the ability to work or operate machinery safely. The main thing is to understand and follow your company's substance abuse policy.

> **CAUTION**
>
> If your doctor prescribes any medication that you think might affect your job performance, ask about its effects. Your safety and the safety of your co-workers depend on everyone being alert on the job.

1.2.4 Lack of Skill

Every worker needs to learn and practice new skills under careful supervision. Never perform new tasks alone until you have been checked out by a supervisor. Lack of skill can cause incidents and accidents quickly. For example, suppose you are told to cut some boards with a circular saw, but you aren't skilled with that tool. A basic rule of circular saw operation is never to cut without a properly functioning guard. Because you haven't been trained, you don't know this. You find that the guard on the saw is slowing you down, so you jam the guard open. The result could be a serious accident. Proper training can prevent this from happening.

Never operate a power tool or machine until you have been trained to use it. Some power tools require that a worker be trained and certified in their use. You can greatly reduce the chances of incidents and accidents by learning the safety rules for each task you perform.

1.2.5 Intentional Acts

When someone purposely causes property damage or injury, it is called an intentional act. An angry or dissatisfied employee may purposely create a situation that leads to an incident or accident. If someone you are working with threatens to get even or pay back someone, let your supervisor know at once.

Did You Know?

Stress Effects

Stress creates a chemical change in your body. Although stress may heighten your hearing, vision, energy, and strength, long-term stress can harm your health. Not all stress is job-related; some stress develops from the pressures of dealing with family and friends and daily living. In the end, your ability to handle and manage your stress determines whether stress hurts or helps you. Use common sense when you are dealing with stressful situations. For example, consider the following:

- Keep daily occurrences in perspective. Not everything is worth getting upset, angry, or anxious about.
- When you have a particularly difficult workday scheduled, get plenty of rest the night before.
- Manage your time. The feeling of always being behind creates a lot of stress. Waiting until the last minute to finish an important task adds unnecessary stress.
- Talk to your supervisor. Your supervisor may understand what is causing your stress and may be able to suggest ways to manage it better.

Unfortunately, terrorism must also be a consideration on the job site. The level of concern is typically dependent on the site itself and the nature of the structure. Pipelines, for example, may represent a significant target due to their size, importance, location (sections of pipeline are in remote areas), and potential for destruction. Controversial sites also tend to attract attention, perhaps from terrorist groups operating within a country that oppose the construction or what the structure represents.

As an individual worker, the most important thing you can do is to pay attention. Become familiar with people on the job site and look for strangers that do not seem to fit. A terrorist attack can come from any direction and be delivered in many different ways. It can begin with a single individual gaining access to the job site.

1.2.6 Unsafe Acts

An unsafe act is a change from an accepted, normal, or correct procedure that can cause an incident or accident. It can be any conduct that causes unnecessary exposure to a job-site hazard or that makes an activity less safe than usual. Here are examples of unsafe acts:

- Failing to use required **personal protective equipment (PPE)**
- Failing to warn co-workers of hazards
- Lifting improperly
- Loading or placing equipment or supplies improperly
- Making safety devices (such as saw guards) inoperable
- Operating equipment at improper speeds
- Operating equipment without authority
- Servicing equipment in motion
- Taking an improper working position
- Using defective equipment
- Using equipment improperly

1.2.7 Rationalizing Risk

Everybody takes risks every day. When you get in your car to drive to work, you know there is a risk of being involved in an accident. Yet when you drive using all the safety practices you have learned, you know that there is a very good chance you will arrive at your destination safely. Driving is an acceptable risk because you have some control over your own safety and that of others.

Some risks are not acceptable. On the job, you must never take risks that endanger yourself or others just because you think you can justify doing so. This is called rationalizing risk which means ignoring safety warnings and practices.

For example, because you are late for work, you might decide to run a red light. Perhaps you feel the risk is worth the time saved. This is not only a dangerous driving habit, but this kind of thinking is unacceptable on the job site.

The following are common examples of rationalized risks on the job:

- Crossing barricades or boundaries because there is no apparent activity
- Not wearing gloves because it's easier to do the job with bare hands
- Removing your hard hat because you are hot and you cannot see anyone working overhead
- Not tying off your fall protection because you only have to lean over by about a foot

Think about the job before you do it. If you think that it is unsafe, then it is unsafe. Stop working until the job can be done safely. Bring your concerns to the attention of your supervisor. Your health and safety, and that of your co-workers, make it worth taking extra care.

1.2.8 Unsafe Conditions

An unsafe condition is a physical state that is different from the acceptable, normal, or correct condition found on the job site. It usually results in an incident or accident. An unsafe condition can be anything that reduces the degree of safety normally present. The following are some examples of unsafe conditions:

- Congested workplace
- Defective tools, equipment, or supplies
- Excessive noise
- Fire and explosive hazards
- Hazardous atmospheric conditions (such as gases, dusts, fumes, and vapors)
- Inadequate supports or guards
- Inadequate warning systems
- Cluttered work area
- Poor lighting
- Poor ventilation
- Radiation exposure
- Unguarded moving parts such as pulleys, drive chains, and belts

All employees should be given the authority to stop work when an unsafe condition is observed.

1.2.9 Poor Housekeeping

Housekeeping means keeping your work area clean and free of scraps, clutter, or spills. It also means being orderly and organized. Store your materials and supplies safely and label them properly. Arranging your tools and equipment to permit safe, efficient work practices, and easy cleaning is also important.

If the work site is indoors, make sure it is well-lit and ventilated. Don't allow aisles and exits to be blocked by materials and equipment. Make sure that flammable liquids are stored in safety cans. Oily rags must be placed only in approved, self-closing metal containers (*Figure 3*).

Remember that the major goal of housekeeping is to prevent incidents. Good housekeeping reduces the chances for slips, fires, explosions, and falling objects. Here are some good housekeeping rules:

- Remove from work areas all scrap material and lumber with nails protruding.
- Clean up spills to prevent falls.
- Remove all combustible scrap materials regularly.
- Make sure you have containers for the collection and separation of refuse. Containers for flammable or harmful refuse must have covers.
- Dispose of wastes often.
- Store all tools and equipment in the proper location and condition when you are finished with them.
- Keep all aisles and walkways clear of materials and tools.

Another term for good housekeeping is pride of workmanship. If you take pride in what you are doing, you won't let trash build up around you. The saying, "a place for everything and everything in its place" is the right attitude on the job site.

1.2.10 Management System Failure

Sometimes the cause of an incident or accident is failure of the management system. The management system should be designed to prevent or correct the acts and conditions that can create safety hazards. If the safety management system is not functioning properly, problems are likely to occur.

Tool Blades

Dull blades cause more accidents than sharp ones. If you do not keep your cutting tools sharpened, they won't cut very easily. When you have a hard time cutting, you exert more force on the tool. When that happens, something is bound to slip. And when something slips, you can be injured.

00101-15_F03.EPS

Figure 3 Container for oily rags and waste.

What traits could mean the difference between a management system that fails and one that succeeds? A company implementing a good management system will do the following:

- Put safety policies and procedures in writing
- Distribute written safety policies and procedures to each employee
- Review safety policies and procedures periodically
- Enforce all safety policies and procedures fairly and consistently
- Evaluate supplies, equipment, and services to see whether they are safe
- Provide regular, periodic safety training for employees

1.3.0 Hazard Recognition, Evaluation, and Control

The process of hazard recognition, evaluation, and control is the foundation of an effective safety program. When hazards are identified and assessed, they can be addressed quickly, reducing the hazard potential. Simply put, the more aware you are of your surroundings and the dangers in them, the less likely you are to be involved in an incident.

There is a standard rule in the United States that affects every worker in most industries. It is often referred to as HAZCOM, which is short for **Hazard Communication Standard**. The US HAZCOM also aligns with the Globally Harmonized System of Classification and Labeling of Chemicals (GHS). HAZCOM may also be called the Right-To-Know requirement. It requires all US contractors to educate their employees about the hazardous chemicals they may be exposed to on the job site. Employees must be taught how to work safely around these materials. Other countries have similar programs.

Many people think that there are very few hazardous chemicals on construction job sites. That is simply not true. In practice, the term *hazardous chemical* applies to solvents, paint, concrete, and even wood dust, along with more obvious substances such as acids.

As an employee, you have the following responsibilities under HAZCOM:

- Know where SDSs are on your job site.
- Report any hazards you spot on the job site to your supervisor.
- Know the physical and health hazards of any hazardous materials on your job site, and know and practice the precautions needed to protect yourself from these hazards.
- Know what to do in an emergency.
- Know the location and content of your employer's written hazard communication program.

The final responsibility for your safety rests with you. Your employer must provide you with information about hazards, but you must know this information and follow safety rules.

1.3.1 Hazard Recognition

There are many potential hazard indicators. The best approach in determining if a situation or equipment is potentially hazardous is to ask these questions:

- How can this situation or equipment cause harm?
- What types of energy sources are present that can cause an incident?
- What is the magnitude of the energy?
- What could go wrong to release the energy?
- How can the energy be eliminated or controlled?
- Will I be exposed to any hazardous materials?

Before you can fully answer these questions, you need to know the different types of incidents and accidents that can happen and the energy sources behind them. Some of the different types of events that can cause injuries include the following:

- Falls on the same elevations or falls from elevations
- Being caught in, on, or between equipment
- Being struck by falling objects
- Contact with acid, electricity, heat, cold, radiation, pressurized liquid, gas, or toxic substances

- Being cut by tools or equipment
- Exposure to high noise levels
- Repetitive motion or excessive vibration

You might recognize the first four high-hazard conditions as the so-called fatal four previously discussed. Remember, these four types of accidents cause more than half of all construction fatalities.

When equipment is the cause of an incident or accident, it is usually because there was an uncontrolled release of energy. The different types of energy sources that can be released include the following:

- Mechanical
- Pneumatic
- Hydraulic
- Electrical
- Chemical
- Thermal (heat or cold)
- Radioactive
- Gravitational
- Stored energy

There are a number of ways to recognize hazards and potential hazards on a job site. Some techniques are more complicated than others. In order to be effective, they all must answer this question: What could go wrong with this situation or operation? No matter what hazard recognition technique you use, answering that question in advance will save lives and prevent equipment damage.

1.3.2 Job Safety Analysis (JSA) and Task Safety Analysis (TSA)

Performing a job safety analysis (JSA), also known as job hazard analysis (JHA), is one approach to hazard recognition. Another common technique is performing a task safety analysis (TSA), also called a task hazard analysis (THA).

In a JSA, the task at hand is broken down into its individual parts or steps and then each step is analyzed for its potential hazards. Once a hazard is identified, certain actions or procedures are recommended that will correct that hazard. For example, during a JSA, it is determined that using a chain hoist to install a pump motor in a tight space would be safer than having a worker do it manually. By using the chain hoist, the chance that the worker's hand would get crushed during installation is reduced. Using the JSA process saved the worker from injury. *Figure 4* shows an example of a form used to conduct a JSA.

JOB SAFETY ANALYSIS

TITLE OF JOB OR TASK

TASK	START	END	HAZARDS	CONTROLS
1.				
2.				
3.				
4.				
5.				
6.				
7.				
8.				

Required Training: _____   Required Personal Protective Equipment (PPE): _____

Job Name: _____

Job Number: _____

Supervisor: _____

Date: _____

Weekly Vehicle Check List: _____   Tire Pressure _____   Transmission Fluid _____

_____ Oil _____ Lights _____

_____ Air Filter _____ Wkly Mileage _____

Names of Employees:	PRINT NAME	SIGN NAME	TOTAL HOURS

00101-15_F04.EPS

Figure 4 Job safety analysis form.

JSAs can also be used as pre-planning tools. This helps to ensure that safety is planned into the job. You may be asked to take part in a JSA during job planning. When JSAs are used as pre-planning tools, they contain the following information:

- Tools, materials, and equipment needs
- Staffing or manpower requirements
- Duration of the job
- Quality concerns

Task safety analysis is similar to job safety analysis in that both require workers to identify potential hazards and needed safeguards associated with a job they are about to do. The difference is the form used to report the hazards. During a TSA, a pre-printed, fill-in-the-blank checklist is often used to document any hazard found during analysis. Before work begins, the first-line supervisor or team leader should discuss the conclusions found during the TSA with the crew. Some companies require workers to sign the completed TSA forms or checklists before they start work. This helps companies document the hazards and ensure that workers have been told of the potential hazards and safety procedures.

1.3.3 Risk Assessment

Whether an action is considered safe is often a matter of evaluating risk. Risk is a measure of the probability, consequences, and exposure related to an event. Probability is the chance that a given event will occur. Consequences are the results of an action, condition, or event. Exposure is the amount of time and/or the degree to which someone or something is exposed to an unsafe condition, material, or environment.

A safe operation is one in which there is an acceptable level of risk. This means there is a low probability of an incident and that the consequences and exposure risk are all acceptable. For example, climbing a ladder has risk that is considered to be acceptable if the proper ladder is being used as intended, if it is set up correctly, and if it is in good condition. The probability of exposure to a hazard and its potential consequences are all low. If any one of these conditions were different, climbing the ladder would have an unacceptable level of risk.

1.3.4 Reporting Injuries, Incidents, and Near-Misses

All on-the-job incidents and accidents, no matter how minor, must be reported to your supervisor.

Some workers think they will get in trouble if they report minor injuries, but that is not the case. Small injuries, like cuts and scrapes, can later become big problems because of infection and other complications.

US employers with more than 10 employees are required to maintain a log of significant work-related injuries and illnesses using specific OSHA forms and documents. Employee names can be kept confidential in certain circumstances. A summary of these injuries must be posted at certain intervals, although employers do not need to submit it to OSHA unless requested. Employers can calculate the total number of injuries and illnesses and compare the result with the average national rates for similar companies. By analyzing incidents and accidents, companies and OSHA can improve safety policies and procedures. By reporting an incident, you can help keep similar events from happening in the future. For details on the operation of OSHA and its important mission, refer to the *Appendix*.

1.3.5 Safety Data Sheets

SDSs are fact sheets prepared by the chemical manufacturer or importer. Each product used on a construction site must have an SDS. An SDS describes the substance, along with its hazards, safe handling, first aid, and emergency spill procedures. The HAZCOM standard requires new SDSs (formerly known as material safety data sheets or MSDSs) to be in a standardized format. The sections of the form include the following:

- *Section 1, Product identification, manufacturer contact information, recommended uses and restrictions*
- *Section 2, Hazard identification*
- *Section 3, Composition/information on ingredients*
- *Section 4, First aid measures*
- *Section 5, Firefighting information*
- *Section 6, Incidental release measures*
- *Section 7, Handling and storage*
- *Section 8, Exposure controls/personal protection*
- *Section 9, Physical and chemical properties*
- *Section 10, Stability and reactivity*
- *Section 11, Toxicological properties*
- *Section 12, Ecological properties*
- *Section 13, Disposal considerations*
- *Section 14, Transport information*
- *Section 15, Regulatory information*
- *Section 16, Other information*

Figure 5 shows a sample SDS for a PVC solvent-cement. The most important things to look for on an SDS are the specific hazards, personal protection, handling procedures, and first aid

GHS SAFETY DATA SHEET

WELD-ON® 705™ Low VOC Cements for PVC Plastic Pipe

Date Revised: **DEC 2011**
Supersedes: **FEB 2010**

SECTION 1 - PRODUCT AND COMPANY IDENTIFICATION

PRODUCT NAME: WELD-ON® 705™ Low VOC Cements for PVC Plastic Pipe

PRODUCT USE: Low VOC Solvent Cement for PVC Plastic Pipe

SUPPLIER:

MANUFACTURER: IPS Corporation
17109 South Main Street, Carson, CA 90248-3127
P.O. Box 379, Gardena, CA 90247-0379
Tel. 1-310-898-3300

EMERGENCY: Transportation: CHEMTEL Tel. 800.255.3924, 813-248-0585 (International)   **Medical:** Tel. 800.451.8346, 760.602.8703 3E Company (International)

SECTION 2 - HAZARDS IDENTIFICATION

GHS CLASSIFICATION:

Health		Environmental		Physical	
Acute Toxicity:	Category 4	Acute Toxicity:	None Known	Flammable Liquid	Category 2
Skin Irritation:	Category 3	Chronic Toxicity:	None Known		
Skin Sensitization:	NO				
Eye:	Category 2B				

GHS LABEL: ⬥ ⬥   OR   ⬥ ✖   **Signal Word:** **Danger**   **WHMIS CLASSIFICATION:** CLASS B, DIVISION 2

Hazard Statements	Precautionary Statements
H225: Highly flammable liquid and vapor	P210: Keep away from heat/sparks/open flames/hot surfaces – No smoking
H319: Causes serious eye irritation	P261: Avoid breathing dust/fume/gas/mist/vapors/spray
H332: Harmful if inhaled	P280: Wear protective gloves/protective clothing/eye protection/face protection
H335: May cause respiratory irritation	P304+P340: IF INHALED: Remove victim to fresh air and keep at rest in a position comfortable for breathing
H336: May cause drowsiness or dizziness	P403+P233: Store in a well ventilated place. Keep container tightly closed
EUH019: May form explosive peroxides	P501: Dispose of contents/container in accordance with local regulation

SECTION 3 - COMPOSITION/INFORMATION ON INGREDIENTS

	CAS#	EINECS #	REACH Pre-registration Number	CONCENTRATION % by Weight
Tetrahydrofuran (THF)	109-99-9	203-726-8	05-2116297729-22-0000	25 - 50
Methyl Ethyl Ketone (MEK)	78-93-3	201-159-0	05-2116297728-24-0000	5 - 36
Cyclohexanone	108-94-1	203-631-1	05-2116297718-25-0000	15 - 30

All of the constituents of this adhesive product are listed on the TSCA inventory of chemical substances maintained by the US EPA, or are exempt from that listing.
* Indicates this chemical is subject to the reporting requirements of Section 313 of the Emergency Planning and Community Right-to-Know Act of 1986 (40CFR372).
# indicates that this chemical is found on Proposition 65's List of chemicals known to the State of California to cause cancer or reproductive toxicity.

SECTION 4 - FIRST AID MEASURES

Contact with eyes:	Flush eyes immediately with plenty of water for 15 minutes and seek medical advice immediately.
Skin contact:	Remove contaminated clothing and shoes. Wash skin thoroughly with soap and water. If irritation develops, seek medical advice.
Inhalation:	Remove to fresh air. If breathing is stopped, give artificial respiration. If breathing is difficult, give oxygen. Seek medical advice.
Ingestion:	Rinse mouth with water. Give 1 or 2 glasses of water or milk to dilute. Do not induce vomiting. Seek medical advice immediately.

SECTION 5 - FIREFIGHTING MEASURES

			HMIS	NFPA	
Suitable Extinguishing Media:	Dry chemical powder, carbon dioxide gas, foam, Halon, water fog.				0-Minimal
Unsuitable Extinguishing Media:	Water spray or stream.	Health	2	2	1-Slight
Exposure Hazards:	Inhalation and dermal contact	Flammability	3	3	2-Moderate
Combustion Products:	Oxides of carbon, hydrogen chloride and smoke	Reactivity	0	0	3-Serious
		PPE	B		4-Severe
Protection for Firefighters:	Self-contained breathing apparatus or full-face positive pressure airline masks.				

SECTION 6 - ACCIDENTAL RELEASE MEASURES

Personal precautions:	Keep away from heat, sparks and open flame.
	Provide sufficient ventilation, use explosion-proof exhaust ventilation equipment or wear suitable respiratory protective equipment. Prevent contact with skin or eyes (see section 8).
Environmental Precautions:	Prevent product or liquids contaminated with product from entering sewers, drains, soil or open water course.
Methods for Cleaning up:	Clean up with sand or other inert absorbent material. Transfer to a closable steel vessel.
Materials not to be used for clean up:	Aluminum or plastic containers

SECTION 7 - HANDLING AND STORAGE

Handling: Avoid breathing of vapor, avoid contact with eyes, skin and clothing.
Keep away from ignition sources, use only electrically grounded handling equipment and ensure adequate ventilation/fume exhaust hoods.
Do not eat, drink or smoke while handling.

Storage: Store in ventilated room or shade below 44°C (110°F) and away from direct sunlight.
Keep away from ignition sources and incompatible materials: caustics, ammonia, inorganic acids, chlorinated compounds, strong oxidizers and isocyanates.
Follow all precautionary information on container label, product bulletins and solvent cementing literature.

SECTION 8 - PRECAUTIONS TO CONTROL EXPOSURE / PERSONAL PROTECTION

EXPOSURE LIMITS:

Component	ACGIH TLV	ACGIH STEL	OSHA PEL	OSHA STEL:
Tetrahydrofuran (THF)	50 ppm	100 ppm	200 ppm	
Methyl Ethyl Ketone (MEK)	200 ppm	300 ppm	200 ppm	
Cyclohexanone	20 ppm	50 ppm	50 ppm	

Engineering Controls: Use local exhaust as needed.

Monitoring: Maintain breathing zone airborne concentrations below exposure limits.

Personal Protective Equipment (PPE):

Eye Protection: Avoid contact with eyes, wear splash-proof chemical goggles, face shield, safety glasses (spectacles) with brow guards and side shields, etc. as may be appropriate for the exposure.

Skin Protection: Prevent contact with the skin as much as possible. Butyl rubber gloves should be used for frequent immersion.
Use of solvent-resistant gloves or solvent-resistant barrier cream should provide adequate protection when normal adhesive application practices and procedures are used for making structural bonds.

Respiratory Protection: Prevent inhalation of the solvents. Use in a well-ventilated room. Open doors and/or windows to ensure airflow and air changes. Use local exhaust ventilation to remove airborne contaminants from employee breathing zone and to keep contaminants below levels listed above. With normal use, the Exposure Limit Value will not usually be reached. When limits approached, use respiratory protection equipment.

00101-15_F05A.EPS

Figure 5A Solvent cement SDS. (1 of 2)

GHS SAFETY DATA SHEET

WELD-ON® 705™ Low VOC Cements for PVC Plastic Pipe

Date Revised: **DEC 2011**
Supersedes: **FEB 2010**

SECTION 9 - PHYSICAL AND CHEMICAL PROPERTIES

Appearance:	Clear or gray, medium syrupy liquid		
Odor:	Ketone	**Odor Threshold:**	0.88 ppm (Cyclohexanone)
pH:	Not Applicable		
Melting/Freezing Point:	-108.5°C (-163.3°F) Based on first melting component: THF	**Boiling Range:**	66°C (151°F) to 156°C (313°F)
Boiling Point:	66°C (151°F) Based on first boiling component: THF	**Evaporation Rate:**	> 1.0 (BUAC = 1)
Flash Point:	-20°C (-4°F) TCC based on THF	**Flammability:**	Category 2
Specific Gravity:	0.9611 @23°C (73°F)	**Flammability Limits:**	LEL: 1.1% based on Cyclohexanone
Solubility:	Solvent portion soluble in water. Resin portion separates out.		UEL: 11.8% based on THF
Partition Coefficient n-octanol/water:	Not Available	**Vapor Pressure:**	129 mm Hg @ 20°C (68°F)based on THF
Auto-ignition Temperature:	321°C (610°F) based on THF	**Vapor Density:**	>2 (Air = 1)
Decomposition Temperature:	Not Applicable	**Other Data: Viscosity:**	Medium bodied
VOC Content:	When applied as directed, per SCAQMD Rule 1168, Test Method 316A,VOC content is: $\leq$ 510 g/l.		

SECTION 10 - STABILITY AND REACTIVITY

Stability:	Stable
Hazardous decomposition products:	None in normal use. When forced to burn, this product gives off oxides of carbon, hydrogen chloride and smoke.
Conditions to avoid:	Keep away from heat, sparks, open flame and other ignition sources.
Incompatible Materials:	Oxidizers, strong acids and bases, amines, ammonia

SECTION 11 - TOXICOLOGICAL INFORMATION

Likely Routes of Exposure:	Inhalation, Eye and Skin Contact
Acute symptoms and effects:	
Inhalation:	Severe overexposure may result in nausea, dizziness, headache. Can cause drowsiness, irritation of eyes and nasal passages.
Eye Contact:	Vapors slightly uncomfortable. Overexposure may result in severe eye injury with corneal or conjunctival inflammation on contact with the liquid.
Skin Contact:	Liquid contact may remove natural skin oils resulting in skin irritation. Dermatitis may occur with prolonged contact.
Ingestion:	May cause nausea, vomiting, diarrhea and mental sluggishness.
Chronic (long-term) effects:	None known to humans

Toxicity:	LD_{50}	LC_{50}
Tetrahydrofuran (THF)	Oral: 2842 mg/kg (rat)	Inhalation 3 hrs. 21,000 mg/m^3 (rat)
Methyl Ethyl Ketone (MEK)	Oral: 2737 mg/kg (rat), Dermal: 6480 mg/kg (rabbit)	Inhalation 8 hrs. 23,500 mg/m^3 (rat)
Cyclohexanone	Oral: 1535 mg/kg (rat), Dermal: 948 mg/kg (rabbit)	Inhalation 4 hrs. 8,000 PPM (rat)

Reproductive Effects	Teratogenicity	Mutagenicity	Embryotoxicity	Sensitization to Product	Synergistic Products
Not Established	Not Established	Not Established	Not Established	Not Established	Not Established

SECTION 12 - ECOLOGICAL INFORMATION

Ecotoxicity:	None Known
Mobility:	In normal use, emission of volatile organic compounds (VOC's) to the air takes place, typically at a rate of $\leq$ 510 g/l.
Degradability:	Biodegradable
Bioaccumulation:	Minimal to none.

SECTION 13 - WASTE DISPOSAL CONSIDERATIONS

Follow local and national regulations. Consult disposal expert.

SECTION 14 - TRANSPORT INFORMATION

Proper Shipping Name:	Adhesives		
Hazard Class:	3	**EXCEPTION for Ground Shipping**	
Secondary Risk:	None	**DOT Limited Quantity:** Up to 5L per inner packaging, 30 kg gross weight per package.	
Identification Number:	UN 1133	**Consumer Commodity:** Depending on packaging, these quantities may qualify under DOT as "ORM-D" .	
Packing Group:	PG II		
Label Required:	Class 3 Flammable Liquid	**TDG INFORMATION**	
Marine Pollutant:	NO	TDG CLASS:	FLAMMABLE LIQUID 3
		SHIPPING NAME:	ADHESIVES
		UN NUMBER/PACKING GROUP:	UN 1133, PG II

SECTION 15 - REGULATORY INFORMATION

Precautionary Label Information:	Highly Flammable, Irritant	Ingredient Listings: USA TSCA, Europe EINECS, Canada DSL, Australia
Symbols:	F, Xi	AICS, Korea ECL/TCCL, Japan MITI (ENCS)
Risk Phrases:	R11: Highly flammable.	
	R20: Harmful by inhalation.	R66: Repeated exposure may cause skin dryness or cracking
	R36/37: Irritating to eyes and respiratory system.	R67: Vapors may cause drowsiness and dizziness
Safety Phrases:	S9: Keep container in a well-ventilated place.	S26: In case of contact with eyes, rinse immediately with plenty of water and seek medical advice.
	S16: Keep away from sources of ignition - No smoking.	S33: Take precautionary measures against static discharges.
	S25: Avoid contact with eyes.	S46: If swallowed, seek medical advise immediately and show this container or label.

SECTION 16 - OTHER INFORMATION

Specification Information:		
Department issuing data sheet:	IPS, Safety Health & Environmental Affairs	All ingredients are compliant with the requirements of the European
E-mail address:	<EHSinfo@ipscorp.com>	Directive on RoHS (Restriction of Hazardous Substances).
Training necessary:	Yes, training in practices and procedures contained in product literature.	
Reissue date / reason for reissue:	12/14/2011 / Updated GHS Standard Format	
Intended Use of Product:	Solvent Cement for PVC Plastic Pipe	

This product is intended for use by skilled individuals at their own risk. The information contained herein is based on data considered accurate based on current state of knowledge and experience. However, no warranty is expressed or implied regarding the accuracy of this data or the results to be obtained from the use thereof.

00101-15_F05B.EPS

Figure 5B Solvent cement SDS. (2 of 2)

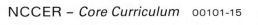

information. Most SDSs have a 24-hour emergency-response number.

Using *Figure 5*, try to find the information you would need to use the cement described on the sample SDS. First locate the hazards; section 2 of the SDS shows that the adhesive is flammable, and also an eye and skin irritant that can cause respiratory irritation, dizziness, and drowsiness.

Next, find out how to minimize these hazards. Section 7 gives general handling and storage information. It indicates that ventilation is needed to reduce hazardous vapors, which can be a fan in an open window. If ventilation is not enough, respiratory protection is needed. Section 8 tells you how to protect your eyes and skin.

Section 4 lists the first aid measures for eye contact, skin contact, or inhalation. Section 5 explains fire hazards and firefighting measures. Now you have the information you need in case of an emergency.

The SDSs must be kept in the work area and be readily accessible to all workers. The company's safety officer or **competent person** should review the SDS before a hazardous material is used. Ask your supervisor to point out where the SDSs are located if you are not sure. Have him or her point out the sections that relate to your job; the health and safety of you and your co-workers may depend on it.

Additional Resources

US Occupational Safety and Health Administration. Numerous safety videos are available on line at **www.osha.gov/video**.

Construction Safety, Jimmie W. Hinze. 2006. Upper Saddle River, NJ: Pearson Education, Inc.

DeWalt Construction Safety/OSHA Professional Reference, Paul Rosenberg; American Contractors Educational Services. 2006. DEWALT

Basic Construction Safety and Health, Fred Fanning. 2014. CreateSpace Independent Publishing Platform.

1.0.0 Section Review

1. The person primarily responsible for your safety is _____.

   a. your foreman
   b. your instructor
   c. yourself
   d. your employer

2. The color commonly used for informational signs is _____.

   a. green
   b. red
   c. yellow
   d. blue

3. The SDS for any chemical used at a job site must be available _____.

   a. at the job site
   b. on line
   c. at the contractor's office
   d. at the nearest hospital

SECTION TWO

2.0.0 ELEVATED WORK AND FALL PROTECTION

Objectives

Describe the safe work requirements for elevated work, including fall protection guidelines.

   a. Identify and describe various fall hazards.

   b. Identify and describe equipment and methods used in fall prevention and fall arrest.

   c. Identify and describe the safe use of ladders and stairs.

   d. Identify and describe the safe use of scaffolds.

Performance Tasks

1. Properly set up and climb/descend an extension ladder, demonstrating proper three-point contact.

2. Inspect the following PPE items and determine if they are safe to use:
   - Fall arrest harnesses
   - Lanyards
   - Connecting devices

3. Properly don, fit, and remove the following PPE:
   - Fall arrest harness

Trade Terms

Cross-bracing: Braces (metal or wood) placed diagonally from the bottom of one rail to the top of another rail that add support to a structure.

Excavation: Any man-made cut, cavity, trench, or depression in an earth surface, formed by removing earth. It can be made for anything from basements to highways. Also see *trench*.

Guarded: Enclosed, fenced, covered, or otherwise protected by barriers, rails, covers, or platforms to prevent dangerous contact.

Hand line: A line attached to a tool or object so a worker can pull it up after climbing a ladder or scaffold.

Lanyard: A short section of rope or strap, one end of which is attached to a worker's safety harness and the other to a strong anchor point above the work area.

Maximum intended load: The total weight of all people, equipment, tools, materials, and loads that a ladder can hold at one time.

Midrail: Mid-level, horizontal board required on all open sides of scaffolds and platforms that are more than 14 inches (35 cm) from the face of the structure and more than 10 feet (3 m) above the ground. It is placed halfway between the toeboard and the top rail.

Planked: Having pieces of material 2 inches (5 cm) thick or greater and 6 inches (15 cm) wide or greater used as flooring, decking, or scaffold decks.

Scaffold: An elevated platform for workers and materials.

Six-foot rule: A rule stating that platforms or work surfaces with unprotected sides or edges that are 6 feet (≈2 m) or higher than the ground or level below it require fall protection.

Toeboard: A vertical barrier at floor level attached along exposed edges of a platform, runway, or ramp to prevent materials and people from falling.

Top rail: A top-level, horizontal board required on all open sides of scaffolds and platforms that are more than 14 inches (36 cm) from the face of the structure and more than 10 feet (3 m) above the ground.

Falls from elevated areas are the leading cause of fatalities in the workplace. Falls from elevated heights account for about one-third of all deaths in the construction trade. Approximately 85 percent of the injured workers lose time from work; approximately 33 percent require hospitalization; some never return to the job.

While the risk of falls is high in construction and some other trades, there are many things that workers can do to safeguard themselves. Using the appropriate PPE; following proper safety procedures; practicing good housekeeping habits; and staying alert at all times will help you stay safe when working at an elevation. Employers are required to provide for both fall prevention and fall arrest, and to make sure workers are trained and certified in the use of fall protection equipment. Fall prevention consists of covered floor openings, climbing aids, barricades, and guardrails that are designed to protect against falls. Fall arrest consists of equipment such as body harnesses, lanyards, connection devices, lifelines, and safety nets that are intended to protect a worker in case a fall occurs. All these topics are covered in this section.

2.1.0 Fall Hazards

Falls are classified into two groups: falls from an elevation and falls from the same level. Falls from an elevation can happen during work from scaffolds, work platforms, decking, concrete forms, ladders, stairs, and work near excavations. Falls from elevation often result in death unless the fall is arrested. Falls on the same level are usually caused by tripping or slipping. Sharp edges and pointed objects, such as exposed concrete reinforcing bars (rebar), could cut and otherwise harm a worker. Other bodily injuries are also common results of tripping or slipping.

In the United States, fall protection is required for platforms or work surfaces with unprotected sides or edges that are 6 feet (≈2 meters) or above the ground or the level below. This is commonly referred to as the six-foot rule. However, some international regulations and company policies may require fall protection for heights less than 6 feet.

2.1.1 Walking and Working Surfaces

Slips, trips, and falls on walking and working surfaces cause 15 percent of all incidental deaths in the construction industry. Some incidents occur due to environmental conditions, such as snow, ice, or wet surfaces. Others happen because of poor housekeeping and careless behavior, such as leaving tools, materials, and equipment out and unattended. You can avoid slips, trips, and falls by being aware of your surroundings and following the rules on your site. Remember these general walking and working surface guidelines to avoid incidents: *Fall prevention*

- Keep all walking and working areas clean and dry. If you see a spill or ice patch, clean it up, or barricade the area until it can be properly attended to.
- Keep all walking and working surfaces clear of clutter and debris.
- Run cables, extension cords, and hoses overhead or through crossover plates so that they will not become tripping hazards.
- Do not run on scaffolds, work platforms, decking, roofs, or other elevated work areas.

2.1.2 Unprotected Sides, Wall Openings, and Floor Holes

Any opening in a wall or floor is a safety hazard. There are two types of protection for these openings: (1) they can be guarded or (2) they can be covered. Cover any hole in the floor when pos-

sible. Hole covers must be clearly marked. When it is not practical to cover the hole, use barricades. If the bottom edge of a wall opening is less than 39 inches (1 m) above the floor and would allow someone to fall 6 feet (≈2m) or more, then place guards around the opening.

The types of barriers and barricades used vary from one job site to another. There may also be different procedures for when and how barricades are put up. Learn and follow the policies at your job site. Several different types of guards are commonly used:

- Railings are used across wall openings or as a barrier around floor openings to prevent falls (*Figure 6A*).
- Warning barricades alert workers to hazards but provide no real protection (*Figure 6B*). Warning barricades can be made of plastic tape or rope strung from wire or between posts. The tape or rope is color-coded as follows:
  - *Red means danger.* No one may enter an area with a red warning barricade. A red barricade is used when there is danger from falling objects or when a load is suspended over an area.
  - *Yellow means caution.* You may enter an area with a yellow barricade, but be sure you know what the hazard is, and be careful. Yellow barricades are used around wet areas or areas containing loose dust. Yellow with black lettering warns of physical hazards such as bumping into something, stumbling, or falling.
  - *Yellow and purple together mean radiation warning.* No one may pass a yellow and purple barricade without authorization, training, and the appropriate PPE. These barricades are often used where piping welds are being X-rayed.
- Protective barricades give both a visual warning and protection from injury (*Figure 6C*). They can be wooden posts and rails, posts and chain, or steel cable. People should not be able to get past protective barricades.
- Blinking lights are placed on barricades so they can be seen at night (*Figure 6D*).
- Hole covers are used to cover open holes in a floor or in the ground (*Figure 6E*).

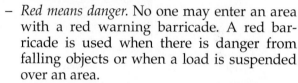

WARNING! Never remove a barricade unless you have been authorized to do so. Follow your employer's procedures for putting up and removing barricades.

(A)

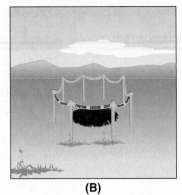
(B)

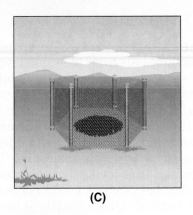
(C)

(D)

(E)

Red = risk
Yellow = Caution

00101-15_F06.EPS

Figure 6 Common types of barriers and barricades.

Follow these guidelines when working near unprotected sides, floor holes, and wall openings:

- Hole covers must be cleated, wired, or otherwise secured to prevent them from slipping sideways or horizontally beyond the hole.
- Covers must extend adequately beyond the edge of the hole.
- Hole covers must be strong enough to support twice the weight of anything that may be placed on top of them. They must also be clearly marked. Use ¾-inch (2 cm) plywood as a hole cover, provided that one dimension of the opening is less than 18 inches (46 cm); otherwise, 2-inch (5-cm) lumber is required.
- Never store material or equipment on a hole cover.
- Guard all stairway floor openings, with the exception of the entrance, with standard railing and **toeboards**.
- Guard all wall openings from which there is a drop of more than 6 feet (≈2m) and for which the bottom of the opening is less than 39 inches (1 m) above the working surface.
- Guard all open-sided floors and platforms 6 feet (≈2m) or more above adjacent floor or ground level, using a standard railing or the equivalent.

2.2.0 Fall Arrest

The key to preventing serious injury or death should a fall occur is the personal fall arrest system (PFAS). A complete PFAS (*Figure 7*) consists of anchor points, a body harness, and connecting devices. Anchor points are related to the structure, and the type and availability of anchor points help determine what other equipment should be chosen. The body harness comprises the system of belts, rings, or hooks worn by the worker. Connecting devices and lanyards are used to maintain attachment between anchor points and the PFAS, and include lanyards and various pieces of hardware.

> **NOTE**
>
> This training alone does not provide any level of certification in the use of fall arrest or fall restraint equipment. Trainees should not assume that the knowledge gained in this module is sufficient to certify them to use fall arrest equipment in the field.

2.2.1 Anchor Points

There are both permanent anchor points and temporary, reusable anchor points (*Figure 8*). Anchor points must be rated at or equal to 5,000 pounds (2,267 kg) breaking or tensile strength, or twice the intended load.

PERSONAL FALL ARREST SYSTEM

ANCHORAGE

ANCHORAGE CONNECTOR

CONNECTING DEVICE

BODY HARNESS

00101-15_F07.EPS

Figure 7 Personal fall-arrest system.

Ideally, the fall arrest anchor point will be located directly above the back D-ring (*Figure 9*), in order to minimize any swing-zone hazards,

Case Histories

Here are some examples of fatal incidents that resulted from failure to provide the proper fall protection:

- A worker taking measurements was killed when he fell backward from an unguarded balcony to the concrete below.
- A roofer handling a piece of fiberboard backed up and tripped over a 7½-inch (19 cm) parapet. He fell more than 50 feet (15 m) to the ground level and died of severe head injuries.

as well as the free-fall distance. The maximum free-fall distance is 6 feet (≈2 m). To understand swing zones, picture a tied-off worker standing three feet (1 m) away from a point directly under the anchor point. If the worker falls, gravity will cause his body to swing toward the anchor point axis and momentum will cause the body to swing past that point. If there is a wall or other solid object close by, he may strike it and be injured. Swing zones are minimized when the anchor point is directly above the worker when he falls. Serious injury and damage can occur when a human body strikes an immoveable object while swinging as a pendulum. Although the PFAS may do its job by preventing the worker from falling a great distance, serious injury or death can still occur by striking an object in the swing zone.

(A) PERMANENT ANCHOR

(B) CONCRETE ANCHOR

(C) BEAM ANCHOR

(D) TIE-BACK LANYARD

00101-15_F08.EPS

Figure 8 Anchor points.

Anchor points are often needed to secure the position of the worker, leaving the hands free to accomplish a task. Ideally, workers will select anchor points to maintain a potential fall distance of no more than 2 feet (61 cm). For example, a worker placing reinforcing bars in a concrete wall form may use two short positioning lanyards connected to the hip rings on the harness. The fall protection lanyard would be connected to the rear D-ring.

Positioning connections cannot be considered the primary anchor. Positioning anchor points are required to be rated at a 3,000 pound (1,360 kg) strength instead of the 5,000 pound (2,268 kg) rating for primary fall arrest. Remember that positioning lanyards, connected to D-rings on the harness other than the back or front chest D-ring, are fall restraints rather than fall arresting connections.

A positioning lanyard does not take the place of a fall arrest lanyard or anchor point.

2.2.2 Harnesses

Full body harnesses, like the one shown in *Figure 10,* are available in sizes that are based on the height and weight of the user. The back D-ring is the only one used to connect the harness to the anchor point for primary fall arrest purposes unless you are climbing a ladder. When climbing a ladder, the front chest D-ring is the likely choice for connecting the fall arrest harness. D-rings located at the hips are used for positioning and fall restraint only. D-rings mounted to shoulders are often used for rescue situations. All of them can be used for fall restraint, but the back D-ring is the primary connection for fall arrest.

Figure 9 Back D-ring.

A body harness must fit correctly to ensure that it will provide proper protection. Do not place additional holes or openings in harness components under any circumstances. No field modifications to a body harness or lanyard should be attempted. Installation, maintenance, and inspection instructions are provided for every harness, and it is the responsibility of the worker to read and understand the details regarding his or her personal equipment. Workers must inspect their body harness each day that is in use.

Harness straps are generally designed with some stretch to help absorb some of the potential force of a fall. This means good, taut installation on the body is essential so that the worker cannot fall out of the harness in the event of a fall.

Figure 11 shows the proper procedure for donning a common full body harness. The most important adjustments to be made include the chest straps, the groin straps, and the final position of the back D-ring. The related details that follow must be considered during the fitting and wearing of a full body harness:

> NOTE
>
> These guidelines are general in nature and the instructions provided for specific equipment by the manufacturer must always take precedence. It is the worker's responsibility to be intimately familiar with the duty of each and every ring and strap on a given harness.

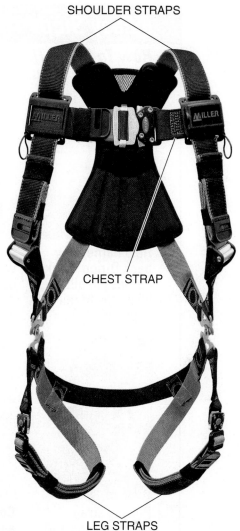

SHOULDER STRAPS

CHEST STRAP

LEG STRAPS

Figure 10 Full body harness.

- The back D-ring location is vital to proper fall arrest. Position this ring between the shoulder blades. If it is too low, it will tend to cause the body to hang in a more horizontal position during fall arrest, increasing pressure on the diaphragm and affecting breathing. If it is positioned too high or with too much slack, the D-ring may strike the worker's head at the base of the fall, and the shoulder straps may be pulled too tightly into the neck and restrict blood flow. The impact at the base of the fall arrest can be dramatic, so the force must be spread all around the body to prevent injury to any one portion.

- Chest straps generally form either an "H" pattern or an "X" pattern. Adjust "H" pattern straps to land between the bottom of the sternum and the belly button. This helps ensure the horizontal portion of the "H" does not contact the throat during a fall, choking the worker. Some harness designs may not allow

How To Don A Harness

1

Hold harness by back D-ring. Shake harness to allow all straps to fall in place.

2

If chest, leg and/or waist straps are buckled, release straps and unbuckle at this time.

3

Slip straps over shoulders so **D-ring is located in middle of back between shoulder blades.**

4

Pull leg strap between legs and connect to opposite end. Repeat with second leg strap. If belted harness, connect waist strap after leg straps.

5

Connect chest strap and position in midchest area. Tighten to keep shoulder straps taut.

6 Snug Fit

After all straps have been buckled, **tighten all buckles so that harness fits snug but allows full range of movement.** Pass excess strap through loop keepers.

00101-15_F11.EPS

Figure 11 Installing the body harness.

for this adjustment, with the final position of the "H" being based solely on a properly sized harness.

- The position of the chest straps is also crucial for "X" pattern harnesses. Position the "X" at or just below the sternum.
- Leg straps are an integral and required part of a PFAS. Adjust the groin straps for a good, snug fit. Too much slack here will cause extreme discomfort in a fall, when the impact snatches them up tight and you are left suspended this way.

- A suspension trauma strap (*Figure 12*) is recommended as part of the PFAS gear. The suspension trauma strap is stored in a pouch connected to the harness within easy reach. This is done by either one end of the strap being permanently sewn to the harness (by the manufacturer); one end of the strap attached to the harness with a carabiner (*Figure 13*); or by choking the pouch around a harness strap or hip D-ring. The strap can then be quickly removed and used without any possibility of the user dropping it. Once connected, the strap allows

personal fall

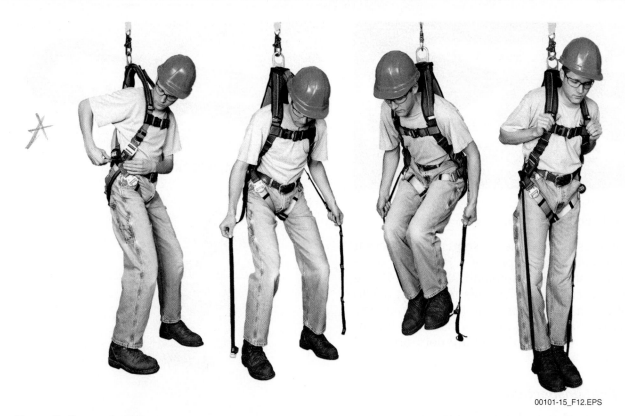

00101-15_F12.EPS

Figure 12 Suspension trauma strap use.

00101-15_F13.EPS

Figure 13 A carabiner.

the worker to stand up in the harness, relieving suspended weight and pressure from the hips and groin. This helps open the path for blood flow from the legs back to the heart, preventing blood from pooling in the lower extremities.

- A separate waist or tool belt, while not considered a necessary component of the fall-arrest system, must be fitted properly. It is best used for body positioning with the D-rings precisely located at the hips, rather than in the front or rear. Do not adjust it in a way that could apply pressure to the kidneys or lower back. If the waist belt is not an integral part of the harness, it must not be worn on the outside of the

harness—put it on first, then add the harness over it. This is also true of any added tool belts.
- Saddles, like waist belts, are also not considered an integral or required part of the PFAS. They are optional and often detachable. They are generally used by workers to allow a seated, suspended position when the task may require long periods in the same location.

It is important to note that not all hardware will qualify as a component of a PFAS. Some hardware is to be used only for attaching tools and equipment to the worker or to structures. Hardware used as connecting devices as part of the PFAS must be drop-forged steel, and have a corrosion-resistant finish resistant to salt spray per ANSI standards. Any type of hook or carabiner must be equipped with safety gates or keepers to prevent the hooked object from being disconnected incidentally. In most cases, these safety gates will be required to be two-step, also called double action. Designs for these features vary, but those designed so that both movements required to open the gate can be done with a single hand are generally better. Using both hands to manipulate a single connector can be a hazard in itself.

Double-locking snap hooks (*Figure 14*) are usually curved and have an opening to allow connection to a line that can then be securely closed. They are usually not as consistent in appearance

Figure 14 Double-locking snap hook.

(A) SHOCK-ABSORBING

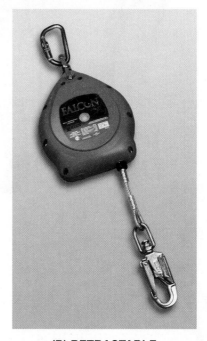

(B) RETRACTABLE

00101-15_F15.EPS

Figure 15 Fall arrest lanyards.

as carabiners, and come in somewhat different shapes. When used as part of a PFAS, the security closure should be automatic. Snap hooks are designed to connect to D-rings primarily, not to each other. In most cases, snap hooks are already connected to a lanyard or rope to ensure the integrity of the connection, and are not purchased separately.

2.2.3 PFAS Inspection

Always inspect your lanyard, harness, and any other fall protection gear prior to use. Never use damaged equipment. Treat a safety harness as if your life depends on it, because it does! To maintain their service life and high performance, all belts and harnesses should be inspected frequently. Damage to fall arrest systems includes burns, hardening due to chemical contact, and excessive wear. When inspecting a harness, check that the buckles and D-ring are not bent or deeply scratched. Check the harness for any cuts or rough spots.

The PFAS should be inspected monthly by a competent person. This requirement should be established through your company's safety program. The competent person has the authority to impose prompt corrective measures to eliminate any hazards. If there is any question about a defect, no matter how small it may seem to be, err on the side of caution. Take the fall arrest component(s) out of service for testing or replacement.

2.2.4 Lanyards

Lanyards consist of some primary material of construction (rope, webbing, aircraft cable, etc.) with a connecting device attached to the ends. Lanyards used for fall arrest are either shock absorbing or self-retracting (*Figure 15*). Non-shock absorbing lanyards, such as the one shown in *Figure 16*, are used for positioning and fall restraint. Lanyards used for positioning are not considered part of the

fall-arrest system; they are fall restraints and will be attached to D-rings on the harness other than the back D-ring. Since fall restraint is all about preventing a fall from happening, lanyards used for positioning should not allow a fall or movement greater than 2 feet (0.61 m). Two such lanyards are usually required to permit movement from one place to another.

The retractable fall arrest lanyard shown in *Figure 15* is rapidly becoming the preferred device. Since it automatically retracts or feeds lanyard as the worker moves about, there is never a great deal of slack that can pose a risk in itself. In addition, the potential fall distance is significantly reduced.

Lanyards for fall restraint or arrest must never be field-fabricated, and must never be connected together to increase their length. Depending on the use, padding may be added during fabrication or added in the field to protect against sharp edges. Never tie a knot in a lanyard, as knots

24   NCCER – *Core Curriculum*   00101-15

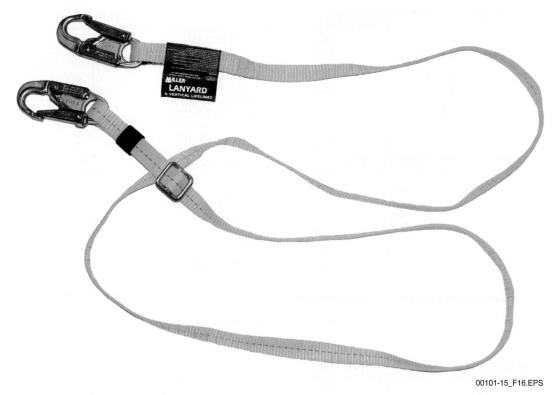

Figure 16 Non-shock absorbing lanyard.

can severely reduce the load limit. Never wrap a lanyard around a structure and then choke it back into itself unless it is designed for this use. A special large D-ring is usually attached to the lanyard for this purpose. It is important to remember that whenever you are wearing a PFAS, 100 percent tie-off is required. A PFAS is absolutely useless unless you are tied off.

The D-ring used on the PFAS for fall arrest may only be connected to one live connection at a time. This can be challenging when trying to move from one point to another, especially horizontally. The user must be able to reach back and disconnect a lanyard, while maintaining one lanyard connected at all times. A Y-configured lanyard (*Figure 17*) can be used for this purpose. A Y-configured lanyard has a single point of attachment at the D-ring that is used to accommodate two lanyards. They are also referred to as double-leg or tieback lanyards.

Before using a shock absorbing or self-retracting lanyard, the potential fall distance is determined. Then the proper equipment is selected to meet available fall clearance (*Figure 18*). Note in the figure that the fall distance with the self-retracting lanyard is significantly shorter than it is with a shock absorbing lanyard; hence its increasing popularity. Failure to select proper equipment and calculate fall distance may result in serious personal injury or death. These calculations must be done by experienced and qualified personnel on the job site. If a personal fall arrest system is actuated due to a fall, it must be inspected by a competent person before it can be used again. In most cases, replacement will be necessary.

2.2.5 Lifelines

A lifeline is a flexible line such as a cable or rope connected vertically to an anchorage at one end (vertical lifeline), or horizontally to an anchorage at both ends (horizontal lifeline). It serves as a means for connecting other components of a personal fall-arrest system to the available anchorage.

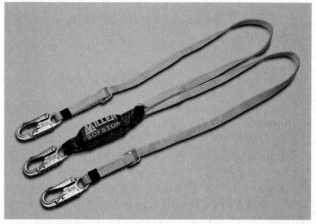

Figure 17 Y-configured shock absorbing lanyard.

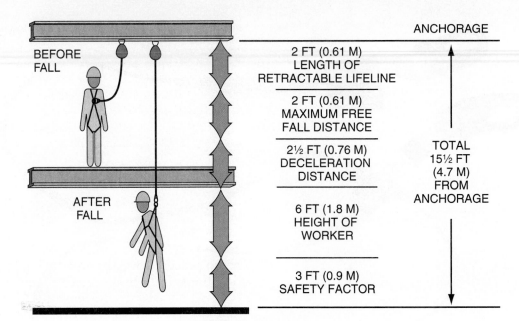

ANCHORAGE

2 FT (0.61 M)
LENGTH OF
RETRACTABLE LIFELINE

2 FT (0.61 M)
MAXIMUM FREE
FALL DISTANCE

2½ FT (0.76 M)
DECELERATION
DISTANCE

6 FT (1.8 M)
HEIGHT OF
WORKER

3 FT (0.9 M)
SAFETY FACTOR

TOTAL
15½ FT
(4.7 M)
FROM
ANCHORAGE

BEFORE
FALL

AFTER
FALL

USING A SELF-RETRACTING LANYARD

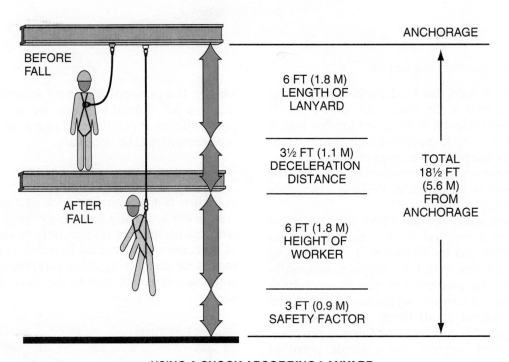

ANCHORAGE

6 FT (1.8 M)
LENGTH OF
LANYARD

3½ FT (1.1 M)
DECELERATION
DISTANCE

6 FT (1.8 M)
HEIGHT OF
WORKER

3 FT (0.9 M)
SAFETY FACTOR

TOTAL
18½ FT
(5.6 M)
FROM
ANCHORAGE

BEFORE
FALL

AFTER
FALL

USING A SHOCK ABSORBING LANYARD

00101-15_F18.EPS

Figure 18 Potential fall distances for self-retracting and shock absorbing lanyards.

Vertical lifelines are suspended from a fixed anchorage. A fall arrest device such as a rope grab (*Figure 19*) or retractable lanyard is attached to the lifeline. A beam grab, or beamer (*Figure 20*), is sometimes used as an anchorage for a lifeline. Vertical lifelines must have a minimum breaking strength of 5,000 pounds (2,267 kg). Workers must use separate vertical lifelines. A vertical lifeline must have a termination on the end unless it extends to the ground or to the next lower level of the structure.

Horizontal lifelines (*Figure 21*) are connected between two fixed anchorages. A lanyard is attached to the lifeline. Horizontal lifelines must be designed, installed, and used under the supervision of a competent person. The required breaking strength of any horizontal lifeline is determined by the number of workers that will be attached to it.

Figure 19 Rope grab.

00101-15_F19.EPS

00101-15_F20.EPS

Figure 20 Beam grab.

2.2.6 Guardrails

Guardrails (*Figure 22*) are a common type of fall prevention. They protect workers by providing a barrier between the work area and the ground or lower work areas. They may be made of wood, pipe, steel, or wire rope and must be able to support 200 pounds (90 kg) of force applied in any direction to the top rail and 150 pounds (68 kg) for the midrail. A guardrail must be 42" ±3" (106 ±8 cm) high to the top rail and have a toeboard that is a minimum of 4" (10 cm) high. This helps to prevent the inadvertent loss of tools or material through the bottom rail. The toeboard must be securely fastened with not more than ¼" (5 mm) clearance above the floor level.

2.2.7 Safety Nets

Safety nets are used for fall protection on bridges and similar projects. They must be installed not more than 30 feet (9 m) beneath the work area. There must be enough clearance under a safety net to prevent a worker who falls into it from hitting the surface below it. There must also be no obstruction between the work area and the net.

Depending on the actual vertical distance between the net and the work area, the net must extend 8 to 13 feet (2 to 4 m) beyond the edge of the work area. Mesh openings in the net must be limited to 36 square inches (232 sq cm) and 6 inches (15 cm) from the side. The border rope must have a 5,000-pound (2,267 kg) minimum breaking strength, and connections between net panels must be as strong as the nets themselves. Safety nets must be inspected at least once a week and after any event that might have damaged or weakened them. Worn or damaged nets must be removed from service.

2.3.0 Ladders and Stairs

Ladders are used daily to perform work in elevated locations. Any time work is performed above ground level, there is a risk of incidents. You can reduce this risk by carefully inspecting ladders before you use them and by using them properly.

Overloading means exceeding the maximum intended load of a ladder. Overloading can cause ladder failure, which means that the ladder could buckle, break, or topple. The maximum intended load is the total weight of all people, equipment, tools, materials, loads that are being carried, and other loads that the ladder can hold at any one time. Check the manufacturer's specifications to determine the maximum intended load. Ladders are usually given a duty rating that indicates their load capacity, as shown in *Table 2*. Note that ladders designed for the metric market are not usually direct equivalents to ladders built for the American market. Capacity ratings of 130 kg (286 lbs) and 150 kg (330 pounds) are the most common for the trades. Ladder heights will also be stated in metric units and may not be exactly the same size as those designed for the American market.

00101-15_F21.EPS

Figure 21 Horizontal lifeline.

00101-15_F22.EPS

Figure 22 Guardrails.

Drop-Testing Safety Nets

Safety nets should be drop-tested at the job site after the initial installation, whenever relocated, after a repair, and at least every six months if left in one place. The drop test consists of a 400-pound (181 kg) bag of sand of 29" to 31" (74 to 79 cm) in diameter that is dropped into the net from at least 42" (107 cm) above the highest walking/working surface at which workers are exposed to fall hazards. If the net is still intact after the bag of sand is dropped, it passed the test.

Table 2 Ladder Duty Ratings and Load Capacities

Duty Rating	Load Capacities
Type IAA	375 lbs., extra-heavy duty/professional use
Type IA	300 lbs., extra-heavy duty/professional use
Type I	250 lbs., heavy duty/industrial use
Type II	225 lbs., medium duty/commercial use
Type III	200 lbs., light duty/household use

There are different types of ladders to use for different jobs (*Figure 23*). Selecting the right ladder for the job at hand is important to complete a job as safely and efficiently as possible. Ladder types include portable straight ladders, extension ladders, and stepladders.

> **WARNING!**
>
> Use ladders only for their intended purposes. Ladders are not interchangeable; incorrect use of a ladder can result in injury or damage.
>
> When using a ladder, be sure to maintain three-point contact with the ladder while ascending or descending. Three-point contact means that either two feet and one hand or one foot and two hands are always touching the ladder.

2.3.1 Straight Ladders

Straight ladders consist of two rails, rungs between the rails, and safety feet on the bottom of the rails (*Figure 24*). Straight ladders are generally made of aluminum, wood, or fiberglass.

Metal ladders conduct electricity and should never be used around electrical equipment. Any portable metal ladder must have "Danger! Do Not Use Around Electrical Installations" stenciled on the rails in two-inch red letters. Ladders made of dry wood or fiberglass, neither of which conducts electricity, should be used around electrical equipment. Check that any ladder, especially a wooden ladder, is completely dry before using it; even a small amount of water will conduct electricity.

Case History

A worker was climbing a 10-foot (3.05 m) ladder to access a landing, which was 9 feet (2.74 m) above the adjacent floor. The ladder slid down, and the worker fell to the floor, sustaining fatal injuries. Although the ladder had slip-resistant feet, it was not secured, and the railings did not extend 3 feet (0.91 m) above the landing.

Different types of ladders are intended for use in specific situations. Aluminum ladders are corrosion-resistant and can be used where they might be exposed to the elements. They are also lightweight and can be used where they must often be lifted and moved. Fiberglass ladders are very durable, so they are useful where some amount of rough treatment is unavoidable. Wooden ladders, which are heavier and sturdier than fiberglass or aluminum ladders, can be used where heavy loads must be moved up and down. However, wooden ladders are subject to more rapid deterioration as the wood swells and shrinks. Both fiberglass and aluminum are easier to clean than wood.

Wooden ladders should never be painted. The paint could hide cracks in the rungs or rails. Clear varnish, shellac, or a preservative oil finish will protect the wood without hiding defects.

Figure 25 shows the safety feet attached to a straight ladder. Make sure the feet are securely attached and that they are not damaged or worn down. Do not use a ladder if its safety feet are not in good working order.

It is very important to place a straight ladder at the proper angle before using it. A ladder placed at an improper angle will be unstable and could cause a fall. *Figure 26* shows a properly positioned straight ladder.

The distance between the foot of a ladder and the base of the structure it is leaning against must be one-fourth of the distance between the ground and the point where the ladder touches the structure. Stated another way, there should be a 4-to-1 ratio between the distances. For example, if the height of the wall shown in *Figure 26* is 16 feet (4.9 m), the base of the ladder should be 4 feet (1.2 m) from the base of the wall. If you are going to step off a ladder onto a platform or roof, the top of the ladder should extend at least 3 feet (0.9 m) above the point where the ladder touches the platform, roof, side rails, etc.

Ladders should be used only on stable and level surfaces unless they are secured at both the bottom and the top to prevent any incidental movement (*Figure 27*). Never try to move a ladder while you are on it. If a ladder must be placed in front of a door that opens toward the ladder, the door should be locked or blocked open. Otherwise, the door could be opened into the ladder.

Ladders are made for vertical use only. Never use a ladder as a work platform by placing it horizontally. Make sure the ladder you are about to climb or descend is properly secure before you do so. Check to make sure the ladder's feet are solidly positioned on firm, level ground. Also check

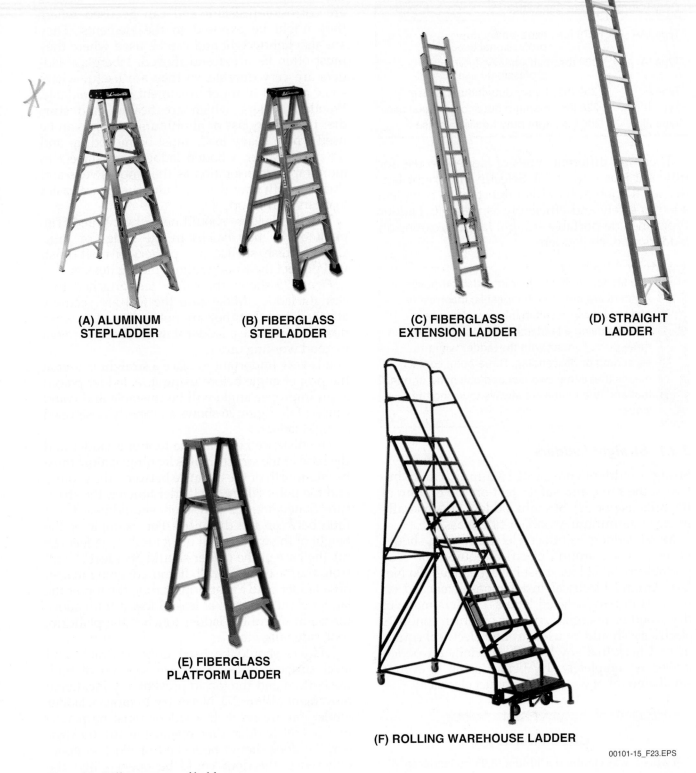

(A) ALUMINUM STEPLADDER

(B) FIBERGLASS STEPLADDER

(C) FIBERGLASS EXTENSION LADDER

(D) STRAIGHT LADDER

(E) FIBERGLASS PLATFORM LADDER

(F) ROLLING WAREHOUSE LADDER

00101-15_F23.EPS

Figure 23 Different types of ladders.

to make sure the top of the ladder is firmly positioned and in no danger of shifting once you begin your climb. Remember that your own weight will affect the ladder's steadiness once you mount it. It is important to test the ladder first by putting some of your weight on it without actually beginning to climb. This way, you can be sure that the ladder will remain steady as you climb.

When climbing a straight ladder, keep both hands on the rails or rungs (*Figure 28*). Maintain

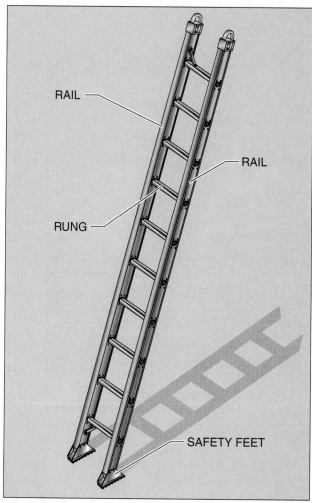

RAIL

RAIL

RUNG

SAFETY FEET

00101-15_F24.EPS

Figure 24 Portable straight ladder.

00101-15_F25.EPS

Figure 25 Ladder safety feet.

AT LEAST 3 FEET (0.9 M)

16 FEET
(4.9 M)

4 FEET
(1.2 M)
(¼ OF THE HEIGHT)

00101-15_F26.EPS

Figure 26 Proper positioning of a straight ladder.

three points of contact at all times, as shown in the figure. This can be two feet and one hand, or one foot and two hands. Always keep your body's weight in the center of the ladder between the rails. Face the ladder at all times. Never go up or down a ladder while facing away from it.

To carry a tool while you are on the ladder, use a **hand line** or tagline attached to the tool. Climb the ladder and then pull up the tool. Don't carry tools in your hands while you are climbing a ladder.

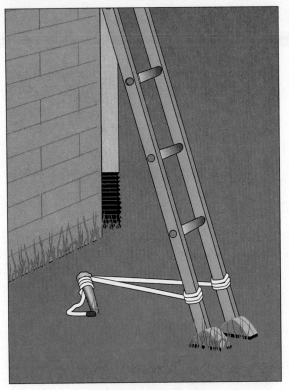

BOTTOM SECURED

TOP SECURED

00101-15_F27.EPS

Figure 27 Securing a ladder.

00101-15_F28.EPS

Figure 28 Moving up or down a ladder.

 ### 2.3.2 *Extension Ladders*

An extension ladder is actually two straight ladders connected so that the overlap between them can be altered to increase or decrease the length of the ladder (*Figure 29*).

Extension ladders are positioned and secured following the same rules as straight ladders. When adjusting the length of an extension ladder, always reposition the movable section from the bottom, not the top, to ensure that the rung locks

ALUMINUM   **FIBERGLASS**

00101-15_F29.EPS

Figure 29 Examples of extension ladders.

(*Figure 30*) are properly engaged after you make the adjustment. Check to make sure the section locking mechanisms are fully hooked over the desired rung. Also check to make sure that all ropes used for raising and lowering the extension are clear and untangled.

Extension ladders are positioned and secured following the same rules as straight ladders. There are, however, some safety rules that are unique to extension ladders:

> **WARNING!**
>
> Extension ladders have a built-in extension stop mechanism. Do not remove this mechanism. If the mechanism is removed, it could cause the ladder to collapse under a load.
>
> Haul materials up on a line rather than hand carrying them up an extension ladder.

- Make sure the extension ladder overlaps between the two sections (*Figure 31*). For ladders up to 36' (10.5 m) long, the overlap must be at least 3' (0.9 m). For ladders 36' to 48' (10.8 to 14.6 m) long, the overlap must be at least 4' (1.2 m). For ladders 48' to 60' (14.6 to 18.3 m) long, the overlap must be at least 5' (1.5 m).

00101-15_F30.EPS

Figure 30 Rung locks.

- Never stand above the highest safe standing level on a ladder. On an extension ladder, this is the fourth rung from the top. If you stand higher, you may lose your balance and fall. Some ladders have colored rungs to show where you should not stand.
- Avoid carrying anything on a ladder, because it will affect your balance and may cause you to fall. Haul materials up on a line instead.
- Keep yourself centered on the ladder. Do not over-reach, lean to one side, or try to move a ladder while standing on it.

2.3.3 Stepladders

Stepladders are self-supporting ladders made of two sections hinged at the top (*Figure 32*). The section of a stepladder used for climbing consists of rails and rungs like those on straight ladders. The other section consists of rails and braces. Spreaders are hinged arms between the sections that keep the ladder stable and keep it from folding while in use.

When positioning a stepladder, be sure that all four feet are on a hard, even surface. Otherwise, the ladder can rock from side to side or corner to corner when you climb it. With the ladder in position, be sure the spreaders are locked in the fully open position. The following safety precautions must be followed when using a stepladder:

- Never stand on the top step or the top of a stepladder. Putting your weight this high will make the ladder unstable. The top of the ladder is made to support the hinges, not to be used as a step.
- Although the rear braces may look like rungs, they are not designed to support your weight. Never use the braces for climbing or climb the back of a stepladder. However, there are specially designed two-person ladders available with steps on both sides.
- Check the load capacity of the ladder and do not exceed it.

> **WARNING!**
>
> Never stand on a step with your knees higher than the top of a stepladder. You need to be able to hold on to the ladder with your hand. Keep your body centered between the side rails.

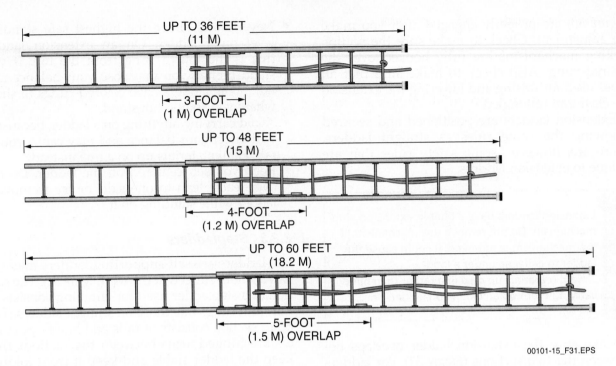

UP TO 36 FEET
(11 M)

3-FOOT
(1 M) OVERLAP

UP TO 48 FEET
(15 M)

4-FOOT
(1.2 M) OVERLAP

UP TO 60 FEET
(18.2 M)

5-FOOT
(1.5 M) OVERLAP

00101-15_F31.EPS

Figure 31 Overlap lengths for extension ladders.

LADDER TOP

TOP STEP

FRONT SIDE
RAILS

SPREADER

REAR SIDE
RAILS

BRACE

SAFETY FEET

00101-15_F32.EPS

Figure 32 Typical fiberglass stepladder.

2.3.4 Inspecting Ladders

Always inspect a ladder before you use it. Check the rails and rungs for cracks or other damage. Also, check for loose rungs. If you find any damage, do not use the ladder. Check the entire ladder for loose nails, screws, brackets, or other hardware. If you find any hardware problems, tighten the loose parts or have the ladder repaired before you use it. OSHA requires regular inspections of all ladders and an inspection just before each use.

> **CAUTION**
>
> Wooden ladders should never be painted. The paint could hide cracks in the rungs or rails. Clear varnish, shellac, or a preservative oil finish will protect the wood without hiding defects.

The same rules for inspecting straight ladders apply to extension ladders. In addition, the rope that is used to raise and lower the movable section of the ladder should be inspected. If the rope is frayed or has worn spots, it should be replaced before the ladder is used.

The rung locks support the entire weight of the movable section and the person climbing the ladder. Inspect them for damage before each use. If they are damaged, they should be repaired or replaced before the ladder is used.

Inspect stepladders the way you inspect straight and extension ladders. Pay special attention to the hinges and spreaders to be sure they are in good repair. Also, be sure the rungs are clean. The rungs of a stepladder are usually flat, so oil, grease, or dirt can build up on them and make them slippery.

2.3.5 Stairways

Stairways are also routinely used on construction sites where there is a break in elevation of 19 inches (46 cm) or more, and no ramp, runway, sloped embankment, or personnel hoist is provided. Observe the following regulations, based on OSHA standards, when using stairways on a job site:

- Stairways having four or more risers or rising more than 30 inches (76 cm), whichever is less, must be equipped with at least one handrail and one stair railing system along each unprotected side.
- Winding and spiral stairways must be equipped with a handrail offset sufficiently to prevent walking on those portions of the stairways where the tread width is less than 6 inches (15 cm).
- Stair railings must be not less than 36 inches (91 cm) from the upper surface of the stair railing system to the surface of the tread, in line with the face of the riser at the forward edge of the tread.

To reduce the likelihood of slips, trips, or falls, keep stairways clean and clear of debris. Do not store any tools or materials on stairways, and clean up liquid spills, rain water, or mud immediately.

Stairways must have adequate lighting. This can sometimes be a problem because permanent lighting is usually installed after stairway construction is completed. If the lighting is inadequate, temporary lighting should be installed in the stairway. Each bulb should be equipped with a protective cover and the string should be inspected daily for burned out or broken bulbs.

Whenever possible, avoid using stairways to transport materials between floors. Carrying small materials and tools is fine, as long as the materials do not block your vision. Going up or down a stairway while carrying large items is physically demanding and increases the chance of injuries and falls. Use the building elevator or crane service to transport large materials from one floor to another.

2.4.0 Scaffolds

Scaffolds provide safe elevated work platforms for people and materials. They are designed and built to comply with high safety standards, but normal wear and tear or incidentally putting too much weight on them can weaken them and make them unsafe. That's why it is important to inspect every part of a scaffold before each use. Personnel who assemble scaffolds must be certified to do so.

> **CAUTION**
>
> Only a competent person has the authority to supervise setting up, moving, and taking down scaffolds. Only a competent person can approve the use of scaffolds on the job site after inspecting the scaffolds.

2.4.1 Types of Scaffolds

Two basic types of scaffolds—self-supporting scaffolds and suspended scaffolds—are used in the construction industry. The rules for safe use apply to both of them. Self-supporting scaffolds can be manufactured units or can be assembled at the site.

Manufactured scaffolds (*Figure 33*) are made of painted steel, stainless steel, or aluminum. They are stronger and more fire-resistant than wooden scaffolds. They are supplied in ready-made, individual units, which are assembled on site. A rolling scaffold has wheels on its legs so that it can be easily moved. The scaffold wheels have brakes so the scaffold will not move while workers are standing on it.

> **WARNING!**
>
> Never unlock the wheel brakes of a rolling scaffold while anyone is on it. People on a moving scaffold can lose their balance and fall.

Built-up scaffolds (*Figure 34*) are built from the ground up at a job site using steel framework sections and lumber. Swing (suspended) scaffolds (*Figure 35A*) are suspended by ropes or cables in a manner that allows it to be raised or lowered as needed. Another type of suspended scaffolding is a work cage (*Figure 35B*). A work cage is typically suspended with rigging devices that attach to I-beams with various sizes of clamps and rollers.

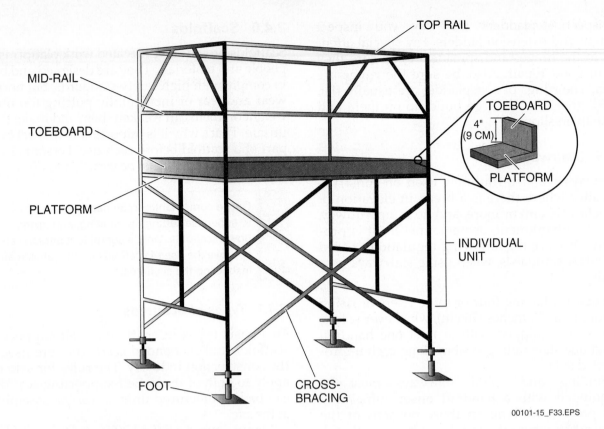

Labels on figure:
TOP RAIL
MID-RAIL
TOEBOARD
PLATFORM
FOOT
CROSS-BRACING
TOEBOARD
4" (9 CM)
PLATFORM
INDIVIDUAL UNIT

00101-15_F33.EPS

Figure 33 Typical manufactured scaffold.

2.4.2 Inspecting Scaffolds

Any scaffold that is assembled on the job site must be tagged to indicate whether the scaffold meets OSHA standards and is safe to use. Three colors of tags are used: green, yellow, and red (*Figure 36*).

- A green tag means the scaffold meets all OSHA standards and is safe to use.
- A yellow tag means the scaffold does not meet all OSHA standards. An example is a scaffold on which a railing cannot be installed because of equipment interference. To use a yellow-tagged scaffold, you must wear a safety harness attached to a lanyard. You may have to take other safety measures as well.
- A red tag means a scaffold is being put up or taken down. Never use a red-tagged scaffold.

Don't rely on the tags alone; inspect all scaffolds before you use them. Check for bent, broken, or badly rusted tubes. Also check for loose joints where the tubes are connected. Any of these problems must be corrected before the scaffold is used.

Make sure you know the weight limit of any scaffold you will be using. Compare this weight limit to the total weight of the people, tools, equipment, and material you expect to put on the scaffold. Scaffold weight limits must never be exceeded.

If a scaffold is more than 10 feet (3.1 m) high, check to see that it is equipped with top rails, midrails, and toeboards; otherwise, use a PFAS. All connections must be pinned. That means they must have a piece of metal inserted through a hole to prevent connections from slipping. Cross-bracing must be used. A handrail is not the same as cross-bracing. The walking area must be completely planked.

If it is possible for people to walk under a scaffold, the space between the toeboard and the top rail must be screened. This prevents objects from falling off the work platform and injuring those below.

When you examine a rolling scaffold, check the condition of the wheels and brakes. Be sure the brakes are working properly and can stop the scaffold from moving while work is in progress. Be sure all brakes are locked before you use the scaffold.

2.4.3 Using Scaffolds

Be sure that a competent person inspects the scaffold before you use it. There should be firm footing under each leg of a scaffold before putting any weight on it. If you are working on loose or soft soil, you can put planks or matting under the scaffold's legs or wheels, as shown in *Figure 34*. When moving a rolling scaffold, first unlock the brakes and then move the scaffold. Once the scaffold is repositioned, don't forget to relock the brakes.

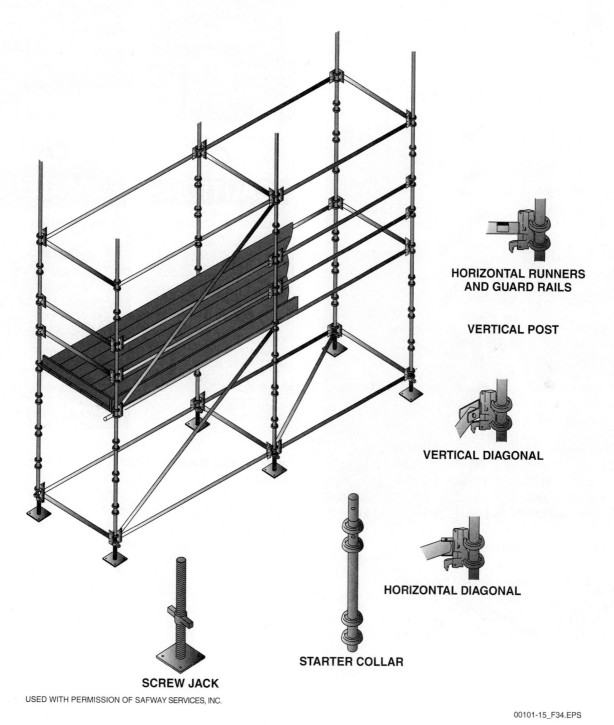

HORIZONTAL RUNNERS AND GUARD RAILS

VERTICAL POST

VERTICAL DIAGONAL

HORIZONTAL DIAGONAL

STARTER COLLAR

SCREW JACK

USED WITH PERMISSION OF SAFWAY SERVICES, INC.

00101-15_F34.EPS

Figure 34 Built-up scaffold.

(A) SUSPENDED SCAFFOLD

(B) WORK CAGE

00101-15_F35.EPS

Figure 35 Suspended scaffold and work cage.

Figure 36 Typical scaffold tags.

00101-15_F36.EPS

NCCER – *Core Curriculum*   00101-15

Additional Resources

US Occupational Safety and Health Administration. Numerous safety videos are available on line at **www.osha.gov/video**.

Construction Safety, Jimmie W. Hinze. 2006. Upper Saddle River, NJ: Pearson Education, Inc.

DeWalt Construction Safety/OSHA Professional Reference. Paul Rosenberg; American Contractors Educational Services. 2006. DEWALT.

Basic Construction Safety and Health, Fred Fanning. 2014. CreateSpace Independent Publishing Platform.

2.0.0 Section Review

1. A barrier with yellow and purple markings indicates a _____.
   a. fire hazard
   b. fall hazard
   c. radiation hazard
   d. confined space hazard

2. Positioning lanyards are used only as fall restraint devices and may *not* be used as fall arrest devices.
   a. True
   b. False

3. When positioning a straight ladder against a wall, how far from the wall should the base of the ladder be?
   a. Four feet (1.2 m)
   b. One-fourth the distance from the ground to the point where the ladder touches the wall
   c. The height of the wall minus 4 feet (1.2 m)
   d. One-half the distance from the ground to the point where the ladder touches the wall

4. If a scaffold has a yellow tag, it means _____.
   a. the scaffold may only be used by one person at a time
   b. the scaffold is condemned and cannot be used
   c. a safety harness must be worn when using it
   d. the scaffold is under assembly and cannot be used

SECTION THREE

3.0.0 STRUCK-BY AND CAUGHT-IN-BETWEEN HAZARDS

Objectives

Identify and explain how to avoid struck-by and caught-in-between hazards.

  a. Identify and explain how to avoid struck-by hazards.

  b. Identify and explain how to avoid caught-in and caught-between hazards.

Trade Terms

Shielding: A structure used to protect workers in trenches.

Shoring: A support system designed to prevent a trench or excavation cave-in.

Signaler: A person who is responsible for directing a vehicle when the driver's vision is blocked in any way.

Spoil: Material such as earth removed while digging a trench or excavation.

There is an apparent overlap in struck-by and caught-in-between hazards that needs to be clarified, as they can involve the same kinds of equipment. *Struck-by* means being hit by a moving object, while *caught-in-between* means being trapped between a moving object and a solid surface. Assume a worker is working near a crane that is swinging a load and the load hits the worker. That is an example of a struck-by incident. However, if the worker is behind the crane and the crane backs up, pinning the worker against a wall, it becomes a caught-in-between incident.

3.1.0 Struck-By Hazards

On any job site, there is a risk of being struck by falling objects, such as tools dropped from above. Flying objects such as debris from grinding and chipping metal are another struck-by hazard. On any site where there is moving equipment, or where workers are near roadways, there is a also the danger of being struck by a vehicle.

 ### 3.1.1 Falling Objects

Workers are at risk from falling objects when they are beneath machinery and equipment such as cranes and scaffolds; where overhead work is being performed; or when working around stacked materials. To protect against struck-by injuries from falling objects, always wear an approved hard hat. Employers generally require workers to wear hard hats at all times on construction sites.

When working near machinery and equipment such as cranes, never stand or work beneath the load or in the fall zone. Barricade hazard areas where rigging equipment is in use, and post warning signs to inform other workers of falling object hazards. Inspect cranes and rigging components before use and do not exceed the rated load capacity.

When performing overhead work, be sure all tools, materials, and equipment are secured to prevent them from falling on people below. Use protective measures such as toeboards, debris nets, catch platforms, or canopies to catch or deflect falling objects. Use tool lanyards to prevent tools from falling.

Many workers are hurt or killed by falling stacks of material. Do not stack materials higher than 4:1 height-to-base ratio. Secure all loads by blocking and interlocking them. Interlocking means placing alternate layers at right angles to each other. Be aware of changing weather conditions, such as wind, that may lift and shift loads.

3.1.2 Flying Objects

There is a danger from flying objects when power tools or activities such as pushing, pulling, or prying causes objects to become airborne. Chipping, grinding, brushing, or hammering are all examples of job tasks that may cause flying objects. Tools that move at very high speed, like pneumatic and powder-actuated tools, can be very dangerous. Injuries from flying objects can range from minor abrasions to concussions, blindness, or death. The workers shown in *Figure 37* are using a pneumatic chipping hammer and a pneumatic grinder, and are properly using protective face shields.

To protect against flying object hazards, follow these guidelines:

- Use eye protection, such as safety glasses, goggles, or face shields where machines or tools may cause flying particles.
- Inspect tools and machines to ensure that protective guards are in place and in good condition.
- Make sure you are trained in the proper operation of pneumatic and powder-actuated tools.
- Use shielding devices such as welding screen or similar equipment to block flying debris.

3.1.3 Vehicle and Roadway Hazards

A common cause of incidents for workers on large job sites and highway projects are vehicle-related hazards such as being run over by vehicles or equipment (especially backing equipment) or by equipment tip-over. The most common cause of death for equipment operators is equipment roll-over. If vehicle safety practices are not observed at your site, you risk being struck by swinging backhoes, moving vehicles, or swinging crane loads. If you work near public roadways (*Figure 38*), you risk being struck by passenger or commercial vehicles. When working near moving vehicles and equipment, follow these guidelines:

- Stay alert at all times and keep a safe distance from vehicles and equipment.
- Maintain eye contact with vehicle or equipment operators to ensure that they see you.
- Never get into blind spots of equipment operators.
- Keep off equipment unless authorized.
- Wear reflective or high-visibility vests or other suitable garments.
- Never stand between pieces of equipment unless they are secured.
- Never stand under loads handled by lifting or digging equipment, or near vehicles being loaded or unloaded.

Operators must also use caution when driving vehicles. The operator of any vehicle is responsible for the safety of passengers and the protection of the load. Follow these safety guidelines when you operate a vehicle on a job site:

- Always wear a seat belt.
- Be sure that each person in the vehicle has a firmly secured seat and seat belt.
- Obey all speed limits. Reduce speed in crowded areas.
- Look to the rear and sound the horn before backing up. If your rear vision is blocked, get a **signaler** to direct you.

00101-15_F37.EPS

Figure 37 Protection from flying particles.

Case Histories

- A worker was standing under a suspended scaffold that was hoisting a workman and three sections of ladder. Sections of that ladder became unlatched and fell 50 feet (≈15m), striking the worker in the skull. The worker was not wearing any head protection and died from his injuries.
- A carpenter was using a powder-actuated tool to anchor a plywood form in preparation for pouring a concrete wall. The nail passed through the hollow wall, traveled 72 feet (≈22m), and struck an apprentice in the head, killing him. The tool operator had never been trained in the proper use of the tool, and none of the employees in the area, including the victim, was wearing PPE.

Tool Lanyards

When working at elevations, a dropped tool can become a serious hazard. If the tool has moving parts, like battery-powered drills, the fall will likely destroy it.

Tool lanyards specifically designed for work in the elevated environment have been introduced by Snap-on®. Tethering tools to the tool belt or wrist can prevent injuries, save tools, prevent component damage, and prevent lost time in retrieval.

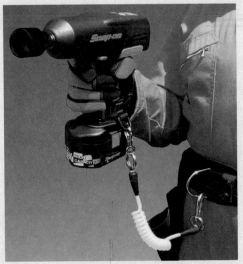

00101-15_SA01.EPS

- Every vehicle must have a backup alarm. Make sure the backup alarm works.
- Always turn off the engine when fueling.
- Turn off the engine and set the brakes before leaving the vehicle.
- Never stay on or in a truck that is being loaded by excavating equipment.

- Keep windshields, rearview mirrors, and lights clean and functional.
- Carry road flares, fire extinguishers, and other standard safety equipment at all times.
- Never use a cell phone while operating a motor vehicle.

00101-15_F38.EPS

Figure 38 A busy job site.

NCCER – *Core Curriculum*   00101-15

3.2.0 Caught-In and Caught-Between Hazards

Congested work sites, heavy equipment, and the presence of multiple trades can contribute to caught-in-between hazards. The primary causes of caught-in-between fatalities include trench/excavation collapse, rotating equipment, and unguarded parts.

One of the most disastrous caught-in hazards on a construction job site is being trapped by a cave-in of the walls of a trench or excavation. Other caught-in or caught-between hazards involve protective guards on power tools and machines, as well as the risk of being crushed by heavy equipment.

3.2.1 Trenches and Excavation

Trenches and excavations are common hazards, especially in construction and pipeline work. Anyone working in or around a trench or excavation must know and follow safety procedures aimed at protecting workers from cave-ins.

An excavation is any man-made cut, cavity, trench, or depression formed by removal of earth or soil. Sometimes the terms *excavation* and *trench* are used interchangeably, but there is a difference. A trench is an excavation that is deeper than it is wide, and usually not wider than 15 feet (4.6 m). Nearly all trenches are dangerous if not protected. Because trenches are narrow, workers can easily become trapped.

Hazards involved with trench and excavation work include the following:

- Cave-ins
- Water accumulation
- Falling objects
- Collapse of nearby structures
- Hazardous atmospheres produced by toxic gases in the soil

> **WARNING!**
>
> Just 2 to 3 feet (0.61 to 0.91 m) of soil can put enough pressure on your lungs to prevent you from breathing. In as little as 4 to 6 minutes without oxygen, you can sustain considerable brain damage.

Cave-ins are the most common and deadly hazard in excavation work. When dirt is removed from an excavation, the surrounding soil becomes unstable and gravity can force it to collapse. Cave-ins occur when soil or rock falls, or slides, into an excavation. Most cave-ins occur in trenches 5 to 15 feet (1.53 to 4.6 m) deep, and happen suddenly with little or no warning. On average, about 1,000 trench collapses occur each year in the United States.

Soil conditions can change, so they must be constantly evaluated. There are certain factors that could change the surroundings of the site, making a cave-in more likely. These factors include the following:

- Changes in weather conditions, such as freezing, thawing, or sudden heavy rain
- An excavation dug in unstable or previously disturbed soil
- Excessive vibration around the excavation
- Water accumulation in an excavation

Sometimes there are visible warning signs around the excavation that can be spotted before a cave-in occurs. Being aware of the warning signs increases your chances of getting out before a collapse occurs. Visible warning signs of a potential cave-in include the following:

- Ground settlement or narrow cracks in the sidewalls, slopes, or surface next to the excavation

Did You Know?

Struck-by Fatalities

Nearly one in four struck-by vehicle deaths involve construction workers – more than any other occupation.

Did You Know?

Excavation Fatalities

The fatality rate for excavation work is 112 percent higher than the rate for general construction.

- Flakes, pebbles, or clumps of soil separating and falling into the excavation
- Changes or bulges in the wall slope

If you notice any of these signs, get out of the excavation immediately and alert your co-workers as well.

Soil type is a key factor in determining the type of protective system needed to ensure that the trench will be safe. Solid rock is the most stable, while sandy soil is the least stable. The four types of soil are shown in *Table 3*.

To be safe, treat soil as if it is Type C soil, per *Table 3*, unless proven otherwise. It is better to over-prepare for a stronger soil than to not prepare enough for a weaker one.

A competent person must inspect excavations daily and decide whether cave-ins or failures of protective systems could occur, and whether there are any other hazardous conditions present. The competent person must conduct inspections before any work begins, as needed throughout the shift, and after every rainstorm or other hazard-increasing incident.

If the inspection reveals indications of protective system failure, hazardous atmospheres, or a possible cave-in, workers must be removed from the hazardous area and may not return until corrective action has been taken. Always ask the competent person on site or your immediate supervisor if you have questions about proper safety practices.

Once visual and manual tests are performed and the soil type is determined, a protective system must be chosen. Protective systems are required in nearly all excavations. There are various types of trench protective systems to meet each type of soil condition. Selecting a protective system for an excavation depends on soil conditions, the depth of the trench, and the environmental conditions surrounding the site.

Regardless of what type of system is used, if the excavation is more than 20 feet (6.1 m) deep, the entire excavation protective system has to be

designed by a registered professional engineer. There are two basic systems of trench protection: sloping and benching systems and support systems. The method to be used is determined by the engineer and is based on the types of soil and the site conditions.

Sloping and benching are forms of trench protection that cut away and slant the excavation face. A sloping system is a method in which the sides of an excavation are cut back to a safe angle using relatively smooth inclines (*Figure 39*, top). A benching system is similar to a sloping system, but instead of smooth inclines, the sides of the trench wall are cut back using a series of steps (*Figure 39*, bottom). Benching systems cannot be used with Type C soil.

Trenches are often located in narrow places, so sloping and benching are not options. In these situations, support systems like shoring or shielding must be used. Shoring structures are typically made of metal or wood and are used to support the sides of a trench and prevent soil from caving in. They consist of plating held firmly in place with expandable braces (*Figure 40*). There are many types of shoring systems. Some of them are easy to install, and others require experience and engineering.

Shielding structures, also known as trench boxes, are placed inside trenches or excavations, and are strong enough to protect workers in the event of a cave-in, so long as the workers are within the confines of the box (*Figure 41*). Trench shields are used only to provide a protected space for workers. Shoring not only protects workers, but also prevents the trench walls from collapsing.

Table 3 Soil Types

Name	Type/Characteristics
Solid Rock	Excavation walls stay vertical as long as the excavation is open.
Type A Soil	Fine-grained, cohesive: clay, hardpan, and caliche. Particles too small to see with the naked eye.
Type B Soil	Angular rock, silt, and similar soil.
Type C Soil	Coarse-grained, granular: sand, gravel, and loamy sand. Particles are visible to the naked eye.

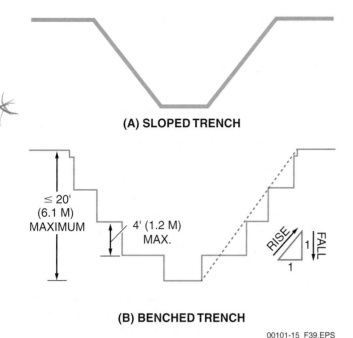

(A) SLOPED TRENCH

≤ 20'
(6.1 M)
MAXIMUM

4' (1.2 M)
MAX.

RISE

FALL

1

1

(B) BENCHED TRENCH

00101-15_F39.EPS

Figure 39 Sloped and benched trenches for Type B soil.

00101-15_F40.EPS

Figure 40 Shoring structure.

Case History

A worker was in a trench installing forms for concrete footers when it caved in, causing fatal injuries. The trench, which was 7½ feet (2.29 m) deep, was in loose, sandy (Type C) soil, and no inspection was conducted before the start of the shift.

Spoil piles, comprised of the material removed from an excavation, and other materials represent a hazard if not handled properly. Loose rock, soil, materials, and equipment on the face or near the excavation can fall or roll into the excavation, or overload and possibly collapse excavation walls. Keep all spoil, materials, and heavy equipment at least 2 feet (0.6 m) away from the edge of an excavation, or set up barricades to contain falling material. Scale the excavation face to remove loose material, and place spoil so that rainwater runs away from the excavation. Use a retaining device strong enough and high enough to resist expected loads. If the spoil cannot be safely stored on site, remove it to a temporary site.

When working in a trench, there must be a safe means of entry and exit for workers, such as a stairway, ladder, or ramp. There must be an exit every 25 feet (7.6 m) for every trench over 4 feet (1.2 m) deep. Lifting equipment such as loader buckets and backhoe shovels are not safe means for entering or exiting a trench. Once you are in the trench and before you begin work, take a moment to look around and find the nearest ladder so that you can plan your exit, if necessary.

00101-15_F41.EPS

Figure 41 Shielding structure.

- Two workers were installing pipe in a trench. No means of protection was provided in the vertical wall trench. A cave-in occurred, fatally injuring one worker and causing serious injury to the other.
- Four workers were in an excavation, boring a hole under a road. Eight-foot (2.44 m) high steel plates used as shoring were placed against the side walls of the excavation at about 30-degree angles, but were not supported by horizontal bracing. One of the plates tipped over, crushing one worker.

Your company is required to have an emergency action plan that must be communicated to every worker. If you are not sure what the emergency action plan is for your site, don't be afraid to ask questions. Your knowledge could help prevent serious injury or even death, and your supervisor wants you and everybody working with you to be as safe as possible.

Most importantly, try to prevent emergencies before they happen. When in doubt, get out! If you notice potentially dangerous conditions while working in an excavation, get yourself and your co-workers out of danger immediately, and inform your supervisor or the competent person of your concerns.

3.2.2 Tool, Machine, and Equipment Guards

Almost all tools and machines used in construction and industrial work are equipped with guards that protect workers from rotating parts. *Figure 42* show the guards on a grinding machine. All tools and machines that could harm workers must have a guard shielding the hazard. The following types of tools and machines must have guards:

- Grinding tools
- Shearing tools
- Presses
- Punches
- Cutting tools
- Rolling machines
- Tools or machines with pinch points
- Tools or machines with sharp edges

Machine guards should prevent moving parts of the machine from coming into contact with your arms, hands, or any other part of the body, while allowing you to use the machine comfortably and efficiently. Some workers find machine guards to be aggravating and try to remove them from machines. Guards should be secure and should not be easily removed. They should be maintained in good condition, made of durable material, and bolted or screwed to the machine so that tools are needed for their removal. *Figure 43* shows a coupling guard installed over the coupling that connects a motor shaft to a pump shaft. The guard is bolted in place and should only be removed when the motor is not running. The guard is there to prevent someone from being caught in the coupling, which rotates at high speed. Its highly visible markings indicate that it is a hazardous location.

Case History

A spoil pile had been placed on top of a curb, which formed the west face of a trench. A backhoe was working on top of the spoil pile. The west face of the trench collapsed on two workers who were installing sewer pipe. One worker was killed; the other received back injuries. The trench was 8 feet (2.4 m) deep with vertical walls. No cave-in protection was provided. The superimposed loads of the spoil pile and backhoe may have caused the collapse.

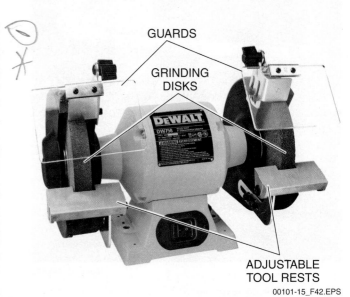

Figure 42 Bench grinder with guards.

COUPLING GUARD

00101-15_F43.EPS

Figure 43 Coupling guard.

Follow these guidelines for using and caring for tool and machine guards:

- Do not remove a guard from a tool or machine except for cleaning purposes or to change a blade or perform other service. Make sure the machine is turned off and tagged out.
- When you are finished with cleaning or maintenance, replace the guard immediately.
- Do not use any material to wedge a guard open.
- Guards and attachments are designed for the specific tool or machine you are using. Use only attachments that are specifically designed for that tool or machine.

3.2.3 *Cranes and Heavy Equipment*

Motorized equipment is used in many different jobs including construction, mining, plant maintenance and operations, road maintenance, equipment transportation, and snow removal. Working with motorized equipment can be dangerous. Workers can be crushed by falling loads, fall from equipment, be electrocuted by power lines, or be struck or trapped by vehicles. The swing radius of equipment can also be a hazard if the job is not carefully planned and properly barricaded. Most heavy equipment has pinch points. *Figure 44* shows some of the pinch point hazards on an excavator.

Dangers exist for both equipment operators and other workers on the site. In one example, a contractor was operating a backhoe when another employee attempted to walk between the swinging back end of the backhoe and a concrete wall. As the employee approached the backhoe from the operator's blind side, the back end hit the victim, crushing him against the wall. A similar problem can occur with a crane or excavator, as shown in *Figure 45*. As the cab swings, the rear end extends beyond the base of the machine. For that reason, barricades must be placed around the working perimeter of the machine.

Because working with or near motorized equipment is dangerous, you must understand that your first responsibility on a job is safety. This includes your own safety, the safety of others on the site, and the safe use of equipment on the site. You must know the hazards and safety procedures of every job you are on, regardless of the work you are doing.

Figure 44 Pinch and crush points.

Figure 45 Example of a caught-between hazard.

Case History

Two employees were attempting to adjust the brakes on a backhoe. The victim told the backhoe operator to raise the wheels off the ground with the front bucket and the outriggers so that he could get to the brakes. The victim then crawled under the machine and began to adjust the brakes. He did this without considering that there was only a 36" (0.9 m) space from the ground to the drive shaft. While adjusting the brakes, the hood of his rain jacket wrapped around the drive shaft and broke his neck. He died instantly.

The Bottom Line: Loose clothing can be caught in moving parts of machinery. You must consider all possible dangers when working on equipment.

Source: The Occupational Safety and Health Administration (OSHA)

Additional Resources

US Occupational Safety and Health Administration. Numerous safety videos are available on line at **www.osha.gov/video**.

Construction Safety Jimmie W. Hinze. 2006. Upper Saddle River, NJ: Pearson Education, Inc.

DeWalt Construction Safety/OSHA Professional Reference. Paul Rosenberg; American Contractors Educational Services. 2006. DEWALT.

Basic Construction Safety and Health, Fred Fanning. 2014. CreateSpace Independent Publishing Platform.

3.0.0 Section Review

1. Which of these activities is most likely to produce flying objects?

   a. Using a chipping hammer
   b. Using a screwdriver
   c. Painting a wall
   d. Climbing a ladder

2. The type of soil most likely to result in a trench cave-in is _____.

   a. solid rock
   b. Type A
   c. Type B
   d. Type C

Section Four

4.0.0 Energy Release Hazards

Objectives

Identify common energy-related hazards and explain how to avoid them.

  a. Describe basic job-site electrical safety guidelines.

  b. Explain the importance of lockout/tagout and describe basic procedures.

Performance Task

4. Inspect a typical power cord and GFCI to ensure their serviceability.

Trade Terms

Ground: The conducting connection between electrical equipment or an electrical circuit and the earth.

Ground fault circuit interrupter (GFCI): A device that interrupts and de-energizes an electrical circuit to protect a person from electrocution.

Lockout/tagout (LOTO): A formal procedure for taking equipment out of service and ensuring that it cannot be operated until an authorized person has removed the lock and/or warning tag.

Proximity work: Work done near a hazard but not actually in contact with it.

Whenever possible, you will de-energize equipment before you begin working on it. Your employer will have written guidelines that you must follow to de-energize the equipment. It is important to follow all guidelines because some circuits receive power from multiple sources. To place equipment in a safe work condition, all energy sources must be removed. Once equipment is de-energized, lockout/tagout (LOTO) devices must be attached to all sources to prevent someone from unknowingly restoring the energy before the work has been completed.

4.1.0 Electrical Safety Guidelines

Electrical safety is a concern for all workers, not just for electricians. On many jobs, no matter what your trade, you will use or work around electrical equipment. Extension cords, power tools, portable lights, and many other pieces of equipment use electricity.

Not all electrical incidents result in death. There are different types of electrical incidents. Any of the following can happen:

- Burns
- Electric shock
- Explosions
- Falls caused by electric shock
- Fires

If the human body comes in contact with an electrically energized conductor and is also in contact with a **ground** at the same time, the body becomes an additional path of resistance for the electrical current to flow through. The electricity flows through the body in less than the blink of an eye without warning. That's why safety precautions are so important when working with and around electrical circuits. When a body conducts electrical current, and the current is high enough, the person can be electrocuted (killed by electric shock). *Table 4* shows the effects of different amounts of electrical current on the human body and lists some common tools that operate using those currents.

Here's an example: A craft worker is operating a portable power drill while standing on damp ground. The power cord inside the drill has become frayed, and the electric wire inside the cord touches the metal drill frame. Three amps of current pass from the wire through the frame, then through the worker's body and into the ground. *Table 4* shows that this worker will probably die.

There are specific policies and procedures to keep the workplace safe from electrical hazards. You can do many things to reduce the chance of an electrical incident. If you ever have any questions about electrical safety on the job site, ask your supervisor.

> **WARNING!**
>
> Less than one amp of electrical current can kill. Always take precautions when working around electricity.

> **Did You Know?**
>
> # Electrocution
>
> Electric shocks or burns are a major cause of incidents in the construction industry. According to the Bureau of Labor Statistics, electrocution is the fourth leading cause of death among construction workers.

Table 4 Effects of Electrical Current on the Human Body

Current	Common Item/Tool	Reaction to Current
0.001 amps	Watch battery	Faint tingle.
0.005 amps	9-volt battery	Slight shock.
0.006 – 0.025 amps (women) 0.009 – 0.030 amps (men)	Christmas tree bulb	Painful shock. Muscular control is lost.
0.050 – 0.9 amps	Small electric radio	Extreme pain. Breathing stops; severe muscular contractions occur. Death may result.
1.0 – 9.9 amps	Jigsaw (4 amps); Sawsall® or Port-a-Band® saw (6 amps); portable drill (3 – 8 amps)	Ventricular fibrillation and nerve damage occur. Death may result.
10 amps and above	ShopVac® (15-gallon); circular saw	Heart stops beating; severe burns occur. Death may result.

4.1.1 Grounding

Grounding is a method of protecting humans from electric shock; however, it is normally a secondary protective measure. The term *ground* refers to a conductive body, usually the earth. A ground is a conductive connection, whether intentional or incidental, by which an electric circuit or equipment is connected to earth or to an engineered grounding system. By grounding a tool or electrical system, a low-resistance path to the earth is intentionally created. When properly done, this path offers low resistance and has enough current-carrying capacity to prevent the buildup of voltages that could create a personnel hazard. This does not guarantee that no one will receive a shock. It will, however, greatly reduce the possibility of such incidents, especially when used in combination with your company's safety program.

Use three-wire extension cords for portable power tools and make sure they are properly connected (*Figure 46*, top). The three-wire system is one of the most common safety grounding systems used to protect you from incidental electrical shock. The third wire is connected to a ground system. If the insulation in a tool fails, the current will pass to ground through the third wire—not through your body. Double-insulated tools (*Figure 46*, bottom) are also very effective in preventing shocks. In fact, it has become more common to use double-insulated tools because they are safer than relying on a three-wire cord alone. *Figure 46* also shows the double-insulated symbol that can be found on double-insulated tools. Double-insulated tools use a two-wire power cord with no ground pin. One prong of the plug is larger than the other so it can only be connected to a polarized receptacle.

4.1.2 Ground Fault Circuit Interrupters

A **ground fault circuit interrupter (GFCI)** is a fast-acting circuit breaker that senses small imbalances in the circuit caused by current leakage to ground. A GFCI continually matches the amount of current going to an electrical device against the amount of current returning from the device. Whenever the two values differ by more than 5 milliamps, the GFCI interrupts the electric power within 1/40th of a second. *Figure 47* shows an extension cord with a GFCI.

A GFCI must be used on all receptacles that are not part of the building's permanent wiring, such as temporary power and extension cords. A GFCI provides protection against a ground fault, which is the most common form of electrical shock. It also provides protection from fires, overheating, and wiring insulation deterioration.

Tripping of GFCIs—interruption of circuit flow—is sometimes caused by wet connectors and tools. Limit the amount of water that tools and connectors come into contact with by using watertight or sealable connectors. Tripping may also be caused by cumulative leakage from several tools or from extremely long circuits. GFCIs should be periodically inspected in accordance with the manufacturer's recommendations or site safety practices. GFCIs have a TEST button that can be pressed to verify that the GFCI is working. The GFCI should be tested before any use.

CAUTION

Do not plug a GFCI-protected device into a GFCI-protected circuit.

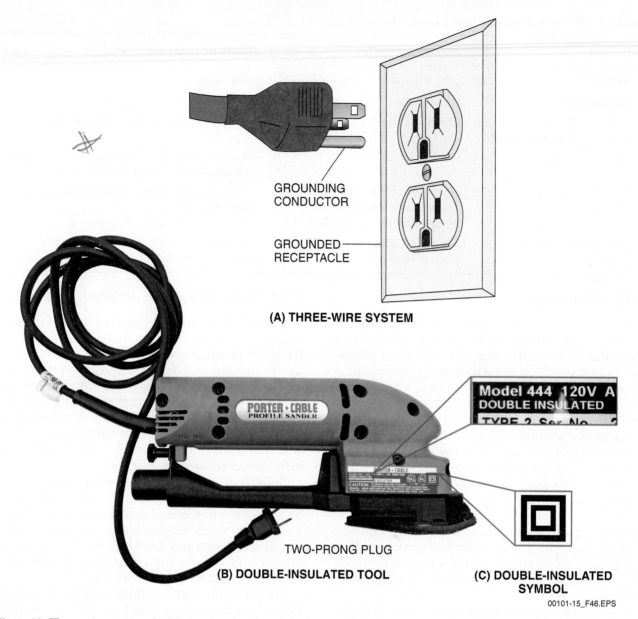

(A) THREE-WIRE SYSTEM

GROUNDING
CONDUCTOR

GROUNDED
RECEPTACLE

Model 444 120V A
DOUBLE INSULATED
TYPE 2 Ser No

TWO-PRONG PLUG

(B) DOUBLE-INSULATED TOOL

**(C) DOUBLE-INSULATED
SYMBOL**

00101-15_F46.EPS

Figure 46 Three-wire system, double-insulated tool, and double insulated symbol.

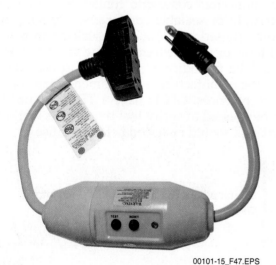

00101-15_F47.EPS

Figure 47 Extension cord with a GFCI.

Case History

A self-employed builder was using a metal cutting tool on a metal carport roof and was not using GFCI protection. The male and female plugs of his extension cord partially separated, and the active pin touched the metal roofing. When the builder grounded himself on the gutter of an adjacent roof, he received a fatal shock.

The Bottom Line: Always use GFCI protection and be aware of potential hazards.

Source: The Occupational Safety and Health Association (OSHA)

4.1.3 Summary of Electrical Safety Guidelines

Here are some basic job-site electrical safety guidelines:

- All power tools used in construction should be ground-fault protected.
- Make sure that panels, switches, outlets, and plugs are grounded.
- Never use bare electrical wire.
- Never use metal ladders near any source of electricity.
- Inspect electrical power tools before you use them.
- Never operate any piece of electrical equipment that has a danger tag or lockout device attached to it.

> **WARNING!**
>
> All work on electrical equipment should be done with circuits de-energized, locked out, and confirmed. All conductors, buses, and connections should be considered energized unless proven otherwise.

- Never use worn or frayed cables (*Figure 48*). If the cord is frayed or worn, disconnect the power and dispose of the cord.
- Make sure light bulbs have protective guards to prevent incidental contact (*Figure 49*).

4.1.4 Working Near Energized Electrical Equipment

No matter what your trade, your job may include working near exposed electrical equipment or conductors. This is one example of **proximity work**. Often, electrical distribution panels, switch enclosures, and other equipment must be left open during construction. This leaves the wires and components in them exposed. Some or all of the wires and components may be energized. Working

00101-15_F49.EPS

Figure 49 Work light with protective guard.

near exposed electrical equipment can be safe, but only if you keep a safe working distance.

Regulations and company policies tell you the minimum safe working distances from exposed conductors. The safe working distance can be very small or span a significant distance, depending on the voltage. The higher the voltage, the greater the required safe working distance.

You must learn the safe working distance for each situation. Make sure you never get any part of your body or any tool you are using closer to exposed conductors than that distance. You can get information on safe working distances from your instructor, your supervisor, company safety policies, and regulatory documents. This subject is also covered in greater detail in other craft curricula where it is relevant.

One of the common causes of electrical shock is coming into contact with overhead wires with metal ladders, cranes, or excavating equipment.

00101-15_F48.EPS

Figure 48 Never use damaged cords.

A distance of at least 10 feet (3 meters) must be maintained from any conductor carrying 50,000 volts or less. Greater distances are required for higher voltages.

4.1.5 If Someone Is Shocked

If you are there when someone gets an electrical shock, you can save a life by taking immediate action. The best thing to do is to immediately shut off the power to the circuit. Do not touch the victim while he or she remains in contact with the power source. Once the circuit is disconnected, call an ambulance or the emergency number at your job site. Render first aid only if you are trained to do so.

> **WARNING!**
>
> Do not touch the victim or the electrical source with your hand, foot, or any other part of your body or with any object or material. You could become part of the circuit—and another victim.

4.2.0 Lockout/Tagout Requirements

Failure to disable machinery before working on it is a major cause of injury and death on job sites. Lockout/tagout procedures safeguard workers against unexpected releases from various energy sources. An energy source can be electrical, mechanical, hydraulic, pneumatic, chemical, or thermal. After the power has been turned off, energy can still be stored in a device or component. For example, even when a hydraulic system has been turned off, hydraulic pressure may remain in all or part of the system. This high-pressure energy can be released if the system is opened during service or repairs, creating an extremely hazardous situation. Additional dangers exist if the device or system contains chemicals, flammable liquids, high-temperature liquids, or gases.

When anyone is working on or around active systems, all equipment that could release energy must be shut down, drained, de-energized, or otherwise rendered harmless whenever possible. Switches, circuit breakers, valves and other components are switched off or closed, and then locks or tags (*Figure 50*) are applied so they cannot be re-energized while work is ongoing.

Generally, each lock has its own key, and the individual who applies the lock keeps the key. A variety of tags are used, depending on the circumstances and the organization. Tags (*Figure 51*) typically have the word *DANGER* on them, along with other printed information, writing space for comments, and a line for the installer's signature and date. The authorized person who applied the lock must be the one to remove it. In an emergency, a supervisor or other person authorized by the employer may remove a lockout/tagout.

Electrical Cord Safety

Electrical cords are frequently seen on construction sites, yet they are often overlooked. Use the following safety guidelines to ensure your safety and the safety of other workers.

- Every electrical cord should have an Underwriters Laboratory (UL) label attached to it. Check the UL label for specific wattage. Do not plug more than the specified number of watts into an electrical cord.
- A cord set not marked for outdoor use is to be used indoors only. Check the UL label on the cord for an outdoor marking.
- Do not remove, bend, or modify any metal prongs or pins of an electrical cord.
- Do not run a cord through doorways or through holes in ceilings, walls, and floors that might pinch the cord. Also, check to see that there are no sharp corners along the cord's path. Any of these situations will lead to cord damage.
- Extension cords are a tripping hazard. They should never be left unattended and should always be put away when not in use.

(A) ELECTRICAL LOCKOUT

(B) VALVE LOCK

(C) PNEUMATIC LOCKOUT

00101-15_F50.EPS

Figure 50 Lockout/tagout devices.

In a lockout, an energy-isolating device such as a disconnect switch or circuit breaker is placed in the Off position and a lock is applied. Padlocks are popular and versatile, but other styles (*Figure 52*) are often specifically suited for the equipment being serviced. Multiple lockout devices (*Figure 53*) are used when more than one person is accessing the equipment. When a worker needs to service a system and sees that another person's lock has been applied, it is tempting to accept this as a safe situation. However, the lock may be removed by the first worker and the circuit re-energized while the second worker is still vulnerable. In these cases, the multiple lockout devices allow several workers to apply locks, and each is ensured that the system cannot be restored until theirs is removed.

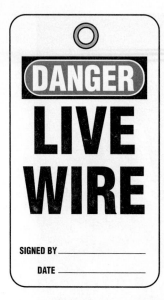

00101-15_F51.EPS

Figure 51 Typical safety tags.

In a tagout, components that control power to equipment and machinery are set to a safe position and a written warning is attached. This method may be more appropriate for valves and other energy-controlling devices that are difficult or impossible to lock, since tags alone cannot prevent a control device or switch from being repositioned.

It is important to emphasize that LOTO procedures are not solely related to electrical energy sources. Technicians cannot forget that other forms of energy may need to be rendered safe as well.

The exact procedures for LOTO will vary by organization and site. Check with your instructor or site supervisor regarding the detailed LOTO procedure at your site and be sure you are familiar with the proper steps.

4.2.1 Pressurized or High-Temperature Systems

Many jobs require that workers be close to tanks, piping systems, and pumps that contain pressurized or high-temperature fluids. Be aware that touching a container of high-temperature fluid can cause burns. Many industrial processes involve fluids that are as hot as several thousand degrees. Also, if a container holding pressurized fluids is damaged, it may leak and spray dangerous fluids. Any work around pressurized or high-temperature systems is considered proximity work. Barricades, a monitor, or both may be needed to ensure the safety of those working nearby.

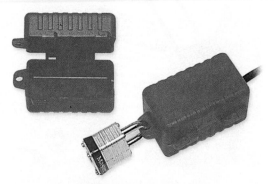

(A) ELECTRICAL PLUG LOCKOUT

(B) CIRCUIT BREAKER LOCK

(C) BALL VALVE LOCKOUT

(C) ELECTRICAL SWITCH LOCKOUT

00101-15_F52.EPS

Figure 52 Lockout devices.

00101-15_F53.EPS

Figure 53 Multiple lockout/tagout device.

Additional Resources

US Occupational Safety and Health Administration. Numerous safety videos are available on line at **www.osha.gov/video**.

DeWalt Construction Safety/OSHA Professional Reference, Paul Rosenberg; American Contractors Educational Services. 2006. DEWALT.

Basic Construction Safety and Health, Fred Fanning. 2014. CreateSpace Independent Publishing Platform.

4.0.0 Section Review

1. The purpose of double-insulating an electrically powered tool is to _____.

    a. prevent the tool from overheating
    b. protect users from electrical shock
    c. keep the tool from getting too cold
    d. allow it to use a three-prong receptacle

2. When a lockout is performed, the key to the lock is _____.

    a. left in the lock
    b. kept by the person who applied the tag
    c. given to the foreman for safekeeping
    d. hidden in a convenient place

SECTION FIVE

5.0.0 PERSONAL PROTECTIVE EQUIPMENT

Objectives

Identify and describe the proper use of personal protective equipment.

a. Identify and describe the PPE used to protect workers from bodily injury.

b. Identify potential respiratory hazards and the basic respirators used to protect workers against those hazards.

Performance Tasks

2. Inspect the following PPE items and determine if they are safe to use:
   - Eye protection
   - Hearing protection
   - Hard hat
   - Gloves
   - Approved footwear

3. Properly don, fit, and remove the following PPE items:
   - Eye protection
   - Hearing protection
   - Hard hat
   - Gloves

Trade Terms

Arc welding: The joining of metal parts by fusion, in which the necessary heat is produced by means of an electric arc.

Every worker is responsible for wearing appropriate PPE on the job. When worn correctly, PPE is designed to protect you from injury. You must keep it in good condition and use it when you need to. When workers are injured on the job, it is often because they are not using required PPE.

You will not see all the potentially dangerous conditions just by looking around a job site. It is important to stop and consider what type of incidents could happen on any job that you are about to do. Knowing how to use PPE will greatly reduce your chance of getting hurt. Keep in mind that PPE requirements are usually established by the job site. Some job sites will have different requirements than others, so always make sure you check. For example, some sites require high-top safety boots instead of the more common standard 6-inch tops.

5.1.0 PPE Items

Remember, while PPE is the last line of defense against personal injury, using it properly and taking care of it are the first steps toward protecting yourself on the job.

The first line of defense is represented by the engineering and administrative controls used by employers to eliminate or reduce the need for some PPE. Engineering controls are implemented by making physical changes, such as reducing noise by the use of mufflers on equipment engines; installing guards on moving equipment; and making sure equipment is properly maintained. An example of an administrative control is operating noisy equipment on a shift when fewer workers are present.

The best protective equipment is of no use unless you follow these rules:

- Regularly inspect it.
- Properly care for it.
- Use it properly when it is needed.
- Never alter or modify it in any way.

The sections that follow describe protective equipment commonly used on construction sites, and tell how to use and care for each piece of equipment. Be sure to wear the equipment according to the manufacturer's specifications.

> **WARNING!**
>
> Your clothing must comply with good general work and safety practices. Do not wear clothing or jewelry that could get caught in machinery or otherwise cause an incident, such as loose clothing, baggy shirts, or dragging pants. You must wear a shirt at all times; some tasks will require long-sleeved shirts. Your shirt should always be tucked in unless you are performing welding.

Protective equipment commonly worn by craft workers on any job site usually includes the following:

- Hard hats
- Eye protection
- Gloves

Sharing PPE

It is not a good idea for workers to share PPE. If one person has an infection, it can be passed to another through the shared equipment.

- Safety footwear
- Hearing protection
- Respiratory protection
- High-visibility clothing

5.1.1 Hard Hats

Figure 54 shows a typical hard hat. The outer shell of the hat can protect your head from a hard blow. The webbing inside the hat keeps space between the shell and your head. Adjust the headband so that the webbing fits your head and there is at least one inch of space between your head and the shell.

Do not alter your hard hat in any way. Inspect your hard hat every time you use it. If there are any cracks or dents in the shell, or if the webbing straps are worn or torn, get a new hard hat. Wash the webbing and headband with soapy water as often as needed to keep them clean. Wear the hard hat only as the manufacturer recommends. Never wear anything but approved clothing under your hard hat.

WARNING!
No articles should be worn under the hard hat that could interfere with fit and visibility. That includes ball caps or hoodies that obscure peripheral vision. Only employer-approved gear is to be worn under the hard hat.

00101-15_F54.EPS

Figure 54 Typical hard hat.

5.1.2 Eye and Face Protection

Wear eye protection (*Figure 55*) wherever there is even the slightest chance of an eye injury. In general, eye protection is required any time you are on a job site. Face shields (*Figure 56*) are added over safety glasses for certain tasks, such as grinding. Appropriate eye or face protection is required when there is a possibility of exposure to hazards from flying particles, molten metal, liquid chemicals, acids or caustic liquids, chemical gases or vapors, or potentially injurious light radiation. The following are examples of tasks requiring eye protection:

- Grinding and chipping metal
- Using power saws and other tools/equipment that can throw out solid material

Did You Know?

Hard Hats

Hard hats were once made of metal. However, metal conducts electricity, so most hard hats are now made of reinforced plastic or fiberglass.

Case History

A training specialist traveled nearly three hours to reach a remote job site in order to observe and photograph the installation of high-voltage transmission lines. On arriving, he parked his car, put on his work boots, hard hat, reflective vest and safety glasses and walked over to the location where the work was being done. He was introduced to the project manager who shocked him by saying: "Sir, I have to ask you to leave my job site because you are not wearing electrically insulated boots". Given no choice, the training specialist headed toward his car for the long trip home. Fortunately, the project manager needed to go back to the office for a meeting and offered the loan of his boots, thus preventing six hours of wasted time, as well as the loss of a rare opportunity.

The Bottom Line: Project managers and superintendents in construction and industrial work are serious about safety. If you are not fully prepared, you may find yourself sitting on the sidelines. A worker who shows up at a job site without the required PPE could lose a day's pay and possibly face disciplinary action.

  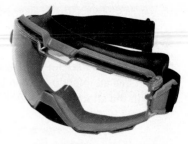

(A) SAFETY GLASSES

(B) TINTED SAFETY GLASSES

(C) SAFETY GOGGLES

00101-15_F55.EPS

Figure 55 Typical safety glasses and goggles.

(A) CLEAR FACE SHIELD

(B) TINTED FACE SHIELD

00101-15_F56.EPS

Figure 56 Full-face shields.

- Working with molten lead, tar pots, and other molten materials
- Working with chemicals, acids, and corrosive liquids
- Arc welding

On the job site, potential eye hazard areas are usually identified, but always be on the lookout for possible hazards.

> **CAUTION**
>
> Standard eyeglasses are not adequate protection in high-hazard work zones. Shatter-proof lenses and side shields are required for eye protection in those areas.

Regular safety glasses will protect you from falling objects or from objects flying toward your face. Side shields provide added protection and are required whenever there is risk of flying debris. Safety goggles provide the best protection from all directions. A face shield is required, in addition to safety glasses or goggles, when there is likely to be flying debris. Grinding and chipping activities are good examples of such activities.

Handle your safety glasses and goggles with care. If they get scratched, replace them; the scratches will interfere with your vision. Clean the lenses regularly with lens tissues or a soft cloth.

Welders must use tinted goggles or welding hoods. The tinted lenses protect the eyes from the bright welding arc or flame. Welders must use filter lenses as specified by job site requirements. Welder's helpers and all employees working in the vicinity of arc welding should not look directly at the welding process and are also required to use eye protection with the prescribed level of shading. Oxyacetylene welding and burning also require the use of a filter lens.

Follow these general precautions for eye care:

- Always report all eye injuries and suspected foreign material in your eye to your supervisor immediately. Do not try to remove foreign matter yourself.
- Keep your hands away from your eyes.

- Keep materials out of your eyes by regularly clearing debris from your hard hat brim, the top of your goggles, and your face shield. When removing a cap, hard hat, or face shield, be careful to remove it in a way that prevents accumulated dirt and debris from falling into your eyes.
- Flood your eyes with water if you feel something in them. Never rub them, as this can make the problem worse.
- Know the location of eyewash stations and how to use them.

5.1.3 Hand Protection

On many construction jobs, you must wear heavy-duty gloves to protect your hands (*Figure 57*). Construction work gloves are usually made of cloth, canvas, or leather. Make sure your work gloves are a good fit. Wearing gloves that are too big for your hands can lead to injury. Never wear gloves around rotating or moving equipment. They can easily get caught in the equipment. Replace gloves when they become worn, torn, or soaked with oil or chemicals.

Gloves help prevent cuts and scrapes when you handle sharp or rough materials. Heat-resistant gloves are needed for handling hot materials. Craftworkers should be familiar with standards governing hand protection, such as ANSI/SEA Standard 105-2011, *Hand Protection Selection Criteria* and European Standard EN388, *Protective Gloves Against Mechanical Risks*. These standards establish a rating system for gloves tested for their resistance to various hazards, including chemical burns, abrasions, cuts, punctures, resistance to heat and flame, and resistance to cold. Additional European standards govern protection from chemicals and microorganisms (EN374); thermal hazards (EN407 and EN511); and ionizing radiation and radioactive contamination (EN421).

Electricians use special rubber-insulated gloves when they work on or around live circuits. When working with solvents or other chemicals, it may be necessary to wear chemical-resistant gloves.

> **WARNING!**
>
> Only specially trained employees are allowed to use dielectrically tested rubber gloves to work on energized equipment. Never attempt this work without proper training and authorization.

5.1.4 Foot and Leg Protection

Approved safety footwear must be worn on all job sites. The best shoes to wear on a construction site are shoes with a safety toe to protect toes from falling objects (*Figure 58*). This type of shoe is generally required on every construction-related job site.

Some specialized work will require different footwear or gear. For example, safety-toed rubber boots are needed on job sites that are subject to chemically hazardous conditions or standing water. When climbing a ladder, your shoes or boots must have a well-defined heel to prevent your feet from slipping off the rungs. You will

(A) FITTED WORK GLOVES

(B) GLOVES WITH CUFFS

(C) RUBBERIZED COTTON

00101-15_F57.EPS

Figure 57 Work gloves.

Figure 58 Safety shoe.

need foot guards when using jackhammers and similar equipment.

Never wear canvas shoes or sandals on a construction site. They do not provide adequate protection. Always replace boots or shoes when the sole tread becomes worn or the shoes have holes, even if the holes are on top. Because of the risk of fire, do not wear oil-soaked shoes when you are welding or cutting metal.

Remember these general guidelines relating to leg protection:

- Never carry pointed tools, such as scissors or screwdrivers, in your pants pockets. Use a canvas or leather tool sheath with all sharp ends pointing down.
- When using certain special equipment, such as chainsaws and brush hooks, use shin guards.
- Consider stability when stepping into or onto locations where materials are stored. Materials may shift and pinch your legs and/or feet.

5.1.5 Hearing Protection

Unlike damage to most parts of the body, ear damage does not always cause pain. Exposure to loud noise over a long period of time can cause hearing loss, even if the noise is not loud enough to cause pain. Vibrations caused by sound waves enter the ear and are received by the cochlea, which consists of tiny hair cells inside the ear that convert the vibrations into sound signals. Exposure to intense sound waves over time can damage the cochlea and reduce the ability to hear. Damage to the cochlea can also cause tinnitus, which is a permanent condition characterized by ringing or buzzing in the ears that never stops. Hearing loss reduces your quality of life and makes simple, daily tasks more complicated. Save your hearing by using hearing protection whenever you are working in a noisy environment.

Here is a good rule to follow: If the noise level is so great that you have to raise your voice to be heard by someone who is less than 2 feet (61 cm) away, you need to wear hearing protection.

Most construction companies follow rules defined in official safety standards in deciding when hearing protection must be used. One type of hearing protection is specially designed earplugs that fit into your ears and filter out noise (*Figure 59*, top). Another type of hearing protection is earmuffs, which are large padded covers for the entire ear (*Figure 59*, bottom). The headband on earmuffs must be adjusted for a snug fit. If the noise level is very high, you may need to wear both earplugs and earmuffs.

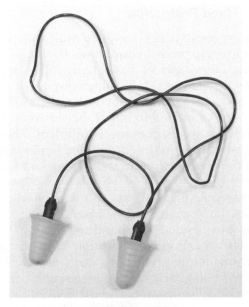

(A) EAR PLUGS

(B) EAR MUFFS

Figure 59 Hearing protection.

Noise-induced hearing loss can be prevented by using noise control measures and personal protective devices. *Table 5* shows the recommended maximum length of exposure to sound levels rated 90 decibels and higher. When noise levels exceed those outlined in *Table 5*, an effective hearing conservation program is required. A company-appointed program administrator will oversee this program. If you have questions about the hearing conservation program on your site, see your supervisor or the program administrator.

5.2.0 Respiratory Hazards and Protection

Wherever there is danger of an inhalation hazard, you must use a respirator. There are a number of job-site conditions that require workers to wear respirators, including the following:

- Dust from metal grinding
- Toxic fumes from welding or flame cutting of some metals
- Working with cleaning solvents
- Working in low-oxygen environments such as confined spaces
- Spray painting
- Sand blasting
- Drilling concrete
- Working with chemicals such as chlorine or ammonia

Silica is a mineral found in concrete, masonry, and rock. During construction, silica may be found in a dust form. Prolonged exposure to silica dust could cause silicosis and lung cancer. Silicosis is an incurable, and sometimes fatal, lung disease. The time it takes for silicosis to develop varies depending on how long the exposure lasted and how much silica the person was exposed to. When you are working in an area where silica dust is present, you must use the appropriate respiratory protection. Never work around silica dust without proper training, authorization, and PPE.

Asbestos is another hazardous material that can be harmful to your lungs. Prolonged exposure can cause lung cancer, asbestosis (scarring of the lung tissue), and a cancer called mesothelioma. It may take more than 20 years for these diseases to develop. Smoking significantly increases the risk of lung disease. If you smoke, tar from cigarettes will stick to the asbestos fibers in the lungs, making it more difficult for your body to get rid of the asbestos.

Sources of asbestos include older insulation and other building materials, such as floor tiles, pipe insulation drywall compounds, and reinforced plaster, along with some roofing and siding materials. The United States banned production of most asbestos products in the 1970s, meaning that asbestos-containing materials (ACMs) are generally found in structures built before 1980. During renovation or maintenance operations, asbestos may be dislodged and the fibers become airborne. Asbestos fibers are particularly hazardous and must not be disturbed unless special techniques and procedures are used to remove it safely. Although you will not feel the effects immediately, this exposure can be a serious health hazard. Never handle asbestos without proper training, authorization, and PPE.

Table 5 Maximum Noise Levels

Sound Level (decibels)	Maximum Hours of Continuous Exposure per Day	Examples
90	8	Power lawn mower
92	6	Belt sander
95	4	Tractor
97	3	Hand drill
100	2	Chain saw
102	1.5	Impact wrench
105	1	Spray painter
110	0.5	Power shovel
115	0.25 or less	Hammer drill

Around the World
Asbestos Use

Asbestos was identified as a health risk many years ago. However, countries have reacted to the hazard in different ways. While some countries banned its use in products and construction decades ago, other countries, such as India, continue to use asbestos without regulation of any form. Indeed, asbestos use in India has risen over 80 percent in the last decade.

The World Health Organization (WHO) estimates that 125 million people worldwide are exposed to asbestos each year in the workplace. Even in the United States, asbestos may still be encountered in old buildings, flooring, and insulation components. Be aware of the regulations concerning the use of asbestos In your region and stay alert to sources that might do you harm if not avoided.

Only workers who have been trained and licensed are authorized to perform work related to asbestos including, but not limited to, the following tasks:

- Demolition
- Removal
- Alteration
- Repair
- Maintenance
- Installation
- Cleanup
- Transportation
- Disposal
- Storage

If asbestos is encountered or suspected on the job site, a supervisor must be notified and all work must stop until the asbestos is either sealed off or removed by licensed professionals.

5.2.1 Types of Respirators

Federal law specifies which type of respirator to use for different types of hazards. There are four general types of respirators (*Figure 60*):

- Self-contained breathing apparatus (SCBA)
- Supplied air mask
- Full facepiece mask with chemical canister (gas mask)
- Half mask or mouthpiece with mechanical filter

A SCBA carries its own air supply in a compressed air tank. It is used where there is not enough oxygen or where there are dangerous gases or fumes in the air.

A supplied-air mask uses a remote compressor or air tank to provide breathable air in oxygen-deficient atmospheres. Supplied-air masks can generally be used under the same conditions as SCBAs.

A full-facepiece mask with chemical canisters is used to protect against brief exposure to dangerous gases or fumes. A half mask or mouthpiece with a mechanical filter is used in areas where you might inhale dust or other solid particles.

5.2.2 Wearing a Respirator

If you need to use a respirator, your employer must provide you with the appropriate training to select, test, wear, and maintain this equipment. Your employer is also responsible for having you medically evaluated to ensure you are fit enough to wear respiratory protection equipment without being harmed. You must also be tested to ensure a proper fit and a good seal.

Wearing a respirator generally places a burden on the employee. A self-contained breathing apparatus is heavy and may be difficult for some workers to carry; negative-pressure respirators restrict breathing; and other respirators may cause claustrophobia. For these and other reasons, workers must be medically evaluated by a physician or other licensed health care professional to determine under what conditions they may safely wear respirators.

Respirators are ineffective unless properly fit-tested to the user. To obtain the best protection from a respirator, perform positive (breathing out) and negative (breathing in) fit checks each time it is worn. These fit checks must be repeated until a good face seal is obtained. A respirator must be clean, in good condition, and all of its components must be in place in order to provide adequate protection.

When a respirator is required, a personal monitoring device is usually also required. This device samples the air to measure the concentration of hazardous chemicals.

5.2.3 Selecting a Respirator

Follow company and government procedures when choosing the type of respirator for a particular job. Also be sure that it is safe for you to wear a respirator. Under current regulations, workers must fill out a questionnaire to identify potential problems in wearing a respirator. Depending on the answers, a medical exam may be required. When a respirator is not required, workers may voluntarily use a dust or particle mask for general protection. These masks do not require fit testing or a medical examination.

Always use the appropriate respiratory protective device for the hazardous material involved and the extent and nature of the work to be performed. Before using a respirator, you must determine the type and concentration level of the contaminant, and whether the respirator can be properly fitted on your face. If you wear contact lenses, practice wearing a respirator with your contact lenses to see whether you have any problems. That way, you will identify any problems before you use the respirator under hazardous conditions.

> CAUTION
>
> All respirator instructions, warnings, and use limitations contained on each package must be read and understood by the wearer before use.

**(A) SELF-CONTAINED
BREATHING APPARATUS (SCBA)**

(B) SUPPLIED AIR MASK

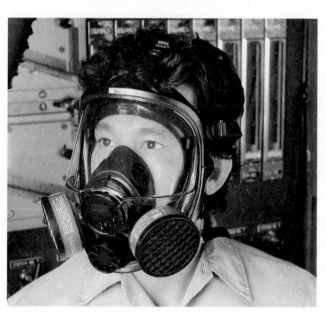

(C) FULL FACEPIECE MASK

(D) HALF MASK

00101-15_F60.EPS

Figure 60 Examples of respirators.

Additional Resources

US Occupational Safety and Health Administration. Numerous safety videos are available on line at **www.osha.gov/video**.

Construction Safety, Jimmie W. Hinze. 2006. Upper Saddle River, NJ: Pearson Education, Inc.

DeWalt Construction Safety/OSHA Professional Reference. Paul Rosenberg; American Contractors Educational Services. 2006. DEWALT.

Basic Construction Safety and Health, Fred Fanning. 2014. CreateSpace Independent Publishing Platform.

5.0.0 Section Review

1. Personnel working in the vicinity of arc welding work must wear eye protection with tinted lenses.

   a. True
   b. False

2. Drilling concrete requires use of a respirator because it produces _____.

   a. asbestos
   b. lead
   c. toxic fumes
   d. silica dust

6.0.0 Job-Site Hazards

Objectives

Identify and describe specific job-site safety hazards.

   a. Identify various exposure hazards commonly found on job sites.
   b. Identify hazards associated with environmental extremes.
   c. Identify hazards associated with hot work.
   d. Identify fire hazards and describe basic firefighting procedures.
   e. Identify confined spaces and describe the related safety considerations.

Trade Terms

Brazing: A process using heat in excess of 800°F (427°C) to melt a filler metal that is drawn into a connection. Brazing is commonly used to join copper pipe.

Flash burn: The damage that can be done to eyes after even brief exposure to ultraviolet light from arc welding. A flash burn requires medical attention.

Flash point: The temperature at which fuel gives off enough gases (vapors) to burn.

Permit-required confined space: A confined space that has been evaluated and found to have actual or potential hazards, such as a toxic atmosphere or other serious safety or health hazard. Workers need written authorization to enter a permit-required confined space. Also see *confined space*.

Qualified person: A person who, by possession of a recognized degree, certificate, or professional standing, or by extensive knowledge, training, and experience, has demonstrated the ability to solve or prevent problems relating to a certain subject, work, or project.

Welding curtain: A protective screen set up around a welding operation designed to safeguard workers not directly involved in that operation.

Wind sock: A cloth cone open at both ends mounted in a high place to show which direction the wind is blowing.

It is impossible to list all the hazards that can exist on a construction or industrial job site. This section describes some of the more common hazards and explains how to deal with them. For your safety, you must know the specific hazards where you are working and how to prevent incidents and injuries. If you have questions specific to a job site, ask your supervisor.

6.1.0 Job-Site Exposure Hazards

The term *exposure* refers to contact with a chemical, biological, or physical hazard. Exposure can be chronic or acute. Chronic exposure is long-term and repeated, and may be mild or severe. Acute exposure is short-term and intense.

According to the HAZCOM standard, your employer must inform you of any hazards to which you might be exposed. In order to better protect you, your employer may require pre-employment and periodic medical examinations to ensure that your health is not being negatively affected by chemical exposure during work. Exams are usually conducted annually, and generally involve targeted organ testing. Because most medical tests cannot directly check for specific chemical exposure, blood tests are used to check the status of the particular organs (called target organs) known to be most affected by the chemicals to which a person has been exposed.

The results of these tests are compared to pre-employment test results and any other annual or periodic exams you may have had. Employers also frequently request a final screening when a worker leaves the company.

Workers can be exposed to chemicals and other hazardous materials in a variety of ways. Routes of exposure include breathing (inhalation), eating or drinking (ingestion), and skin contact (absorption). Chemical exposure is subject to permissible exposure limits (PELs). In other words, some level of exposure to certain chemicals and fumes is considered acceptable and does not represent a hazard. A PEL is the maximum concentration of a substance that a worker can be exposed to in an eight-hour shift. When a worker approaches or reaches the PEL, he or she must be removed from that environment.

There are various types of hazardous materials you may come in contact with during various types of construction work, including the following:

- Lead
- Bloodborne pathogens
- Chemicals

Taking the time to understand each of these hazards will help you stay healthy and safe. Working around these hazards requires special training and PPE. Never handle any of these materials without proper training, authorization, and PPE.

6.1.1 Lead

Lead occurs naturally in the Earth's crust and is spread through human activities, such as mining and the burning of fossil fuels. As an element, lead is indestructible—it does not break down. Once it is released into the environment, it can move from one medium to another. For example, lead in dust can be carried long distances, dissolve in slightly acidic water, and find its way into soil where it can remain for years. Lead is difficult to detect because it has no distinctive taste or smell.

Lead has many useful properties. It is soft and easily shaped, durable, resistant to some chemicals, and fairly common. Because lead is so versatile, it is used in the production of piping, batteries, and casting metals. However, lead is a toxic metal that can cause serious health problems. You can be exposed to lead by breathing air, drinking water, eating food, and swallowing or touching dust or dirt that contains lead.

You may encounter lead-based paints during demolition or renovation of structures built before 1978, when lead-based paint was banned. Dust created from sanding lead-based paints is hazardous. Protective clothing and equipment must always be used when lead levels are above the PEL. If you are unsure whether lead is present, consult your supervisor.

All waste contaminated with lead is considered hazardous. Those who handle hazardous waste must have special training. Never handle hazardous materials or waste without proper training, authorization, and PPE.

6.1.2 Bloodborne Pathogens

Bloodborne pathogens are another health hazard you may encounter on the job site. Sometimes you will hear them referred to as bloodborne infectious diseases. They are diseases that can be transmitted by contact with an infectious person's blood or other bodily fluids, the most common being HIV (the virus that causes AIDS) and hepatitis B and C.

You could be exposed to another person's blood on the job site when administering first aid to an injured person or being involved in a multi-victim incident. In order to safeguard yourself, you must know the universal precautions to prevent exposure and follow them closely:

- Always use appropriate gloves, eye protection, and a mask when administering first aid.
- If you come in contact with someone's blood, immediately wash the affected areas with soap and water. Otherwise, use antiseptic hand cleanser or antiseptic towelettes until washing with soap and running water is possible.
- Notify your supervisor of the contact right away.

You may need to seek medical attention for precautionary screening, particularly if that person has any bloodborne infectious diseases. Your employer should have a written exposure control plan and procedure for such occurrences.

6.1.3 Chemical Splashes

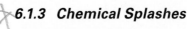

Chemicals such as acids and solvents can pose physical hazards, including acid reactions or burns. Acids can create toxic vapors or react violently when mixed with other chemicals. Other chemicals are flammable, combustible, or explosive. For example, solvents and compressed gases are often flammable. These materials can catch fire in their liquid or gaseous state. Some materials, such as blasting caps, are dangerous solids. These items should be kept away from open flames, sparks, intense heat, or other ignition sources to prevent fires or explosions. Know the physical hazards of the chemicals being used in order to prevent fires and chemical reactions.

Workers should always check the SDS for a chemical before working with or around it. That way, you will be prepared to take appropriate corrective action if you or another worker is exposed.

WARNING!

Chemical splashes can become medical emergencies. Before working with any chemicals, ensure that you know where the nearest shower and eyewash station are, and verify that they are functioning properly.

Remember to wear prescribed clothing and PPE on the job site at all times. This will help guard against your skin and eyes being splashed with chemicals. If you are not sure of the PPE needed on your job site or for a certain task, consult your supervisor before beginning work.

6.1.4 Container Labeling

On a construction site, any material in a container must have a label. Labels describe what is in a container and warn of chemical hazards. The HAZCOM Standard states that hazardous material containers must be labeled, tagged, or marked. The label must include the name of the material, the appropriate hazard warnings, and the name and address of the manufacturer. On December 1, 2015, OSHA will require that all shipments be labeled as outlined by the Global Harmonization System (GHS), which has been adopted as a worldwide standard to enhance awareness and improve the safety and health of workers exposed to these substances.

In today's global marketplace, chemicals and similar hazardous materials are in constant movement between countries. With many countries having their own hazardous material labeling system, products must often be labeled multiple times to support the systems of both the shipping and receiving countries. The GHS will solve this problem by providing some standardized labeling for various materials. There are nine pictograms that workers must become familiar with are shown here. Those that are related to flammable, combustible, and explosive materials are shown in *Figure 61*, and those that represent other hazards are shown in *Figure 62*.

Pictograms have been in use for a number of years to depict a variety of hazards or to communicate information. The advantage of

Process Safety Management

One way to avoid the hazards associated with chemical spills is to evaluate those hazards and plan ahead for dealing with unwanted releases. Forward-thinking companies adopt a method known as process safety management for this purpose. Process safety management involves training employees and contractors; making information available; establishing formal procedures; analyzing potential hazards; and taking steps to mitigate the hazards.

pictograms is that they cross any language barrier. Regardless of language, pictograms such as those used in the GHS tell the user what type of hazard to be concerned about. Although other systems of container labeling will be encountered for years to come, the pictograms all share common features and visually represent the hazard of concern.

If a material is transferred from a labeled container to a new container, the new container must be labeled with all of the information from the original label. Make sure that any materials being worked with are labeled. Be sure that you understand your company's labels.

FLAME

- FLAMMABLES
- PYROPHORICS
- SELF-HEATING
- EMITS FLAMMABLE GAS
- SELF-REACTIVES
- ORGANIC PEROXIDES

FLAME OVER CIRCLE

- OXIDIZERS

EXPLODING BOMB

- EXPLOSIVES
- SELF-REACTIVES
- ORGANIC PEROXIDES

00101-15_F61.EPS

Figure 61 GHS labels for flammable, combustible, and explosive materials.

HEALTH HAZARD

- CARCINOGEN
- MUTAGENICITY
- REPRODUCTIVE TOXICITY
- RESPIRATORY SENSITIZER
- TARGET ORGAN TOXICITY
- ASPIRATION TOXICITY

EXCLAMATION MARK

- IRRITANT (SKIN AND EYE)
- SKIN SENSITIZER
- ACUTE TOXICITY (HARMFUL)
- NARCOTIC EFFECTS
- RESPIRATORY TRACT IRRITANT
- HAZARDOUS TO OZONE LAYER (NON-MANDATORY)

GAS CYLINDER

- GASES UNDER PRESSURE

CORROSION

- SKIN CORROSION/BURNS
- EYE DAMAGE
- CORROSIVE TO METALS

ENVIRONMENT (NON-MANDATORY)

- AQUATIC TOXICITY

SKULL AND CROSSBONES

- ACUTE TOXICITY (FATAL OR TOXIC)

00101-15_F62.EPS

Figure 62 GHS labels for various materials.

6.1.5 Radiation Hazards

Radioactive materials are used in construction during radiographic testing of welds in piping, vessels, and pumps. These hazards are also found at nuclear power plants. Radioactive materials must be properly labeled and warning signs must be posted in the work area. Only trained workers are allowed in these areas. If you see signs containing symbols like those shown in *Figure 63*, stay away from that area unless properly trained.

Radioactive materials are a special type of physical hazard. Excessive radiation exposure can cause skin burns, nausea, vomiting, infertility, and cancer. Employees can minimize radiation exposure by limiting the amount of time they are exposed and/or increasing their distance from the source. Always use the proper shielding or PPE. Check with your supervisor to find out if there are any radioactive hazards and how to avoid them.

6.1.6 Biological Hazards

Biological hazard signs are used to warn workers of the actual or potential presence of a biological hazard (*Figure 64*). Biological hazards, or biohazards, can be any infectious agents that create a real or potential health risk. Biological hazard signs are commonly used to identify contaminated equipment, containers, rooms, materials, and areas housing experimental animals.

NCCER – Core Curriculum 00101-15

00101-15_F64.EPS

Figure 64 Biological hazard symbol.

00101-15_F63.EPS

Figure 63 Radioactivity warning symbols.

Background colors on signs may vary, but must contrast enough for the symbol to be easily identified. Wording is used on the sign to indicate the nature of the hazard or to identify the specific hazard.

6.1.7 Evacuation

In many work environments, specific evacuation procedures are needed. These procedures go into effect when dangerous situations arise, such as fires, chemical spills, and gas leaks. In an emergency, you must know the evacuation procedures. You must also know the signal (usually a horn or siren) that tells workers to evacuate.

When the evacuation signal sounds, follow the evacuation procedures precisely. That usually means taking a certain route to a designated assembly area and telling the person in charge that you are there. If hazardous materials are released into the air, you may have to determine which way the wind is blowing. Some sites may have a **wind sock** that indicates the wind direction. Different evacuation routes are planned for

different wind directions. Taking the right route will keep you from being exposed to the hazardous material.

6.2.0 Environmental Extremes

Workers on construction and industrial sites are often required to work outdoors in extremes of heat and cold. It is important to know how to protect yourself in such environments and to recognize the symptoms of conditions that could be caused by environmental extremes.

6.2.1 Heat Stress

Heat stress occurs when abnormally hot air and/or high humidity, or extremely heavy exertion, prevents your body from cooling itself fast enough. When this happens, you may suffer heat cramps, heat exhaustion, or a heat stroke. To prevent heat stress, take the following precautions:

- Drink plenty of water.
- Avoid alcoholic or caffeinated drinks.
- When possible, perform the most strenuous work during cooler parts of the day.
- Wear lightweight, light-colored clothing.
- Wear loose-fitting cotton clothing if it does not create a hazard.
- Keep your head covered and your face shaded.
- Take frequent, short breaks.
- Rest in the shade whenever possible.

Workers are encouraged to drink 4 cups (≈1 liter) of water per hour when the heat index is 103°F (39.5°C) or higher when working in hot conditions. Start by having one or two glasses of water before beginning work, and then drink one cup (237 milliliters) every 15 to 20 minutes when working in hot conditions. It is possible to drink too much water as well. Water intake should not exceed 6 cups (1.4 l) per hour or 3 gallons (11.4 l) per day. Drinking water should be cool but not cold.

6.2.2 Heat Cramps

Heat cramps are muscular pains and spasms caused by heavy exertion. Any muscles can be affected, but most often it is the muscles that have been used the most. Loss of water and electrolytes from heavy sweating causes these cramps. Symptoms of heat cramps include the following:

- Painful muscle spasms and cramping
- Pale, sweaty skin
- Normal body temperature
- Abdominal pain
- Nausea

As noted above, body temperature is not normally elevated during the onset of heat cramps. An increase in body temperature usually indicates that the problem is escalating to a more serious stage. If you experience heat cramps, take the following steps: move to a cool area, drink some water, and gently stretch and massage cramped muscles.

6.2.3 Heat Exhaustion

Heat exhaustion typically occurs when people exercise heavily or work in a warm, humid place where body fluids are lost through heavy sweating. When it is humid, sweat does not evaporate fast enough to cool the body properly.

Symptoms of heat exhaustion include the following:

- Cool, pale, and moist skin
- Heavy sweating
- Headache, nausea, vomiting
- Dilated pupils
- Dizziness
- Possible fainting
- Fast, weak pulse
- Slight elevation in body temperature

If someone is showing the symptoms of heat exhaustion, get the victim to a shaded area and have him or her lie down. Try to cool the victim by applying cold wet cloths, fanning, removing heavy clothing, and giving small amounts of cool water if the victim is conscious. If the victim's condition does not improve within a few minutes, call emergency medical services.

6.2.4 Heat Stroke

Heat stroke is life threatening. The body's temperature-control system, which produces sweat to cool the body, stops working. Body temperature can rise so high that brain damage and death may result if the body is not cooled quickly.

WARNING!
If you suspect someone has heat stroke, call emergency medical services immediately.

The symptoms of heat stroke include the following:

- Hot, dry, or spotted skin
- Extremely high body temperature
- Very small pupils
- Mental confusion
- Headache
- Vision impairment
- Convulsions
- Loss of consciousness

While waiting for help to arrive, try to cool the victim using the methods described for heat exhaustion. Ice packs placed in the armpits and the groin should be used, if possible.

6.2.5 Cold Stress

When your body temperature drops even a few degrees below normal, which is about 98.6°F (37°C), you can begin to shiver uncontrollably and become weak, drowsy, disoriented, unconscious, or even fatally ill. This loss of body heat is known as cold stress, or hypothermia.

People who work outdoors during the winter need to learn how to protect against loss of body heat. The following guidelines can help you keep your body warm and avoid the dangerous consequences of hypothermia, frostbite, and overexposure to the cold.

WARNING!
Always seek immediate medical attention if you suspect hypothermia or frostbite.

Outdoors, indoors, in mild weather, or in cold, it is a good idea to dress in layers. Layering your clothes allows you to adjust what you are wearing to suit changing temperature conditions. In cold weather, wear cotton, polypropylene, or lightweight wool next to your skin, and wool layers over your undergarments. For outdoor activities, choose clothing made of waterproof, wind-resistant fabrics such as nylon. Since a great deal of body heat is lost through the head, always wear a hat for added protection.

Water chills the body far more rapidly than air or wind. Always take along an extra set of clothing whenever working outdoors. If clothes become damp, change to dry clothes to prevent your body temperature from dropping in cold weather. Wear waterproof boots in damp or snowy weather, and always pack raingear even if the forecast calls for sunny skies.

6.2.6 Frostbite

Frostbite is a dangerous condition that can have lifelong effects on your body. It usually affects the hands, fingers, feet, toes, ears, and nose. Symptoms of frostbite include a pale, waxy-white skin color and hard, numb skin.

Remember the following when providing first aid for frostbite:

- Never rub the affected area. This can damage the skin and tissue.
- After the affected area has been warmed, it may become puffy and blister.
- The affected area(s) may have a burning feeling or numbness. When normal feeling, movement, and skin color have returned, the affected area(s) should be dried and wrapped to keep it warm.
- If there is a chance the affected area may get cold again, do not warm the skin. If the skin is warmed and then becomes cold again, it will cause severe tissue damage. Seek medical attention as soon as possible. First aid for frostbite includes moving the victim to a warm area,

removing wet clothing, and applying warm water of about 105°F (41°C) to the affected area. Do not pour the warm water directly onto the area, however.

6.2.7 Hypothermia

Hypothermia is a serious, potentially fatal condition caused by loss of body heat. You do not have to be in below-freezing temperatures to be at risk for hypothermia. Hypothermia can happen on land or in water.

The effects of hypothermia can be gradual and often go unnoticed until it is too late. If you know you'll be working outdoors for an extended period of time, work with a buddy. At the very least, let someone know where you'll be and what time you expect to return. Ask your buddy to check you for overexposure to the cold, and do the same for your buddy. Check for shivering, slurred speech, mental confusion, drowsiness, and weakness. If anyone shows any of these signs, call for emergency medical assistance and see that the victim is moved indoors as soon as possible to warm up.

> **WARNING!**
>
> If a co-worker exhibits uncontrolled shivering, slurred speech, clumsy movements, fatigue, and confused behavior, immediately call for medical assistance.

Symptoms of hypothermia include the following:

- A drop in body temperature
- Fatigue or drowsiness
- Uncontrollable shivering
- Slurred speech
- Clumsy movements
- Irritable, irrational, confused behavior

If a person is showing symptoms of hypothermia from exposure to cold air, immediately call for medical assistance. The victim should be moved to a warm, dry area. Any wet clothing should be removed and replaced with dry

Around the World
Temperature Extremes

Some places in the world reach incredibly hot and cold temperatures. Although work may come to a halt in the worst conditions, the job goes on between the extremes. In Chita, Russia for example, the average low temperature in January is roughly –30°C (≈–22°F). The average high during this period is around –17°C (≈1°F). Since these are average temperatures, people in Chita must adapt and continue working. The admiration of the world's workers should go to those who must work in temperature extremes like this without hesitation.

clothing or blankets. If possible, give the victim warm sweet drinks; avoid drinks with caffeine, as well as alcohol. Have the victim move his arms and legs to create muscle heat. If he is unable to do this, place warm bottles or hot packs around the armpits, groin, neck, and head. Do not rub the victim's body or place him in a warm bath. This could cause additional harm.

Body heat is lost up to 25 times faster in water than on land and the methods used to help victims while waiting for medical help to arrive is different. If you are a victim, first of all, do not remove any clothing. Close up and tighten all clothing to create a layer of trapped water close to the body. This provides insulation that slows heat loss. Keep your head out of the water and put on a hat or hood if possible. Get out of the water as quickly as possible, or climb onto a floating object. Do not attempt to swim unless you can reach a floating object or another person. Swimming or other physical activity uses the body's heat and reduces survival time by about 50 percent.

If it is not possible to get out of the water, wait quietly, and conserve your body heat by folding your arms across your chest, keeping your thighs together, bending your knees and crossing your ankles. If another person is in the water, huddle together with your chests pressed tightly together.

6.3.0 Hot Work Hazards

Welding and torch cutting of metals is done using heat that is high enough to melt steel. In addition, most flame cutting and some welding processes are done with flammable gases. For those reasons, such processes are classified as hot work and are subject to special safety precautions. Grinding with a pneumatic or electric grinding machine (*Figure 65*) is also classified as hot work, as it gives off sparks that could cause a fire or explosion in an environment containing flammable gases or concentrated dust. Hot work always carries a risk of severe burns as well as fire hazards. Being aware of these risks is of great importance and will help keep you safe.

6.3.1 Arc Welding Hazards

Arc welding is a process in which metals are joined using a high-intensity electric arc (*Figure 66*) at temperatures in the range of 1,500 to 3,000°F (815 to 1,648°C). The arc melts and fuses the base metals. In most cases, a filler metal is melted along with the base metal to strengthen the joint. Welders are required to wear specialized PPE

00101-15_F65.EPS

Figure 65 Grinding creates sparks.

00101-15_F66.EPS

Figure 66 Arc welding.

that protects their eyes from damage and protects their bodies from burns caused by sparks and molten metal. Such eye protection includes tinted goggles as well as a full face shield with a tinted viewing window. Anyone working in the vicinity of arc welding must also be protected, especially from damage to their eyes caused by looking at the welding arc. Never look at an arc welding operation without wearing the proper tinted eye protection, because the ultraviolet light from the arc will burn your eyes. Even a reflected arc can harm your eyes. It is extremely important to follow proper safety procedures at and around all welding operations. Serious eye injury or even blindness, as well as burns, can result from unsafe conditions.

When people are working near a welding operation, welding curtains (*Figure 67*) must be set up and everyone in the vicinity must wear tinted protective eyewear.

Even a brief exposure to the ultraviolet light from arc welding can cause a flash burn and damage your eyes badly. You may not notice the symptoms until sometime after the exposure. Here are some symptoms of flash burns to the eye:

- Headache
- Feeling of sand in your eyes
- Red or weeping eyes
- Trouble opening your eyes
- Impaired vision
- Swollen eyes

Welded material is dangerously hot. Mark it with a sign and stay clear for a while after the welding has been completed.

6.3.2 Oxyfuel Cutting, Welding, and Brazing

Oxygen and acetylene (oxyfuel) gases are combined to produce a flame with a temperature high enough to melt steel. So-called oxyfuel is commonly used to cut and weld steel and to join copper pipe by brazing.

Figure 68 shows a worker using an oxyfuel torch to cut steel. Persons working with or around welding, brazing, or metal cutting equipment can be severely burned from the intense heat and flames generated in these processes. In addition,

00101-15_F68.EPS

Figure 68 Oxyfuel cutting.

the gases used in these processes pose an additional danger because they are stored under pressure in metal cylinders (*Figure 69*).

Many hazards are involved in the handling, storage, and use of compressed gases. It takes energy to compress and confine the gas. That energy is stored until purposely released to perform useful work or until incidentally released by container failure or other causes.

00101-15_F69.EPS

Figure 69 Compressed-gas cylinders used for oxyfuel cutting.

00101-15_F67.EPS

Figure 67 Welding curtain.

Some gases, such as acetylene, are highly flammable. Oxygen is an explosion hazard if it comes into contact with grease or oil. Flammable compressed gases have additional stored energy besides simple compression-released energy. In case of a fire, for example, escaping oxygen will make the fire more intense. Other compressed gases, such as nitrogen, can cause asphyxiation simply because they displace oxygen.

The cylinders containing oxygen and acetylene must be transported, stored, and handled very carefully. Always follow these safety guidelines:

- Keep the work area clean and free from potentially hazardous items such as combustible materials and petroleum products.

> **WARNING!**
> Keep oxygen away from sources of flame and combustible materials, especially substances containing oil, grease, or other petroleum products. Compressed oxygen mixed with oil or grease will explode. Never use petroleum-based products around fittings that serve compressed oxygen lines.

- Use great caution when you handle compressed gas cylinders.
- Store cylinders in an upright position where they will not be struck, where they will be away from corrosives, and where they cannot tip over or fall. Cylinders must be stored in the upright position in a secure container or secured to a wall with chains. When in storage, oxygen and fuel cylinders must be separated by 20 feet (6.1 m) or by a 5-foot (1.5 m) high ½-hour fire-rated barrier (*Figure 70*).

> **WARNING!**
> Do not remove the protective cap unless a cylinder is secured. If the cylinder falls over and the nozzle breaks off, the cylinder could shoot off like a rocket, injuring or killing anyone in its path.

6.3.3 Transporting and Securing Cylinders

Always handle cylinders with care. They are under high pressure and should never be dropped, knocked over, rolled, or exposed to heat in excess of 140°F (60°C). When moving cylinders, always be certain that the valve caps are in place. Cylinders must be transported to the workstation in the upright position on a hand truck or bottle cart such as the one shown in *Figure 69*. Oxygen and cutting gases should not be transported on the same cart unless the cart has a divider, as shown.

Be aware that some job sites will not permit oxygen and cutting gases to be transported on the same cart under any circumstances.

Never attempt to lift a cylinder using the holes in a safety cap; use an approved lifting cage (*Figure 71*). Make sure that the cylinder is secured in the cage. Cages of various sizes are available for high-pressure cylinders and cylinders containing liquids.

6.3.4 Hot Work Permits

A hot work permit (*Figure 72*) is an official authorization from the site manager to perform work that may pose a fire hazard. The permit includes information such as the time, location, and type of work being done. The hot work permit system promotes the development of standard fire safety guidelines. Permits also help managers keep records of who is working where and at what time. This information is essential in the event of an emergency at times when personnel need to be evacuated.

Most sites require the use of hot work permits and fire watches. When these requirements are violated, severe penalties may be imposed because of the risk it represents. Before a hot work permit is issued, a competent person must inspect the work area to ensure that no flammable or combustible materials are present. This inspection includes the areas above and below where the work will be performed, as well as the surrounding area.

A fire watch is posted when welding or cutting work is being done in many areas. One person other than the welding or cutting operator must constantly scan the work area for fires. Fire watch personnel should have ready access to fire extinguishers and alarms and know how to use them. Welding and cutting operations should never be performed without a fire watch. The area where welding is done must be monitored afterwards until there is no longer a risk of fire. The person on fire watch must be dedicated to that activity and may not perform other work during the time when hot work is being performed.

6.4.0 Fire Hazards and Firefighting

Fire is always a hazard on construction job sites. Many of the materials used in construction are flammable. In addition, welding, grinding, and many other construction activities create heat or sparks that can cause a fire. Fire safety involves two elements: fire prevention and firefighting.

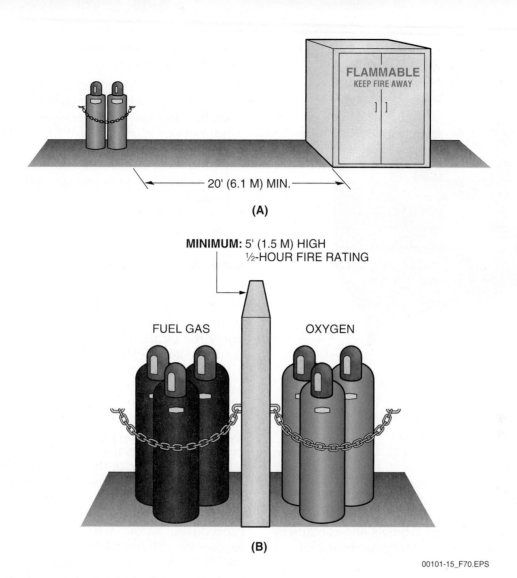

(A)

MINIMUM: 5' (1.5 M) HIGH
½-HOUR FIRE RATING

FUEL GAS   OXYGEN

(B)

00101-15_F70.EPS

Figure 70 Cylinder storage.

6.4.1 How Fires Start

For a fire to start, three things are needed in the same place at the same time: fuel, heat, and oxygen. If one of these three is missing, a fire will not start.

Fuel is anything that will combine with oxygen and heat to burn. Oxygen is always present in the air. When pure oxygen is present, such as near a leaking oxygen hose or fitting, material that would not normally be considered fuel (including some metals) will burn.

Heat refers to a source of ignition. It may be as simple as a single spark. Heat is anything that will raise a fuel's temperature to the **flash point**. The flash point is the temperature at which a fuel is encouraged to burn. The flash points of some fuels are quite low—room temperature or less. When the burning gases raise the temperature of a fuel to the point at which it ignites, the fuel itself will burn—and keep burning—even if the original source of heat is removed.

What is needed for a fire to start can be shown as a fire triangle (*Figure 73*). If one element of the triangle is missing, a fire cannot start. If a fire has started, removing any one element from the triangle will put it out.

6.4.2 Combustibles

Combustibles are categorized as liquid, gas, or ordinary combustibles. The term *ordinary combustibles* means paper, wood, cloth, and similar fuels. Liquids can be flammable or combustible. Flammable liquids have a flash point below 100°F (38°C). Combustible liquids have a flash point at or above 100°F (38°C). Flammable gases used on construction sites include acetylene, hydrogen, ethane, and propane. To save space, these gases are compressed so that a large amount is stored in a small cylinder or bottle. As long as the gas is kept in the cylinder, oxygen cannot get to it and start a fire. The cylinders should be stored away

00101-15_F71.EPS

Figure 71 Cylinder lifting cage.

from sources of heat. If oxygen is allowed to escape and mix with a flammable gas, the resulting mixture will explode under certain conditions.

The easiest way to prevent fire in ordinary combustibles is to keep a neat, clean work area. If there are no scraps of paper, cloth, or wood lying around, there will be no fuel to start a fire. Establish and maintain good housekeeping habits. Use approved storage cabinets and containers for all waste and other ordinary combustibles.

6.4.3 Fire Prevention

The best way to ensure fire safety is to prevent a fire from starting. Fire can be prevented by the following actions:

- *Removing the fuel* – Liquid does not usually burn. What generally burns are the gases (vapors) given off as the liquid evaporates. Keeping liquids in an approved, sealed container prevents evaporation. If there is no evaporation, there is no fuel to burn.
- *Removing the heat* – If the liquid is stored or used away from a heat source, it will not be able to ignite.
- *Removing the oxygen* – The vapor from a liquid will not burn if oxygen is not present. Keeping safety containers tightly sealed prevents oxygen from coming into contact with the fuel.

The best way to prevent a fire is to make sure that fuel, oxygen, and heat are never present in the same place at the same time.

The following are some basic safety guidelines for fire prevention:

- Always work in a well-ventilated area, especially when using flammable materials such as shellac, lacquer, paint stripper, or construction adhesives.
- Never smoke or light matches when working with or near flammable materials.
- Keep oily rags in approved, self-closing metal containers.
- Store combustible materials only in approved containers.

6.4.4 Basic Firefighting

You are not expected to be an expert firefighter, but you may have to deal with a fire to protect your safety and the safety of others. You need to know the locations of firefighting equipment on your job site as well as which equipment to use on different types of fires. However, only a **qualified person** is placed in a position where firefighting skills are a required activity.

Most companies tell new employees where fire extinguishers are kept. If you have not been told, be sure to ask. Also, ask how to report fires. The telephone number of the nearest fire department should be clearly posted in your work area. If your company has a fire brigade, learn how to contact them. Learn your company's fire safety procedures. Know what type of extinguisher to use for different kinds of fires and how to use them. Make sure all extinguishers are fully charged. Never remove the tag from an extinguisher—it shows the date the extinguisher was last serviced and inspected.

A fire watch is required to have a fire extinguisher while on duty. A portable extinguisher such as the one shown in *Figure 74* is commonly provided for that purpose.

The function of a fire extinguisher is to remove one of the three elements (oxygen, heat, fuel) needed to sustain a fire. A fire extinguisher cannot remove fuel, so it is designed to remove either heat or oxygen. Heat is removed by using a coolant such as water; oxygen is removed by smothering the fire with dry chemicals or CO_2.

Fire extinguishers are rated for specific types of fires. Class A extinguishers, for example, are intended for fires involving ordinary combustibles such as paper, wood, and fabric. An extinguisher that is rated only for Class A fires might contain water, which could not be used on electrical or grease fires. Using water on a Class B fire (flammable liquids, grease, or gases) could spread the fire or splash burning fuel. Using water on a

HOT WORK PERMIT

For Cutting, Welding, or Soldering with Portable Gas or ARC Equipment

(References: 1997 Uniform Fire Code Article 49 & National Fire Protection Association Standard NFPA 51B.)

Job Date_____ Start Time_____ Expiration_____ WO #_____

Name of Applicant_____ Company_____ Phone_____

Supervisor_____ Phone_____

Location / Description of work _____

IS FIRE WATCH REQUIRED?

1. _____ (yes or no) Are combustible materials in building construction closer than 35 feet to the point of operation?

2. _____ (yes or no) Are combustibles more than 35 feet away but would be easily ignited by sparks?

3. _____ (yes or no) Are wall or floor openings within a 35 foot radius exposing combustible material in adjacent areas, including concealed spaces in floors or walls?

4. _____ (yes or no) Are combustible materials adjacent to the other side of metal partitions, walls, ceilings, or roofs which could be ignited by conduction or radiation?

5. _____ (yes or no) Does the work necessitate disabling a fire detection, suppression, or alarm system component?

YES to any of the above indicates that a qualified fire watch is required.

Fire Watcher Name(s) _____ Phone_____

NOTIFICATIONS

Notify the following groups at least 72 hours prior to work and 30 minutes after work is completed.

Write in names of persons contacted.

Notify in person OR by phone ONLY if question #5 above is answered "yes":

• Facilities Management Fire Alarm Supervisor

Notify by phone or in person: (If by phone, write down name of person and send them a completed copy of this permit.)

• Facilities Management Fire Protection Group
• Environmental Health & Safety Industrial Hygiene Group

SIGNATURES REQUIRED

University Project Manager_____ Date _____ Phone_____

I understand and will abide by the conditions described in this permit. I will implement the necessary precautions which are outlined on both sides of this permit form. Thirty minutes after each hot work session, I will reinspect work areas and adjacent areas to which spark and heat might have spread to verify that they are fire safe, and contact Facilities Management Alarm Technicians to have any disabled fire protection systems reactivated.

_____   _____ Date _____ Phone_____
        Permit Applicant   Company or Department

1/17/03

Figure 72 Hot work permit.

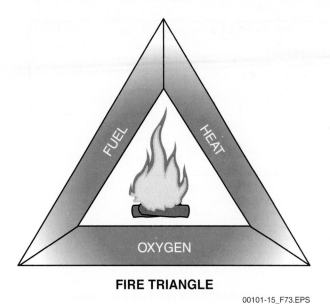

FIRE TRIANGLE

00101-15_F73.EPS

Figure 73 The fire triangle.

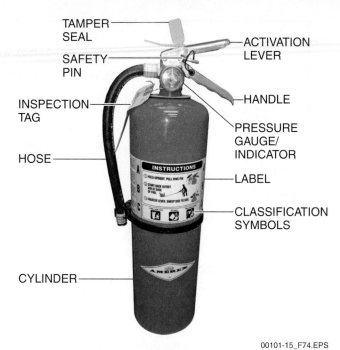

00101-15_F74.EPS

Figure 74 A portable fire extinguisher.

Class C (electrical) fire could cause you to be electrocuted because water conducts electricity. Class D fire extinguishers are required for fires involving reactive metals such as sodium, potassium, magnesium, titanium, zirconium, and the metal hydrides. A Class D fire extinguisher contains a powder designed to coat the metal to either smother the fire or keep oxygen from reaching the fire. If the powder coat fails to extinguish the fire, the best solution is to reapply the powder to keep the fire from spreading.

A pictogram label (*Figure 75*) on the extinguisher denotes the type(s) of fire for which the extinguisher is intended. The extinguisher shown in the figure is a dry-chemical extinguisher, which can be used on a Class A, B, or C fire. This type is the most common because it can be used to fight most fires. Some CO_2 extinguishers can also be used on Class A, B, or C fires, but others are designed for use on only class B and C fires. This can be seen in *Figure 76*.

CO_2 is heavier than oxygen, so it smothers the fire by displacing the available oxygen. It also cools as it expands, so it can serve as a coolant. Some companies permit only CO_2 extinguishers to be used because the dry chemical in a dry-chemical extinguisher is a corrosive that can damage electrical systems. In addition, the fine powder emitted by the material in a dry-chemical extinguisher can cause eye and respiratory irritation when it becomes airborne. One advantage of the dry-chemical extinguisher is that its working range is 10 to 20 feet (3.1 to 6.1 m), while the range of the CO_2 extinguisher is 3 to 8 feet (0.9 to 2.4 m). A pressurized-water extinguisher, which is used only for Class A fires, has a range of about 50 feet (15 m).

Prevention and Preparation Are the Keys to Fire Safety

Any fire in the workplace can cause serious injury or property damage. When chemicals are involved, the risks are even greater. Prevention is the key to eliminating the hazards of fire in the workplace. Preparation is the key to controlling any fires that do start. Take the following precautions to ensure safety from fire in the workplace:

- Keep work areas clean and clutter-free.
- Know how to handle and store chemicals.
- Know what you are expected to do in case of a fire emergency.
- Should a fire start, call for professional help immediately. Don't let a fire get out of control.
- Know what chemicals you work with. You might have to tell firefighters at a chemical fire what kinds of hazardous substances are involved.
- Make sure you are familiar with your company's emergency action plan for fires.
- Use caution when using power tools near flammable substances.

Figure 75 Fire extinguisher pictograms.

00101-15_F75.EPS

Fire extinguishers must be periodically inspected to make sure they are properly charged. The needle on the extinguisher's pressure gauge must register in the green area (*Figure 77*). The inspection must be documented on the tag attached to the extinguisher. Whenever you check out an extinguisher, make sure its inspection tag (*Figure 78*) is up to date and that the pressure gauge registers the correct charge. Not all extinguishers have gauges, however. Extinguishers without gauges must be weighed to determine if they are fully charged.

6.4.5 Using a Fire Extinguisher

Use the PASS method (*Figure 79*) when attacking a fire:

- **P**ull the pin from the handle, breaking the tamper seal in the process (*Figure 80*).
- **A**im the nozzle at the base of the fire while 8 to 10 feet away.
- **S**queeze the discharge handle.
- **S**weep the nozzle back and forth at the base of the fire.

Note that the method of attacking the fire can vary from one extinguisher to another. The instructions for a specific fire extinguisher are printed on the label, as shown in *Figure 81*. Be sure you are familiar with the type of extinguishers that are commonly provided on your current job site.

As the fire is extinguished, move in closer until it is completely out. Stop shooting every three or four sweeps to check progress. Try not to use all of the contents on the initial attack, in case the fire flares up. Once the extinguisher has been used, it must be re-charged.

> **WARNING!**
>
> Do not use a CO$_2$ extinguisher on a Class D fire, as it can cause the fire to spread. Also note that CO$_2$ is heavier than air, so it will concentrate in low areas, displacing oxygen. For that reason, a CO$_2$ extinguisher must never be taken into, or used in, a confined space. Before it can be used in a confined space, all personnel in the space must first be evacuated. After the extinguisher has been used in the confined space, the space must be cleared by a competent person before it can be re-entered.
>
> Finally, CO$_2$ is discharged from the extinguishers at a temperature of −109°F (−78°C), so it can cause frostbite if it contacts the skin. Never discharge it while it is pointed at someone and do not hold it by the activation lever unless the activation lever is insulated.

> **WARNING!**
>
> When using a CO$_2$ extinguisher do not hold the activation lever with your bare hand, as it will become extremely cold.

Ordinary Combustibles

Fires in paper, cloth, wood, rubber, and many plastics require a water-type extinguisher labeled A.

⚠ TRASH•WOOD•PAPER

CO₂

OR

Dry Chemical

Flammable Liquids

Fires in oils, gasoline, some paints, lacquers, grease, solvents, and other flammable liquids require an extinguisher labeled B.

Ⓑ LIQUIDS

Electrical Equipment

Fires in liquids, fuse boxes, energized electrical equipment, computers, and other electrical sources require an extinguisher labeled C.

Ⓒ ELECTRICAL EQUIP

Ordinary Combustibles, Flammable Liquids, or Electrical Equipment

Multi-purpose dry chemical extinguishers are suitable for use on Class A, B, and C fires.

⚠ TRASH•WOOD•PAPER   Ⓑ LIQUIDS   Ⓒ ELECTRICAL EQUIP

Metals

Combustible metals such as magnesium and sodium require special extinguishers labeled D.

Ⓓ

00101-15_F76.EPS

Figure 76 Fire extinguisher applications.

00101-15_F77.EPS

Figure 77 Fire extinguisher pressure gauge.

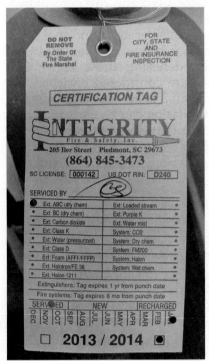

00101-15_F78.EPS

Figure 78 Fire extinguisher inspection tag.

00101-15_F79.EPS

Figure 79 PASS Method.

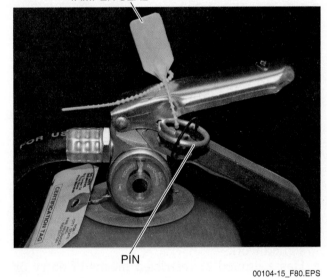

00104-15_F80.EPS

Figure 80 Pin and tamper seal.

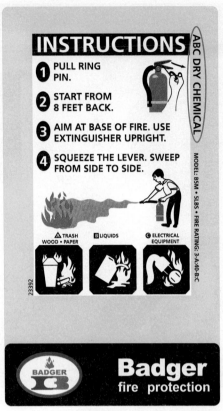

Figure 81 Example of fire extinguisher instructions.

Figure 81 Example of fire extinguisher instructions.

00101-15_F81.EPS

6.5.0 Confined Spaces

Construction and maintenance work is not always done outdoors; a lot of it is done in confined spaces. A confined space is a space that is large enough to work in but that has limited means of entry or exit. A confined space is not designed for human occupancy and it has limited ventilation. Examples of confined spaces are tanks, vessels, silos, storage bins, hoppers, vaults, pits, and certain compartments on ships and barges (*Figure 82*).

OSHA defines a confined space as a space that has the following characteristics:

- Is large enough and so configured that employees can bodily enter and perform their assigned work.
- Has a limited or restricted means of entry or exit.
- Is not designed for continuous employee occupancy.

OSHA further defines a **permit-required confined space** as a space that has one or more of the following characteristics:

- Contains or has the potential to contain a hazardous atmosphere.
- Contains a material that has the potential for engulfing an entrant.
- Has an internal configuration such that an entrant could be trapped or asphyxiated by inwardly converging walls or by a floor that slopes downward and tapers to a smaller cross-section.
- Contains any other recognized serious safety or health hazard.

Atmospheric hazards are a concern in a confined space. The air in the space can contain flammable or explosive vapors or toxic gases. The space can also contain too much or too little oxygen. For safe working conditions, the oxygen level in a confined space atmosphere must range between 19.5 and 23.5 percent by volume as measured with an oxygen analyzer, with 21 percent being considered the normal level. Oxygen concentrations below 19.5 percent by volume are considered deficient; those above 23.5 percent by volume are considered enriched. Special meters are available to test atmospheric hazards. Forced ventilation can be used to overcome these hazards, but that does not eliminate the need to satisfy other confined space safety requirements, including inspection and permitting.

> **WARNING!**
>
> If too much oxygen is introduced into a confined space, it can be absorbed by a worker's clothing and ignite. If too little oxygen is present, it can lead to death in minutes. For this reason, the following precautions apply:
>
> - Make sure confined spaces are ventilated properly for cutting or welding purposes.
> - Never use oxygen in confined spaces for ventilation purposes.
> - Always remain aware of the work going on around you, as conditions can change.

A permit-required confined space is a type of confined space that has been evaluated by a qualified person and found to have actual or potential hazards. Written authorization, commonly known as an entry permit, is required in order to enter a permit-required confined space. Once the entry permit is issued, it must be posted at the entry to the confined space.

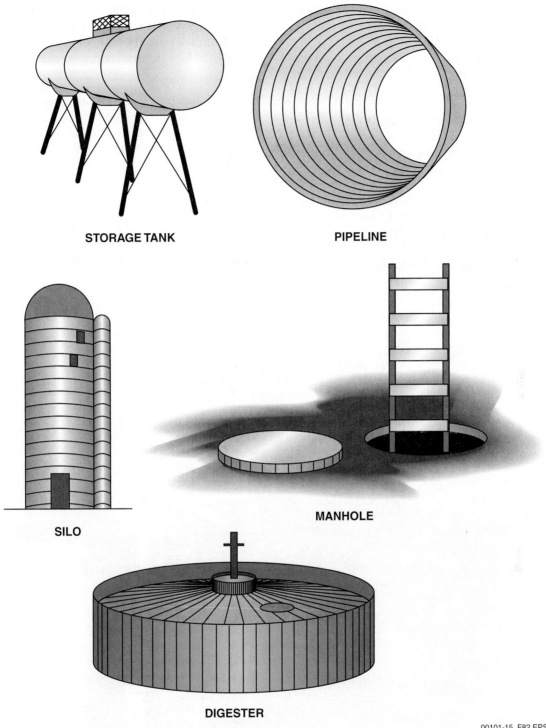

STORAGE TANK

PIPELINE

SILO

MANHOLE

DIGESTER

00101-15_F82.EPS

Figure 82 Examples of confined spaces.

When equipment is operating, confined spaces may contain hazardous gases or fluids, or may be oxygen-deficient. In addition, the work you are doing may introduce hazardous fumes into the space. Welding and metal cutting are examples of such work. For safety, you must take special precautions both before you enter and leave a confined space, and while you work there.

Until you have been trained to work in permit-required confined spaces and have taken the needed precautions, you must stay out of them. If you are not sure whether a confined space requires a permit, ask your supervisor. You must always follow your employer's procedures and your supervisor's instructions. Confined space procedures may include getting clearance from a safety representative before starting the work. You will be told what kinds of hazards are involved and what precautions you need to take. You will also be shown how to use the required PPE. Remember, it is better to be safe than sorry, so ask!

> **WARNING!**
>
> Without proper training, no employee is allowed to enter a permit-required or non-permit-required confined space. Employers are required to have programs to control entry to, and hazards in, both types of confined spaces.

Never work in a confined space without an attendant. An attendant must remain outside a permit-required confined space. The attendant monitors entry, work, and exit (*Figure 83*).

Figure 83 Permit-required confined space.

00101-15_F83.EPS

Additional Resources

US Occupational Safety and Health Administration. Numerous safety videos are available on line at **www.osha.gov/video**.

Construction Safety, 2006. Jimmie W. Hinze. Upper Saddle River, NJ: Pearson Education, Inc.

DeWalt Construction Safety/OSHA Professional Reference, Paul Rosenberg; American Contractors Educational Services. 2006. DEWALT.

Basic Construction Safety and Health, Fred Fanning. 2014. CreateSpace Independent Publishing Platform.

6.0.0 Section Review

1. The term PEL refers to a worker's _____.

   a. personal energy level
   b. maximum exposure limit
   c. the amount of lead in the bloodstream
   d. product efficiency level

2. A condition in which body temperature drops a few degrees below normal is known as _____.

   a. heat stress
   b. cold stroke
   c. cold stress
   d. heat exhaustion

3. Swollen eyes and a sandy feeling in the eyes are symptoms of _____.

   a. a flash burn
   b. heat stroke
   c. asbestos exposure
   d. frostbite

4. The three elements needed for a fire to start are fuel, combustible liquid, and oxygen.

   a. True
   b. False

5. A confined space is an enclosed space that has _____.

   a. only one door
   b. limited access
   c. hazardous gases
   d. locked doors

SUMMARY

Although the typical job site has many hazards, it does not have to be a dangerous place to work. Your employer has programs to deal with potential hazards. Basic rules and regulations help protect you and your co-workers from unnecessary risks.

This module has presented many of the basic guidelines you must follow to ensure your safety and the safety of your co-workers. These guidelines fall into the following categories:

- Following safe work practices and procedures
- Inspecting safety equipment before use
- Using safety equipment properly

The basic approach to safety is to eliminate hazards in the equipment and the workplace; to learn the rules and procedures for working safely with and around the remaining hazards; and to apply those rules and procedures. The information covered here provides the foundation for a safe, productive, and rewarding career.

1. The four leading causes of death in the construction industry include electrical incidents, struck-by incidents, caught-in or caught-between incidents, and _____.
   a. vehicular incidents
   b. falls
   c. radiation exposure
   d. chemical burns

2. A sign that has a white background with a green panel with white lettering is a _____.
   a. general information sign
   b. safety instruction sign
   c. caution sign
   d. danger sign

3. To properly dispose of oily rags, they must be _____.
   a. stored in a container designed for the purpose
   b. washed thoroughly and returned to use
   c. taken outdoors and thrown into a dumpster
   d. burned at the end of the shift

4. Keeping your work area clean and free of scraps or spills is referred to as _____.
   a. managing
   b. organizing
   c. housekeeping
   d. stacking and storing

5. HAZCOM classifies all paint, concrete, and wood dust as _____.
   a. hazardous materials
   b. common materials
   c. inexpensive materials
   d. nonhazardous materials

6. Under HAZCOM, if you spot a hazard on your job site you must _____.
   a. report it to your supervisor
   b. leave immediately
   c. notify your co-workers
   d. correct the problem

7. Which of the following must be reported to your supervisor?
   a. Only major injuries
   b. Only incidents and major injuries
   c. All injuries, incidents, and accidents
   d. Only incidents in which a death occurred

8. Metal ladders should *not* be used near _____.
   a. stairways
   b. scaffolds
   c. electrical equipment
   d. windows

9. If you lean a straight ladder against the top of a 16-foot (4.8 m) wall, the base of the ladder should be _____.
   a. 3 feet (0.9 m) from the base of the wall
   b. 4 feet (1.2 m) from the base of the wall
   c. 5 feet (1.5 m) from the base of the wall
   d. 6 feet (1.8 m) from the base of the wall

10. The two basic types of scaffolds are _____.
    a. self-supporting and suspended scaffolds
    b. fixed and portable scaffolds
    c. metal and wooden scaffolds
    d. assembled and deliverable scaffolds

11. Interlocking stacked material is done by _____.
    a. applying a chain and padlock to it
    b. driving stakes around the stack
    c. securing the objects with rope
    d. placing the objects at right angles

12. To reduce the risk of workers being hurt or killed by falling materials, the maximum height-to-base ratio of a stack of materials should be _____.
    a. 2:1
    b. 4:1
    c. 8:1
    d. 10:1

13. The most common cause of death for equipment operators is _____.
    a. hit and run
    b. equipment rollover
    c. brake malfunction
    d. head-on collisions

14. Most cave-ins happen suddenly with little or no warning and occur in trenches 5 feet (1.5 m) to _____ .

    a. 10 feet (3.4 m) deep
    b. 15 feet (4.6 m) deep
    c. 20 feet (6.1 m) deep
    d. 25 feet (7.6 m) deep

15. In every trench over 4 feet (1.2 m) deep, there must be an exit every _____.

    a. 10 feet (3.4 m)
    b. 12 feet (3.7 m)
    c. 18 feet (5.5 m)
    d. 25 feet (7.6 m)

16. The minimum distance that a spoil pile must be located from the edge of an excavation is _____.

    a. 6 inches (15 cm)
    b. 2 feet (61 cm)
    c. 5 feet (1.5 m)
    d. 100 yards (91.4 m)

17. The type of trench protection designed to prevent trench wall cave-ins is _____.

    a. trench shield
    b. trench box
    c. spoil pile
    d. shoring

18. Protective guards are provided on power tools and machines in order to keep _____.

    a. dirt out of the tool or machine
    b. workers from being caught in rotating or moving parts
    c. the tool or machine from being struck by moving equipment
    d. unauthorized personnel from using the tool or machine

19. The minimum safe working distance from exposed electrical conductors _____.

    a. depends on the voltage
    b. is 6 inches (15 cm)
    c. is one foot (30 cm)
    d. is unlimited

20. Work that is performed near a hazard but not in direct contact with it is called _____.

    a. close call work
    b. near miss work
    c. proximity work
    d. barricade work

21. Circuit breakers and disconnect switches are examples of _____.

    a. energy-isolating devices
    b. energy-removal devices
    c. lockout/tagout devices
    d. multiple lockout devices

22. Which of the following provide the best eye protection?

    a. Welding hoods
    b. Face shields
    c. Safety goggles
    d. Strap-on glasses

23. The type of respirator that has its own clean air supply is the _____.

    a. half mask
    b. mouthpiece with mechanical filter
    c. self-contained breathing apparatus
    d. full facepiece mask

24. A worker can be exposed to hazardous materials by inhalation, ingestion, and _____.

    a. osmosis
    b. absorption
    c. radiation
    d. proximity

25. Hypothermia is a condition brought on by _____.

    a. excessive alcohol consumption
    b. prolonged exposure to cold
    c. excessive sweating
    d. prolonged exposure to heat

26. When in storage, oxygen and fuel cylinders used in oxyfuel cutting must be separated by _____.

    a. 20 feet (6.1 m)
    b. a metal barrier
    c. a 20-foot (6.1 m) wall
    d. 10 feet (3.4 m)

27. Grease and oil must be kept away from oxygen tanks because _____.

    a. grease or oil will contaminate the oxygen
    b. oxygen causes oil to freeze
    c. grease or oil can cause oxygen to explode
    d. oxygen will contaminate the oil or grease

28. Which of these must be present in the same place at the same time for a fire to occur?

   a. Oxygen, carbon dioxide, and heat
   b. Oxygen, heat, and fuel
   c. Hydrogen, oxygen, and wood
   d. Grease, liquid, and heat

29. A fire extinguisher labeled C would be used to fight a(n) _____.

   a. electrical fire
   b. magnesium fire
   c. paper fire
   d. gasoline fire

30. Which of the following characteristics is typical of a confined space?

   a. It has a limited amount of ventilation.
   b. There is no means of escape.
   c. It is too small to work in.
   d. It may be entered by untrained employees.

Trade Terms Quiz

Fill in the blank with the correct term that you learned from your study of this module.

1. The formal procedure for taking equipment out of service and ensuring it cannot be operated until an authorized person has returned it to service is ___Lock out tagout___

2. A(n) ___excavation___ is any man-made cut, cavity, trench, or depression in an earth surface, formed by removing earth.

3. A(n) _____ identifies unsanitary, hazardous, or dangerous working conditions and has the authority to correct or eliminate them.

4. A(n) _____ is large enough to work in but has limited means of entry or exit.

5. Liquids that are _____ must be stored in safety cans to avoid the risk of fire.

6. To save lives, prevent injuries, and protect the health of America's workers, _____ publishes rules and regulations that employees and employers must follow.

7. Even a brief exposure to the ultraviolet light from arc welding can damage the eyes, causing a(n) _____.

8. _____ is an OSHA rule requiring all contractors to educate their employees about the hazardous chemicals they may be exposed to on the job site.

9. The temperature at which a fuel gives off enough gases to burn is called the _____.

10. A(n) ___Ground___ is the conducting connection between electrical equipment or an electrical circuit and the earth.

11. When climbing a ladder or scaffold, a tagline or ___Hand line___ should be used to pull up tools.

12. If a work area is ___planked___, that means it has pieces of material at least 2 inches (5 cm) thick and 6 inches (15 cm) wide used as flooring, decking, or scaffolds.

13. When doing work more than 6 feet (≈2 m) above the ground, a worker must wear a safety harness with a(n) ___six foot___ that is attached to a strong anchor point.

14. A structure that prevents trench walls from collapsing is known as _____.

15. Information on how to handle hazardous substances is located in the _____.

16. Overloading, which means exceeding the _____ of a ladder, can cause ladder failure.

17. If you are operating a vehicle on a job site and cannot see to your rear, get a(n) _____ to direct you.

18. Before working in a(n) _____ you must be trained, obtain written authorization, and take the necessary precautions.

19. A(n) _____ is a narrow excavation made below the surface of the ground that is generally deeper than it is wide.

20. The _____ on a scaffold is placed halfway between the toeboard and the top rail.

21. _____ for welding includes a face shield, ear plugs, and gloves.

22. A(n) _____ has proven his or her extensive knowledge, training, and experience and has successfully demonstrated the ability to solve problems relating to the work.

23. A(n) _____ provides clean air for breathing.

24. A(n) _____ is an elevated working platform for workers and material.

25. Working near a hazard, but not actually in contact with it is known as _____.

26. A vertical barrier called a(n) ___Toe board___ is used at floor level on scaffolds to prevent materials from falling.

27. A(n) _____ shows which way the wind is blowing.

28. A horizontal board called a(n) ___Top rail___ is used at top-level on all open sides of scaffolds and platforms.

29. The _____ is a general construction rule that states that fall protection is required any time you are working 6 feet (≈2 m) above a lower level.

30. _____ is the company organization that includes reporting procedures and supervisory responsibility.

31. The process of joining metal parts using an electric arc is known as _____.

32. An equipment or process powered by compressed air is _____.

33. Wood or metal braces placed diagonally from the bottom of one rail to the top of another rail in order to provide support are referred to as Cross bracing

34. When a whole company sees the value of safety it is said to have a(n) Culture.

35. Material such as soil removed from a trench or excavation is known as _____.

36. An area that is enclosed, fenced, covered, or otherwise protected by barriers, rails, covers, or platforms to prevent dangerous contact is said to be Guarded

37. A(n) _____ is an unintentional, electrically conducting connection between an ungrounded conductor of an electrical circuit and the normally noncurrent-carrying conductors, metal objects, or the earth.

38. A(n) Ground _____ is a device that interrupts and de-energizes an electrical circuit to protect a person from electrocution.

Ground fault circuit

39. The type of equipment that is powered by fluid pressure is _____.

40. An unplanned event that results in personal injury is referred to as a(n) _____.

41. A protective screen set up around a welding operation designed to safeguard workers not directly involved in that operation is a(n) _____.

42. A process using heat in excess of 800°F (427°C) to melt base metal, along with a filler metal in order to join copper pipe is called _____.

43. A material that is capable of easily igniting and rapidly burning with a flash point at or above 100°F (38°C) is considered _____.

44. A structure used to protect workers in trenches, but lacking the ability to prevent cave-ins is _____.

45. A near miss that occurs but does not result in personal injury is considered to be a(n) _____.

Trade Terms

Accident -40
Arc welding 31
Brazing -42
Combustible -43
Competent person 3
Confined space 4
Cross-bracing 33
Excavation _
Flammable 5
Flash burn 7
Flash point 9
Ground 10
Ground fault 37

Ground fault circuit interrupter (GFCI) 38
Guarded
Hand line 11
Hazard communication Standard (HAZCOM) 8
Hydraulic 39
Incident 45
Lanyard 13
Lockout/tagout (LOTO)
Management system 30
Maximum intended load 16
Midrail 20

Occupational Safety and Health Administration (OSHA) 6
Permit-required confined space 18
Personal protective equipment (PPE) 21
Planked 12
Pneumatic 32
Proximity work 25
Qualified person 22
Respirator 23
Safety culture 34

Safety data sheet (SDS) 15
Scaffold 24
Shielding 44
Shoring 14
Signaler 12
Six-foot rule 29
Spoil 35
Toeboard 26
Top rail 28
Trench 19
Welding curtain 41
Wind sock 27

Bob Fitzgerald
Southern Company
Manager – Project Safety and Health

How did you choose a career in the Safety and Health field?
After many years working as an emergency medical technician (EMT) in the public and private sectors, I was hired as a medic on a construction site. What sparked my interest in safety and health was the notion of not just treating the injured, but actually being able to do something to prevent the incident from happening in the first place. When working in the medical field, you have to really care and have compassion for people. I think that is one of the major attributes of a successful safety professional.

What types of training have you been through?
I have taken all sorts of safety and health classes through the years, including trenching, shoring, electrical, fall protection, respiratory protection, and more. I was privileged to attend the NCCER Safety Academy at Clemson University early in my safety career and I think that was a turning point in my philosophy and thinking about safety management. Since then I have earned my Bachelor's of Science and a Master's degree in Safety and Health. I also hold the Certified Safety Professional and Construction Health and Safety Technician designation.

What kinds of work have you done in your career?
I have had two jobs in my professional career: EMT and Safety Professional.

Tell us about your present job.
I serve as the Manager for Project Safety & Health for Southern Company Operations Engineering and Construction Services, presently based in Birmingham, Alabama. Southern Company is the premier energy company serving the Southeast through its subsidiaries. Our firm provides engineering and project management services for most major capital work at Southern Company Electric Generating facilities as well as new generation projects.

What do you enjoy most about your job?
I enjoy helping people and making a difference in their lives. I feel very gratified to know that something I may say or some influence I may have could impact a person getting home safely to their families. That truly humbles me and is my motivation.

What factors have contributed most to your success?
You really have to care about people. Safety is a people business! My early experiences in the medical field helped me to have empathy for people and to see the pain and suffering that could happen if things go wrong.

What advice would you give to those new to Safety and Health field?
Realize that the value for safety and health is fundamental in all crafts or whatever role you perform in construction or industrial work. Watch out for each other and don't be afraid to speak up if you see an at-risk situation or action. Take the risk of intervening to help someone. Learn all you can about safety and put it to use.

Interesting career-related fact or accomplishment:
In 1995, I had the privilege to work with a wonderful team of construction professionals and craftworkers to earn the OSHA Voluntary Protection Program (VPP) STAR designation at a site in Stevenson, Alabama. This particular construction site was the first in the state of Alabama to earn the OSHA STAR. The effort that went into achieving OSHA STAR still echoes today in folks that I meet that worked on that site. They have instilled a culture of safety and a lifelong safety commitment. It was, and is, great!

OSHA Mission and Standards

The mission of the Occupational Safety and Health Administration (OSHA) is to save lives, prevent injuries, and protect the health of America's workers. To accomplish this, federal and state governments work in partnership with millions of working men and women who are covered by the Occupational Safety and Health Act (OSH Act) of 1970.

Nearly every worker in the US comes under OSHA's jurisdiction. There are some exceptions, such as miners, transportation workers, many public employees, and the self-employed. These specific groups are not covered by OSHA standards, but most are covered by standards developed by the specific industry.

The Code of Federal Regulations

The *Code of Federal Regulations (CFR)* Part 1910 covers the OSHA standards for general industry. *CFR* Part 1926 covers the OSHA standards for the construction industry. Either or both may apply to you, depending on where you are working and what you are doing. If a job-site condition is covered in the CFR book, then that standard must be used. However, if a more stringent requirement is listed in *CFR* 1910, it should also be met. Check with your supervisor to find out which standards apply to your job.

29 *CFR* 1926 is divided into subparts A through Z. As you progress in task-specific training, you will learn about all the subparts applicable to your work. *Subpart C of* 29 *CFR* 1926 applies to all construction and maintenance work. It outlines the general safety and health provisions for the construction industry. It covers the following topics:

- Safety training and education
- Injury reporting and recording
- First aid and medical attention
- Housekeeping
- Illumination
- Sanitation
- PPE
- Standards incorporated by reference
- Definitions
- Access to employee exposure and medical records
- Means of egress
- Employee emergency action plans

For assistance in identifying parts, sections, paragraphs, and subparagraphs of an OSHA standard, refer to Appendix *Figure 1*.

All of OSHA's safety requirements in the Code of Federal Regulations apply to residential as well as commercial construction. In the past, OSHA enforced safety only at commercial sites. The increasing rate of incidents at residential sites led OSHA to enforce safety guidelines for the building of houses and townhomes. Today, however, OSHA still focuses its enforcement efforts on commercial construction.

The General Duty Clause

If a standard does not specifically address a hazard, the general duty clause must be invoked. Failing to adhere to the general duty clause can result in heavy fines for your employer. The general duty clause reads as follows:

In practice, OSHA, court precedent, and the review commission have established that if the following elements are present, a general duty clause citation may be issued:

- The employers failed to keep the workplace free of a hazard to which employees of that employer were exposed.
- The hazard was recognized. (Examples might include: through your safety personnel, employees, organization, trade organization, or industry customs.)
- The hazard was causing or was likely to cause death or serious physical harm.
- There was a feasible and useful method to correct the hazard.

Employee Rights and Responsibilities

While it is the employer's responsibility to keep workers safe by complying with the General Duty Clause and all other OSHA regulations, workers have certain rights and responsibilities on the job site as well. First and foremost, workers must follow their employers' safety rules. While workers cannot be cited or fined by OSHA, they can be disciplined for violating their employer's safety rules. Workers must also wear the provided personal protective equipment. Workers should also inform their foreman about health and safety concerns on the job.

An OSHA Standard reference may look like this:

29 CFR 1926.501 (a)(1)(i)(A)

and breaks down like this:

29	=	Title (Labor)
CFR	=	Code of Federal Regulations
1926	=	Part (Construction)
.501	=	Section
(a)	=	Paragraph
(1)	=	Subparagraph
(i)	=	Subparagraph
(A)	=	Subparagraph

00101-15_A01.EPS

Figure 1 Reading OSHA standards.

Section 11(c) of the OSH Act prohibits employers from disciplining or discriminating against any worker for practicing their rights under OSHA, including filing a complaint. You have the right to file a complaint if you do not think that your employer is protecting your health and safety at work. You may submit a written request to OSHA asking for an inspection of your worksite. Workers who file a complaint have the right to have their names withheld from their employers, and OSHA will not reveal this information.

Workers who would like an on-site inspection must submit a written request. You have the following rights when job site inspection is conducted:

- You must be informed of imminent dangers. An OSHA inspector must tell you if you are exposed to an imminent danger. An imminent danger is one that could cause death or serious injury now or in the near future. The inspector will also ask your employer to stop any dangerous activity.
- You have the right to accompany the OSHA inspector in the walk-around inspection. Walk-around activities include all opening and closing conferences related to the conduct of the inspection.
- You have the right to be told about citations issued at your workplace. Notices of OSHA citations must be posted in the workplace near the site where the violation occurred and must remain posted for three days or until the hazard is corrected, whichever is longer.

After an inspection has been performed, OSHA will give the employer a date by which any hazards cited must be fixed. Employers can appeal these dates, and appeals must be filed within 15 days of the citation. Workers have the right to meet privately with the OSHA inspector to discuss the results of the inspection.

If you have been discriminated against for asserting your OSHA rights, you have the right to file a complaint with the OSHA area office within 30 days of the incident. Make sure you file your complaint as soon as possible, as the time limit is strictly enforced.

You also have the right to see and copy any medical records about you that the employer has obtained. Your employer is required by OSHA 29 *CFR* 1926.33 and OSHA 29 *CFR* 1910.1020 to maintain your medical records for 30 years after you leave employment. If you are employed for less than one year, the employer can maintain your records or give them to you when you leave the job.

INSPECTIONS

OSHA conducts six types of inspections to determine if employers are in compliance with standards:

- *Imminent danger inspections* – OSHA's top priority for inspection, conducted when workers face an immediate risk of death or serious physical harm.
- *Catastrophe inspections* – Performed after an incident that requires hospitalization of three or more workers. Employers are required to report fatalities and catastrophes to OSHA within eight hours.
- *Worker complaint and referral inspections* – Conducted due to complaints by workers or a worker representative, or a referral from a recognized professional.
- *Programmed inspection* – Aimed at high-risk areas based on OSHA's targeting and priority methods.
- *Follow-up inspection* – Completed after citations to assure employer has corrected violations.
- *Monitoring inspection* – Used for long-term abatement follow-up or to assure compliance with variances.

Before beginning an inspection, OSHA staff must be able to determine from the complaint that there are reasonable grounds to believe that a violation of an OSHA standard or a safety or health hazard exists. If OSHA has information indicating the employer is aware of the hazard and is correcting it, the agency may not conduct an inspection after obtaining the necessary documentation from the employer.

Complaint inspections are typically limited to the hazards listed in the complaint, although other violations in plain sight may be cited as well. The inspector may decide to expand the inspection based on professional judgment or conversations with workers.

Complaints are not necessarily inspected in first-come, first-served order. OSHA ranks complaints based on the severity of the alleged hazard and the number of workers exposed. That is why lower-priority complaints can often be handled more quickly using the phone/fax method than through on-site inspections.

Inspections are typically performed by conducting a walk-around. During a walk-around inspection, the inspector typically does the following:

- Observes conditions of the job site.
- Talks to workers.
- Inspects records.
- Examines posted hazard warnings and signs.
- Points out hazards and suggests ways to reduce or eliminate them.

After the walk-around inspection, there is typically a closing conference held between the inspector and the site contractor or company managers. During this conference, inspectors discuss their findings, citing specific violations and suggested abatement methods. Inspectors may also conduct interviews with the employers, workers, and representatives at this point.

VIOLATIONS

Employers who violate OSHA regulations can be fined. The fines are not always high, but they can harm a company's reputation for safety. Fines for serious safety violations can cost up to $7,000. Fines for each violation that was done willfully can cost up to $70,000. In 2012, roughly $260,000,000 in fines were issued against employers. In the decade prior to 2012, annual fines averaged approximately $150,000,000. Significant increases and decreases in annual fines tend to be dependent on the current federal administration and OSHAs budget provided by federal lawmakers.

COMPLIANCE

Just as employers are responsible to OSHA for compliance, employees must comply with their company's safety policies and rules. Employers are required to identify hazards and potential hazards within the workplace and eliminate them, control them, or provide protection from them. This can only be done through the combined efforts of the employer and employees. Employers must provide written programs and training on hazards, and employees must follow the procedures. You, as the employee, must read and understand the OSHA poster at your job site explaining your rights and responsibilities. If you are unsure where the OSHA poster is, ask your supervisor.

To help employers provide a safe workplace, OSHA requires companies to provide a competent person to ensure the safety of the employees. In *OSHA 29 CFR 1926*, OSHA defines a competent person as follows:

> Competent person means one who is capable of identifying existing and predictable hazards in the surroundings or working conditions which employees, and who has authorization to take prompt corrective measures to eliminate them.

In comparison, *OSHA 29 CFR 1926* defines a qualified person as follows:

> Someone who, by possession of a recognized degree, certificate, or professional standing, or who by extensive knowledge, training, and experience, has successfully demonstrated his ability to solve or resolve problems relating to the subject matter, work, or the project.

In other words, a competent person is experienced and knowledgeable about the specific operation and has the authority from the employer to correct the problem or shut down the operation until it is safe. A qualified person has the knowledge and experience to handle problems. A competent person is not necessarily a qualified person.

These terms will be an important part of your career. It is important for you to know who the competent person is on your job site. OSHA requires a competent person for many of the tasks you may be assigned to perform, such as confined space entry, ladder use, and trenching. Different individuals may be assigned as a competent person for different tasks, according to their expertise. To ensure safety for you and your co-workers, work closely with your competent person and supervisor.

Trade Terms Introduced in This Module

Accident: Per the US Occupational Safety and Health Administration (OSHA), an unplanned event that results in personal injury or property damage.

Arc welding: The joining of metal parts by fusion, in which the necessary heat is produced by means of an electric arc.

Brazing: A process using heat in excess of 800°F (427°C) to melt a filler metal that is drawn into a connection. Brazing is commonly used to join copper pipe.

Combustible: Capable of easily igniting and rapidly burning; used to describe a fuel with a flash point at or above 100°F (38°C).

Competent person: A person who is capable of identifying existing and predictable hazards in the surroundings or working conditions that are unsanitary, hazardous, or dangerous to employees, and who has authorization to take prompt corrective measures to eliminate them.

Confined space: A work area large enough for a person to work in, but with limited means of entry and exit and not designed for continuous occupancy. Tanks, vessels, silos, pits, vaults, and hoppers are examples of confined spaces. Also see *permit-required confined space*.

Cross-bracing: Braces (metal or wood) placed diagonally from the bottom of one rail to the top of another rail to add support to a structure.

Excavation: Any man-made cut, cavity, trench, or depression in an earth surface, formed by removing earth. It can be made for anything from basements to highways. Also see *trench*.

Flammable: Capable of easily igniting and rapidly burning; used to describe a fuel with a flash point below 100°F (38°C).

Flash burn: The damage that can be done to eyes after even brief exposure to ultraviolet light from arc welding. A flash burn requires medical attention.

Flash point: The temperature at which fuel gives off enough gases (vapors) to burn.

Ground: The conducting connection between electrical equipment or an electrical circuit and the earth.

Ground fault: An unintentional, electrically conducting connection between an ungrounded conductor of an electrical circuit and the normally noncurrent-carrying conductors, metal objects, or the earth.

Ground fault circuit interrupter (GFCI): A device that interrupts and de-energizes an electrical circuit to protect a person from electrocution.

Guarded: Enclosed, fenced, covered, or otherwise protected by barriers, rails, covers, or platforms to prevent dangerous contact.

Hand line: A line attached to a tool or object so a worker can pull it up after climbing a ladder or scaffold.

Hazard Communication Standard (HAZCOM): The standard that requires contractors to educate employees about hazardous chemicals on the job site and how to work with them safely.

Hydraulic: Powered by fluid under pressure.

Incident: Per the US Occupational Safety and Health Administration (OSHA), an unplanned event that does not result in personal injury but may result in property damage or is worthy of recording.

Lanyard: A short section of rope or strap, one end of which is attached to a worker's safety harness and the other to a strong anchor point above the work area.

Lockout/tagout: A formal procedure for taking equipment out of service and ensuring that it cannot be operated until a qualified person has removed the lock and/or warning tag.

Management system: The organization of a company's management, including reporting procedures, supervisory responsibility, and administration.

Maximum intended load: The total weight of all people, equipment, tools, materials, and loads that a ladder can hold at one time.

Midrail: Mid-level, horizontal board required on all open sides of scaffolds and platforms that are more than 14 inches (35 cm) from the face of the structure and more than 10 feet (3.05 m) above the ground. It is placed halfway between the toeboard and the top rail.

Occupational Safety and Health Administration (OSHA): An agency of the US Department of Labor. Also refers to the Occupational Safety and Health Act of 1970, a law that applies to more than more than 111 million workers and 7 million job sites in the country.

Permit-required confined space: A confined space that has been evaluated and found to have actual or potential hazards, such as a toxic atmosphere or other serious safety or health hazard. Workers need written authorization to enter a permit-required confined space. Also see *confined space*.

Personal protective equipment (PPE): Equipment or clothing designed to prevent or reduce injuries.

Planked: Having pieces of material 2 inches (5 cm) thick or greater and 6 inches (15 cm) wide or greater used as flooring, decking, or scaffold decks.

Pneumatic: Powered by air pressure, such as a pneumatic tool.

Proximity work: Work done near a hazard but not actually in contact with it.

Qualified person: A person who, by possession of a recognized degree, certificate, or professional standing, or by extensive knowledge, training, and experience, has demonstrated the ability to solve or prevent problems relating to a certain subject, work, or project.

Respirator: A device that provides clean, filtered air for breathing, no matter what is in the surrounding air.

Safety culture: The culture created when the whole company sees the value of a safe work environment.

Safety data sheet (SDS): A document that must accompany any hazardous substance. The SDS identifies the substance and gives the exposure limits, the physical and chemical characteristics, the kind of hazard it presents, precautions for safe handling and use, and specific control measures.

Scaffold: An elevated platform for workers and materials.

Shielding: A structure used to protect workers in trenches but lacking the ability to prevent cave-ins.

Shoring: Using pieces of timber, usually in a diagonal position, to hold a wall in place temporarily.

Signaler: A person who is responsible for directing a vehicle when the driver's vision is blocked in any way.

Six-foot rule: A rule stating that platforms or work surfaces with unprotected sides or edges that are 6 feet (≈2 m) or higher than the ground or level below it require fall protection.

Spoil: Material such as earth removed while digging a trench or excavation.

Toeboard: A vertical barrier at floor level attached along exposed edges of a platform, runway, or ramp to prevent materials and people from falling.

Top rail: A top-level, horizontal board required on all open sides of scaffolds and platforms that are more than 14 inches (36 cm) from the face of the structure and more than 10 feet (3 m) above the ground.

Trench: A narrow excavation made below the surface of the ground that is generally deeper than it is wide, with a maximum width of 15 feet (4.6 m). Also see *excavation*.

Welding curtain: A protective screen set up around a welding operation designed to safeguard workers not directly involved in that operation.

Wind sock: A cloth cone open at both ends mounted in a high place to show which direction the wind is blowing.

Additional Resources

This module presents thorough resources for task training. The following reference material is suggested for further study.

US Occupational Safety and Health Administration. Numerous safety videos are available on line at **www.osha.gov/video**.

Construction Safety, Jimmie W. Hinze. 2006. Upper Saddle River, NJ: Pearson Education, Inc.

DeWalt Construction Safety/OSHA Professional Reference, Paul Rosenberg; American Contractors Educational Services. 2006. DEWALT.

Basic Construction Safety and Health, Fred Fanning. 2014. CreateSpace Independent Publishing Platform.

Figure Credits

Honeywell Safety Products, Module opener, Figures 8B–D, 21, 50B, 52, 57C

Accuform Signs, Figures 2, 51

Courtesy of Justrite Mfg. Co. LLC, Figure 3

L.J. LeBlanc, Figure 4

Weld-On Adhesives, Inc., a division of IPS Corporation, Figure 5

Fall protection materials provided courtesy of Miller Fall Protection, Franklin, PA, Figures 7, 8A, 9-11, 13, 15–17, 19

Photo Courtesy of Capital Safety (DBI-SALA and PROTECTA), Figure 12

Topaz Publications Inc., Figures 14, 20, 37, 43, 46B, 47–49, 50A, 57A, 58, 59A, 65, 75, 77, 78, 80

Guardian Fall Protection, Figure 22

Courtesy of Louisville Ladders, Figures 23, 32

Werner Co., Figures 29, 30

Courtesy of Safway Services LLC, Figure 34

Spider Staging, Figure 35

Courtesy of Snap-on Industrial - Tools at Height Program, SA01

Carolina Bridge Co., Figure 38

Kundel Industries, Figures 40, 41

DeWALT Industrial Tool Co., Figure 42

Courtesy of Atlas Copco, Figure 44

Panduit Corp., Figure 50B

Photo courtesy of Bullard, Figure 54

MSA The Safety Company, Figures 55, 56, 59B, 60A-B, 60D

North Safety Products USA, Figures 57B, 60C

The Lincoln Electric Company, Cleveland, OH, USA, Figures 66, 68

Sellstrom Manufacturing, Figure 67

Vestil Manufacturing, Figure 69

Courtesy of Saf-T-Cart, Figure 71

Courtesy of Amerex Corporation, Figures 74, 76 (photos), 79

Badger Fire Protection, Figure 81

Section Review Answer Key

Answer	Section Reference	Objective
Section One		
1. c	1.0.0	1a
2. d	1.2.1	1b
3. a	1.3.5	1c
Section Two		
1. c	2.1.2	2a
2. a	2.2.1	2b
3. b	2.3.1	2c
4. c	2.4.2	2d
Section Three		
1. a	3.1.2	3a
2. d	3.2.1	3b
Section Four		
1. b	4.1.1	4a
2. b	4.2.0	4b
Section Five		
1. a	5.1.2	5a
2. d	5.2.0	5b
Section Six		
1. b	6.1.0	6a
2. c	6.2.5	6b
3. a	6.3.1	6c
4. b	6.4.1	6d
5. b	6.5.0	6e

NCCER CURRICULA — USER UPDATE

NCCER makes every effort to keep its textbooks up-to-date and free of technical errors. We appreciate your help in this process. If you find an error, a typographical mistake, or an inaccuracy in NCCER's curricula, please fill out this form (or a photocopy), or complete the online form at **www.nccer.org/olf**. Be sure to include the exact module ID number, page number, a detailed description, and your recommended correction. Your input will be brought to the attention of the Authoring Team. Thank you for your assistance.

Instructors – If you have an idea for improving this textbook, or have found that additional materials were necessary to teach this module effectively, please let us know so that we may present your suggestions to the Authoring Team.

NCCER Product Development and Revision

13614 Progress Blvd., Alachua, FL 32615

Email: curriculum@nccer.org
Online: www.nccer.org/olf

❏ Trainee Guide ❏ Lesson Plans ❏ Exam ❏ PowerPoints Other _____

Craft / Level: _____ Copyright Date: _____

Module ID Number / Title: _____

Section Number(s): _____

Description: _____

_____

_____

_____

_____

Recommended Correction: _____

_____

_____

_____

_____

Your Name: _____

Address: _____

_____

Email: _____ Phone: _____

00102-15

Introduction to Construction Math

Overview

In the construction trades, workers must use math day in and day out. Plumbers use math to calculate pipe length, read plans, and lay out fixtures. Carpenters use math to lay out floor systems and frame walls and ceilings. In some cases, algebra, geometry, and even trigonometry may be required. This module reviews basic mathematical procedures and provides the opportunity to practice mathematical tasks related to construction activities.

Module Two

Trainees with successful module completions may be eligible for credentialing through the NCCER Registry. To learn more, go to **www.nccer.org** or contact us at **1.888.622.3720**. Our website has information on the latest product releases and training, as well as online versions of our *Cornerstone* magazine and Pearson's product catalog.

Your feedback is welcome. You may email your comments to **curriculum@nccer.org,** send general comments and inquiries to **info@nccer.org**, or fill in the User Update form at the back of this module.

This information is general in nature and intended for training purposes only. Actual performance of activities described in this manual requires compliance with all applicable operating, service, maintenance, and safety procedures under the direction of qualified personnel. References in this manual to patented or proprietary devices do not constitute a recommendation of their use.

Objectives

When you have completed this module, you will be able to do the following:

1. Identify whole numbers and demonstrate how to work with them mathematically.
   a. Identify different whole numbers and their place values.
   b. Demonstrate the ability to add and subtract whole numbers.
   c. Demonstrate the ability to multiply and divide whole numbers.
2. Explain how to work with fractions.
   a. Define equivalent fractions and show how to find lowest common denominators.
   b. Describe improper fractions and demonstrate how to change an improper fraction to a mixed number.
   c. Demonstrate the ability to add and subtract fractions.
   d. Demonstrate the ability to multiply and divide fractions.
3. Describe the decimal system and explain how to work with decimals.
   a. Describe decimals and their place values.
   b. Demonstrate the ability to add, subtract, multiply, and divide decimals.
   c. Demonstrate the ability to convert between decimals, fractions, and percentages.
4. Identify various tools used to measure length and show how they are used.
   a. Identify and demonstrate how to use rulers.
   b. Identify and demonstrate how to use measuring tapes.
5. Identify and convert units of length, weight, volume, and temperature between the Imperial and metric systems of measurement.
   a. Identify and convert units of length measurement between the Imperial and metric systems.
   b. Identify and convert units of weight measurement between the Imperial and metric systems.
   c. Identify and convert units of volume measurement between the Imperial and metric systems.
   d. Identify and convert units of temperature measurement between the Imperial and metric systems.
6. Identify basic angles and geometric shapes and explain how to calculate their area and volume.
   a. Identify various types of angles.
   b. Identify basic geometric shapes and their characteristics.
   c. Demonstrate the ability to calculate the area of two-dimensional shapes.
   d. Demonstrate the ability to calculate the volume of three-dimensional shapes.

Performance Tasks

This is a knowledge-based module; there are no performance tasks.

Trade Terms

Acute angle	Equilateral triangle	Positive numbers
Adjacent angles	Equivalent fractions	Product
Angle	Force	Quotient
Area	Formula	Radius
Base	Fraction	Rectangle
Bisect	Improper fraction	Remainder
Circle	Invert	Right angle
Circumference	Isosceles triangle	Right triangle
Cube	Joist	Scalene triangle
Decimal	Loadbearing	Solid geometry
Degree	Mass	Square
Denominator	Mixed number	Straight angle
Diagonal	Negative numbers	Stud
Diameter	Obtuse angle	Sum
Difference	Opposite angles	Triangle
Digit	Perimeter	Unit
Dividend	Pi	Vertex
Divisor	Place value	Volume
Equation	Plane geometry	Whole numbers

Industry Recognized Credentials

If you are training through an NCCER-accredited sponsor, you may be eligible for credentials from NCCER's Registry. The ID number for this module is 00102-15. Note that this module may have been used in other NCCER curricula and may apply to other level completions. Contact NCCER's Registry at 888.622.3720 or go to **www.nccer.org** for more information.

Note

You should do the math problems without using a calculator, except when the text specifically calls for calculator use or when you need to check the answers to your problems.

Contents

Topics to be presented in this module include:

Contents (continued)

Figures and Tables

Figures and Tables (continued)

1.0.0 WHOLE NUMBERS

Objective

Identify whole numbers and demonstrate how to work with them mathematically.

a. Identify different whole numbers and their place values.
b. Demonstrate the ability to add and subtract whole numbers.
c. Demonstrate the ability to multiply and divide whole numbers.

Trade Terms

Decimal: A part of a number represented by digits to the right of a point, called a decimal point. For example, in the number 1.25, .25 is the decimal portion of the number. In this case, it represents 25 percent of the whole number 1.

Difference: The result of subtracting one number from another. For example, in the problem $8 - 3 = 5$, 5 is the difference between the two numbers.

Digit: Any of the numerical symbols 0 to 9.

Dividend: In a division problem, the number being divided is the dividend.

Divisor: In a division problem, the number that is divided into another number is called the divisor.

Equation: A mathematical statement that indicates that the value of two mathematical expressions, such as 2×2 and 1×4, are equal. An equation is written using the equal sign in this manner: $2 \times 2 = 1 \times 4$.

Fraction: A portion of a whole number represented by two numbers. The upper number of a fraction is known as the numerator and the bottom number is known as the denominator.

Negative numbers: Numbers less than zero. For example, −1, −2, and −3 are negative numbers.

Place value: The exact value a digit represents in a whole number, determined by its place within the whole number or by its position relative to the decimal point. In the number 124, the number 2 represents 20, since it is in the tens position.

Positive numbers: Numbers greater than zero. For example, 1, 2, and 3 are positive numbers. Any number without a negative (−) sign in front of it is considered to be a positive number.

Product: The answer to a multiplication problem. For example, the product of 6×6 is 36.

Quotient: The result of a division problem. For example, when dividing 6 by 2, the quotient is 3.

Remainder: The amount left over in a division problem. For example, in the problem $34 \div 8$, 8 goes into 34 four times ($8 \times 4 = 32$) with 2 left over; in other words, 2 is the remainder.

Sum: The resulting total in an addition problem. For example, in the problem $7 + 8 = 15$, 15 is the sum.

Whole numbers: Complete number units without fractions or decimals.

No matter which construction trade you enter, you can be sure that you will be required to work with whole numbers in both your written and oral communication. The ability to read, write, and communicate whole numbers to others accurately is extremely important on any job site. Carpenters will often use whole numbers during the early phases of planning to quickly estimate the square feet of drywall, or linear feet of baseboard necessary to finish a room. Sheet metal workers will use whole numbers during the planning phase for installing an air handling system and to estimate, with relative accuracy, trunk line dimensions, lengths, and the amount of air it will be required to handle.

Be aware that a mistake made when communicating any number to others can result in wasted time, effort, and materials, all of which negatively affect the bottom line.

In this section, you will learn how to work with whole numbers. Whole numbers are complete number units without fractions or decimals. This section presents whole numbers only. Working with fractions and decimals will be covered in other sections.

Did You Know?

Whole Numbers

The following are whole numbers...

1 5 67 335 2,654

... but the following are *not* whole numbers:

½ ¾ 7⅛ 0.45 4.25

1.1.0 Place Values of Whole Numbers

Each **digit** that makes up a given whole number has a specific **place value**. A digit is any of the numerical symbols from 0 to 9. *Figure 1* provides an example of a whole number with seven digits. To read this seven-digit whole number out loud, say "five million, three hundred sixteen thousand, two hundred forty-seven."

Each of this whole number's seven digits represents a specific place value. Each digit has a value that depends on its place, or location, in the whole number. In this whole number, for example, the place value of the 5 is five million, while the place value of the 2 is two hundred.

Other important points to keep in mind about whole numbers include the following:

- Numbers larger than zero are called **positive numbers** (such as 1, 2, 3 …). Except for zero, all numbers without a minus sign in front of them are positive.
- Numbers less than zero are called **negative numbers** (such as −1, −2, −3 …). Negative numbers are preceded by a negative (−) sign. A number does not need to be positive to still be considered a whole number; negative numbers can also be whole numbers.
- Zero (0) is neither positive nor negative.

Some whole numbers may contain the digit zero. For example, the whole number 7,093 has a zero in the hundreds place. When you read that number out loud, you would say "seven thousand ninety-three."

$$5 \quad , \quad 3 \quad 1 \quad 6 \quad , \quad 2 \quad 4 \quad 7$$

| MILLIONS | HUNDRED THOUSANDS | TEN THOUSANDS | THOUSANDS | HUNDREDS | TENS | ONES |

00102-15_F01.EPS

Figure 1 Place values.

1.1.1 Study Problems: Place Values of Whole Numbers

1. Look at the following description of a number. This number would be written as _____.
   Digit in the hundreds place: 9
   Digit in the ones place: 4
   Digit in the thousands place: 3
   Digit in the tens place: 6
   a. 3,964
   b. 4,693
   c. 30,964
   d. 39,064

2. In the number 25,718, the numeral 5 is in the _____.
   a. tens place
   b. thousands place
   c. ones place
   d. hundreds place

3. An estimate for a commercial flooring job requires one thousand, six hundred ninety-three square meters of carpet to complete the first floor. How would you write this amount as a whole number?
   a. 163
   b. 1,693
   c. 10,693
   d. 16,093

4. A supervisor estimates that a commercial building will require sixteen thousand, five hundred feet of copper piping to complete all of the restroom facilities. How would you write this value as a whole number?
   a. 1,650
   b. 16,500
   c. 160,500
   d. 16,000,500

5. An engineer estimates that the total cost to install a building's HVAC system will be three hundred twenty-two thousand, nine hundred and seven dollars. How would you write this cost using digits?
   a. $3,297.00
   b. $300,297.00
   c. $322,907.00
   d. $322,000,907.00

1.2.0 Adding and Subtracting Whole Numbers

To add means to combine the values of two or more numbers together into one **sum** or total. To add whole numbers, use the following steps:

Step 1 Line up the digits in the top number and the bottom number by place value columns.

$$723$$
$$+\ 84$$

Step 2 Beginning at the right side, add the numbers in the ones column (3 and the 4) together first.

$$723$$
$$+\ 84$$
$$\overline{7}$$

Step 3 Continue to add the digits in each column, moving from right to left, one column at a time. In this example, when you add the 2 and the 8 in the tens column you get 10. This requires you to carry the 1 in the tens column over to the next column to the left. To do so, place the 0 in the tens column and carry the 1 over to the top of the hundreds column as shown. That carried-over number is now added to the rest of the digits in that column.

$$\overset{1}{7}23$$
$$+\ 84$$
$$\overline{07}$$

Step 4 Add the 7 already in the hundreds column to the 1 carried over. The resulting sum is 807.

$$\overset{1}{7}23\ \rangle\ \textit{ADDENDUM}$$
$$+\ 84\ \rangle$$
$$\overline{807} \rightarrow \textit{Sum = Answer}$$

Note that you may need to carry-over digits several times in the same problem. The following is an example of such a problem:

$$\overset{1\ 1\ 1}{66,}723$$
$$+\ 5,784$$
$$\overline{72,507}$$

Minuend ← 1250
Subtrahend 1034
Difference ← $216

To subtract means to take away a given amount of one number from the total amount of a second number to find the **difference**. To subtract whole numbers, use the following steps:

Step 1 Line up the digits in the top number and the bottom number by place value columns. Generally, position the larger number on the top. If not, the result will be a negative number. That is appropriate for some calculations, but it is extremely rare in math used on the job site.

$$12,766$$
$$-\ 1,483$$

Step 2 As in addition, start with the right column—the ones column. Subtract the 3 from the 6 to get 3, and record it under the ones column. As you work your way left into the tens column, note that you are unable to subtract 8 from 6. This will require you to borrow a 1 from the hundreds column. To borrow, cross out the 7 in the hundreds column, change it to a 6, and carry a 1 over to the 6 in the tens column. This makes it 16 instead of 6. Subtract 8 from 16 to get 8, and record it in the tens column.

$$12,\overset{6\ \ 1}{\cancel{7}}66$$
$$-\ 1,483$$
$$\overline{83}$$

Step 3 Continue to work towards the left, column by column, completing each subtraction for the remaining place values.

$$12,\overset{6\ \ 1}{\cancel{7}}66$$
$$-\ 1,483$$
$$\overline{11,283}$$

Note that you may need to borrow several times in one problem to calculate the difference. The following problem provides an example of this situation:

$$9\overset{3\ 12\ 10\ 15}{\cancel{4}\cancel{3},\cancel{1}\cancel{5}}3$$
$$-\ 436,372$$
$$\overline{506,781}$$

1.2.1 Study Problems: Adding and Subtracting Whole Numbers

Use addition and subtraction to solve the following problems. Read each question carefully to determine the appropriate procedure. Be sure to show all of your work.

1. In calculating a bid for a roof restoration, a contractor estimates that he will need $847 for lumber, $456 for roofing shingles, and $169 for hardware. What is the total cost for the materials portion of the bid?

   $ __1472.00__

2. Brazil's currency is called the real, and is denoted with this symbol: R$. A plumbing contractor allotted R$10,236 in his bank account to complete three residential jobs. If he estimates Job 1 to cost R$2,477, Job 2 to cost R$2,263, and Job 3 to cost R$3,218, how much money will he have left in the account for unexpected costs?

   R$ __2278__

3. An HVAC contracting company sent out three work crews to complete three installations over the past week. If Crew One worked 10-, 9-, 11-, 12-, and 9-hour days, Crew Two worked 9-, 12-, 12-, 9-, and 9-hour days, and Crew Three worked 12-, 12-, 10-, 9-, and 11-hour days, how many total hours did the three crews work for the week?

   __155__ hours

4. A general contractor ordered three different sized windows to complete a job on a residential home. She ordered a bow window that cost $874; one 36" × 36" double-hung window that cost $67; and one 36" × 54" double-hung window that cost $93. If she had set aside $1,250 to purchase the windows in her estimate, how much will she have left after buying them?

   $ __216__

5. Russia's currency is called the ruble, and is denoted with this symbol: Rub. An electrical contractor has Rub850,360 in his business's bank account. If at the end of the week he deposits Rub119,980 in payments made from clients and then pays out Rub79,205 in wages, how much money will he have left in his account?

   Rub __891135__

(handwritten work)
850360
+ 119980
970340
79205

1.3.0 Multiplying and Dividing Whole Numbers

Multiplication is a quick way to add the same number to itself numerous times. For example, assume four different people each gave you $8 to pick up lunch for the crew. You could add $8 together four times. However, it is easier to multiply 4 times 8 to get a **product** of 32 than it is to add 8 together four times. With larger numbers, it could take a long time to complete all the addition. To multiply whole numbers, use the following steps:

Step 1 Line up the digits in the top number and the bottom number by place value columns. Position the greater number on the top. This is a matter of convenience and makes the task easier. However, note that it will have no effect at all on the accuracy of the result.

$$\begin{array}{r} 374 \\ \times\ 26 \\ \hline \end{array}$$

Step 2 Start with the ones column (the right column) and multiply the 6 by the 4 (6 × 4 = 24). Write down the 4 in the ones column and carry the 2 up to the tens column. Unlike addition, you will multiply the 6 by every digit in the top number before proceeding to the 2.

$$\begin{array}{r} \overset{2}{374} \\ \times\ 26 \\ \hline 4 \end{array}$$

Step 3 Multiply the 6 by the 7 next, and then add the 2 that was carried over ($6 \times 7 = 42 + 2 = 44$). Write down the 4 in the tens column and carry the 4 to the hundreds column.

$$
\begin{array}{r}
{\scriptstyle 4\ 2} \\
374 \\
\times\ 26 \\
\hline
44
\end{array}
$$

Step 4 Next multiply the 6 by the 3, and then add the 4 that you previously carried over ($6 \times 3 = 18 + 4 = 22$). Write down a 2 in the hundreds column and a 2 in the thousands column, since there are no more places in the upper number.

$$
\begin{array}{r}
{\scriptstyle 4\ 2} \\
374 \\
\times\ 26 \\
\hline
2,244
\end{array}
$$

Step 5 You will then need to multiply each digit in 374 by the 2 in 26. Begin by multiplying by the 2 by the four in the ones column ($2 \times 4 = 8$). Write the 8 down under the tens column, directly under the 2. Notice that you do not write the 8 down in the ones column because the 2 in the number 26 is in the tens column. A zero (0) can be placed in the ones column to help keep the columns aligned if it helps.

$$
\begin{array}{r}
374 \\
\times\ 26 \\
\hline
2,244 \\
80
\end{array}
$$

Step 6 Now multiply the 2 by the 7 in the tens column ($2 \times 7 = 14$). Write down a 4 in the hundreds column and carry the 1.

$$
\begin{array}{r}
{\scriptstyle 1} \\
374 \\
\times\ 26 \\
\hline
2,244 \\
480
\end{array}
$$

Step 7 Multiply the 2 by the 3, and then add the 1 that you carried over ($2 \times 3 = 6 + 1 = 7$). Write a 7 in the thousands column.

$$
\begin{array}{r}
{\scriptstyle 1} \\
374 \\
\times\ 26 \\
\hline
2,244 \\
+7,480
\end{array}
$$

Step 8 The next step requires addition. Add the two products to get a final product of 9,724. There will be as many products to add together as there are numbers in the bottom portion of the problem. In this example, there are two numbers in the bottom portion of the problem (the 2 and the 6). Thus, there are two products to add together to arrive at the final product.

$$
\begin{array}{rl}
{\scriptstyle 1} & \\
374 & \\
\times\ 26 & \\
\hline
2,244 & \text{product of } 6 \times 374 \\
+7,480 & \text{product of } 20 \times 374 \\
\hline
9,724 & \text{final product}
\end{array}
$$

Division is the opposite of multiplication. Instead of adding a number several times ($5 + 5 + 5 = 15$, or $5 \times 3 = 15$), you subtract a number several times to find a **quotient** in division. The two numbers in a division problem have their own names. The number being divided is known as the **dividend**. The number the dividend is being divided by is called the **divisor**.

When adding, subtracting, or multiplying whole numbers, the result of the work always results in a whole number. However, in division, a given number may not be neatly divisible by another given number. In this case, the result will be a whole number and/or part of a number. For example, dividing 6 by 3 results in the whole number 2. But what happens when we divide 6 by 4? The four can only be divided into the 6 one time, with 2 left over. This left over portion is referred to as the **remainder**.

Construction Estimating

Creating accurate estimates is an essential part of the job for most contractors. A job that is engineered typically has a budget for the project that the engineer has to work within. For example, granite cannot be specified for a countertop unless the project owner has the money in the budget. Once a design is completed, various contractors can then estimate the cost of the construction, including materials, labor, and the overhead—costs that are associated with putting the labor on the job, keeping the lights and telephones on, and paying the accounting department for the work they do. Then the desired amount of profit is added. The job usually goes to the lowest bidder.

Long division, as will be demonstrated here, requires a sequence of mathematical operations. Not only is division required, but multiplication and subtraction are also required repeatedly.

To divide whole numbers, use the following steps:

Step 1 Begin by setting up the division problem using the division bar as shown below. The ÷ symbol can also be used, writing the problem as 2,638 ÷ 24. However, the division bar provides much better organization when working a problem with pencil and paper. Position the divisor on the left side of the division bar and place the dividend on the right side of the division bar. In this example, 24 is the divisor and 2,638 is the dividend.

$$24\overline{)2{,}638}$$

Step 2 Unlike addition, subtraction, and multiplication, where you start the procedure in the ones column, with division you start with the place value of the dividend farthest to the left. In this example, you would first try to divide 24 into 2. However, 2 cannot be divided by 24 at least one full time. As a result, the first two digits must be considered together. You then divide 24 into the 26 of 2,638. If, for example, the divisor was 124 instead of 24, you would need to work with the first three digits. 24 does go into 26 one full time, so write a 1 above the 6 and write 24 under the 26 in 2,638. The 24 is brought down because it is the result of 1 × 24. Then subtract 24 from 26 to get 2. Remember to use zeros (0) to help keep your place values straight.

```
     0,1??
24 )2,638
   - 24
     02
```

Step 3 Bring down the next number in 2,638 (the 3) and place it next to the 2. Next, determine if 24 can go into 23. The answer is no, so place a 0 above the 3 on the answer line. 24 times 0 is 0, so write zeros in the appropriate place value columns and subtract the two numbers to get 23.

```
     0,10?
24 )2,638
   - 24
     023
   - 000
     023
```

Step 4 Bring down the next number in 2,638 (the 8) and place it next to the 3. Then determine if 24 can go into 238. Yes it can, but how many times? This is where the value of learning multiplication tables can be very valuable. To figure this, think about numbers that are easy to work with. 10 times 24 is 240. That is almost the same as 238, but just a little over. Therefore, you quickly see that 24 will go into 238 nine times. Write a 9 in the answer line next to the 0. Write 216 below the 238 and subtract to get a remainder of 22. So the final answer is that 24 will go into 2,638 one hundred nine times with a remainder of 22. The answer is written as 109 r22, with 109 being the quotient for the problem.

```
 *              010?? quotient      0109  r22
 divisor    24 )2,638            24 )2,638   Dividend
              -24                   -24
              023                   023
            -000                  -000
             0238                  0238
                                 -0216
                                    22
```

Numerators and Denominators

Working with fractions will also be presented in this module. Think of a fraction as just another way to write a division problem. The number on top of a fraction is known as the numerator, but it is the same thing as the dividend in a long division problem. The lower number, called the denominator, is the divisor. The biggest difference in fractions and long division is that most fractions you will encounter on the job site result in a quotient that is less than one. In the average long division problem, the quotient is usually greater than one. But, in the end, fractions simply represent another way to write a division problem. In the case of many fractions you will encounter, the quotient of the division problem is not important—the fraction itself provides the needed information.

Refer back to Step 4. Note the discussion about how many times the number 24 can be divided into 238. The following fact is important to remember in long division problems like these: the answer to each division step should always be a number from 0-9. If the answer is 10 or more, then something went wrong. Either a multiplication error has been made, or 24 could actually be divided into the digits (in this case, the 23 of 238) at least one full time after all.

1.3.1 The Order of Operations

Many math problems cannot be solved without using more than one operation. Problems with multiple operations are most often found on one or both sides of an **equation**. Although no complex equations are presented in this mod, even the simplest equations must be completed using a specific order of operations.

In order for an answer to be accurate, operations must be done in the proper order. The order of operations for math was actually established in the 1500s. For simple equations, the order is: multiplication, division, addition, and subtraction. A simple acronym—MDAS—can remind you of this order. To remember this acronym, think of it as "My Dear Aunt Sally."

The following equation is a good example:

$$6 + 3 \times 5 = A$$

Adding 6 and 3, and then multiplying the sum by 5 results in an answer of 45. However, multiplying 3 times 5 first, and then adding 6 results in an answer of 21. Following the correct order of

operations shows that 21 is, in fact, the correct answer. As you can see, following the proper order makes a big difference in the result.

A second acronym—PEMDAS— can be used for slightly more complicated equations. The P represents parentheses, and the E represents exponents. Therefore, if an equation has an operation in parentheses, such as (6×3), that operation is done first. A number with an exponent, such as 6^2, would be completed next. Some remember this acronym through the phrase, "Please Excuse My Dear Aunt Sally."

1.3.2 Study Problems: Multiplying and Dividing Whole Numbers

Use multiplication and division to solve the following practical problems. Read each question carefully to determine the appropriate procedure. Be aware that addition or subtraction may also be required. Be sure to show all of your work.

1. Your supervisor sends you to the truck for 180 special fasteners. When you get there, you find that the fasteners come in bags of 15. How many bags of fasteners will you need to bring back?

   __12__ bags

2. If the following amounts of lumber need to be delivered to each of 2 different staging areas at 4 different job sites, how many total boards of each size will you need?

   a. (65) 2 × 4s __260__
   b. (45) 2 × 8s __180__
   c. (25) 2 × 10s __100__

3. If one plumbing job requires 45 meters of PVC pipe, and a second job requires 30 meters, how many lengths of pipe will you need if it comes in 6-meter lengths? Remember that you cannot order a partial length of pipe; only orders for whole lengths are generally accepted.

   __13__ lengths of pipe

   How much pipe will be left over, assuming there are no errors?

   __3__ meters

4. If a crane rental company charges $800 per day, $3,400 per week (5 days), and $10,500 per month:

   a. How much would it cost to rent the crane for 3 days? __2400$__
   b. How much would it cost to rent the crane for 12 days? __$8400__
   c. How much would it cost for one month and 11 days? __$18100.00__

Applied Construction Math: A Novel Approach, NCCER. 2006. Upper Saddle River, NJ: Prentice Hall.

Mathematics for Carpentry and the Construction Trades, Alfred P. Webster; Kathryn B. Judy. 2001. Upper Saddle River, NJ: Prentice Hall.

Mathematics for the Trades: A Guided Approach, Robert A. Carman; Hal Saunders. 2014. Pearson Learning.

1.0.0 Section Review

1. Given the number 92475, which digit is in the ten-thousands column of place value?

   a. 2
   b. 4
   c. 7
   d. 9

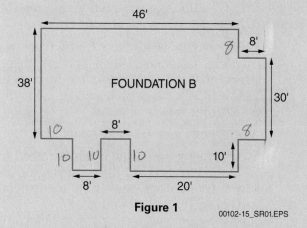

Figure 1

00102-15_SR01.EPS

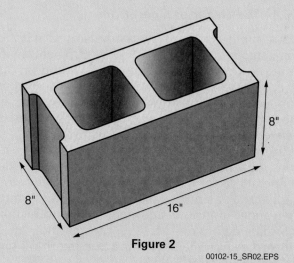

Figure 2

00102-15_SR02.EPS

2. What is the distance around the foundation shown in Section Review Question *Figure 1*?

   a. 206 feet
   b. 214 feet
   c. 224 feet
   d. 234 feet

3. How many concrete blocks like the one shown in Section Review Question *Figure 2* are required to erect a 2-foot high wall for the foundation shown in *Figure 1*? Assume there is no waste, and round your answer *up* to the nearest whole number (since only whole blocks can be purchased).

   a. 336 blocks
   b. 504 blocks
   c. 514 blocks
   d. 528 blocks

2.0.0 FRACTIONS

Objective

Explain how to work with fractions.
a. Define equivalent fractions and show how to find lowest common denominators.
b. Describe improper fractions and demonstrate how to change an improper fraction to a mixed number.
c. Demonstrate the ability to add and subtract fractions.
d. Demonstrate the ability to multiply and divide fractions.

Trade Terms

Denominator: The part of a fraction below the dividing line. For example, the 2 in ½ is the denominator. It is equivalent to the divisor in a long division problem.

Equivalent fractions: Fractions having different numerators and denominators but still have equal values, such as the two fractions ½ and ¾.

Improper fraction: A fraction whose numerator is larger than its denominator. For example, ¾ and ⅔ are improper fractions.

Invert: To reverse the order or position of numbers. In fractions, inverting means to reverse the positions of the numerator and denominator, such that ¾ becomes ⅓. When you are dividing by fractions, one fraction is inverted.

Mixed number: A combination of a whole number with a fraction or decimal. Examples of mixed numbers are 3⁷⁄₁₆, 5.75, and 1¼.

Numerator: The part of a fraction above the dividing line. For example, the 1 in ½ is the numerator. It is the equivalent of the dividend in a long division problem.

A fraction divides whole numbers into parts. Common fractions are written as two numbers, separated by a slash or by a horizontal line, like this: ½

The slash or horizontal line means the same thing as the ÷ sign. So think of a fraction as another way to write a division problem. The fraction ½ means 1 divided by 2, or one divided into two equal parts. This fraction is spoken as one-half.

The lower number of the fraction, known as the denominator, tells you the number of parts by which the upper number is being divided.

In a typical division problem, it is the divisor. The upper number, known as the numerator, is a whole number that tells you how many parts are going to be divided. It is the dividend of this division problem. In the fraction ½, the 1 is the upper number, or numerator, and the 2 is the lower number, or denominator. A fraction in which both the numerator and the denominator are the same number (²⁄₂, ⁸⁄₈, ¹⁶⁄₁₆) is always equal to the number 1.

2.1.0 Equivalent Fractions and Lowest Common Denominators

Equivalent fractions, such as ½, ²⁄₄, and ⁴⁄₈, have the same value. *Figure 2* shows the individual units of ½, ¼, and ⅛. It is often easier and more practical to work with fractions that have been reduced to their lowest common denominator. For the three equivalent fractions shown above, ½ represents the fraction with the lowest common denominator. Equivalent fractions and finding the lowest common denominator are presented in this section.

2.1.1 Finding Equivalent Fractions

If you are used to working with fractions, it is instantly apparent that ½, ²⁄₄, and ⁴⁄₈ are equivalent fractions. However, there is a mathematical way to determine this. Each fraction's value can be determined by dividing the denominator by the numerator. So 2 divided by 1 equals 2; 4 divided by 2 equals 2; and 8 divided by 4 also equals 2. Since the quotient for all three of these is the same (2), they can be proven to be equivalent fractions. It can be proven physically as well. If pieces of wood are cut at ½-inch long, ¾-inch long, and ⁴⁄₈-inch long, all three pieces of wood are the same length.

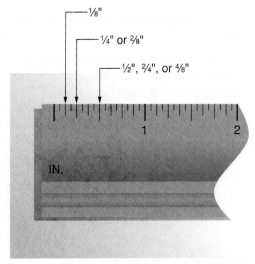

00102-15_F02.EPS

Figure 2 Equivalent fractions.

Fractions on the Job

Those of you who use the inch-pound system of measurement will realize how important it is to understand fractions from the first day you report to work on a construction site. In the real world, most measurements are not whole numbers. Typically, pipe, lumber, and other materials will be measured and cut to fractional lengths such as ⅜, ⁵⁄₁₆, or ¾ of an inch or foot. Being comfortable working with fractions is an essential job skill for all, but especially for users of the inch-pound system. The metric system of measurement eliminates the need for fractions in day-to-day measurements.

When you measure objects, it is often best to record all measurements as the same fractions—in sixteenths of an inch, for example. Technically, this means using the same denominator for each measurement. Doing this allows you to easily compare, add, and subtract fractional measurements.

Suppose that you are taking measurements and want all of them to be recorded in sixteenths of an inch. To find out how many sixteenths of an inch are equal to ½ inch, for example, you need to multiply both the numerator and the denominator by the same number. As long as the numerator and denominator of a given fraction are being multiplied by the same number, you are creating an equivalent fraction.

For this example, ask yourself what number you would multiply by 2 to get 16. The answer is 8, so you multiply both numbers (the numerator and the denominator) by 8.

$$\frac{1}{2} \times \frac{8}{8} = \frac{8}{16}$$

The answer is ⁸⁄₁₆ inch, equivalent to ½ inch.

2.1.2 Reducing Fractions to Their Lowest Terms

If you find that the measurement of something is ⁴⁄₁₆", you may want to reduce the measurement to its lowest terms so the number is easier to work with. In fact, if you call out a measurement like ⁴⁄₁₆" on a construction job site, there will likely be some puzzled listeners. Although they will know what you mean, they will not understand why you chose to communicate the measurement that way. To find the lowest terms of ⁴⁄₁₆", use division as follows:

Step 1 To reduce a fraction, determine the largest number that you can divide evenly into both the numerator and the denominator. With a little practice, this is a mental exercise that should not require pencil and paper. If there is no number (other than 1) that will divide evenly into both numbers, the fraction is already in its lowest term.

Step 2 Divide the numerator and the denominator by the number determined in Step 1; both must be divided by the same number. In this example, divide both the numerator and the denominator by 4.

$$\frac{4}{16} \div \frac{4}{4} = \frac{1}{4}$$

This shows that the lowest term of ⁴⁄₁₆ is ¼. Again, the number divided into the numerator and denominator must be the same number, and it must divide into both numbers evenly. In other words, a whole number must result from both operations.

2.1.3 Comparing Fractions and Finding Lowest Common Denominators

Which measurement is larger: ¾" or ⅝"?

This question may be best answered this way: Would you have a longer piece of lumber if you had three sections from a board that was cut up into four equal sections or if you had five sections of a board that was cut up into eight equal sections? *Figure 3* provides a visual reference for this question.

As you can see, it is hard to compare fractions that do not have common denominators, just as it is to mentally compare pieces of lumber cut into

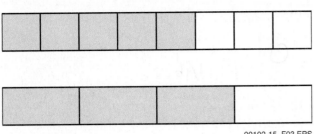

Figure 3 Lumber divided into equal sections.

00102-15_F03.EPS

sections of different lengths. The visual reference in *Figure 3* makes it easy to see that ¾ of a board is more than ⅝ of a board, but a visual reference like this is rarely available on the job. There is however, a mathematical way to make the task easier. If the fractions had the same denominator, then you would only need to look at the numerators of the two fractions to know which of them was larger.

$$\frac{3}{4} \text{ or } \frac{5}{8}$$

To easily compare them, you need to find a common denominator for the board sections. The common denominator is a number that both denominators can go into evenly.

Step 1 Multiply the two denominators together (4 × 8 = 32). 32 is a common denominator between the two fractions, because each of the two denominators can be divided into it evenly. So a common denominator has been found that allows the two fractions to be compared more easily.

Step 2 Now finish converting the two fractions so that they will both have the same denominator of 32.

$$\frac{3}{4} \times \frac{8}{8} = \frac{24}{32}$$

$$\frac{5}{8} \times \frac{4}{4} = \frac{20}{32}$$

Now it is easy to compare the two fractions to see which is larger. You would have a longer section of lumber if you choose ¾ because you would have ²⁴⁄₃₂ instead of ²⁰⁄₃₂ of the lumber piece.

You have found a common denominator for this lumber problem. However, working with fractions like ²⁴⁄₃₂ or ²⁰⁄₃₂ can still be a little difficult. To make working with them easier, find the lowest common denominator possible between them, which means reducing the fractions to their lowest terms. Although 32 is a common denominator for the two fractions, it is not the lowest common denominator.

To find the lowest common denominator, follow these steps:

Step 1 Reduce each fraction to its lowest terms.

Step 2 Find the lowest common multiple of the denominators. Sometimes this is as simple as one denominator already being a multiple of the other. If this is the case, all you have to do is find the equivalent fraction for the term with the smaller denominator.

Step 3 If neither of the denominators is a multiple of the other, you can multiply the denominators together to get a common denominator.

Let's look at the 2 × 4 example again where ¾ and ⅝ are already in their lowest terms. Upon looking at the denominators, you see that 8 is a multiple of 4. So the equivalent fraction for ¾ that has a denominator of 8 can be found this way:

$$\frac{3}{4} \times \frac{2}{2} = \frac{6}{8}$$

You can now compare ⁶⁄₈ to ⅝ and see that ⁶⁄₈ is the larger fraction. You could see this same basic result as ²⁴⁄₃₂ and ²⁰⁄₃₂ were compared. However, in this case, you are now working with the lowest common denominator.

Whether the lowest common denominator is determined or the denominators are simply multiplied together to find an acceptable common denominator will depend on the situation. In some applications, you may want all fractions involved to have a particular denominator. In the carpentry trade for example, working to within ¹⁄₁₆" is common, so carpenters may tend to think of most fractional measurements in sixteenths.

Did You Know?
Nominal Measurements

A nominal 8" × 8" × 16" concrete block is actually 15⅝" × 7⅝" × 7⅝" in dimension. It is referred to as an 8" × 8" × 16" block because of the addition of the ⅜-inch mortar joint between the blocks. This measurement has been adopted as the trade standard for masonry joints. Adding the ⅜ inch to the block's actual dimensions causes the block's measurements to come out in even inches, rather than as fractions. This holds true for other sizes and types of blocks.

2.1.4 Study Problems: Finding Equivalent Fractions

Find the equivalents of the following fractions:

1. ¼ equals how many sixteenths?
   a. ²⁄₁₆
   b. ⁴⁄₁₆
   c. ⁶⁄₁₆
   d. ⁸⁄₁₆

2. ²⁄₁₆ equals how many thirty-seconds?
   a. ¹⁄₃₂
   b. ²⁄₃₂
   c. ⁴⁄₃₂
   d. ⁸⁄₃₂

3. ¾ equals how many eighths?
   a. ²⁄₈
   b. ⁴⁄₈
   c. ⁵⁄₈
   d. ⁶⁄₈

4. ¾ equals how many sixty-fourths?
   a. ⁴⁸⁄₆₄
   b. ⁵⁰⁄₆₄
   c. ⁵²⁄₆₄
   d. ⁵⁴⁄₆₄

5. ³⁄₁₆ equals how many thirty-seconds?
   a. ²⁄₃₂
   b. ⁴⁄₃₂
   c. ⁶⁄₃₂
   d. ⁸⁄₃₂

Find the lowest term for each of the following fractions without using a calculator.

6. ²⁄₁₆ = __1/8__

7. ²⁄₈ = __1/4__

8. ¹²⁄₃₂ = __3/8__

9. ⁴⁄₈ = __1/2__

10. ⁴⁄₆₄ = __1/16__

Find the lowest common denominator for the following pairs of fractions.

11. ²⁄₆ and ¾.
   a. 6
   b. 10
   c. 12
   d. 16

12. ¼ and ³⁄₈.
   a. 4
   b. 8
   c. 12
   d. 18

13. ⅛ and ½.
   a. 3
   b. 5
   c. 7
   d. 8

14. ¼ and ³⁄₁₆.
   a. 8
   b. 16
   c. 18
   d. 20

15. ⁴⁄₃₂ and ⁵⁄₈.
   a. 64
   b. 32
   c. 16
   d. 8

2.2.0 Improper Fractions and Mixed Numbers

An **improper fraction** is defined as one in which the numerator is larger than the denominator. The fractions ⁴³⁄₆, ¹⁹⁄₁₆, and ⁵⁵⁄₈ are all improper fractions.

Improper fractions make better sense and are often more practical to work with when they are converted into a **mixed number**. A mixed number is one in which a whole number has been combined with a fraction. Based on the examples of improper fractions above, writing them as mixed numbers would result in 7⅙, 3³⁄₁₆, and 6⅞. Converting some improper fractions, such ⁴²⁄₆, will result in a whole number alone, without an additional fraction. In this case, the fraction ⁴²⁄₆ converts to the whole number 7.

As noted previously, fractions are really just another way of writing a division problem. To convert an improper fraction to a mixed number, the division problem is worked out. For this example, we will work with the improper fraction ⁶⁷⁄₃₂. Here are the steps to convert an improper fraction to a mixed number:

Step 1 The numerator of the fraction (dividend) must be divided by the denominator (divisor). If the denominator cannot be divided into the numerator at least one full time, it is not an improper fraction. Set up a division problem if the numbers are difficult to handle mentally:

$$32 \overline{)67}$$

Step 2 Complete the long division.

$$
\begin{array}{r}
02 \text{ r3} \\
32 \overline{)67} \\
-64 \\
\hline
3
\end{array}
$$

Step 3 The quotient of the problem, 2, is the whole number of the mixed number. Any remainder becomes the numerator of the fraction, using the original denominator.

$$2^{3}/_{32}$$

2.2.1 Study Problems: Changing Improper Fractions to Mixed Numbers

Change these improper fractions to mixed numbers:

1. $^{35}/_{8}$ = _4 3/8_
2. $^{97}/_{16}$ = _6 1/16_
3. $^{13}/_{4}$ = _3 1/4_
4. $^{14}/_{5}$ = _2 4/5_
5. $^{48}/_{16}$ = _3_

2.3.0 Adding and Subtracting Fractions

To add fractions, it is necessary to find a common denominator and convert them to equivalent fractions. Once the denominators are the same, simple addition is used to add the numerators together. To add fractions, proceed as follows:

Step 1 Find the common denominator of the fractions you wish to add. A common denominator for ¾ and ⅝ is 32. The lowest common denominator is 8. The addition can be done using either one; it will not affect the accuracy of the result.

Step 2 Convert the fractions to equivalent fractions with the same denominator as shown.

$$\frac{3}{4} \times \frac{8}{8} = \frac{24}{32}$$

$$\frac{5}{8} \times \frac{4}{4} = \frac{20}{32}$$

Step 3 Add the numerators of the fractions. Place this sum over the denominator.

$$\frac{24}{32} + \frac{20}{32} = \frac{44}{32}$$

Step 4 Reduce the fraction to its lowest terms. In this example, both the numerator and denominator can be divided by 4. Doing so reduces the fraction to ¹¹⁄₈. As you now know, this is an improper fraction. The improper fraction ¹¹⁄₈ can be changed to the mixed number 1⅜.

As noted in *Step 1*, any common denominator can be used. However, when the lowest common denominator is used, the resulting sum of the fractions is already reduced to its lowest terms. Of course, the result may still be an improper fraction that must be changed to a mixed number.

Subtracting fractions is very much like adding fractions. You must find a common denominator before you can subtract. For example, assume you have ⅞ of a liter of paint left. If another worker asks for ¼ of it and you agree to share, how much will you have left?

Around the World

The Roots of Fractions

The word *fraction* originated from the Latin word *fractio* which means "to break." Other common words in the English language, such as *fracture*, were also derived from this Latin word. A number of different cultures, including the Egyptians, Babylonians, and the Indians all devised their own ways of considering and writing fractions over the years. It was the Arabs who added the line we now draw, sometimes on a slant and sometimes horizontally, between the numerator and the denominator.

Step 1 Find a common denominator. In this case, 8 is a common denominator and is also the lowest common denominator.

$$\frac{7}{8} \quad \frac{1}{4}$$

Step 2 Since one fraction is already in eighths, only the $\frac{1}{4}$ needs to converted to eighths. Multiply each term (numerator and denominator) by 2 to get a fraction with the denominator of 8.

$$\frac{1}{4} \times \frac{2}{2} = \frac{2}{8}$$

Step 3 With both fractions using the same denominator, the numerators of the two fractions can be subtracted. This results in a difference of $\frac{5}{8}$.

$$\frac{7}{8} - \frac{2}{8} = \frac{5}{8}$$

Sometimes a fraction must be subtracted from a whole number. This requires that the whole number be changed to a fraction as well. Once it is a fraction, find a common denominator and subtract as before. In this example, $\frac{2}{7}$ will be subtracted from 5.

Changing any whole number into a fraction is very simple. By simply adding a 1 as a denominator, a whole number becomes a fraction. In this case, the whole number 5 becomes the fraction $\frac{5}{1}$. The lowest common denominator between $\frac{5}{1}$ and $\frac{2}{7}$ is 7. Converting $\frac{5}{1}$ into sevenths results in the fraction $\frac{35}{7}$. Now the two fractions can be subtracted easily, resulting in a difference of $\frac{33}{7}$. This improper fraction can be changed to the mixed number of $4\frac{5}{7}$.

2.3.1 Study Problems: Adding and Subtracting Fractions

Find the answers to the following addition problems. Remember to reduce the sum to the lowest terms and change any improper fractions to mixed numbers.

1. $\frac{1}{8} + \frac{4}{16} =$ 3/8
2. $\frac{4}{8} + \frac{6}{16} =$ 7/8
3. $\frac{2}{4} + \frac{3}{4} =$ 5/4
4. $\frac{3}{4} + \frac{2}{8} =$ 1
5. $\frac{14}{16} + \frac{3}{8} =$ 20/16

Find the answers to the following subtraction problems. Remember to reduce the differences to the lowest terms.

6. $\frac{3}{8} - \frac{5}{16} =$ 1/16
7. $\frac{11}{16} - \frac{5}{8} =$ 1/16
8. $\frac{3}{4} - \frac{2}{6} =$ 5/12
9. $\frac{11}{12} - \frac{4}{8} =$ 5/12
10. $\frac{11}{16} - \frac{1}{2} =$ 3/16

Find the answers to the following subtraction problems involving fractions and whole numbers. Reduce the fractions to their lowest terms and change any improper fractions to mixed numbers.

11. $8 - \frac{3}{4} =$ 7 + 1/4 = 7 1/4
12. $12 - \frac{5}{8} =$ 11 8/8 − 11 3/8
13. Two punches are made from steel bar stock $9\frac{7}{16}$ inches long. If one punch is $4\frac{1}{64}$ inches long and the other is $4\frac{3}{32}$ inches long, how many inches of stock are wasted?
    a. $1\frac{1}{16}$ inches
    b. $1\frac{5}{16}$ inches
    c. $1\frac{15}{64}$ inches
    d. $1\frac{21}{64}$ inches
14. If you saw $12\frac{1}{16}$ inches off a board that is $20\frac{3}{4}$ inches long, the length of the remaining board will be _____.
    a. $8\frac{1}{4}$ inches
    b. $8\frac{11}{16}$ inches
    c. $11\frac{1}{4}$ inches
    d. $17\frac{3}{8}$ inches
15. A rough opening for a window measures $36\frac{3}{8}$ inches. The window to be placed in the rough opening measures $35\frac{15}{16}$ inches. The total clearance that should exist between the window and the rough opening will be _____.
    a. $\frac{7}{16}$ inch
    b. 1 inch
    c. $1\frac{7}{16}$ inches
    d. $1\frac{3}{4}$ inches

2.4.0 Multiplying and Dividing Fractions

Multiplying and dividing fractions is very different from adding and subtracting fractions. You do not have to find a common denominator when you multiply or divide fractions.

In a word problem, the words used let you know if you need to multiply. If a problem asks

"What is $\frac{2}{8}$ of 9?" then think of the problem this way: $\frac{2}{8} \times \frac{9}{1}$. Remember that any number (except 0) over 1 equals itself. Therefore, $\frac{9}{1}$ is simply a fractional way to write the whole number 9.

Using $\frac{4}{8} \times \frac{5}{6}$ as an example, follow these steps:

Step 1 Multiply the numerators together to get a new numerator. Multiply the denominators together to get a new denominator.

$$\frac{4}{8} \times \frac{5}{6} = \frac{20}{48}$$

Step 2 Multiplication often results in fractions that are not in their lowest terms. Reduce the product if possible. In this example, $\frac{20}{48}$ can be reduced to $\frac{5}{12}$, since both the numerator and denominator are divisible by 4.

Although you can multiply fractions without first reducing them to their lowest terms, you can reduce them before you multiply. This will sometimes make the multiplication easier, since you will be working with smaller numbers. It will also make it easier to reduce the product to the lowest terms. What may seem like an extra step can save you time in the long run. Regardless of which way is chosen, the same product will result.

Dividing fractions is very much like multiplying fractions, with one difference. You must **invert**, or flip, the fraction you are dividing by. Using $\frac{1}{2} \div \frac{3}{4}$ as an example, follow these steps:

Step 1 Invert the fraction that is the divisor. In our example, $\frac{3}{4}$ is the divisor. Therefore:

$$\frac{3}{4} \text{ becomes } \frac{4}{3}$$

Step 2 Now change the division sign (÷) to a multiplication sign (×).

$$\frac{1}{2} \div \frac{3}{4} = \frac{1}{2} \times \frac{4}{3}$$

Step 3 Multiply the fraction as described earlier.

$$\frac{1}{2} \times \frac{4}{3} = \frac{4}{6}$$

Step 4 Reduce to the lowest terms when possible.

$$\frac{4}{6} \text{ reduce to } \frac{2}{3}$$

Thus, $\frac{3}{4}$ will go into $\frac{1}{2}$ two-thirds of a time. A division problem can be checked with multiplication to ensure its accuracy. Multiplying $\frac{3}{4}$ times $\frac{2}{3}$ results in a product of $\frac{1}{2}$—the original dividend of the problem.

If you are working with a mixed number, $2\frac{1}{3}$ for example, it must be converted into a fraction before it is inverted. Do this by multiplying the denominator of the fraction (3) by the whole number (2). Then add the numerator of the fraction and place the result over the denominator. That portion looks like this:

$$[(3 \times 2) + 1]$$

The result of the math above, 7, becomes the numerator of the fraction. The original denominator remains in place. The complete process of changing the mixed number $2\frac{1}{3}$ to a fraction looks like this:

$$2^1/_3 = \frac{(3 \times 2) + 1}{3} = \frac{7}{3}$$

When dividing a fraction by a whole number, first place the whole number over a 1 to convert it to a fraction. Remember that $\frac{4}{1}$ is the same as 4. Then invert the fraction. For example:

$$^1/_2 \div 4$$
$$^1/_2 \div {}^4/_1$$
$$^1/_2 \times {}^1/_4 = {}^1/_8$$

2.4.1 Study Problems: Multiplying and Dividing Fractions

Find the answers to the following multiplication problems without using a calculator. Reduce the products to their lowest terms and change improper fractions to mixed numbers.

1. $\frac{4}{16} \times \frac{5}{8}$ = 5/32
2. $\frac{3}{4} \times \frac{7}{8}$ = 21/32
3. $\frac{2}{8} \times 15$ = 15/4
4. $\frac{3}{7} \times 49$ = 21
5. $\frac{8}{16} \times \frac{32}{64}$ = 1/4

Find the answers to the following division problems without using a calculator. Reduce the quotients to their lowest terms and change improper fractions to mixed numbers.

6. $\frac{3}{8} \div 3$ = 1/8
7. $\frac{5}{8} \div \frac{1}{2}$ = 5/4
8. $\frac{3}{4} \div \frac{3}{8}$ = 2

9. On construction drawings, smaller dimensions are often used to represent larger ones. This allows the large object or structure to fit on the paper. On such a drawing, if ¼-inch represents a distance of 1 foot, then a line on the drawing measuring 8½ inches would represent how many feet?

a. 34
b. 36
c. 38
d. 40

10. How many ⅞-inch long strips can be cut from a single 7-inch long strip of material?

a. 5
b. 6
c. 7
d. 8

Additional Resources

Applied Construction Math: A Novel Approach, NCCER. 2006. Upper Saddle River, NJ: Prentice Hall.

Mathematics for Carpentry and the Construction Trades, Alfred P. Webster; Kathryn B. Judy. 2001. Upper Saddle River, NJ: Prentice Hall.

Mathematics for the Trades: A Guided Approach, Robert A. Carman; Hal Saunders. 2014. Pearson Learning.

2.0.0 Section Review

1. Which of the following fraction pairs are equivalent fractions?

a. ½ and ¹⁷⁄₃₂
b. ½ and ³²⁄₆₄
c. ½ and ¾
d. ½ and ⁶³⁄₁₂₈

2. The improper fraction ⁷⁶⁄₆₄ converts to the mixed number 1⁹⁄₃₂.

a. True
b. False

3. ¹¹⁄₁₆ + ⅜ = _____.

a. ⁹⁄₁₆
b. ⅞
c. 1¹⁄₁₆
d. 1⅛

4. 1¼ − ⁷⁄₁₆ = _____.

a. ⁵⁄₁₆
b. ⅝
c. ¾
d. ¹³⁄₁₆

5. ⅛ × ¹⁄₁₀ = _____.

a. ¹⁄₈₀
b. ¹⁄₄₀
c. ⅘
d. 1¼

6. ⁷⁄₁₆ ÷ ⅞ = _____.

a. ⁴⁹⁄₁₂₈
b. ⅜
c. ½
d. 2

NCCER – *Core Curriculum* 00102-15

SECTION THREE

3.0.0 THE DECIMAL SYSTEM

Objective

Describe the decimal system and explain how to work with decimals.

a. Describe decimals and their place values.
b. Demonstrate the ability to add, subtract, multiply, and divide decimals.
c. Demonstrate the ability to convert between decimals, fractions, and percentages.

Decimals are based entirely on the number 10. Through learning the place values of digits in a given number earlier in this module, you have learned half of the decimal system already. Whole numbers, such as 26,493, do not typically have a decimal point written beside them. However, that does not mean a decimal point is not actually there. When a decimal point is not visible, one is always assumed to be to the right of the last number. For the number 26,493, a decimal point is assumed to be on the right side of the number 3.

3.1.0 Decimals

By using digits placed to the right of a decimal point, a part of a number can be written. Proper fractions represent part of the number 1. When using decimals instead of fractions, all numbers written to the right of a decimal point also represent part of the number 1. In fact, decimals are often referred to as decimal fractions, since they represent part of a whole number.

Figure 4 shows the place values of digits placed on the left and on the right of the decimal point.

Of course, place values extend far beyond what is shown, in both directions. For applications requiring extreme precision, extending a number five places to the right of the decimal point usually provides the required precision. For more common applications, the precision offered by extending a number one or two places to the right of the decimal point is often sufficient. This depends entirely on the application. In some cases, even a whole number is sufficient.

To read a decimal, say the number as it is written and then the name of the place value for the right-most digit. For example, 0.56 is spoken as fifty-six hundredths. The right-most digit, the 6, is located in the hundredths place.

Note that the number written for this example (0.56) begins with a zero on the left of the decimal point. Although it is not required to properly read the decimal value, it is common to fill the first position to the left of the decimal with a zero for clarity when the decimal represents a value less than one.

Mixed numbers also appear in decimals. The number 15.7 is an example. It is read as fifteen and seven-tenths. Notice the use of the word "and" to separate the whole number from the decimal.

3.1.1 Rounding Decimals

Sometimes an answer is a bit more precise than required. As an example, if tubing costs €3.76 (the € symbol represents the euro, currency of the European Union) per meter and €800 has been budgeted for the tubing, how much tubing can be purchased?

The precise mathematical answer is 212.7659574 meters. However, it is quite unlikely that 0.7659574 meters of tubing can be purchased. A value to the nearest tenth is sufficient for this application. For this exercise, 212.7659574 will be rounded to the nearest tenth. This represents only one position to the right of the decimal point.

5	,	3	1	6	,	2	4	7	.	4	2	9	6	3	5
MILLIONS		HUNDRED THOUSANDS	TEN THOUSANDS	THOUSANDS		HUNDREDS	TENS	ONES		TENTHS	HUNDREDTHS	THOUSANDTHS	TEN-THOUSANDTHS	HUNDRED-THOUSANDTHS	MILLIONTHS

00102-15_F04.EPS

Figure 4 Place values on both sides of the decimal point.

Step 1 Underline the place to which you are rounding.

212.7659574

Step 2 Look at the digit one place to its right.

212.7659574

Step 3 If the digit to the right is 5 or more, you will round up by adding 1 to the underlined digit. If the digit is 4 or less, leave the underlined digit the same. In this example, the digit to the right is 6, which is more than 5, so you round up by adding 1 to the underlined digit. Then drop all the remaining digits to the right of it.

212.8

3.1.2 Comparing Decimals with Decimals

It is relatively simple to tell if one decimal value is larger or smaller than another. It is really no different than determining if one whole number is larger than another. You should be able to tell which of these numbers is larger on sight:

4,214 or 4,217

Obviously 4,217 is the larger of the two numbers. A decimal point can be added at any position to make this number a decimal.

0.4214 or 0.4217

The result is the same. The number with the digits 4217 is larger than the other.

3.1.3 Study Problems: Working With Decimals

For the following problems, identify the words that represent the proper way to speak the decimal value shown.

1. 0.4 = _____.
   a. four
   b. four-tenths
   c. four-hundredths
   d. four-thousandths

2. 0.05 = _____.
   a. five
   b. five-tenths
   c. five-hundredths
   d. five-thousandths

3. 2.5 = _____.
   a. two and five-tenths
   b. two and five-hundredths
   c. two and five-thousandths
   d. twenty-five-hundredths

For the following problems, identify the written numerical value of the spoken number.

4. Eighteen-hundredths = _____.
   a. 0.0018
   b. 0.018
   c. 0.18
   d. 1.8

5. Five and eight-tenths = _____
   a. 5.0
   b. 5.008
   c. 5.08
   d. 5.8

For the following problems, select the answer that places the decimals in order from smallest to largest.

6. 0.400, 0.004, 0.044, and 0.404
   a. 0.400, 0.004, 0.044, 0.404
   b. 0.004, 0.044, 0.404, 0.400
   c. 0.004, 0.044, 0.400, 0.404
   d. 0.404, 0.044, 0.400, 0.004

7. 0.567, 0.059, 0.56, and 0.508
   a. 0.508, 0.56, 0.567, 0.059
   b. 0.059, 0.56, 0.508, 0.567
   c. 0.567, 0.059, 0.56, 0.508
   d. 0.059, 0.508, 0.56, 0.567

8. 0.320, 0.032, 0.302, and 0.003
   a. 0.003, 0.032, 0.302, 0.320
   b. 0.320, 0.302, 0.032, 0.003
   c. 0.302, 0.320, 0.003, 0.032
   d. 0.003, 0.032, 0.320, 0.302

9. 0.867, 0.086, 0.008, and 0.870
   a. 0.870, 0.867, 0.086, 0.008
   b. 0.008, 0.086, 0.867, 0.870
   c. 0.086, 0.008, 0.867, 0.870
   d. 0.008, 0.870, 0.867, 0.086

10. 0.626, 0.630, 0.616, and 0.641
    a. 0.616, 0.641, 0.630, 0.626
    b. 0.616, 0.626, 0.630, 0.641
    c. 0.641, 0.616, 0.626, 0.630
    d. 0.630, 0.616, 0.626, 0.641

3.2.0 Adding, Subtracting, Multiplying, and Dividing Decimals

In many ways, mathematical operations with decimals are not significantly different than working with whole numbers. Placing the decimal point in the proper location is however, essential to finding the correct answer.

3.2.1 Adding and Subtracting Decimals

The most important rule to remember when adding and subtracting decimals is to keep the decimal points aligned as the problem is written.

In this example, 4.76 and 0.834 will be added together. Note that these two numbers have a different number of digits to the right of the decimal. Line up the problem as shown here, adding a 0 if needed to help keep the numbers lined up.

$$\begin{array}{r} 4.760 \\ + \ 0.834 \\ \hline 5.594 \end{array}$$

Note that the decimal points of the two numbers being added are aligned vertically. The decimal point of the sum is also aligned with them. The same thing is true for subtraction of decimals. To subtract 2.724 from 5.6, line up the decimal points as shown.

$$\begin{array}{r} 5.600 \\ - \ 2.724 \\ \hline 2.876 \end{array}$$

Did You Know?

Digits and Decimals

When a number is written, a symbol representing that number is marked or printed. This symbol is often referred to as a digit. The word *digit* is derived from the Latin word for finger.

Early people (and a lot of us inhabiting the planet today) naturally used their fingers as a means of counting—a base-ten counting system, since we have ten fingers. The decimal system naturally evolved from this. The word *decimal* is also derived from a Latin word. In Latin, decimal means "ten."

Notice that two zeros were added to the end of the first number to make it easier to see where you need to borrow. This is not required, but it does help to ensure that the problem is properly aligned.

3.2.2 Multiplying Decimals

A series of partitions must be set up in an office. One partition is measured to determine its width, which is 4.5 feet. There are seven panels of the same width. How many feet of partition can be erected if the panels are standing side-by-side?

Step 1 Set up the problem just like the multiplication of whole numbers.

$$\begin{array}{r} 4.5 \\ \times \ 7 \end{array}$$

Step 2 Proceed to multiply.

$$\begin{array}{r} 4.5 \\ \times \ 7 \\ \hline 315 \end{array}$$

Step 3 Once you have the answer, count the number of digits to the right of the decimal point in both of the numbers being multiplied. In this example, there is only one with a number to the right of the decimal point (4.5), and there is only one digit to the right of it.

Step 4 In the answer, count over the same number of digits, from right to left, and place the decimal point there.

$$\begin{array}{r} 4.5 \\ \times \ 7 \\ \hline 31.5 \end{array}$$

You may have to add one or more zeros if there are more digits to the right of the decimal points than there are in the answer, as shown in the following example.

$$\begin{array}{r} 0.507 \\ \times \ 0.022 \\ \hline 1014 \\ 10140 \\ 000000 \\ + \ 0000000 \\ \hline 11154 = 0.011154 \end{array}$$

Add the digits to the right of the decimal point in the two numbers. There are six total. Count six digits from right to left in the product. In this case, you'll need to add a zero.

3.2.3 Dividing with Decimals

There are three types of division problems involving decimals:

- Those that have a decimal point in the number being divided (the dividend):

$$22\overline{\smash{\big)}44.5}$$

- Those that have a decimal point in the number you are dividing by (the divisor):

$$0.22\overline{\smash{\big)}4{,}450}$$

- Those that have decimal points in both numbers (the dividend and the divisor):

$$0.22\overline{\smash{\big)}44.5}$$

The examples above will be used to demonstrate each type of decimal division problem.

For the first type of problem where a decimal point is shown only in the dividend, follow these steps:

Step 1 Place a decimal point directly above the decimal point in the dividend.

$$22\overline{\smash{\big)}44.5}$$

Step 2 Divide as usual.

$$
\begin{array}{r}
02.0 \\
22\overline{\smash{\big)}44.5} \\
-44 \\
\hline
00.5r
\end{array}
$$

The answer is 2, with a remainder of 0.5. Note again the alignment of the decimal point in the quotient, directly above the decimal point in the dividend. This is critical for this type of problem.

For the second type of problem, where only a decimal point is found in the divisor, follow these steps:

Step 1 Move the decimal point in the divisor to the right until you have a whole number.

0.22 becomes 22

Step 2 Next, move the decimal point in the dividend the same number of places (two) to the right. Zeros will have to be added, since the dividend is a whole number. Now place the decimal point in the quotient directly above the one in the dividend. Then divide as usual.

$$
\begin{array}{r}
20227.2 \\
22\overline{\smash{\big)}4450.00.0} \\
-44 \\
\hline
0050 \\
-0044 \\
\hline
00060 \\
-00044 \\
\hline
000160 \\
-000154 \\
\hline
0000060 \\
-0000044 \\
\hline
0000016r
\end{array}
$$

Remember that division like this can always be checked by multiplying the quotient times the divisor, and then adding the remainder. The answer should always result in the dividend.

For the third type of problem, where there is a decimal point in both the dividend and the divisor, follow these steps:

Did You Know?

Scientific Notation

Scientific notation (sometimes called exponential notation) is a system that allows you to conveniently write very large or very small decimal-based numbers using an exponent. The exponent represents the number of times you multiply the multiplier by the multiplicand. Scientific notation is commonly used by scientists, mathematicians, and engineers. You may already be familiar with scientific notation if you used ft^2 (feet squared) in a measurement. The following are some examples of scientific notation:

Decimal Notation	Scientific Notation
1	1×10^0
50	5×10^1
7,530,000,000	7.53×10^9
−0.0000000082	-8.2×10^{-9}

Step 1 Move the decimal point in the divisor to the right until you have a whole number.

0.22 becomes 22

Step 2 Move the decimal point in the dividend the same number of places to the right. 0.22 and 44.5 now become 22 and 4,450, respectively. By moving the decimal point in both the divisor and the dividend, the quotient will not be affected.

$$22\overline{)44.50}$$

Step 3 Then divide as usual.

```
          202
    22 ) 4450
       − 44
         005
       − 000
         0050
       − 0044
         0006r
```

The answer is 202 with a remainder of 6.

3.2.4 Using the Calculator to Add, Subtract, Multiply, and Divide Decimals

Performing operations on the calculator using decimals is very much like performing the operations on whole numbers. Follow these steps using the problem 45.6 + 5.7 as an example.

Step 1 Turn the calculator on.

Step 2 Press 4, 5, . (decimal point), and 6. The number 45.6 appears in the display.

> **NOTE**
>
> For this step, press whichever operation key the problem calls for: + to add, − to subtract, × to multiply, ÷ to divide.

Step 3 Press the + key. The 45.6 is still displayed.

Step 4 Press 5, . (decimal point), and 7. The number 5.7 is displayed.

Step 5 Press the = key. After you press the = key, whether you are adding, subtracting, multiplying, or dividing, the answer will appear on your display.

45.6 + 5.7 = 51.3
45.6 − 5.7 = 39.9
45.6 × 5.7 = 259.92
45.6 ÷ 5.7 = 8

Step 6 Press the On or C key to fully clear the calculator. The CE key clears only the most recent entry. A zero (0) appears in the display.

Did You Know?

Other Counting Systems

Ancient Babylonians developed one of the first place-value systems—a base-sixty system. It was based on the number 60, and numbers were grouped by sixties. At the beginning, there was no symbol for zero in the Babylonian system. This made it difficult to perform calculations. It was not always possible to determine if a number represented 24, 204, or 240. Although this system is no longer used today, we still use a base-sixty system for measuring time: 60 seconds in a minute and 60 minutes in an hour.

A base-twelve system requires 12 digits. This system is called a duodecimal system, from the Latin word *duodecim*, meaning "twelve." Since the decimal system has only ten digits, two new digits must be added to the base-twelve system. In a duodecimal system, each place value is 12 times greater than the place to the right. Although this is not a common system, a base-twelve system is used to count objects by the dozen (12) or by the gross (144 = 12 × 12).

A hexadecimal system groups numbers by sixteens. The word *hexadecimal* comes from the Greek word for "six" and the Latin word for "ten." Just as the duodecimal system required two new place numerals, the hexadecimal system requires six additional digits. In a hexadecimal system, each place value is 16 times greater than the place to the right. Computers often use the hexadecimal system to store information. Common configurations of random access memory (RAM) come in multiples of 16. 1,024 kilobytes (kB) equals 1 megabyte (MB).

3.2.5 Study Problems: Decimals

Find the answers to these addition and subtraction problems without using a calculator. Be sure to maintain vertical alignment of the decimal points.

1. 2.50 + 4.20 + 5.00 = _~~11.17~~_

2. 1.82 + 3.41 + 5.25 = _10.48_

3. 6.43 + 86.4 = _92.83_

4. The combined thickness of a piece of sheet metal 0.078 centimeters (cm) thick and a piece of band iron 0.25 cm thick is _____.
   a. 0.308 cm
   † b. 0.328 cm
   c. 3.08 cm
   d. 32.8 cm

5. Yesterday, a lumber yard contained 6.7 tons of wood. Since then, 2.3 tons were removed. How many tons of wood remain?
   a. 3.4 tons
   — b. 4.4 tons
   c. 5.4 tons
   d. 6.4 tons

Use the following information to answer Questions 6 and 7:

A part is being machined. The starting thickness of the part is 6.18 inches. Three cuts are taken. Each cut is three-tenths of an inch.

6. How many inches of material have been removed at this point?
   a. 0.6 inches
   b. 0.8 inches
   c. 0.9 inches
   d. 1.09 inches

   $.3$
   $+.3$
   $.3$
   $.9$

7. The remaining thickness of the part is _____.
   a. 5.28 inches
   b. 6.08 inches
   c. 6.10 inches
   d. 6.15 inches

   6.18
   $-.90$
   5.28

8. Ceramic tile weighs 4.75 pounds per square foot. Therefore, 128 square feet of ceramic tile weighs _____.
   a. 598 pounds
   b. 608 pounds ✗
   c. 908 pounds
   d. 1108 pounds

Find the answers to the following division problems without using a calculator. Answers should be rounded to the nearest hundredth.

9. 45.36 ÷ 18 = _2.52_

10. 4.536 ÷ 18 = _0.252_ .25

11. 0.4536 ÷ 18 = _0.0252_ = 0.3

Round the quotients of these problems to the nearest hundredth.

12. 25) 10.20   0.408 = 41

13. 6) 31.2   5.2

Perform the following division problems on a separate piece of paper without using a calculator. Note that a decimal point is only present in the divisors of these problems. Answers should be rounded to the nearest hundredth.

14. 282 ÷ 14.1 = _20_

15. 694 ÷ 3.2 = _216.88_

16. 99 ÷ 0.45 = _220_

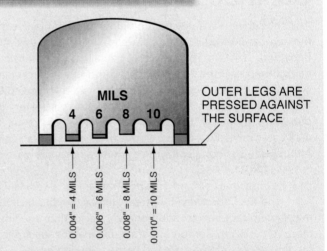

Measuring the Thickness of a Coating

Coating thickness is important on many parts and surfaces because too little or too much can cause problems. A coating such as paint needs a minimum thickness to prevent corrosion, withstand abrasion, and look good. A coating that is too thick may crack, flake, blister, or not cure properly.

The thickness of the coating to be applied to an object or surface is often specified by the buyer or client. To ensure that the specifications are met, periodic checks of the wet-film thickness can be made using a wet-film thickness gauge. Typically these gauges use measurements in mils or microns. As shown here on a typical gauge, a mil is equal to one one-thousandth of an inch (0.001), which can also be written as 10^{-3}. A mil is also equal to 0.0254 millimeters.

MILS
4 6 8 10

OUTER LEGS ARE PRESSED AGAINST THE SURFACE

0.004" = 4 MILS
0.006" = 6 MILS
0.008" = 8 MILS
0.010" = 10 MILS

Thickness is determined by the longest tooth with paint and the shortest tooth without paint. The coating thickness here is 6-8 mils.

00102-15_SA01.EPS

17. $2.5\overline{)102}$ *4.8*

18. $0.6\overline{)312}$ *520*

Find the answers to the following division problems without using a calculator. Note that a decimal point is present in both the divisor and the dividend of these problems. Answers should be rounded to the nearest hundredth.

19. $20.82 \div 4.24 =$ _____   *4.910*

20. $38.9 \div 3.7 =$ _____   *10.51*

21. $9.9 \div 0.45 =$ _____   *22*

22. $0.25\overline{)10.20}$ *= 40.*

23. $0.6\overline{)31.2}$ *= 52*

Use your calculator to find the answers to the following problems, rounding your answers to the nearest hundredth.

24. 45.89
    $+ 7.85$
    53.74

25. 7.6
    $\times 0.12$
    0.912

26. 685.79
    $- 56.266$
    629.524

27. $6.45 \div 3.25 =$ *1.98*

28. 34.76
    $+ 3.64$
    38.40

Solve these problems to practice rounding decimals. Round your answers to the nearest tenth.

29. You need to cut a 90.5-inch pipe into as many 3.75-inch pieces as possible. How many complete 3.75-inch pieces will you be able to cut?
    a. 14
    b. 24
    c. 34
    d. 44

30. If a car traveled 1,001 kilometers on 151.8 liters of gas, how many kilometers per liter would it be achieving, to the nearest tenth of a kilometer?
    a. 0.1
    b. 0.2
    c. 6.5
    d. 6.6

31. If wire costs $4.30 per pound and you pay a total of $120.95, how many pounds of wire were purchased, to the nearest hundredth of a pound?
    a. 0.28
    b. 2.8
    c. 28.1
    d. 28.13

32. One size of vent pipe is on sale at XYZ Supply Company this week for $0.37 per linear foot. How many feet of vent pipe can be purchased with $115.38, to the nearest tenth of a foot?
    a. 308.1
    b. 310.8
    c. 311.8
    d. 311.9

33. Vent pipe at XYZ Supply costs $0.48 per linear foot when it is not on sale. If you spend $115.38, how many feet can be purchased, to the nearest tenth of a foot?
    a. 240.38
    b. 240.4
    c. 241
    d. 241.4

3.3.0 Converting Decimals, Fractions, and Percentages

Sometimes numbers need to be converted from one form to another to make them easier to work with. Some numbers relevant to the job at hand may appear as decimals, some as percentages, and others as fractions. Decimals, percentages, and fractions are all just different ways of expressing the same thing. The decimal 0.25, the percent 25%, and the fraction ¼ all represent the same numerical value or quantity. In order to work with different forms of numbers, they must be converted from one form into another.

3.3.1 Converting Decimals to Percentages and Percentages to Decimals

To understand percentages, think of a whole number divided into 100 parts. Thinking about a $1.00 bill is a perfect way to understand percentages, since it is broken into 100 equal parts (100 pennies). Each penny represents 1% of a dollar.

Any part of a whole can be expressed as a percentage. *Figure 5* provides an example more appropriate for the trades. The tank has a capacity of 100 gallons. It is now filled with 50 gallons. To what percentage is the tank filled?

The correct answer is 50%. The percentage reflects a quantity based on the there being 100 parts to something. How many gallons out of 100 does the tank contain? It contains 50 out of 100, or 50 percent. Percentages are an easy way to express parts of a whole.

Decimals and fractions also express parts of a whole. The tank in *Figure 5* is 50 percent full. If you expressed this as a fraction, you would say it is ½ full. You could also express this as a decimal and say it is 0.50 full. In fact, a percentage can be expressed as a decimal as well. If a half-gallon of liquid is added to the tank in *Figure 5*, it would be 50.5% full.

Sometimes decimals need to be expressed as percentages, or percentages as decimals. Suppose a gallon of cleaning solution needs to be prepared.

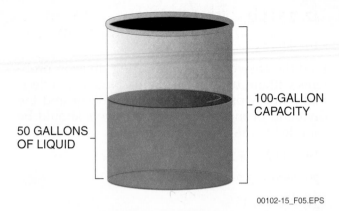

00102-15_F05.EPS

Figure 5 100-gallon tank.

The mixture should be comprised of 10% to 15% of the cleaning agent. The rest should be water. There is 0.12 gallon of cleaning agent concentrate available. Will this be enough to prepare a gallon of the solution? To answer the question, the decimal value of the concentrate (0.12) needs to be converted to a percentage. Follow these steps to make the conversion:

Decimals at Work

The use of decimal measurements is very important to a number of trades, including millwrights. Millwrights are often tasked with aligning pump and drive motor shafts. To eliminate vibration and prevent damage to the drive coupling and pump seal, many shafts must be aligned with extreme accuracy. This pump and motor are being aligned using laser technology. The level of precision is specified by the pump manufacturer, often to within 0.002 to 0.003 of an inch. Shims of various thicknesses are purchased or fabricated in various thicknesses and placed under the feet of the pump and/or motor to achieve alignment.

00102-15_SA02.EPS

00102-15_SA03.EPS

Step 1 Multiply the decimal by 100. (Tip: When multiplying by 100, simply move the decimal point of the number two places to the right.)

$$0.12 \times 100 = 12$$

Step 2 Add a % sign.

12%

Recall that the mixture should be from 10% to 15% cleaning agent. Since there is 12% of a gallon available, there is enough cleaning agent to make a gallon of solution.

You may also need to convert percentages to decimals. Use the following as an example. Suppose a mixture should contain 22% of a certain chemical by weight. One pound of the mixture is to be made. The ingredients are weighed on a digital scale, which cannot be read in percentages of a pound but can be read in decimal portions of a pound. To determine how much chemical is needed in decimal form, convert the percentage (22%) to a decimal by following these steps:

Step 1 Drop the % sign.

22

Step 2 Divide the number by 100. (Tip: When dividing by 100, simply move the decimal point two places to the left.)

$$22 \div 100 = 0.22$$

Therefore, add 0.22 pound of the chemical to 0.78 pound of the other ingredient(s) to make one pound at a 22% mixture.

3.3.2 Converting Fractions to Decimals

You will often need to change a fraction to a decimal. Remember that a fraction is already a division problem that has been written differently. To change a fraction to a decimal value, simply complete the division problem. The following steps show how it is done for the fraction ¾, as in ¾ of a liter:

Step 1 Divide the numerator of the fraction by the denominator.

$$4 \overline{)3.0}$$

In this example, place a decimal point and the zero after the number 3. Since 4 will not divide into 3, more digits will be needed to work the problem.

Step 2 Place the decimal point for the quotient directly above its location in the dividend.

$$4 \overline{)3.0}$$

Step 3 Now divide as you normally would.

$$
\begin{array}{r}
.75 \\
4 \overline{)3.00} \\
-\ 2.8 \\
\hline
0.20 \\
-\ 0.20 \\
\hline
0.00
\end{array}
$$

The decimal equivalent of ¾ is 0.75. As demonstrated earlier, this value can be turned into a percentage by moving the decimal point two positions to the right. So ¾ of a liter is also 0.75 liters or 75% of a liter.

3.3.3 Converting Decimals to Fractions

Converting a decimal to a fraction is relatively simple. Remember that both decimals and fractions are different ways to express the same thing. When a decimal value is spoken, it is spoken like a fraction. When it is written the same way it is spoken, it automatically translates as a fraction.

Drill bits come in fractional sizes, decimal sizes, and even sets identified by numbers and letters alone. Assume a hole that is to be drilled is specified as 0.125 inch in size. However, all you have are fractional drill sizes. It is possible that a fractional drill size is actually the same size. To find out, follow these steps to convert the decimal 0.125 to a fraction:

Step 1 Say the decimal in words.

0.125 is one hundred twenty-five thousandths

Step 2 Write the decimal as a fraction, just like it was spoken.

0.125 written as a fraction is $^{125}/_{1000}$

Step 3 Reduce the fraction to its lowest terms.

$$\frac{125}{1000} = \frac{125}{1000} \div \frac{125}{125} = \frac{1}{8}$$

0.125 converted to a fraction is ⅛. If there is a ⅛" drill in the box, the specification has been met and the hole can be drilled.

3.3.4 Converting Inches to Decimal Equivalents in Feet

Unlike the United States, the vast majority of the world works with the metric system of measurement. The metric system is perfectly matched to the decimal system, since both are based on units of 10. However, for those working with the Imperial or inch-pound system of measurement, measurements may need to be converted to decimals. For example, what decimal part of a foot does 3 inches represent?

First, express the inches as a fraction that has 12 as the denominator. This is accurate because an inch represents ¹⁄₁₂th of a foot. The fraction for 3 inches is written as ³⁄₁₂.

In this example, the fraction ³⁄₁₂ can be reduced to lower terms—¼. Convert the fraction ¼ to a decimal by dividing the 4 into 1.00:

$$
\begin{array}{r}
.25 \\
4\overline{)\,1.00} \\
-\,0.8 \\
\hline
0.20 \\
-\,0.20 \\
\hline
0.00
\end{array}
$$

3 inches is equal to 0.25 foot.

3.3.5 Study Problems: Converting Different Values

Convert these decimals to percentages.

1. 0.62 = _____

2. 0.475 = _____

3. 0.7 = _____

Convert these percentages to decimals.

4. 72% = _____

5. 12.5% = _____

Convert the following fractions to their decimal equivalents without using a calculator.

6. ¼ = _____

7. ¾ = _____

8. ⅛ = _____

9. ⁵⁄₁₆ = _____

10. ²⁰⁄₆₄ = _____

Practical Math Application

When contractors calculate the expenses for building a house, they must pay close attention to the percentages they allow for various factors. Of course, the profit realized by the company is an important consideration. For example, if a contractor is building a house that will have a selling price of $175,000, what will the profit be with the following percentage breakdown?

- Profit =?%
- Overhead = 27%
- Materials = 35%
- Labor = 24%

Solution

Since $175,000 represents 100% of the revenue, the total cost and profit together must equal 100%. Adding together the known percentages above reveals that the total cost of the job was 86% of the revenue. That means the profit must be 14% of the revenue. The profit, in dollars, can now be found this way:

Profit = $175,000 × 14% = $175,000 × 0.14 = $24,500

You can also use your calculator to work with percentages. For example, to solve the profit component of this problem, key it into your calculator as follows:

Step 1 Punch 175000 into the calculator.
Step 2 Press the multiplication button.
Step 3 Punch 14 into the calculator.
Step 4 Press the percent (%) button.

Note that you can also use the calculator to multiply 175,000 by 0.14, the decimal equivalent of 14%.

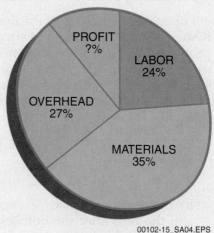

00102-15_SA04.EPS

Convert the following decimals to fractions without using a calculator. Reduce them to their lowest terms.

11. 0.5 = _____

12. 0.12 = _____

13. 0.125 = _____

14. 0.8 = _____

15. 0.45 = _____

Convert the following measurements to a decimal value in feet. Round the answer to the nearest hundredth.

16. 9 inches = _____ foot

17. 10 inches = _____ foot

18. 2 inches = _____ foot

19. 4 inches = _____ foot

20. 17 inches = _____ feet

3.3.6 Practical Applications

The following are examples of some of the practical applications a tradesperson encounters daily on the job site. Use the information provided to solve each conversion problem. Be sure to show all of your work.

1. Find the cost of baseboard needed for the office building shown in the floor plan in *Figure 6*. The lumber company charges $1.19 per linear foot of baseboard, and there is a 12% discount. All door widths are the standard 30 inches. Add tax after you reduce for the sale cost.

   Costs:

   a. Total linear feet needed _____

   b. Initial baseboard cost $ _____

   c. Discount amount $ _____

   d. Baseboard cost after sale reduction
      $ _____

   e. Amount added for 6% tax $ _____

   f. Total baseboard cost $ _____

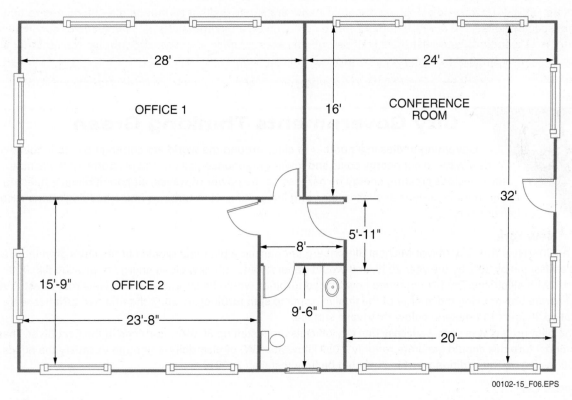

Figure 6 Office floor plan.

00102-15_F06.EPS

2. Use the site plan in *Figure 7* to determine the percentage of the lot that will be used for parking, the building, and for walkways. You may use a calculator. Round the percentages off to the nearest tenth of a percent.

   a. Percentage for parking _____ %

   b. Percentage for building _____ %

   c. Percentage for walkways _____ %

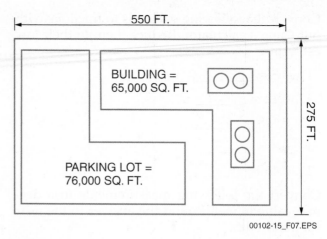

Figure 7 Site plan.

00102-15_F07.EPS

Direct and Indirect Costs

When contractors determine a percentage of profit, they must take into consideration many different costs. These costs can be grouped into one of two categories: direct costs and indirect costs.

Direct costs are expenses that can be directly attributed to completing the job. For example, all of the materials (paint, insulation, lumber) and labor hours required to build a house would be considered direct costs. On larger-scale jobs, direct costs can include not only materials and labor hours, but also subcontracted demolition and cleanup crews, cranes and earth-moving machinery, or even job-site security.

Indirect costs (or overhead costs), on the other hand, are expenses that cannot be directly related to building that specific house, but are required to run the company on a day-to-day basis. Indirect costs can include such items as insurance, administrative staff payroll, marketing, office supplies, and property taxes. What other expenses can you think of that would be considered indirect costs?

City Governments Thinking Green

GOING GREEN

Governing bodies in a number of cities around the world are coming up with innovative new ways to cut energy costs and lower greenhouse-gas emissions. Some of these new ideas include creating energy by harnessing the power of waves, air conditioning a building using rooftop gardens and lake water, and integrating wind turbines into a building's construction to generate energy.

New York

In 2006, New York City Mayor Michael Bloomberg announced a plan that would cut the city's greenhouse-gas emissions by 30% by the year 2030, and would generate enough new clean energy to provide 640,000 homes with electricity. The plan outlined how bladed turbines would be submerged into New York's East River to generate power using the energy of the tidal currents to spin turbines. In 2007, the first five turbines were installed 30 feet (9.14 meters) below the river's surface.

The pilot project was such a success that city officials are installing 30 more turbines in the East River. These additional turbines should generate roughly 1,050 kilowatts (kW) of electricity— enough to satisfy the needs of 9,500 residents. The additional turbines should be fully installed in 2015.

Aspen, CO

Many years ago, Aspen became the first municipality located west of the Mississippi River to make use of hydroelectric power—a renewable energy resource. Today, over 75% of Aspen's energy comes from renewable resources. The city is trying to achieve the goal of deriving 100% of its power needs from renewable resources. Those resources include geothermal energy, hydroelectric power, solar energy, and wind. As of 2014, wind power alone was already providing 26% of the city's energy needs.

NCCER – *Core Curriculum* 00102-15

Applied Construction Math: A Novel Approach, NCCER. 2006. Upper Saddle River, NJ: Prentice Hall.

Mathematics for Carpentry and the Construction Trades, Alfred P. Webster; Kathryn B. Judy. 2001. Upper Saddle River, NJ: Prentice Hall.

Mathematics for the Trades: A Guided Approach, Robert A. Carman; Hal Saunders. 2014. Pearson Learning.

3.0.0 Section Review

1. In the number 135.792, what value is represented by the 9?

    a. 90
    b. nine tenths
    c. nine one-hundredths
    d. nine one-thousandths

2. The number 0.960 is larger than the number 0.0962.

    a. True
    b. False

3. 3.625 + 4.9 = _____.

    a. 7.525
    b. 8.525
    c. 41.15
    d. 52.63

4. 42.58 − 7.577 = _____.

    a. 35.003
    b. 35.523
    c. 36.523
    d. 50.157

5. 9.64 × 12 = _____.

    a. 11.568
    b. 21.64
    c. 115.2
    d. 115.68

6. 123.82 ÷ 6.5 = _____.

    a. 19.049
    b. 18.76
    c. 1.905
    d. 0.190

7. Express the number 0.479 as a percentage.

    a. 0.00479%
    b. 0.479%
    c. 47.9%
    d. 479%

8. The decimal equivalent of the fraction ⅞ is _____.

    a. 0.0875
    b. 0.75
    c. 0.875
    d. 8.75

9. If a fuel tank is ⅔ full, then it is more than 65% full.

    a. True
    b. False

4.0.0 Measuring Length

Objective

Identify various tools used to measure length and show how they are used.

a. Identify and demonstrate how to use rulers.
b. Identify and demonstrate how to use measuring tapes.

Trade Terms

Joists: Lengths of wood or steel that usually support floors, ceiling, or a roof. Roof joists will be at the same angle as the roof itself, while floor and ceiling joists are usually horizontal.

Loadbearing: Carrying a significant amount of weight and/or providing necessary structural support. A loadbearing wall typically carries some portion of the roof weight and cannot be removed without risking structural failure or collapse.

Stud: A vertical support inside the wall of a structure to which the wall finish material is attached. The base of a stud rests on a horizontal baseplate, and a horizontal cap plate rests on top of a series of studs.

In the construction trade, you will need to use a measuring tool to measure the dimensions of various objects. The primary measuring devices you will see on the job are the standard English tape measure or ruler, and the metric tape measure or ruler (*Figure 8*).

CAUTION

Most rulers, especially wooden ones (similar to the metric ruler shown in *Figure 8*) are designed with extra material at the ends in order to maintain accurate measurements in the event that one of the ends is damaged. When using this type of ruler, make sure that you start your measurement at the first marked line and not at the physical end of the ruler.

4.1.0 Reading English and Metric Rulers

Reading English and metric rulers is not difficult. However, those who do not use them often may fail to recall the value of the small marks between the major units. When a measurement is misread, the error can lead to significant mistakes, which can be costly when material is scarce or expensive.

4.1.1 The English Ruler

The English ruler is divided into whole inches and then halves, fourths, eighths, and sixteenths. Some standard rulers may be divided into thirty-seconds, and some into sixty-fourths. These represent fractions of an inch. In this subsection, you will work with a standard ruler and standard fractions. In *Figure 8*, the English tape measure is marked with ⅛-inch increments along the top and ¹⁄₁₆-inch increments along the bottom. It is not uncommon to see tape measures or rulers with different markings along the top and bottom. However, the distances on the ruler shown in *Figure 9* are marked only in ¹⁄₁₆-inch increments.

STANDARD ENGLISH TAPE MEASURE

METRIC RULER

00102-15_F08.EPS

Figure 8 Various measurement tools.

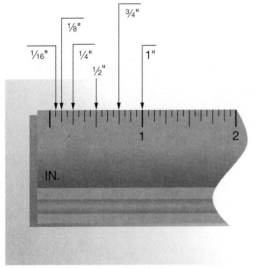

Figure 9 Standard ruler showing ¹⁄₁₆-inch increments and larger values.

In *Figure 9*, the increment between the ⅛-inch and ¼-inch increments would be called out as three-sixteenths (³⁄₁₆) of an inch. Similarly, the increment immediately after the ¾-inch increment would be called out as thirteen-sixteenths (¹³⁄₁₆) of an inch.

4.1.2 The Metric Ruler

Metric tape measures and rulers (*Figure 10*) are typically divided into centimeters and millimeters. The larger lines with numbers printed next to them are centimeters, and the smaller lines represent millimeters. The metric system is known as a base ten system, since each millimeter increment is ¹⁄₁₀ of a centimeter. Therefore, if you measure 7 marks after 2 centimeters as shown in *Figure 10*, it is 2.7 (two point seven) centimeters or 27 millimeters. Both of these numbers represent the same distance.

A measurement less than 1 centimeter, six marks before the 1-centimeter mark for example, would be recorded as 0.4 (point four) centimeters or 4 millimeters. This is also shown in *Figure 10*.

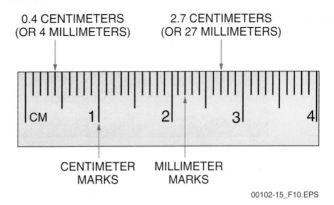

Figure 10 Increments on a metric ruler.

4.1.3 Study Problems: Reading Rulers

Use the information from *Figure 9*, if necessary, to help you identify the marked lengths numbered 1 through 10 in *Figures 11* and *12*. Label each length in the space provided or on a separate sheet of paper.

Identify the marked lengths in *Figure 13*. Record the correct answers for Questions 11, 12, and 13 in centimeters, and the increments for Questions 14 and 15 in millimeters.

4.2.0 The Measuring Tape

Measuring tapes are commonly referred to as tape measures; both names are quite common, depending upon the area of the country or world. The measuring tape's blades show either Imperial markings, metric markings, or possibly both as shown in *Figure 14*. If the blade only has markings from one system, which is most common, they are printed along both edges of the blade so that measurements can be taken accurately from either side.

Some common lengths for measuring tapes using both measurement standards are 16 feet (5 meters) and 25 feet (8 meters). There are many measuring tape lengths available with only one set of markings on the blade. Metric-only tapes are commonly available in 3.5, 5, and 8 meter lengths, for example. Longer tapes are generally

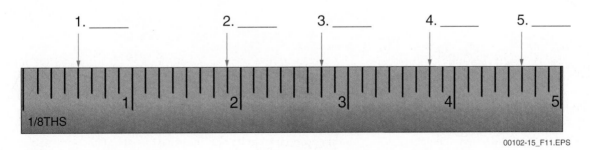

Figure 11 Practice with a ⅛-inch ruler.

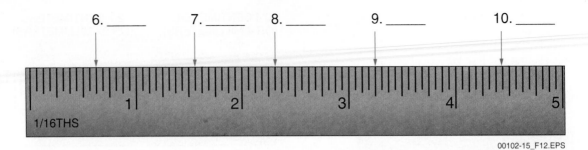

Figure 12 Practice with a ¹⁄₁₆-inch ruler.

used for site layout, while shorter tapes are more compact and convenient for carpentry, pipefitting, and other crafts.

4.2.1 The English Measuring Tape

The English measuring tape is marked similarly to the ruler described earlier with a few additional markings. Along with the ½-inch, ¼-inch, ⅛-inch, and ¹⁄₁₆-inch markings, a measuring tape usually features additional markings that make the task of wall framing easier.

As shown in *Figure 15(A)*, every 24 inches is marked with a contrasting black background. Note that 24-inch spacing on center is used most commonly for the studs in walls that do not carry a heavy load from above. The markings in *Figure 15(B)* are used for the 16-inch on center spacing most commonly used for loadbearing walls. These are highlighted with a red background. This common stud spacing is shown in the photograph in *Figure 15*. With the measuring tape stop pulled tight against the left side of one stud, the left side of the next stud is 16 inches away. Therefore, these studs are placed on 16-inch centers. In *Figure 15(C)*, 19.2 inches is marked with a small black diamond. This spacing is an alternate and less commonly used spacing scheme for special joists. It is not generally used for stud spacing.

Figure 16 shows other markings that may be printed on a standard tape measure. The red foot number (1F) works in conjunction with the red inch number (2) to quickly determine a measurement. The measurement indicated is 1 foot, 2 inches. The number below it (black 14) indicates 14 inches. If you add the top scale numbers together, it will equal the number of inches displayed on the bottom scale (1 foot or 12 inches + 2 inches = 14 inches). The red numbers on the top are marked in the same way along the length of the blade.

Around the World

Nominal and True Dimensions

A 2" × 4" board used in the United States is not actually 2 inches by 4 inches in dimension. The only time the board is truly 2 inches by 4 inches is when it is initially rough-cut from the log. After the board has been dried and planed, it is reduced to a finished size of 1½ inches by 3½ inches. Here are some true dimensions of other common board sizes in the Imperial system:

Nominal Size	Actual Measure	Actual Metric Measure
1" × 10"	¾" × 9¼"	19 × 235 mm
2" × 6"	1½" × 5½"	38 × 140 mm
2" × 8"	1½" × 7¼"	38 × 184 mm
2" × 10"	1½" × 9¼"	38 × 235 mm
4" × 4"	3½" × 3½"	89 × 89 mm

Equivalent lumber sizes in the metric world were generally converted using 25 mm per inch as a base. Note that this is not the precise value of an inch in the metric system, but this value was chosen for simplicity. As a result, the equivalent of a 2" × 4" in the metric system is referred to as a 50 mm × 100 mm board. However, as is the case in the United States, the actual dimensions are smaller; typically 40 mm × 90 mm. Over the years, the actual size of lumber compared to its nominal size has steadily decreased. Continued reductions in lumber sizes may result in changes to structural codes to maintain building strength.

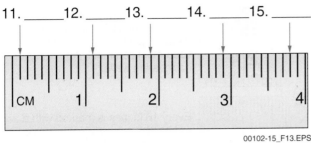

Figure 14 Measuring tape showing Imperial and metric measurements.

Figure 13 Metric ruler practice.

4.2.2 The Metric Measuring Tape

Metric measuring tapes are made in the same basic way as a standard measuring tape. In the United States, it may be difficult to find a measuring tape that uses only metric increments. At best, they will have both systems of measurement marked on the tape. In other countries, metric measuring tapes without inch markings on them are the most common.

Most metric-only measuring tapes do not have the same markings for standard stud spacing and other common spacing values. When they do, the markings are different to accommodate national or regional standards and the common dimensions of metric lumber.

4.2.3 Using a Tape Measure

Using a tape measure is relatively simple. You will note that the tape is not flat, but instead is concave on the top. This strengthens the tape and allows it to be extended a significant distance outside the case while remaining straight, without support along its length. Thicker blade material and a more concave shape allows for longer extension. In marketing literature, manufacturers may refer to this distance as stickout.

Step 1 Pull the tape straight out of its case with one hand.

Step 2 Hook the tape to one end of the object being measured using the metal hook on the end of the tape. If you are unable to hook onto the material, it may be necessary to have another worker hold the end of the tape. Be aware that the hook end is designed to slide a small amount to compensate for its own thickness; it is slightly loose and free to move for a very good reason. It slides in slightly when making an inside measurement, and slides out when the hook is used to make an outside measurement. If the end does not slide freely, your measurements can be off by as much as $\frac{1}{16}$ (1.6 mm) of an inch. A bent or damaged hook can have the same effect.

Step 3 Extend the tape by pulling the case, allowing the tape to extend to the point where the measurement will be taken.

Step 4 When the desired tape length is reached, slide the thumb lock on the case down to hold the tape in place. Not all tape measures have a thumb lock, and some tape measures will lock automatically. On automatic locking tape measures, a thumb hold release is used to retract the tape. Never let the tape retract rapidly back into the case, because this can cause damage to the hook.

It is sometimes necessary to use a tape measure to make an inside measurement. For example, you may need to make a measurement between two walls to determine the length of a shelf. Most measuring tapes have a small notation that indicates the exact size of the case. By firmly bumping the hook against one wall and the back of the case against the other wall, a measurement can be taken. Read the length where the tape meets the case, and then add the length of the case as printed or stamped on it.

Number of Courses

When erecting block walls, you will often hear the term *number of courses*. The number of courses refers to the number of block rows that will need to be stacked in order to establish the correct height of the wall. If you are told to erect a 15-course wall using 8" × 8" × 16" concrete blocks, you would be building a wall that is 10 feet high.

Block height = 8"
Number of courses = 15
12 inches = 1 foot

Therefore:

8" × 15 = 120"
120" ÷ 12" = 10 feet

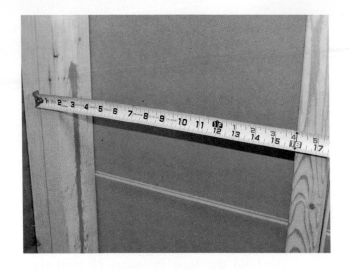

A

Every 24 inches is marked with a contrasting black background. 24-inch spacing on center is used most commonly for nonbearing walls.

B

Every 16 inches is marked with a red background. 16-inch spacing on center is used most commonly for loadbearing walls.

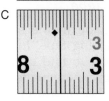

C

Every 19.2 inches is marked with a small black diamond. 19.2-inch spacing on center is an alternate, less-commonly used spacing scheme for loadbearing walls.

00102-15_F15.EPS

Figure 15 Wall-framing markings on a tape measure.

NOTE

Always remember to double-check your measurements before cutting a piece of material. Keep in mind the old saying, "Measure twice, cut once."

4.2.4 *Study Problems: Reading Measuring Tapes*

Identify the marked lengths numbered 1 through 5 in *Figures 17* and *18*. Record your answers on a separate piece of paper. Note that answers related to *Figure 17* should be in inches (using fractions as necessary). The answers related to *Figure 18* should be in centimeters (using tenths of a centimeter as necessary).

00102-15_F16.EPS

Figure 16 Other markings on a standard tape measure.

Around the World

Visualizing Metric Units

For users of the inch-pound system, it may help to visualize some metric measurements, using the examples below.

Metric Examples Visualized (approximation only):

- 1 millimeter = the thickness of the edge of a dime
- 1 centimeter = the width of a standard paperclip
- 1 decimeter = the length of a crayon
- 1 meter = the distance from a door handle to the floor (about 1.1 yards)
- 1 kilometer = the length of 6 city blocks (about 0.6 miles)
- 1 gram = weight of a paperclip
- 1 kilogram = weight of a brick (about 2.2 pounds)

NCCER – *Core Curriculum*   00102-15

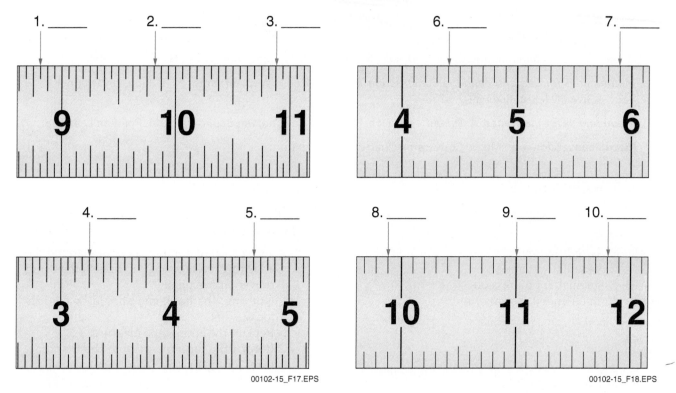

Figure 17 Tape measure practice.

Figure 18 Metric tape measure practice.

Did You Know?

Standard Measure

Determining the distance to a neighbor's farm may have been the earliest measurement of interest to people. Imagine that you wanted to measure the distance to a favorite fishing pond. You could walk, or pace off, the distance and count the number of steps you took. You might find that it was 570 paces between your home and the pond. You could then pace off the distance to another pond. By comparing the paces of one distance to the other, you could tell which distance was longer and by how much. Pace became a unit of measure. Of course, the length of individual paces differs from person to person.

Feet, arms, hands, and fingers were useful for measuring all sorts of things. In fact, the body was such a common basis for measuring that we still have traces of that system in our measurement standards. Lengths are measured in feet. Horses are said to be some number of hands high. An inch roughly matches the width of a person's thumb.

Complications arise when you have to decide whose legs, arms, or hands to use for measuring. If you wanted to mark off a piece of land, you might choose a tall person with long legs in hopes of getting more than your money's worth. If you owed someone a substantial amount of rope, you would want someone with short arms to measure the quantity and possibly save you some rope.

The problem demanded a standardized system of measurement that everyone could agree on. Legend says that the foot measurement used in France was the length of Charlemagne's foot. The yard is supposed to have been the distance from King Henry I of England's nose to the fingertips of his outstretched arm. Kings were not going to travel around making measurements for people, so the solution was to transfer the yard measurement to a stick. The distance marked on the stick became the standard measurement, and the government could send out duplicate sticks to each of the towns and cities as secondary standards. Civil officials could then check local merchants' measurements using the secondary standards.

During the Middle Ages, associations of craftspeople enforced strict adherence to the established standards. The success and reputation of the craft depended on providing accurately measured goods. Violators who used faulty measurements were often punished.

Additional Resources

Applied Construction Math: A Novel Approach, NCCER. 2006. Upper Saddle River, NJ: Prentice Hall.

Mathematics for Carpentry and the Construction Trades, Alfred P. Webster; Kathryn B. Judy. 2001. Upper Saddle River, NJ: Prentice Hall.

Mathematics for the Trades: A Guided Approach, Robert A. Carman; Hal Saunders. 2014. Pearson Learning.

Metric-conversion.org : Metric Conversion Charts and Calculators.

4.0.0 Section Review

1. The metric system can be referred to as a(n) _____.

   a. base 10 system
   b. base 100 system
   c. geometric progression
   d. open-ended system

2. If the hook end of a tape measure is slightly loose, what should you do?

   a. Using a small hammer, carefully pound the attaching rivets enough to tighten them and prevent movement.
   b. Nothing; the hook end should be slightly loose.
   c. Replace the entire tape measure.
   d. Return the tape measure to the manufacturer for repair.

5.0.0 METRIC AND IMPERIAL MEASUREMENT SYSTEMS

Objective

Identify and convert units of length, weight, volume, and temperature between Imperial and metric systems of measurement.

  a. Identify and convert units of length measurement between the Imperial and metric systems.
  b. Identify and convert units of weight measurement between the Imperial and metric systems.
  c. Identify and convert units of volume measurement between the Imperial and metric systems.
  d. Identify and convert units of temperature measurement between the Imperial and metric systems.

Trade Terms

Force: A push or pull on a surface. In this module, force is considered to be the weight of an object or fluid. This is a common approximation.

Mass: The quantity of matter present.

Unit: A definite standard of measure.

Volume: The amount of space contained in a given three-dimensional shape.

More than 95-percent of the world uses the metric system of measurement. Products are manufactured across the world and shipped to global destinations on a daily basis. The dimensions, weights, temperatures, and pressures related to these products may be provided using metric system values, Imperial system values, or both.

The use of the metric system in the United States is becoming more common. Many US manufacturers publish their documentation using values from both systems, regardless of the product destination. For some trades, it is essential to become familiar with both systems and understand how to convert from one to the other. *Figure 19* shows some metric system and Imperial system values that have become familiar to many people through the marketplace.

A great deal of work in science, engineering, and the trades is based on the exact measurement of physical quantities. A measurement is simply a comparison of a quantity to some definite standard measure of dimension called a unit. Whenever a physical quantity is described, the units of the standard to which the quantity was compared, such as a foot, a liter, or a pound, must be specified. A number alone is not enough to describe a physical quantity.

Once it is understood, the metric system is actually simpler to use than the Imperial system. This is because it is a decimal-based system in which unit prefixes are used to denote powers of ten. The Imperial system, on the other hand, requires the use of conversion factors from one Imperial unit to another. These factors must be memorized or determined from tables. For example, one mile is 5,280 feet, and 1 inch is ¹⁄₁₂ of a foot. In contrast, a centimeter is one-one hundredth of a meter and a kilometer is 1,000 meters. Any conversion done within the metric system of measurement involves some power of 10. An additional advantage of the metric system is that it is not necessary to add and subtract fractions for measurement purposes.

Metric system prefixes are listed in *Table 1*. From this table, it can be seen that the metric system is logically arranged, and that the prefix of the unit represents its order of magnitude.

The most common metric system prefixes are mega- (M), kilo- (k), centi- (c), milli- (m), and micro- (µ). Even though these prefixes may seem difficult to understand at first, many people are probably more familiar with them than they realize.

5.1.0 Units of Length Measurement

Sometimes you may need to change from one unit of measurement to another within the same system—for example, from inches to yards or from centimeters to meters. You may also need to convert length measurements from one system to the other. Before considering conversions between the two systems of measurement, we will examine the common units of length in each system.

5.1.1 Imperial System Units of Length

Table 2 shows the most common units of length in the Imperial system and their relationships. The inch is often broken down into fractions, as was presented in the section related to rulers and measuring tapes. As a general rule, the smallest

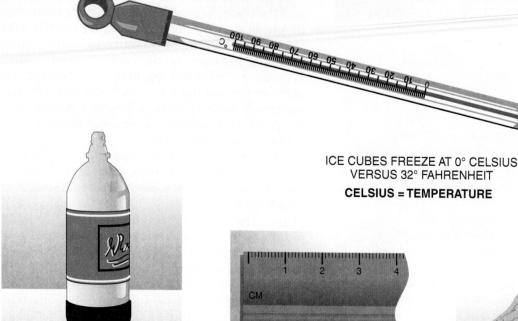

ICE CUBES FREEZE AT 0° CELSIUS
VERSUS 32° FAHRENHEIT

CELSIUS = TEMPERATURE

A 2-LITER BOTTLE OF SODA
INSTEAD OF A HALF-GALLON

LITERS = VOLUME

A METER STICK
INSTEAD OF A YARDSTICK

METERS = LENGTH

A GRAM OF GOLD
INSTEAD OF AN OUNCE

GRAMS = WEIGHT

00102-15_F19.EPS

Figure 19 Common Imperial and metric measured values.

fractional value the inch is broken into is ¹⁄₆₄. For work requiring an even higher level of precision, the inch can also be broken down to decimal values. The increments are usually thousandths (0.001 inches) or ten-thousandths (0.0001 inches) when extreme precision is required.

Each of the units shown here may be abbreviated several different ways:

- An inch or inches may be abbreviated as *in* or identified by the symbol ".
- A foot or feet may be abbreviated as *ft* or identified by the symbol '.
- A yard or yards may be abbreviated as *yd*.
- A mile or miles is rarely abbreviated, but the most likely abbreviation is *mi*.

Converting one unit to another involves multiplication or division by the proper value. This can sometimes be done in steps, but it is usually best to do it in one step when a calculator is available. For example, to change from inches to yards, you may first divide the number of inches by 12 (the number of inches in a foot) to find the number of feet. You would then divide that number by 3 (the number of feet in a yard) to find the number of yards. This method may be simpler when trying to make the conversion without pencil and paper or a calculator. A quicker way though, is to divide

the number of inches by 36. (*Figure 20*). This requires less work and eliminates a step, which reduces the opportunity for an error.

5.1.2 Metric System Units of Length

Table 3 shows some common units of length used in the metric system and their relationships. Notice again that all the units are related to each other by some power of ten.

These units may be abbreviated as follows:

- A millimeter is abbreviated as *mm*.
- The centimeter is abbreviated as *cm*.
- The meter is abbreviated as *m*.
- The kilometer is abbreviated as *km*.

Conversions within the metric system are easier. To make a conversion, you simply move the decimal point, because the system is based on multiples of 10. The keys to success are knowing which direction to move the decimal point, and how far. When converting from a smaller value to a larger one, the decimal point must move to the left. When converting from a larger value to a smaller one, the decimal point must move to the right. Tables and charts provide the information to determine how far to move the decimal point.

Table 1 Metric System Prefixes

Prefix		Unit			
micro- (µ)	¹⁄₁,₀₀₀,₀₀₀	0.000001	10^{-6}	One-millionth
milli- (m)	¹⁄₁,₀₀₀	0.001	10^{-3}	One-thousandth
centi- (c)	¹⁄₁₀₀	0.01	10^{-2}	One-hundredth
deci- (d)	¹⁄₁₀	0.1	10^{-1}	One-tenth
deka- (da)	10	10.0	10^{1}	Tens
hecto- (h)	100	100.0	10^{2}	Hundreds
kilo- (k)	1,000	1,000.0	10^{3}	Thousands
mega- (M)	1,000,000	1,000,000.0	10^{6}	Millions
giga- (G)	1,000,000,000	1,000,000,000.0	10^{9}	Billions

00102-15_T01.EPS

With the above in mind, determine how many meters there are in 72 centimeters (*Figure 21*). Since 1 centimeter = 0.01 meter, the decimal point must move two positions. Further, since a smaller value is being converted to a larger one, the decimal point must move to the left. Therefore, 72 centimeters becomes 0.72 meters.

It is important to note that moving decimal points provides the same result as multiplication, but without the math. For example, 72 centimeters can also be converted to meters by dividing by 100, since 1 meter is equal to 100 centimeters. The result remains the same at 0.72 meters.

Table 2 Common Imperial Units of Length

IMPERIAL LENGTH UNITS		
1 inch	=	¹⁄₁₂ᵗʰ of a foot; 0.0833 feet
1 foot	=	12 inches; ¹⁄₃ʳᵈ of a yard
1 yard	=	36 inches; 3 feet
1 mile	=	5,280 feet; 1,760 yards

00102-15_T02.EPS

5.1.3 Converting Length Units Between Systems

Converting measurements from the Imperial system to the metric system, and vice versa, is much like converting units within the Imperial system; multiplication or division by some factor is required. Since both systems are not based on powers of 10, moving the decimal point alone will not work. Many reference books, including dictionaries, contain charts or tables that show basic equivalents between the Imperial system and metric system units. Examples of some comparison charts are provided in the *Appendix*.

The successful conversion of units from one system to another depends heavily on using the proper conversion value. Even if the mathematical calculation is correct, using the wrong factor will result in an incorrect result. *Table 4* provides some common factors for converting length measurements from one system to another. Once the math is done and the new value has been determined, you can determine if the number can be rounded off to some degree.

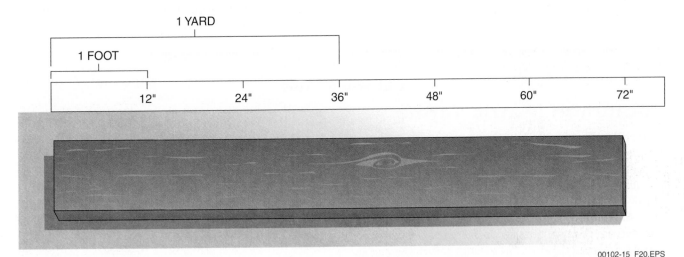

00102-15_F20.EPS

Figure 20 Converting inches to yards.

Table 3 Common Metric Units of Length

METRIC LENGTH UNITS		
1 kilometer	=	1,000 meters
1 meter	=	100 centimeters; 0.001 kilometers
1 centimeter	=	10 millimeters; 0.01 meters
1 millimeter	=	0.1 centimeters; 0.001 meters

00102-15_T03.EPS

For example, according to *Table 4*, 1 centimeter is equal to 0.3937 inches. To convert 13 centimeters to inches, multiply this factor times 13. The result is 5.1181 inches. For most applications, this number can be rounded to 5.12 inches, or even to 5.1 inches. The degree of rounding is dependent upon the level of accuracy required in the measurement. If, for instance, the conversion is related to determining the precise center of a steel plate, then a high level of accuracy may be in order. However, if the conversion is related to measuring how far away from a driveway to set a mailbox post, then a high level of accuracy is far less important.

5.1.4 Study Problems: Converting Measurements

Find the answers to the following conversion problems without using a calculator. Round your answers to the nearest hundredth.

1. 0.45 meter = _____ centimeters

2. 3 yards = _____ inches

3. 36 feet = _____ yards

4. 90 inches = _____ yards

5. 1 centimeter = _____ meters

6. 66 inches = _____ centimeters

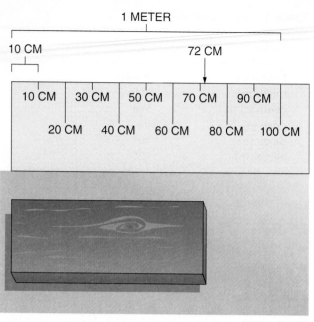

00102-15_F21.EPS

Figure 21 Converting centimeters to meters.

7. 47 feet = _____ meters

8. 54.5 centimeters = _____ feet

9. 19 yards = _____ meters

10. 4.7 meters = _____ inches

5.2.0 Units of Weight Measurement

This section will focus on units used to measure weight in both the Imperial and metric systems.

5.2.1 Imperial Units of Weight

Weight is actually the **force** an object exerts on the surface of the Earth due to its **mass** and the pull of the Earth's gravity. In the Imperial system, common units for weight include the ounce, the pound, and the ton. The relationship between these three units is shown in *Table 5*.

Did You Know?

Common Metric Terms

The most common units of length measure in the metric system are the millimeter, centimeter, meter, and kilometer. Unit names for measurements larger than a kilometer are not commonly used. For example, the term *hectometer* (100 meters) is not often used. In the Olympics, you hear about the 200-meter and the 400-meter races, not the 2-hectometer and the 4-hectometer races.

Measurements smaller than the millimeter (0.001 meter) are usually used by scientists and in precision machining operations. The micrometer (0.000001 meter), nanometer (0.000000001), and the picometer (0.000000000001 meter) are certainly not used in everyday measuring.

Table 4 Length Conversion Factors

Unit	Centimeter	Inch	Foot	Meter	Kilometer
1 millimeter	0.1	0.03937	0.003281	0.001	0.000001
1 centimeter	1	0.3937	0.3281	0.01	0.00001
1 inch	2.54	1	0.08333	0.0254	0.0000254
1 foot	30.48	12	1	0.3048	0.0003048
1 meter	100	39.37	3.281	1	0.001
1 kilometer	100,000	39,370	3,281	1,000	1

00102-15_T04.EPS

Table 5 Common Imperial Units of Weight

IMPERIAL WEIGHT UNITS		
1 ounce	=	⅟₁₆th of a pound; 0.0625 pounds
1 pound	=	16 ounces; 0.0005 tons
1 ton	=	2,000 pounds

00102-15_T05.EPS

These units of measure may be abbreviated as follows:

- An ounce or ounces may be abbreviated as *oz* or *ozs*.
- A pound or pounds may be abbreviated as *lb* or *lbs*.
- The ton may be abbreviated as *t*.

Conversions from one weight unit to another within the Imperial system are done in the same manner as length conversions. For example, to convert 134 ounces to pounds, divide by 16—the number of ounces in one pound. The result is 8.375 pounds. The number can be rounded depending upon the level of precision required for the result.

5.2.2 Metric Units of Weight

The most commonly used metric units of weight are the milligram, the gram, and the kilogram. As is the case with length, these metric units are related by powers of ten. Their relationships are shown in *Table 6*.

Use the following abbreviations with these units:

- A milligram is abbreviated as *mg*.
- The gram is abbreviated as *g*.
- The kilogram is abbreviated as *kg*.

Converting weight units within the metric system can be done through multiplication, or by simply moving the decimal point the proper

Around the World

Metric System as a Modern Standard

As scientific thought developed during the 1500s and later, scholars and scientists had trouble explaining their measurements to one another. The measuring standard varied among the different cultures. By the 1700s, scientists were debating how to establish a uniform system for measurement.

In the 1790s, French scientists created a standard length called a meter (based on the Latin word for measure). This became the basis for the metric system that is still in use throughout most of the world. Not until the 1970s did both the United States and Canada begin to truly acknowledge the metric system. This was partly driven by massive increases in global trade. There is still resistance to fully converting to the metric system in the United States, even though virtually every other country in the world uses it.

The original international standard meter—a solid bar measuring precisely a meter—was made of platinum-iridium and kept in the International Bureau of Weights and Measures near Paris, France. The bar was made from a platinum-iridium alloy because the material would not rust or change over time. That ensured the accuracy of the standard.

In more modern times, scientists have found that a natural standard of measure was more accurate than anything made out of metal. In 1983 the speed of light was calculated as precisely 299,792,458 meters per second. So the distance that light travels in 1/299,792,458 second is now the standard for a meter.

Table 6 Common Metric Units of Weight

METRIC WEIGHT UNITS		
1 metric ton	=	1,000 kilograms
1 kilogram	=	1,000 grams; 0.001 metric tons
1 gram	=	1,000 milligrams; 0.001 kilograms
1 milligram	=	0.001 gram; 0.000001 kilograms

00102-15_T06.EPS

number of places in the correct direction. For example, *Table 6* shows that 1 gram is equal to 1,000 milligrams. To convert 6,439 milligrams to grams, divide 6,439 grams by 1,000—the number of milligrams in a single gram. The alternative is to move the decimal point to the left three places. Moving three decimal places is correct since a milligram is 0.001 grams. Regardless of the method chosen, the result is the same—6.439 grams.

5.2.3 Converting Weight Units Between Systems

Table 7 shows the relationship of weight units between the Imperial and metric systems. The equivalent values in some cases have been carried a number of places beyond the decimal point. You may not always need to make calculations at this level of detail, depending upon the related task. The result of a conversion can also be rounded off to the level of precision required for the application.

Weight-related metric conversions might involve the weight of refrigerant for air-conditioning system charging or the amount of lubricant that must be added to an engine. There are many possible situations where weight conversion may be necessary. Imagine that you must set a new piece of equipment on a rooftop, but the specifications provide the weight in metric units as 1,200

Water and the Gram

In the metric system, the milliliter is a popular unit of liquid measure, equal to $\frac{1}{1000}$th (0.001) of a liter. One milliliter of pure water weighs exactly one gram. This is yet another example of the practicality of the metric system. Other liquids, of course, will have different weights, depending on their density.

kilograms (kg). How do you select the equipment needed to safely raise the unit onto the roof when the rigging equipment is rated in pounds only?

Table 7 shows that 1 kg = 2.205 pounds, so multiplying 1,200 kg by 2.205 yields 2,646 pounds. Therefore, any slings or lifting equipment used to raise the unit must have a rated capacity greater than 2,646 pounds.

When working with smaller weights, ounces or grams are often used as the unit of measure. One ounce is equal to 28.35 grams, so the gram is obviously the smaller unit of measure. Ounces can then be converted to grams by multiplying the number of ounces by 28.35. Grams are converted to ounces by dividing by 28.35.

5.2.4 Study Problems: Converting Weight Units

Convert these weights from imperial to metric weight units, or vice versa. Answers should be stated to the nearest hundredth.

1. 50 pounds = _____ kilograms

2. 50 kilograms = _____ pounds

3. 15.9 ounces = _____ grams

4. 94 grams = _____ ounces

Table 7 Imperial and Metric Weight Conversion Chart

IMPERIAL AND METRIC WEIGHT CONVERSION							
Unit	Metric Ton	Ton	Kilogram	Pound	Ounce	Gram	Milligram
1 metric ton	1	1.102	1,000	2,204.62	35,274	1,000,000	1,000,000,000
1 ton	0.9072	1	907.185	2,000	32,000	907,185	907,200,000
1 kilogram	0.001	0.0011	1	2.205	35.27	1,000	1,000,000
1 pound	0.000454	0.0005	0.4536	1	16	453.6	453,592
1 ounce	0.00002835	0.00003125	0.02835	0.0625	1	28.35	28,349.5
1 gram	0.000001	0.000001102	0.001	0.002205	0.03527	1	1,000
1 milligram	0.000000001	0.000000001102	0.000001	0.000002205	0.00003527	0.001	1

00102-15_T07.EPS

5.3.0 Units of Volume Measurement

In this section, units of volume measurement in the Imperial and metric systems will be examined. Note that volume here applies to three-dimensional objects such as cubes and cylinders. Liquid units of measure are not included in this discussion.

Volume is the amount of space occupied by a three-dimensional object such as a barrel or the piston of an engine. Three separate two-dimensional measurements are typically needed to properly calculate volume: length, width, and height. One of these dimensions, usually height, may be referred to as depth or thickness. Cubic units of measure describe the volume of different spaces. The same units used for length measurement—such as inches or centimeters—are used in units of volume. The result of multiplying the three measurements together to determine volume is indicated by adding the word *cubic* to the unit of measure. For example, if a cube measures 1" × 1" × 1", the volume is shown as 1 cubic inch.

A review of volume calculations for various three-dimensional shapes is provided in Section Six of this module.

5.3.1 Imperial Units of Volume

Table 8 shows the three most common units of volume measurement in the Imperial system. Note that these units are all based on familiar units of length measurement.

These units can be abbreviated as follows:

- The cubic inch may be abbreviated as *cu in* or in^3. It may also be seen as CI but this is not a standard abbreviation.
- The cubic foot may be abbreviated as *cu ft* or ft^3.
- The cubic yard may be abbreviated as *cu yd* or yd^3.

The units shown here can be converted from one to the other in the same manner as other units of measure. For example, to convert 864 cubic inches to cubic feet, multiply 864 by the conversion factor of 0.0005787. The result is 0.4999, or 0.5 cubic feet.

Table 8 Common Imperial Units of Volume

IMPERIAL VOLUME UNITS		
1 cubic inch	=	0.0005787 cubic feet
1 cubic foot	=	1,728 cubic inches; 0.037 cubic yards
1 cubic yard	=	27 cubic feet

5.3.2 Metric Units of Volume

The most common metric units used for volume are the cubic centimeter and the cubic meter. Their relationship to each other is shown in *Table 9*. Like all units of measure in the metric system, there are other volume units such as the cubic millimeter and cubic kilometer, all having a relationship to each other based on powers of ten. These two however, are the most commonly used in the trades. Therefore the conversion from one unit to another within the metric system is done in the same manner as other metric units of measure. The cubic meter represents a three-dimensional object that is 100 centimeters × 100 centimeters × 100 centimeters in size. Therefore, 1 cubic meter is equal to 1,000,000 cubic centimeters.

Cubic centimeters may be abbreviated as *cu cm* or cm^3. Cubic meters may be abbreviated as *cu m* or m^3.

5.3.3 Converting Volume Units Between Systems

Table 10 shows the relationship between Imperial and metric units of volume. Workers may need to convert volume from one system to the other when working with concrete for a foundation or a given amount of soil in site work. Only in rare cases would there be a need to convert volumes between very large units, such as cubic meters, and very small units such as cubic inches or centimeters. In most cases, the conversion will be to a unit in the other system that is relatively close to the same volume. Converting from cubic inches to cubic centimeters, or cubic yards to cubic meters, would be far more likely.

For this example, cubic centimeters will be converted to cubic inches. Assume that an enclosure must be large enough to accommodate a cube-shaped battery that the manufacturer says has a volume of 1,600 cubic centimeters. You know the area allowed for the battery is also cube-shaped and measures 112 cubic inches. Will the battery fit?

To convert cubic centimeters to cubic inches, multiply the number of cubic centimeters by the conversion factor. Since the conversion factor is 0.0610, multiply 1,600 by 0.0610. The result is 97.6 centimeters. So a cube-shaped object of 1,600 centimeters will fit into a space equal to 112 cubic inches, with room to spare.

Table 9 Common Metric Units of Volume

METRIC VOLUME UNITS		
1 cubic centimeter	=	0.000001 cubic meters
1 cubic meter	=	1,000,000 cubic centimeters

Table 10 Imperial and Metric Volume Conversion Chart

IMPERIAL AND METRIC VOLUME CONVERSION					
Unit	Cubic Meter	Cubic Yard	Cubic Foot	Cubic Inch	Cubic Centimeter
1 cubic meter	1	1.308	35.315	61,023.7	1,000,000
1 cubic yard	0.765	1	27	46,656	764,554.86
1 cubic foot	0.0283	0.037	1	1,728	28,316.85
1 cubic inch	0.000016387	0.000021433	0.0005787	1	16.387
1 cubic centimeter	0.000001	0.000001308	0.00003531	0.0610	1

00102-15_T10.EPS

Another example applies to foundation work. You have calculated that a concrete slab for a building requires 59 cubic yards of cement. However, the supplier works in cubic meters. How much cement would you need to order?

To convert cubic yards to cubic meters using the factor from *Table 10*, multiply the number of cubic yards (59) by the conversion factor of 0.765. The result is 45.1 cubic meters. Although slightly more cement may be ordered than necessary to eliminate any possibility of shortage, you are now working with the same volume units as the supplier.

5.3.4 Study Problems: Converting Volume Units

Convert these volumes from the Imperial system to the metric system, or vice versa. Answers should be rounded to the nearest tenth.

1. 11,600 cubic inches = _____ cubic feet

2. 1.9 cubic meters = _____ cubic centimeters

3. 512 cubic meters = _____ cubic yards

4. 7 cubic feet = _____ cubic centimeters

5. 0.2 cubic meters = _____ cubic feet

5.4.0 Temperature Units

Temperature conversions are quite often necessary. Temperature is important to many crafts, and in many different ways. However, temperature is somewhat different from other units of measure, in that most countries of the world are likely to use several temperature scales for different applications.

Temperature can be defined as the intensity level of heat. Temperature is measured in degrees on a temperature scale. In order to establish the scale, a substance is needed that always responds to reproducible conditions in the same manner.

The substance used for this purpose is water. The point at which water freezes at atmospheric pressure is one reproducible condition, and the point at which water boils at standard atmospheric pressure is another.

The four temperature scales commonly used today are the Fahrenheit scale, Celsius scale, Rankine scale, and Kelvin scale (*Figure 22*). Temperature is most often measured in degrees Fahrenheit or degrees Celsius. On the Fahrenheit scale, the freezing temperature of water is 32°F and the boiling temperature is 212°F. On the Celsius scale, the freezing temperature of water is 0°C and the boiling temperature is 100°C. The temperatures at which these fixed points occur were established by the inventors of the scales. The abbreviations for each unit are shown on *Figure 22*. Each is a single capital letter.

The Rankine scale and the Kelvin scale are based on the theory that at some extremely low temperature, no molecular activity occurs. The temperature at which this condition occurs is

Around the World
Coldest Recorded Temperature

In August, 2014, physicists at Yale University succeeded in chilling molecules to the coldest temperature ever recorded. Using a process called magneto-optical trapping, they were able to reduce the temperature of strontium fluoride molecules to a temperature of −459.67°F. The final calculations indicated that the temperature achieved was less than 0.003°F above absolute zero.

The coldest natural temperature recorded at ground level is −128.6°F (−89.2°C) at a Soviet station in Antarctica in 1983.

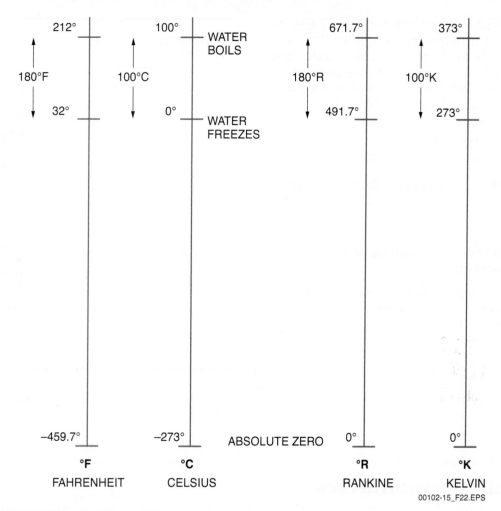

Figure 22 Comparison of temperature scales.

called absolute zero, the lowest temperature possible. Both the Rankine and Kelvin scales have their zero degree points at absolute zero. On the Rankine scale, the freezing point of water is 491.7°R and the boiling point is 671.7°R—a 180° range. This is the same number of degrees between the freezing and boiling points of water found on the Fahrenheit scale. Therefore, the increments on the Rankine scale correspond in size to the increments on the Fahrenheit scale; for this reason, the Rankine scale is sometimes called the absolute Fahrenheit scale.

On the Kelvin scale, the freezing point of water is 273°K and the boiling point is 373°K. The range between the freezing and boiling points of water is 100°. This shows the relationship between the Kelvin and Celsius scales. The increments on the Kelvin scale correspond to the increments on the Celsius scale. For this reason, the Kelvin scale is sometimes called the absolute Celsius scale. Both the Kelvin and Celsius scales are part of the metric system of measurement.

The scales of primary importance in most crafts are the Fahrenheit scale and the Celsius scale. The Rankine scale and the Kelvin scale are better suited for scientific applications.

Charts and tables are readily available from a number of sources to make the conversion of temperatures easy and fast. However, it is also beneficial to understand the math behind them. All such charts and tables are based on the mathematical relationship that follows.

On the Fahrenheit scale, there are 180 degrees between the freezing temperature and boiling temperature of water. On the Celsius scale, there are 100 degrees between the freezing and boiling temperatures of water. The relationship between the two scales can be expressed as follows:

$$\frac{\text{Fahrenheit range (freezing to boiling)}}{\text{Celsius range (freezing to boiling)}} = \frac{180°}{100°} = \frac{9}{5}$$

Therefore, one degree Fahrenheit is ⁵⁄₉ths of one degree Celsius and conversely, one degree Celsius is ⁹⁄₅ths of one degree Fahrenheit. Thus, to convert a Fahrenheit temperature to a Celsius temperature, it is necessary to subtract 32° (since 32° corresponds to 0° on the Celsius scale), and then multiply by ⁵⁄₉. To convert a Celsius temperature to a Fahrenheit temperature, it is necessary to multiply by ⁹⁄₅, and then add 32°C. This can be written mathematically as follows:

$$°C = ⅝ \, (°F - 32°)$$
$$°F = (⅖ × °C) + 32°$$

Practice these calculations to become more comfortable with using them. *Figure 23* shows two examples.

5.4.1 Study Problems: Converting Temperatures

Convert these temperatures from Fahrenheit to Celsius, or vice versa. Answers should be stated to the nearest tenth of a degree.

1. 180°F = __82.2__ °C
2. 66°F = __18.9__ °C
3. −26°C = __−14.8__ °F
4. 71°C = __159.8__ °F

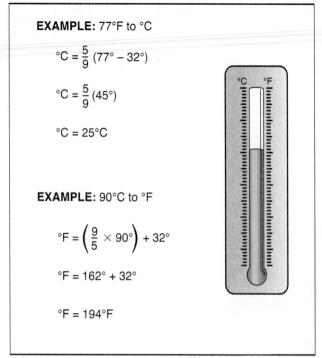

EXAMPLE: 77°F to °C

$$°C = \frac{5}{9} \, (77° - 32°)$$

$$°C = \frac{5}{9} \, (45°)$$

$$°C = 25°C$$

EXAMPLE: 90°C to °F

$$°F = \left(\frac{9}{5} × 90°\right) + 32°$$

$$°F = 162° + 32°$$

$$°F = 194°F$$

00102-15_F23.EPS

Figure 23 Sample temperature conversions.

Digital Thermometers

Thermometers with digital readouts can provide temperature readings in both Celsius and Fahrenheit. The one on the left is a temperature probe with a digital readout. The one on the right is an infrared thermometer that measures temperature based on an infrared signature from the surface. Many analog thermometers are also calibrated for both scales.

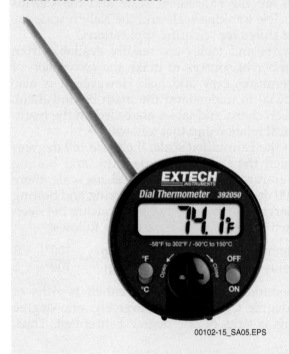

00102-15_SA05.EPS

00102-15_SA06.EPS

Additional Resources

Applied Construction Math: A Novel Approach, NCCER. 2006. Upper Saddle River, NJ: Prentice Hall.

Mathematics for Carpentry and the Construction Trades, Alfred P. Webster; Kathryn B. Judy. 2001. Upper Saddle River, NJ: Prentice Hall.

Mathematics for the Trades: A Guided Approach, Robert A. Carman; Hal Saunders. 2014. Pearson Learning.

Metric-conversion.org : Metric Conversion Charts and Calculators.

5.0.0 Section Review

1. When converting from a larger metric value to a smaller one, the decimal point must move _____.

   a. to the left
   b. to the right
   c. all the way to the left end of the number
   d. all the way to the right end of the number

2. Weight is the force an object exerts on the surface of the Earth due to its mass and _____.

   a. density
   b. volume
   c. the current temperature
   d. the pull of the Earth's gravity

3. The cubic meter represents a three-dimensional object that measures _____.

   a. 10 cm × 10 cm × 10 cm
   b. 100 mm × 100 mm × 100 mm
   c. 100 cm × 100 cm × 100 cm
   d. 10 m × 10 m × 10 m

4. On the Celsius temperature scale, the difference between the freezing and boiling points of water is _____.

   a. 32 degrees
   b. 100 degrees
   c. 180 degrees
   d. 212 degrees

6.0.0 INTRODUCTION TO GEOMETRY

Objective

Identify basic angles and geometric shapes and explain how to calculate their area and volume.

a. Identify various types of angles.
b. Identify basic geometric shapes and their characteristics.
c. Demonstrate the ability to calculate the area of two-dimensional shapes.
d. Demonstrate the ability to calculate the volume of three-dimensional shapes.

Trade Terms

Acute angle: Any angle between 0 degrees and 90 degrees.

Adjacent angles: Angles that have the same vertex and one side in common.

Angle: The shape made by two straight lines coming together at a point. The space between those two lines is measured in degrees.

Area: The surface or amount of space occupied by a two-dimensional object such as a rectangle, circle, or square.

Base: As it relates to triangles, the base is the line forming the bottom of the triangle.

Bisect: To divide into two parts that are often equal. When an angle is bisected for example, the two resulting angles are equal.

Circle: A closed curved line around a central point. A circle measures 360 degrees.

Circumference: The distance around the curved line that forms the circle.

Cube: A three-dimensional square, with the measurements in all the three dimensions being equal.

Degree: A unit of measurement for angles. For example, a right angle is 90 degrees, an acute angle is between 0 and 90 degrees, and an obtuse angle is between 90 and 180 degrees.

Diagonal: Line drawn from one corner of a rectangle or square to the farthest opposite corner.

Diameter: The length of a straight line that crosses from one side of a circle, through the center point, to a point on the opposite side. The diameter is the longest straight line you can draw inside a circle.

Equilateral triangle: A triangle that has three equal sides and three equal angles.

Formula: A mathematical process used to solve a problem. For example, the formula for finding the area of a rectangle is Side A times Side B = Area, or $A \times B = Area$.

Isosceles triangle: A triangle that has two equal sides and two equal angles.

Obtuse angle: Any angle between 90 degrees and 180 degrees.

Opposite angles: Two angles that are formed by two straight lines crossing. They are always equal.

Perimeter: The distance around the outside of a closed shape, such as a rectangle, circle, square, or any irregular shape.

Pi: A mathematical value of approximately 3.14 (or $^{22}\!/_7$) used to determine the area and circumference of circles. It is sometimes symbolized by π.

Plane geometry: The mathematical study of two-dimensional (flat) shapes.

Radius: The distance from a center point of a circle to any point on the curved line, or half the width (diameter) of a circle.

Rectangle: A four-sided shape with four 90-degree angles. Opposite sides of a rectangle are always parallel and the same length. Adjacent sides are perpendicular and are not equal in length.

Right angle: An angle that measures 90 degrees. The two lines that form a right angle are perpendicular to each other. This is the angle used most in the trades.

Right triangle: A triangle that includes one 90-degree angle.

Scalene triangle: A triangle with sides of unequal lengths.

Solid geometry: The mathematical study of three-dimensional shapes.

Square: (1) A special type of rectangle with four equal sides and four 90-degree angles. (2) The product of a number multiplied by itself. For example, 25 is the square of 5; 16 is the square of 4.

Straight angle: A 180-degree angle or flat line.

Triangle: A closed shape that has three sides and three angles.

Vertex: A point at which two or more lines or curves come together.

Geometry might sound complicated, but it is really made up of common things you are already familiar with—**circles**, **triangles**, **squares**, and **rectangles**, for example. The construction industry is based on a world of measurements and shapes. It is important to recognize basic shapes and understand them mathematically in order to make use of geometry in your chosen craft.

The first portion of this section focuses on **plane geometry**. In plane geometry, the shapes are two-dimensional. These shapes have length and width only. In **solid geometry**, also known as 3D geometry, shapes have three dimensions, including height.

6.1.0 Angles

An **angle** is an important term in the construction trades. It is used by all building trades to describe the shape made by two straight lines that meet at a point called the **vertex**. Angles are measured in **degrees** to describe the relationship between the two lines. To measure angles, a tool called a protractor is used. The following are the typical angles (*Figure 24*) you will measure in construction:

- **Acute angle** – An angle that measures between 0 and 90 degrees. The most common acute angles are 30, 45, and 60 degrees.
- **Right angle** – An angle that measures 90 degrees. The two lines that form the right angle are perpendicular to each other. Imagine the shape of a capital letter L. This is a right angle, because the sides of the L are perpendicular to one another. This is the angle used most often in the construction trades. A right angle is indicated in plans or drawings with a square symbol at the vertex, as shown in *Figure 24* .
- **Obtuse angle** – An angle that measures between 90 and 180 degrees.
- **Straight angle** – A straight angle measures precisely 180 degrees (a flat line).
- **Adjacent angles** – These angles have the same vertex and one side in common. Adjacent refers to objects that are next to each other.
- **Opposite angles** – Angles formed by two straight lines that cross are opposite. Opposite angles are always equal.

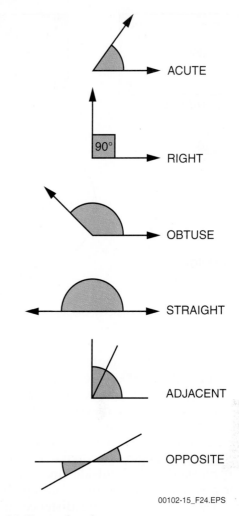

00102-15_F24.EPS

Figure 24 Types of angles.

6.2.0 Shapes

Common shapes that are essential to your work in the trades include rectangles, squares, triangles, and circles (*Figure 25*).

6.2.1 Rectangle

A rectangle is a four-sided shape with four 90-degree angles. The sum of all four angles in any rectangle is 360 degrees. A rectangle has two pairs of equal sides that are parallel to each other. The **diagonals** of a rectangle are always equal: diagonals are lines connecting opposite corners. If you cut a rectangle on the diagonal, you will have two identical **right triangles**, as shown in *Figure 26*. All right triangles have one 90-degree angle.

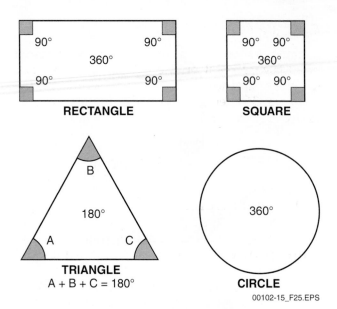

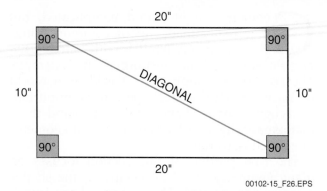

Figure 26 Cutting a rectangle on the diagonal produces two right triangles.

Figure 25 Common plane geometry shapes.

6.2.2 Square

A square is a type of rectangle with four sides of equal length and four 90-degree angles. The sum of all four angles in all squares is 360 degrees. If you cut a square on the diagonal between opposite corners, you will have two right triangles. Each right triangle will have two 45-degree angles and one 90-degree angle, as shown in *Figure 27*.

When measuring the outside lines of a rectangle or a square, you are determining the **perimeter**. The perimeter is the distance around any two-dimensional figure, but it also applies to some three-dimensional figures. You may need to calculate the perimeter of a shape to measure, mark, and cut the right amount of material. For example, if you need to install molding along all four walls of a room, the perimeter measurement must be known to purchase the correct amount of material. If the room is 14 feet by 12 feet, you

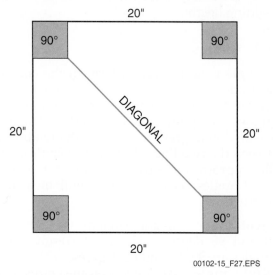

Figure 27 Cutting a square on the diagonal produces two right triangles.

would calculate: 14 + 12 + 14 + 12 = 52 feet of molding. Another way to calculate it would be (2 × 14 feet) + (2 × 12 feet) = 52 feet.

Since a square has four sides of equal length, any side can be measured and multiplied by four to determine the perimeter.

Did You Know?

Rope Stretchers

The word *geometry* comes from two Greek words: *geos*, meaning "land", and *metrein*, meaning "to measure."

In ancient Egypt, most farms were located beside the Nile River. Every year during the flood season, the Nile overflowed its banks and deposited mineral-rich silt over the farmland. But the floodwaters destroyed the markers used to establish property lines between farms. When this happened, the farms had to be measured again. Men called rope stretchers re-marked the property lines each year.

The rope stretchers calculated distances and directions using ropes that had equally spaced knots tied along the length of the rope. Stretching the ropes to measure distances on level land was easy. In many cases, however, the rope stretchers had to measure property lines from one side of a hill to the opposite side or across a pond.

The hills and ponds made this measurement difficult. To adjust for the uneven land or ponds that stood in the way, rope stretchers determined new ways of measuring. Such discoveries became the foundation of geometry.

6.2.3 Triangle

A triangle is a closed shape that has three sides and three angles. Although the angles in a triangle can vary, the sum of the three angles is always 180 degrees (*Figure 28*). The following are different types of triangles you will use in construction:

- *Right triangle* – A right triangle has one 90-degree angle.
- **Equilateral triangle** – An equilateral triangle has three equal angles and three equal sides.
- **Isosceles triangle** – An isosceles triangle has two equal angles and two sides equal in length. A line that **bisects** (runs from the center of the **base** of the triangle to the highest point) an isosceles triangle creates two adjacent right angles.
- **Scalene triangle** – A scalene triangle has three sides of unequal lengths.

6.2.4 Circle

A circle is a closed curved line around a center point. Every point on the curved line is exactly the same distance from the center point. A circle measures 360 degrees. The following terms apply to circles (*Figure 29*):

- **Circumference** – The circumference of a circle is the length of the closed curved line that forms the circle. The **formula** for finding circumference is **pi** (3.14) × **diameter**.
- *Diameter* – The diameter of a circle is the length of a straight line that crosses from one side of the circle through the center point to a point on the opposite side. The diameter is the longest straight line you can draw inside a circle.
- *Pi* or π – pi is a mathematical constant value of approximately 3.14 (or $^{22}/_7$) used to determine the area and circumference of circles.
- **Radius** – The radius of a circle is the length of a straight line from the center point of the circle to any point on the closed curved line that forms the circle. It is equal to half the diameter.

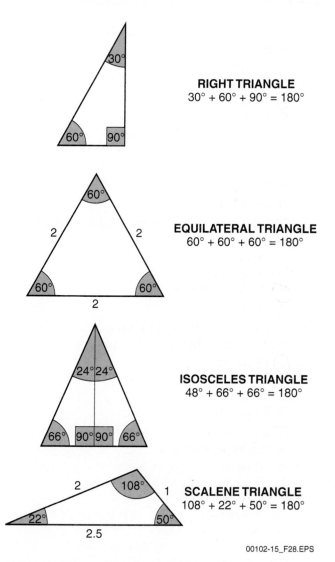

RIGHT TRIANGLE
30° + 60° + 90° = 180°

EQUILATERAL TRIANGLE
60° + 60° + 60° = 180°

ISOSCELES TRIANGLE
48° + 66° + 66° = 180°

SCALENE TRIANGLE
108° + 22° + 50° = 180°

00102-15_F28.EPS

Figure 28 The sum of a triangle's three angles always equals 180 degrees.

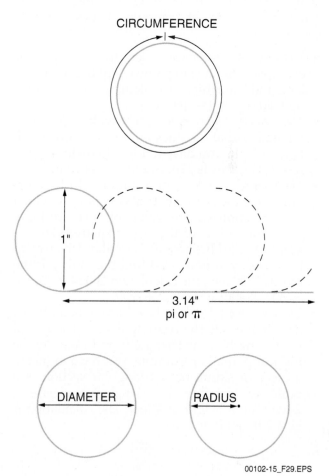

00102-15_F29.EPS

Figure 29 Measurements that apply to circles.

6.3.0 Calculating the Area of Shapes

Area is the measurement of the surface of a two-dimensional object. For example, you must calculate the area of a shape, such as a floor or a wall, to order the proper amount of material, such as carpeting or paint. Squared units of measure describe the amount of surface area. Area measurements in the Imperial system are typically in square inches (sq in or in^2), square feet (sq ft or ft^2), and square yards (sq yd or yd^2). Area measurements in the metric system are often in square centimeters (sq cm or cm^2) and square meters (sq m or m^2). When larger areas are involved, such as land, the units may be in square miles or square kilometers.

- 1 square inch = 1 inch $\times$ 1 inch = 1 inch^2
- 1 square foot = 1 foot $\times$ 1 foot = 1 foot^2
- 1 square yard = 1 yard $\times$ 1 yard = 1 yard^2
- 1 square centimeter = 1 cm $\times$ 1 cm = 1 cm^2
- 1 square meter = 1 m $\times$ 1 m = 1 m^2

You must be able to calculate the area of basic shapes. Mathematical formulas make this very easy to do. In the *Appendix*, you will find formulas for calculating the areas of various shapes. You need to become familiar with these formulas at this stage in your training. The formulas for calculating the most common shapes are presented here:

- *The area of a rectangle* = length $\times$ width. For example, you have to paint a wall that is 20 feet long and 8 feet high. To calculate the area, multiply 20 ft $\times$ 8 ft = 160 sq ft.
- *The area of a square* = length $\times$ width. However, remember that all sides of a square are equal. For example, you have to tile a 12-meter square room. The area is 12m $\times$ 12m = 144 sq m.
- *The area of a circle* = pi $\times$ radius^2. In this formula, use the mathematical constant pi, which has an approximate value of 3.14. Multiply pi by the radius of the circle squared. For example, to find the area of a circular driveway to be sealed, you must first find the radius. If the radius is 20 feet, the calculation is 3.14 $\times$ (20 ft)^2 or 3.14 $\times$ 400 sq ft = 1,256 sq ft.
- *The area of a triangle* = ½ $\times$ base $\times$ height. The base is the side the triangle sits on. The height is the length of the triangle from its base to the highest point. For example, you have to install a piece of siding on a triangular section of a building. You find the triangle has a base of 2 feet and a height of 4 feet. The calculation is ½ $\times$ 2 ft $\times$ 4 ft = 4 sq ft.

Diagonals

Diagonals have a number of uses. If you have to make sure that a surface is a true rectangle with 90-degree corners, you can measure the diagonals to find out. For example, before applying a piece of sheathing, you must make sure it is a true rectangle. Using your tape measure, find the length of the sheathing from one corner to the opposite corner. Now, find the length of the other two opposing corners. Do the diagonals match? If so, the piece of sheathing is a true rectangle. If not, the piece is not a true rectangle and will cause problems when you install it.

6.3.1 *Study Problems: Calculating Area*

1. The area of a rectangle that is 8 feet long and 4 feet wide is _____.
    a. 12 sq ft
    b. 22 sq ft
    c. 32 sq ft
    d. 36 sq ft

2. The area of a 16 cm square is _____.
    a. 256 sq cm
    b. 265 sq cm
    c. 276 sq cm
    d. 278 sq cm

3. The area of a circle with a 14-foot diameter is _____.
    a. 15.44 sq ft
    b. 43.96 sq ft
    c. 153.86 sq ft
    d. 196 sq ft

4. The area of a triangle with a base of 4 centimeters and a height of 6 cm is _____.
    a. 12 sq cm
    b. 24 sq cm
    c. 32 sq cm
    d. 36 sq cm

5. The area of a rectangle that is 14 meters long and 5 meters wide is _____.
    a. 60 sq m
    b. 65 sq m
    c. 70 sq m
    d. 75 sq m

6.4.0 Volume of Three-Dimensional Shapes

Volume is the amount of space occupied in three dimensions. To calculate volume, you must use three measurements: length, width, and height. One dimension, usually height, may be referred to as depth or thickness. Cubic units of measure describe the volume of different spaces. Measurements in the Imperial system are in cubic inches (cu in or in^3), cubic feet (cu ft or ft^3), and cubic yards (cu yd or yd^3). Metric measurements include cubic centimeters (cu cm or cm^3) and cubic meters (cu m or m^3).

- 1 cubic inch = 1 inch × 1 inch × 1 inch = 1 inch^3
- 1 cubic foot = 1 foot × 1 foot × 1 foot = 1 foot^3
- 1 cubic yard = 1 yard × 1 yard × 1 yard = 1 yard^3
- 1 cubic centimeter = 1 centimeter × 1 centimeter × 1 centimeter = 1 cm^3
- 1 cubic meter = 1 meter × 1 meter × 1 meter = 1 m^3

One very important fact to keep in mind is that all three dimensions used in calculating volume must be in the same units. For example, a concrete slab may have dimensions of 4 feet × 6 feet × 6 inches. If these three values are multiplied together to calculate volume, there will be a major error in the result. The calculation would result in 144 ft^3. The final dimension in inches must be converted to feet first, or the other two must be converted to inches. In this case, 6 inches is equal to 0.5 feet. The correct result then would be a volume of 12 ft^3. As you can see, the first answer is incorrect by a factor of 12 due to mismatched units.

Note that there are many opinions about the proper name for a rectangle in its three-dimensional form. The concrete slab just discussed would be such a figure. In a geometry class, you may hear it referred to as a rectangular parallel pipe, rectangular prism, or a three-dimensional orthotope. To avoid confusion over such rarely-heard terms, we will simply refer to them here as three-dimensional rectangles.

You must be able to calculate the volume of common shapes in many different crafts. The following sections provide the mathematical formulas that make this easy to do. In the *Appendix*, you will find a list of the formulas for calculating the volumes of a number of shapes. It is best to become familiar with these formulas at this stage in your training. Remember to ensure that all dimensions are in matching units before multiplying.

6.4.1 Three-Dimensional Rectangles

The volume of a three-dimensional rectangle is calculated by multiplying length × width × depth. For example, you might need to order the right amount of concrete (which is delivered by the cubic yard) for a slab that is 20 feet long and 8 feet wide and 4 inches thick (*Figure 30*). You must know the total volume of the slab. To calculate this, perform the following steps:

Geometry Practical Application

Before a metal building can be built, you must first pour a slab and install anchor bolts. The anchor bolts will be used to attach the structure to the slab. Because cement comes in cubic yards (yd^3), a volume measure, you must determine how much cement you need to pour the slab. To determine volume, multiply the length times the width times the depth, or thickness, of the slab. If the area of the slab is already known, simply multiply the area times the depth.

Solution

Volume = area × depth
Area of slab = 50' × 40' = 2,000 square feet (ft^2)
Slab depth = 6"
(convert inches to feet:
6" = $^6/_{12}$ = ½ foot)
Volume = 2,000 ft^2 × ½ foot = 1,000 ft^3
Volume of slab = 1,000 ft^3

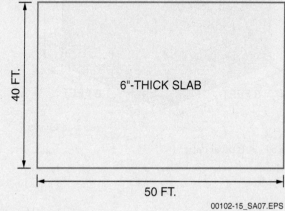

40 FT.

6"-THICK SLAB

50 FT.

00102-15_SA07.EPS

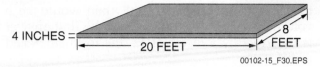

00102-15_F30.EPS

Figure 30 Volume of a proposed concrete slab.

Step 1 Convert inches to feet.

20 ft × 8 ft × (4 in ÷ 12) =

Step 2 Multiply length × width × depth.

20 ft × 8 ft × 0.33 ft = 52.8 cu ft

Step 3 Convert cubic feet to cubic yards.

52.8 cu ft ÷ 27 (cu ft per cu yd)
= 1.96 cu yd of concrete

6.4.2 Cubes

A **cube** is a three-dimensional square. The volume of a cube is basically calculated in the same way as three-dimensional rectangles: length × width × depth. However, all sides of a cube are equal in length (*Figure 31*). To find the volume of a cube, you can cube (multiply the number by itself three times) one of the dimensions. Perform the following steps to determine the volume of a cube:

Step 1 Determine the volume of an 8-foot cube.

8 ft × 8 ft × 8 ft = 512 cu ft

Step 2 If necessary for the task, convert cubic feet to cubic yards.

512 cu ft ÷ 27 = 18.96 cu yd of concrete

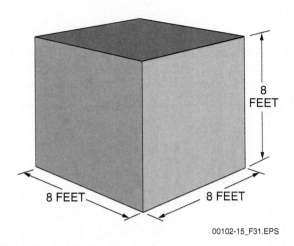

00102-15_F31.EPS

Figure 31 A typical cube.

6.4.3 Cylinders

The volume of a cylinder is calculated using the following formula: pi × radius^2 × height. You may also see this written as area = $\pi r^2 h$, where h represents the height. This is the same as the formula for finding the area of a circle with the added dimension of height. For example, you must find the volume of a cylinder that is 22 feet in diameter and 10 feet high (*Figure 32*):

Step 1 First, calculate the area of the circle using πr^2. Since the diameter is 22 feet, the radius will be 11 feet.

Area of the circle = 3.14 × 11^2 = 379.94 sq ft

Step 2 Then calculate the volume (area × height).

379.94 sq ft × 10 ft = 3,799.4 cu ft

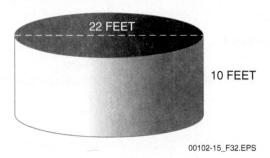

00102-15_F32.EPS

Figure 32 A cylinder.

NCCER – *Core Curriculum* 00102-15

6.4.4 Triangular Prisms

The volume of a triangular prism is calculated using the following formula: 0.5 × base × height × depth (thickness). Note that this is not a pyramid, but a triangle that has depth and is consistent in dimension, such as the one shown in *Figure 33*. In this example, you must fill a triangular shape that has a base of 6 cm, a height of 12 cm, and a depth of 2 cm:

Step 1   Calculate the area of the flat triangle first:

0.5 × 6 × 12 = 36 sq cm area

Step 2   Then calculate the volume of the prism, by adding the factor of depth:

36 sq cm × 2 cm = 72 cu cm

6.4.5 Study Problems: Calculating Volume

1. The volume of a rectangular shape 5 feet high, 6 feet thick, and 13 feet long is _____.
   a. 24 cu ft
   b. 43 ft
   c. 95 ft
   d. 390 cu ft

2. The volume of a 3 cm cube is _____.
   a. 6 cu cm
   b. 9 cu cm
   c. 12 cu cm
   d. 27 cu cm

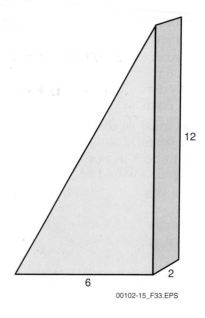

00102-15_F33.EPS

Figure 33   Volume of a triangular prism.

3. The volume of a triangular prism that has a 6-inch base, a 2-inch height, and a 4-inch depth is _____.
   a. 12 sq in
   b. 24 cu in
   c. 36 cu in
   d. 48 sq in

Did You Know?

3-4-5 Rule

The 3-4-5 rule is based on the Pythagorean theorem, and it has been used in building construction for centuries. This simple method for laying out or checking right angles requires only the use of a tape measure. The numbers 3-4-5 represent dimensions in feet that describe the sides of a right triangle. Right triangles that are multiples of the 3-4-5 triangle are commonly used, such as 9-12-15, 12-16-20, and 15-20-25. The specific multiple used is determined by the relative distances involved in the job being laid out or checked.

Refer to the figure for an example of the 3-4-5 theory using the multiples of 15-20-25. In order to square, or check, a corner, first measure and mark 15'-0" down the line in one direction, then measure and mark 20'-0" down the line in the other direction. The distance measured between the 15'-0" and 20'-0" points must be exactly 25'-0" to ensure that the angle is a perfect right (90 degree) angle.

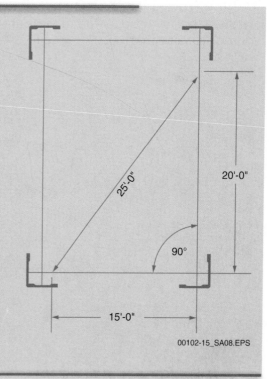

00102-15_SA08.EPS

Handwritten at top: πr^2

Handwritten at top right: Value: $28.26 \times 6 = 16.96$

4. The volume of a cylinder that is 6 meters in diameter and 60 centimeters high is _____.

   a. 16.96 cu m *(circled)*
   b. 18.23 cu m
   c. 1130.4 cu m
   d. 6782.4 cu m

   Handwritten: 60cm = 0.6m
   6 m Dia = 3r
   Area: 3.14 × 3 × 3 = 28.26

5. If gravel must be distributed across an area that measures 17 feet square and the gravel layer is to be 6 inches thick, the volume of gravel needed would be _____.

   a. 3.77 cu yds
   b. 5.35 cu yds
   c. 102 cu yds
   d. 144.5 cu yds *(circled)*

   Handwritten: 17 × 17 = 289
   6 inch = 0.5 feet
   289 × 0.5 = 144.5

6.4.6 Practical Applications Using Volume

The following examples illustrate some of the practical applications a tradesperson may encounter on the job site that requires an understanding of geometry-based principals to solve. Use the information provided to solve each problem. Be sure to show all of your work.

1. Use *Figure 34* to calculate the volume of a section of round sheet metal pipe. The volume of the section shown = _____ cubic inches.

2. To pour the concrete sidewalk shown in *Figure 35*, approximately how many cubic feet of topsoil will you need to remove for the 4"-thick sidewalk if the owner wants the finish surface of the sidewalk to be level with the adjacent topsoil? Round your answer to the nearest cubic foot. _____ cu ft

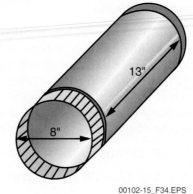

00102-15_F34.EPS

Figure 34 Finding the volume of a pipe section.

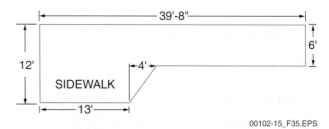

00102-15_F35.EPS

Figure 35 Removing topsoil for a sidewalk.

Additional Resources

Applied Construction Math: A Novel Approach, NCCER. 2006. Upper Saddle River, NJ: Prentice Hall.

Mathematics for Carpentry and the Construction Trades, Alfred P. Webster; Kathryn B. Judy. 2001. Upper Saddle River, NJ: Prentice Hall.

Mathematics for the Trades: A Guided Approach, Robert A. Carman; Hal Saunders. 2014. Pearson Learning.

Metric-conversion.org : Metric Conversion Charts and Calculators.

6.0.0 Section Review

1. An angle of 66 degrees is considered a(n) _____.
   a. right angle
   b. straight angle
   c. obtuse angle
   d. acute angle

2. What is the difference between circumference and perimeter?
   a. The circumference relates only to shapes with angles, while the perimeter relates only to circles.
   b. The perimeter relates to all shapes, while the circumference relates only to circles.
   c. The perimeter represents one-half of the circumference.
   d. The circumference is never larger than the perimeter.

3. Calculate the area of a rectangle that is 27.3 meters × 9.3 meters.
   a. 245.7 m^3
   b. 251.1 m^2
   c. 253.89 m^2
   d. 253.89 m^3

4. Calculate the volume of a shipping container that is 26' long × 13'8" wide × 3' deep.
   a. 355.32 ft^3
   b. 355.32 ft^2
   c. $1,065.48 \text{ ft}^3$
   d. $1,065.48 \text{ ft}^2$

SUMMARY

Mathematics is not just something you need to learn to survive your days in school. The construction environment requires math every day to get a project done. Whether you are cutting stock, charging an air conditioning unit with refrigerant, or installing electrical systems, you will need math skills on the job. Basic operations such as addition, subtraction, multiplication, and division are the keys to completing these tasks. However, more complex mathematical operations will be necessary for a number of tasks, such as planning a piping offset. Being competent and comfortable with math increases your value to an employer, which helps ensure your job security.

6 inch -

1. The number matching the words "two thousand, six hundred eighty-nine" is _____.
   a. 2,286
   b. 2,689
   c. 6,289
   d. 20,689

Solve Questions 2 through 5 without using a calculator.

2. A bricklayer lays 649 bricks the first day, 632 the second day, and 478 the third day. During the three-day period, the bricklayer laid a total of _____.
   a. 1,759
   b. 1,760
   c. 1,769
   d. 1,770

3. A total of 1,478 feet of cable was supplied for a job. Only 489 feet were installed. How many feet of cable remain?
   a. 978
   b. 980
   c. 989
   d. 1,099

4. A worker has been asked to deliver 15 scaffolds to each of 26 different sites. The worker will deliver a total of _____.
   a. 120
   b. 240
   c. 375
   d. 390

5. Your company has 400 rolls of insulation that must be equally distributed to 5 different job sites. How many rolls of insulation will need to go to each site?
   a. 8
   b. 20
   c. 80
   d. 208

6. Which of the following fractions is an equivalent fraction for $\frac{3}{8}$?
   a. $\frac{3}{64}$
   b. $\frac{6}{64}$
   c. $\frac{24}{64}$
   d. $\frac{36}{64}$

7. The lowest common denominator for the fractions $\frac{5}{64}$ and $\frac{8}{32}$ is _____.
   a. 8
   b. 16
   c. 24
   d. 32

Perform the calculations for Questions 8 and 9. Reduce the answers to their lowest terms.

8. $\frac{3}{8} + \frac{9}{16} =$ _____
   a. $\frac{3}{4}$
   b. $\frac{12}{16}$
   c. $\frac{7}{8}$
   d. $\frac{15}{16}$

9. $\frac{11}{32} - \frac{2}{8} =$ _____
   a. $\frac{3}{32}$
   b. $\frac{1}{8}$
   c. $\frac{9}{32}$
   d. $\frac{19}{32}$

10. Put the following decimals in order from smallest to largest: 0.402, 0.420, 0.042, 0.442
    a. 0.420, 0.042, 0.442, 0.402
    b. 0.042, 0.402, 0.420, 0.442
    c. 0.442, 0.420, 0.402, 0.042
    d. 0.042, 0.402, 0.442, 0.420

11. Two coatings have been applied to a pipe. The first coating is 51.5 nanometers thick; the second coating is 89.7 nanometers thick. How thick is the combined coating?
    a. 141.2 nanometers
    b. 142.12 nanometers
    c. 144.2 nanometers
    d. 145.02 nanometers

12. $3.53 \times 9.75 =$ _____
    a. 12.28
    b. 34.42
    c. 36.14
    d. 48.13

13. It costs $2.37 to paint one square foot of wall. You need to paint a wall that measures 864.5 square feet. To paint that wall it will cost _____. (Round your answer to the nearest hundredth.)

   a. $204.88
   b. $2,048.87
   c. $2,848.88
   d. $2,888.86

14. $89.435 \div 0.05 =$ _____

   a. 1788.7
   b. 17.887
   c. 4.47175
   d. 447.175

Solve Questions 15 and 16 without using a calculator.

15. $13.9\% =$ _____

   a. 0.009
   b. 0.013
   c. 0.139
   d. 1.39

16. Convert 14.75 to its equivalent fraction expressed in lowest terms.

   a. $^{1475}/_{100}$
   b. $^{295}/_{20}$
   c. $^{59}/_{4}$
   d. $^{147}/_{4}$

17. The arrow in Review Question *Figure 1* is pointing to _____.

   a. $4\frac{3}{8}$ inches
   b. $4\frac{1}{4}$ inches
   c. $4\frac{1}{2}$ inches
   d. $4\frac{5}{8}$ inches

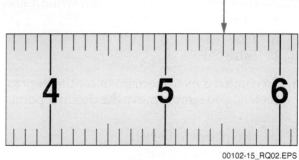

00102-15_RQ02.EPS

Figure 2

18. The arrow in Review Question *Figure 2* is pointing to _____.

   a. 5.2 cm
   b. 5.3 cm
   c. 5.4 cm
   d. 5.5 cm

00102-15_RQ03.EPS

Figure 3

19. The red background around the number 32 in Review Question *Figure 3* indicates _____.

   a. that the tape measure is metric
   b. the Imperial measurement equivalent to one meter
   c. the marking for studs placed on 16-inch centers in a wall structure
   d. the marking for studs placed on 24-inch centers

1/8THS

00102-15_RQ01.EPS

Figure 1

20. A centimeter is ¹⁄₁₀₀th of a meter, while a kilometer is equal to 1000 meters.

   a. True
   b. False

21. To convert a measurement in centimeters to meters, you simply move the decimal point _____.

   a. two places to the left
   b. three places to the right
   c. three places to the left
   d. four places to the right

22. Convert 67 inches to centimeters.

   a. 17.01 cm
   b. 26.38 cm
   c. 170.18 cm
   d. 263.80 cm

23. A pound of water weighs more than a kilogram of water.

   a. True
   b. False

24. Convert the weight of 49 ounces to grams.

   a. 1.73 grams
   b. 3.06 grams
   c. 747.42 grams
   d. 1389.15 grams

25. Convert 15°F to Celsius.

   a. −9.4°C
   b. 9.4°C
   c. −59°C
   d. 59°C

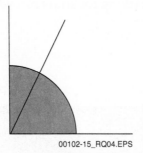

00102-15_RQ04.EPS

Figure 4

26. The two angles shown in Review Question *Figure 4* are both _____.

   a. obtuse and adjacent
   b. obtuse and opposite
   c. right and opposite
   d. acute and adjacent

27. The sum of the four angles in a rectangle is always 180 degrees.

   a. True
   b. False

28. The mathematical constant pi is needed when calculating the area of a _____.

   a. circle
   b. square
   c. rectangle
   d. cube

Use a calculator to answer Questions 29 and 30.

29. The volume of a cylindrical tank that is 23 feet high with a radius of 6.25 feet is _____. (Round your answer to the nearest cubic foot)

   a. 144 cu ft
   b. 898 cu ft
   c. 903 cu ft
   d. 2,821 cu ft

30. The volume of a three-dimensional triangular prism that has a 7 cm base, a 4 cm height, and a 30 mm depth is _____. (Remember to convert all measurements to the same unit before applying the formula.)

   a. 42 cm^3
   b. 42 mm^3
   c. 82 mm^3
   d. 82 cm^3

Trade Terms Quiz

Fill in the blank with the correct term that you learned from your study of this module.

1. The _____ is equal to half the diameter of a circle.

2. A(n) _____ measures between 0 and 90 degrees.

3. When you _____ an angle, you divide it into two equal parts.

4. _____ can be measured in ft², in², cm², as well as other square units.

5. A(n) _____ is the shape made by two straight lines coming together at a point.

6. A(n) _____ is a curved line drawn at a consistent distance around a central point.

7. The _____ is a unit of measurement for angles.

8. The mathematical study of two-dimensional shapes is known as _____.

9. A wall that cannot be moved because it is carrying the weight of the roof is considered a _____ wall.

10. In the fraction ¾, 3 is called the _____.

11. In the same fraction ¾, 4 is called the _____.

12. A line drawn from one corner of a rectangle to the opposite corner is called a(n) _____.

13. The _____ is the longest straight line you can draw inside a circle, representing the distance across it and passing through the center point.

14. The vertical support element inside a wall to which wall finish material is attached is called a _____.

15. In the problem 17 − 8 = 9, the number 9 is the _____.

16. Calculating the volume of three-dimensional objects, such as a cube, relies on the math associated with _____.

17. The numerical symbols from 0 to 9 are called _____.

18. In the number 7,890,342, the _____ of the 3 is three hundred.

19. Fractions having different numerators and denominators, yet they represent the same value, such as ¼ and ⅜, are called _____.

20. The _____ for finding the area of a triangle is ½ × base × height.

21. ¼ and ⅛ are examples of _____.

22. The point at which two or more lines come together to form an angle is called the _____.

23. ⅗ and ¾ are examples of _____.

24. To solve the problem ¾ ÷ ½, you must _____ the second fraction and multiply.

25. In the problem 25 + 25 = 50, the number 50 is the _____.

26. In the problem 86 ÷ 4, the answer is 21 with a(n) _____ of 2.

27. A(n) _____ is one that measures between 90 and 180 degrees.

28. A(n) _____ is a combination of a whole number with a fraction or decimal.

29. _____ are angles that have the same vertex and one side in common.

30. The line that forms the bottom of a triangle is called the _____.

31. To determine the _____, measure the distance around the outside of a closed shape such as a square.

32. Complete numerical units without fractions or decimals are called _____.

33. Numbers that are less than zero are called _____.

34. Numbers that are greater than zero are called _____.

35. The type of triangle that has sides of unequal lengths is a(n) _____.

36. If a triangle has one 90-degree angle, it is a(n) _____ regardless of the other angles.

37. The shape that has three equal sides and three equal angles is called a(n) _____.

38. A(n) _____ has two equal sides and two equal angles.

39. A(n) _____ measures exactly 90 degrees.

40. A(n) _____ measures 180 degrees (basically a flat line).

41. _____ is a mathematical constant that equals approximately 3.14 or ²²⁄₇.

42. A(n) _____ is a geometric figure with three sides and three angles.

43. To calculate the _____ of a cube, multiply the length times the width times the height.

44. The answer to a multiplication problem is called the _____.

45. In a division problem the answer is called the _____.

46. Any definite standard of measurement, such as a centimeter or a foot, is called a(n) _____.

47. The distance around the outside of a circle is called the _____.

48. A specific type of rectangle, with all four sides equal in length, is a(n) _____.

49. When a square is given a third dimension, equal to the other two, it becomes a(n) _____.

50. In a division problem, the number being divided is called the _____.

51. The number divided into the other number in a division problem is the _____.

52. A(n) _____ is a two-dimensional shape that has four sides, with both pairs of opposite sides being equal in length.

53. _____ can be defined as a push or pull on a surface.

54. When a number is written using a whole number plus a portion of one, with the portion separated from the whole number by a dot, it is called a(n) _____.

55. The term _____ represents the quantity of a given material that is present.

56. Two equal angles that are formed when two lines cross over each other are referred to as _____.

57. Pieces of lumber placed horizontally to support a floor or ceiling are called _____.

58. A mathematical statement that indicates that two different expressions are equal, separating them with an equal sign, is called a(n) _____.

Trade Terms

Acute angle	Digit	Mixed number	Right angle
Adjacent angles	Dividend	Negative numbers	Right triangle
Angle	Divisor	Numerator	Scalene triangle
Area	Equation	Obtuse angle	Solid geometry
Base	Equilateral triangle	Opposite angles	Square
Bisect	Equivalent fractions	Perimeter	Straight angle
Circle	Force	Pi	Stud
Circumference	Formula	Place value	Sum
Cube	Fraction	Plane geometry	Triangle
Decimal	Improper fraction	Positive numbers	Unit
Degree	Invert	Product	Vertex
Denominator	Isosceles triangle	Quotient	Volume
Diagonal	Joists	Radius	Whole numbers
Diameter	Loadbearing	Rectangle	
Difference	Mass	Remainder	

Erin M. Hunter
River Valley Technical Center
Carpentry Instructor

How did you become interested in carpentry?
I have always loved working with wood, building things, and being outside. My father has been in concrete construction as long as I can remember. He also did a lot of work renovating the house I grew up in. There was always a project to help with, so I guess you could say I grew up with it. My love of working with wood led me to jobs building furniture. My career in construction started on a historical restoration project as a finish carpenter/project manager. I later began taking on jobs of my own and learning a variety of rough carpentry skills. This led to the start of my own company called E.M. Hunter Construction, where I did everything from roofing to renovations.

What types of training have you been through, and how did it help you get to where you are now?
I had no formal training in carpentry. I had to learn the hard way and from anyone willing to take the time to show me. I wish the opportunities that are available to today's young people were available when I started.

What kinds of work have you done in your career?
I have done custom furniture, historical restoration, finish carpentry, cabinetry, built-ins, tiling, decks, gazebos, timber-framing, roofing, additions; you name it.

What are some of the things you do in your job?
I currently teach a two-year carpentry program to juniors and seniors in a high school career and technical center.

What do you think it takes to be a success in your trade?
Carpentry is a craft that requires integrity, attention to detail, good people skills, adaptability, problem-solving skills, and—most importantly—the willingness to work. To be successful, carpenters need to care about the quality of their work and take pride in what they do.

What do you like about the work you do?
I really enjoy teaching 16- to 18-year-old students. I love helping them learn skills that are relevant and real-life. Carpentry is a great springboard into learning other life skills and disciplines like math. My students get to see themselves as successful in a school setting where they may not have been able to in other classes.

What advice would you give someone just starting out?
Stay open and be willing to learn. Carpentry is an ever-changing field. Sometimes I see young people cut themselves short because they feel that they should already know it all. One of the joys of carpentry is that there is always something you can learn or get better at. I encourage my students to be lifelong learners. Our class motto is "no fear." No fear of learning and no fear of trying something new.

MULTIPLICATION TABLE

Trace across and down from the numbers that you want to multiply, and find the answer. In the example, $7 \times 7 = 49$.

	2	3	4	5	6	7	8	9	10	11	12
2	4	6	8	10	12	14	16	18	20	22	24
3	6	9	12	15	18	21	24	27	30	33	36
4	8	12	16	20	24	28	32	36	40	44	48
5	10	15	20	25	30	35	40	45	50	55	60
6	12	18	24	30	36	42	48	54	60	66	72
7	14	21	28	35	42	49	56	63	70	77	84
8	16	24	32	40	48	56	64	72	80	88	96
9	18	27	36	45	54	63	72	81	90	99	108
10	20	30	40	50	60	70	80	90	100	110	120
11	22	33	44	55	66	77	88	99	110	121	132
12	24	36	48	60	72	84	96	108	120	132	144

00102-15_A01.EPS

CONVERSION FACTORS AND COMMON FORMULAS

COMMON MEASURES

WEIGHT UNITS
- 1 ton = 2,000 pounds
- 1 pound = 16 dry ounces

LENGTH UNITS
- 1 yard = 3 feet
- 1 foot = 12 inches

VOLUMES
- 1 cubic yard = 27 cubic feet
- 1 cubic foot = 1,728 cubic inches
- 1 gallon = 4 quarts
- 1 quart = 2 pints
- 1 pint = 2 cups
- 1 cup = 8 fluid ounces

AREA UNIT
- 1 square yard = 9 square feet
- 1 square foot = 144 square inches

00102-15_A02.EPS

WEIGHT UNITS

1 kilogram	=	1,000 grams
1 hectogram	=	100 grams
1 dekagram	=	10 grams
1 gram	=	1 gram
1 decigram	=	0.1 gram
1 centigram	=	0.01 gram
1 milligram	=	0.001 gram

LENGTH UNITS

1 kilometer	=	1,000 meters
1 hectometer	=	100 meters
1 dekameter	=	10 meters
1 meter	=	1 meter
1 decimeter	=	0.1 meter
1 centimeter	=	0.01 meter
1 millimeter	=	0.001 meter

VOLUME UNITS

1 kiloliter	=	1,000 liters
1 hectoliter	=	100 liters
1 dekaliter	=	10 liters
1 liter	=	1 liter
1 deciliter	=	0.1 liter
1 centiliter	=	0.01 liter
1 milliliter	=	0.001 liter

00102-15_A04.EPS

PREFIX	SYMBOL	NUMBER	MULTIPLICATION FACTOR
giga	G	billion	$1,000,000,000 = 10^9$
mega	M	million	$1,000,000 = 10^6$
kilo	k	thousand	$1,000 = 10^3$
hecto	h	hundred	$100 = 10^2$
deka	da	ten	$10 = 10^1$
			BASE UNITS $1 = 10^0$
deci	d	tenth	$0.1 = 10^{-1}$
centi	c	hundredth	$0.01 = 10^{-2}$
milli	m	thousandth	$0.001 = 10^{-3}$
micro	μ	millionth	$0.000001 = 10^{-6}$
nano	n	billionth	$0.000000001 = 10^{-9}$

00102-15_A03.EPS

US TO METRIC CONVERSIONS

WEIGHTS

1 ounce	=	28.35 grams
1 pound	=	435.6 grams or 0.4536 kilograms
1 (short) ton	=	907.2 kilograms

LENGTHS

1 inch	=	2.540 centimeters
1 foot	=	30.48 centimeters
1 yard	=	91.44 centimeters or 0.9144 meters
1 mile	=	1.609 kilometers

AREAS

1 square inch	=	6.452 square centimeters
1 square foot	=	929.0 square centimeters or 0.0929 square meters
1 square yard	=	0.8361 square meters

VOLUMES

1 cubic inch	=	16.39 cubic centimeters
1 cubic foot	=	0.02832 cubic meter
1 cubic yard	=	0.7646 cubic meter

LIQUID MEASUREMENTS

1 (fluid) ounce	=	0.095 liter or 28.35 grams
1 pint	=	473.2 cubic centimeters
1 quart	=	0.9263 liter
1 (US) gallon	=	3,785 cubic centimeters or 3.785 liters

TEMPERATURE MEASUREMENTS

To convert degrees Fahrenheit to degrees Celsius, use the following formula: $C = 5/9 \times (F - 32)$.

00102-15_A05.EPS

METRIC TO US CONVERSIONS

WEIGHTS

1 gram (G)	=	0.03527 ounces
1 kilogram (kg)	=	2.205 pounds
1 metric ton	=	2,205 pounds

LENGTHS

1 millimeter (mm)	=	0.03937 inches
1 centimeter (cm)	=	0.3937 inches
1 meter (m)	=	3.281 feet or 1.0937 yards
1 kilometer (km)	=	0.6214 miles

AREAS

1 square millimeter	=	0.00155 square inches
1 square centimeter	=	0.155 square inches
1 square meter	=	10.76 square feet or 1.196 square yards

VOLUMES

1 cubic centimeter	=	0.06102 cubic inches
1 cubic meter	=	35.31 cubic feet or 1.308 cubic yards

LIQUID MEASUREMENTS

1 cubic centimeter (cm^3)	=	0.06102 cubic inches
1 liter (1,000 cm^3)	=	1.057 quarts, 2.113 pints, or 61.02 cubic inches

TEMPERATURE MEASUREMENTS

To convert degrees Celsius to degrees Fahrenheit, use the following formula: $F = (9/5 \times C) + 32$.

00102-15_A06.EPS

PRESSURE:

Absolute pressure = gauge pressure + atmospheric pressure. The atmospheric pressure at sea level is typically accepted as 14.7 psi.

Gauge pressure to static pressure: P = hd/144, where P = the pressure in psi; h = the height of the column in feet; d = the density of the liquid in pounds per cubic foot.

CIRCLE:

Area = πr^2, where r is the radius
Circumference = πd, where d is the diameter

SQUARES/RECTANGLES:

Area = length × width
Volume = length × width × height

TEMPERATURE CONVERSION:

°C = 5/9 (°F − 32°)
°F = (9/5 × °C) + 32°

SEQUENCE OF OPERATIONS:

PEMDAS = parenthesis, exponents, multiplication, division, addition, and subtraction

AIR FLOW VOLUME CHANGE:

New cfm = new rpm × existing cfm/existing rpm

TRIANGLES:

Area = (ab)/2, where a is the base length and b is the height

Pythagorean theorem for right triangles:
$c^2 = a^2 + b^2$, or $c = \sqrt{a^2 + b^2}$, where c is the hypotenuse of the triangle. The hypotenuse is the side opposite the right angle. Sides a and b are adjacent to the angle.

PYRAMID:

Volume = (Ah)/3, where A is the area of the base and h is the height

CYLINDER:

Volume = $\pi r^2 h$, where r is the radius of the base and h is the height

CONE:

Volume = $(\pi r^2 h)/3$, where r is the radius of the base and h is the height

SPHERE:

Volume = $(4\pi r^3)/3$, where r is the radius

00102-15_A07.EPS

INCHES CONVERTED TO DECIMALS OF A FOOT

Inches	Decimals of a Foot	Inches	Decimals of a Foot	Inches	Decimals of a Foot
1/16	0.005	2 1/16	0.172	4 1/16	0.339
1/8	0.010	2 1/8	0.177	4 1/8	0.344
3/16	0.016	2 3/16	0.182	4 3/16	0.349
1/4	0.021	2 1/4	0.188	4 1/4	0.354
5/16	0.026	2 5/16	0.193	4 5/16	0.359
3/8	0.031	2 3/8	0.198	4 3/8	0.365
7/16	0.036	2 7/16	0.203	4 7/16	0.370
1/2	0.042	2 1/2	0.208	4 1/2	0.374
9/16	0.047	2 9/16	0.214	4 9/16	0.380
5/8	0.052	2 5/8	0.219	4 5/8	0.385
11/16	0.057	2 11/16	0.224	4 11/16	0.391
3/4	0.063	2 3/4	0.229	4 3/4	0.396
13/16	0.068	2 13/16	0.234	4 13/16	0.401
7/8	0.073	2 7/8	0.240	4 7/8	0.406
15/16	0.078	2 15/16	0.245	4 15/16	0.411
1	0.083	3	0.250	5	0.417
1 1/16	0.089	3 1/16	0.255	5 1/16	0.422
1 1/8	0.094	3 1/8	0.260	5 1/8	0.427
1 3/16	0.099	3 3/16	0.266	5 3/16	0.432
1 1/4	0.104	3 1/4	0.271	5 1/4	0.438
1 5/16	0.109	3 5/16	0.276	5 5/16	0.443
1 3/8	0.115	3 3/8	0.281	5 3/8	0.448
1 7/16	0.120	3 7/16	0.286	5 7/16	0.453
1 1/2	0.125	3 1/2	0.292	5 1/2	0.458
1 9/16	0.130	3 9/16	0.297	5 9/16	0.464
1 5/8	0.135	3 5/8	0.302	5 5/8	0.469
1 11/16	0.141	3 11/16	0.307	5 11/16	0.474
1 3/4	0.146	3 3/4	0.313	5 3/4	0.479
1 13/16	0.151	3 13/16	0.318	5 13/16	0.484
1 7/8	0.156	3 7/8	0.323	5 7/8	0.490
1 15/16	0.161	3 15/16	0.328	5 15/16	0.495
2	0.167	4	0.333	6	0.500

00102-15_A08A.EPS

Inches	Decimals of a Foot	Inches	Decimals of a Foot	Inches	Decimals of a Foot
6 1/16	0.505	8 1/16	0.672	10 1/16	0.839
6 1/8	0.510	8 1/8	0.677	10 1/8	0.844
6 3/16	0.516	8 3/16	0.682	10 3/16	0.849
6 1/4	0.521	8 1/4	0.688	10 1/4	0.854
6 5/16	0.526	8 5/16	0.693	10 5/16	0.859
6 3/8	0.531	8 3/8	0.698	10 3/8	0.865
6 7/16	0.536	8 7/16	0.703	10 7/16	0.870
6 1/2	0.542	8 1/2	0.708	10 1/2	0.875
6 9/16	0.547	8 9/16	0.714	10 9/16	0.880
6 5/8	0.552	8 5/8	0.719	10 5/8	0.885
6 11/16	0.557	8 11/16	0.724	10 11/16	0.891
6 3/4	0.563	8 3/4	0.729	10 3/4	0.896
6 13/16	0.568	8 13/16	0.734	10 13/16	0.901
6 7/8	0.573	8 7/8	0.740	10 7/8	0.906
6 15/16	0.578	8 15/16	0.745	10 15/16	0.911
7	0.583	9	0.750	11	0.917
7 1/16	0.589	9 1/16	0.755	11 1/16	0.922
7 1/8	0.594	9 1/8	0.760	11 1/8	0.927
7 3/16	0.599	9 3/16	0.766	11 3/16	0.932
7 1/4	0.604	9 1/4	0.771	11 1/4	0.938
7 5/16	0.609	9 5/16	0.776	11 5/16	0.943
7 3/8	0.615	9 3/8	0.781	11 3/8	0.948
7 7/16	0.620	9 7/16	0.786	11 7/16	0.953
7 1/2	0.625	9 1/2	0.792	11 1/2	0.958
7 9/16	0.630	9 9/16	0.797	11 9/16	0.964
7 5/8	0.635	9 5/8	0.802	11 5/8	0.969
7 11/16	0.641	9 11/16	0.807	11 11/16	0.974
7 3/4	0.646	9 3/4	0.813	11 3/4	0.979
7 13/16	0.651	9 13/16	0.818	11 13/16	0.984
7 7/8	0.656	9 7/8	0.823	11 7/8	0.990
7 15/16	0.661	9 15/16	0.828	11 15/16	0.995
8	0.667	10	0.833	12	1.000

00102-15_A08B.EPS

Acute angle: Any angle between 0 degrees and 90 degrees.

Adjacent angles: Angles that have the same vertex and one side in common.

Angle: The shape made by two straight lines coming together at a point. The space between those two lines is measured in degrees.

Area: The surface or amount of space occupied by a two-dimensional object such as a rectangle, circle, or square.

Base: As it relates to triangles, the base is the line forming the bottom of the triangle.

Bisect: To divide something into two parts that are often equal. When an angle is bisected for example, the two resulting angles are equal.

Circle: A closed curved line around a central point. A circle measures 360 degrees.

Circumference: The distance around the curved line that forms the circle.

Cube: A three-dimensional square, with the measurements in all the three dimensions being equal.

Decimal: A part of a number represented by digits to the right of a point, called a decimal point. For example, in the number 1.25, .25 is the decimal portion of the number. In this case, it represents 25% of the whole number 1.

Degree: A unit of measurement for angles. For example, a right angle is 90 degrees, an acute angle is between 0 and 90 degrees, and an obtuse angle is between 90 and 180 degrees.

Denominator: The part of a fraction below the dividing line. For example, the 2 in ½ is the denominator. It is equivalent to the divisor in a long division problem.

Diagonal: Line drawn from one corner of a rectangle or square to the farthest opposite corner.

Diameter: The length of a straight line that crosses from one side of a circle, through the center point, to a point on the opposite side. The diameter is the longest straight line you can draw inside a circle.

Difference: The result of subtracting one number from another. For example, in the problem 8 − 3 = 5, 5 is the difference between the two numbers.

Digit: Any of the numerical symbols 0 to 9.

Dividend: In a division problem, the number being divided is the dividend.

Divisor: In a division problem, the number that is divided into another number is called the divisor.

Equation: A mathematical statement that indicates the value of two mathematical expressions, such as 2 × 2 and 1 × 4, are equal. An equation is written using the equal sign in this manner: 2 × 2 = 1 × 4.

Equilateral triangle: A triangle that has three equal sides and three equal angles.

Equivalent fractions: Fractions having different numerators and denominators but still have equal values, such as the two fractions ½ and ¼.

Force: A push or pull on a surface. In this module, force is considered to be the weight of an object or fluid. This is a common approximation.

Formula: A mathematical process used to solve a problem. For example, the formula for finding the area of a rectangle is Side A times Side B = Area, or A × B = Area.

Fraction: A portion of a whole number represented by two numbers. The upper number of a fraction is known as the numerator and the bottom number is known as the denominator.

Improper fraction: A fraction whose numerator is larger than its denominator. For example, ¾ and ⁵⁄₃ are improper fractions.

Invert: To reverse the order or position of numbers. In fractions, inverting means to reverse the positions of the numerator and denominator, such that ¾ becomes ⁴⁄₃. When you are dividing by fractions, one fraction is inverted.

Isosceles triangle: A triangle that has two equal sides and two equal angles.

Joist: Lengths of wood or steel that usually support floors, ceiling, or a roof. Roof joists will be at the same angle as the roof itself, while floor and ceiling joists are usually horizontal.

Loadbearing: Carrying a significant amount of weight and/or providing necessary structural support. A loadbearing wall is typically carrying some portion of the roof weight and cannot be removed without risking structural failure or collapse.

Mass: The quantity of matter present.

Mixed number: A combination of a whole number with a fraction or decimal. Examples of mixed numbers are $3\frac{7}{16}$, 5.75, and $1\frac{1}{4}$.

Negative numbers: Numbers less than zero. For example, –1, –2, and –3 are negative numbers.

Numerator: The part of a fraction above the dividing line. For example, the 1 in ½ is the numerator. It is the equivalent of the dividend in a long division problem.

Obtuse angle: Any angle between 90 degrees and 180 degrees.

Opposite angles: Two angles that are formed by two straight lines crossing. They are always equal.

Perimeter: The distance around the outside of a closed shape, such as a rectangle, circle, square, or any irregular shape.

Pi: A mathematical value of approximately 3.14 (or $\frac{22}{7}$) used to determine the area and circumference of circles. It is sometimes symbolized by π.

Place value: The exact value a digit represents in a whole number, determined by its place within the whole number or by its position relative to the decimal point. In the number 124, the number 2 actually represents 20, since it is in the tens column.

Plane geometry: The mathematical study of two-dimensional (flat) shapes.

Positive numbers: Numbers greater than zero. For example, 1, 2, and 3 are positive numbers. Any number without a negative (–) sign in front of it is considered to be a positive number.

Product: The answer to a multiplication problem. For example, the product of 6 × 6 is 36.

Quotient: The result of a division problem. For example, when dividing 6 by 2, the quotient is 3.

Radius: The distance from a center point of a circle to any point on the curved line, or half the width (diameter) of a circle.

Rectangle: A four-sided shape with four 90-degree angles. Opposite sides of a rectangle are always parallel and the same length. Adjacent sides are perpendicular and are not equal in length.

Remainder: The amount left over in a division problem. For example, in the problem 34 ÷ 8, 8 goes into 34 four times (8 × 4 = 32) with 2 left over; in other words, 2 is the remainder.

Right angle: An angle that measures 90 degrees. The two lines that form a right angle are perpendicular to each other. This is the angle used most in the trades.

Right triangle: A triangle that includes one 90-degree angle.

Scalene triangle: A triangle with sides of unequal lengths.

Solid geometry: The mathematical study of three-dimensional shapes.

Square: (1) A special type of rectangle with four equal sides and four 90-degree angles. (2) The product of a number multiplied by itself. For example, 25 is the square of 5; 16 is the square of 4.

Straight angle: A 180-degree angle or flat line.

Stud: A vertical support inside the wall of a structure to which the wall finish material is attached. The base of a stud rests on a horizontal baseplate, and a horizontal cap plate rests on top of a series of studs.

Sum: The resulting total in an addition problem. For example, in the problem 7 + 8 = 15, 15 is the sum.

Triangle: A closed shape that has three sides and three angles.

Unit: A definite standard of measure.

Vertex: A point at which two or more lines or curves come together.

Volume: The amount of space contained in a given three-dimensional shape.

Whole numbers: Complete number units without fractions or decimals.

Additional Resources

This module presents thorough resources for task training. The following resource material is suggested for further study.

Applied Construction Math: A Novel Approach, NCCER. 2006. Upper Saddle River, NJ: Prentice Hall.

Mathematics for Carpentry and the Construction Trades, Alfred P. Webster; Kathryn B. Judy. 2001. Upper Saddle River, NJ: Prentice Hall.

Mathematics for the Trades: A Guided Approach, Robert A. Carman; Hal Saunders. 2014. Pearson Learning.

Metric-conversion.org : Metric Conversion Charts and Calculators.

Figure Credits

Section Review Answer Key

Answer	Section Reference	Objective
Section One		
1. d	1.1.0	1a
2. c*	1.2.0	1b
3. b*	1.3.0	1c
Section Two		
1. b*	2.1.0	2a
2. b*	2.2.0	2b
3. c*	2.3.0	2c
4. d*	2.3.0	2c
5. a*	2.4.0	2d
6. c*	2.4.0	2d
Section Three		
1. c	3.1.0	3a
2. a	3.1.0	3a
3. b*	3.2.1	3b
4. a*	3.2.1	3b
5. d*	3.2.2	3b
6. a*	3.2.3	3b
7. c	3.3.1	3c
8. c*	3.3.2	3c
9. a*	3.3.2	3c
Section Four		
1. a	4.1.2	4a
2. b	4.2.3	4b
Section Five		
1. b	5.1.2	5a
2. d	5.2.1	5b
3. c	5.3.2	5c
4. b	5.4.0	5d
Section Six		
1. d	6.1.0	6a
2. b	6.2.2; 6.2.4	6b
3. c*	6.3.0	6c
4. c*	6.4.1	6d

*The math calculations for these answers are provided on the following page.

Section Review Calculations

Section 1.0.0

1. $46 + 8 + 8 + 30 + 8 + 10 + 20 + 10 + 8 + 10 + 8 + 10 + 10 + 38 = 224$ feet

3. 224 feet × 12 inches per foot = 2688 inches ÷ 16 inches per block = 168 blocks per course
   A 24-inch-high foundation requires 3 courses of 8-inch block; therefore, 3 × 168 blocks = 504 blocks.

Section 2.0.0

1. $\frac{1}{2} \times \frac{32}{32} = \frac{32}{64}$

2. $\frac{76}{64} = 76 \div 64 = 1$, with a remainder of 12 or $\frac{12}{64}$
   To change the fraction to a decimal: $12 \div 64 = .1875$, making the decimal equivalent of $\frac{76}{64} = 1.1875$

3. Convert $\frac{3}{8}$ to a common denominator with $\frac{11}{16}$: $\frac{3}{8} \times \frac{2}{2} = \frac{6}{16}$
   $\frac{11}{16} + \frac{6}{16} = \frac{17}{16}$, or $1\frac{1}{16}$

4. Convert $1\frac{1}{4}$ to a common denominator with $\frac{7}{16}$: $1\frac{1}{4} = \frac{5}{4}$; $\frac{5}{4} \times \frac{4}{4} = \frac{20}{16}$
   $\frac{20}{16} - \frac{7}{16} = \frac{13}{16}$

5. $\frac{1}{8} \times \frac{1}{10} = 1 \times \frac{1}{8} \times 10 = \frac{1}{80}$

6. Invert the divisor and multiply:
   $\frac{7}{16} \div \frac{7}{8} = \frac{7}{16} \times \frac{8}{7} = 7 \times \frac{8}{16} \times 7 = \frac{56}{112}$
   Using a factor of 56, reduce to lowest terms: $56 \div \frac{56}{112} \div 56 = \frac{1}{2}$

Section 3.0.0

3. $3.625 + 4.9 = 8.525$

4. $42.58 - 7.577 = 35.003$

5. $9.64 \times 12 = 115.68$

6. $123.82 \div 6.5 = 19.049$

8. $\frac{7}{8} = 7 \div 8 = 0.875$

9. $\frac{2}{3} = 2 \div 3 = 0.666.$
   Convert 0.666 to a percentage by moving the decimal 2 places to the right: 66.6%
   66.6% is greater than 65%

Section 6.0.0

3. Area = l × w
   Area = 27.3 meters × 9.3 meters
   Area = 253.89 m^2

4. Volume = l × w × d
   Volume = 26 feet × 13.66 feet × 3 feet
   Volume = 1,065.48 ft^3

NCCER CURRICULA — USER UPDATE

NCCER makes every effort to keep its textbooks up-to-date and free of technical errors. We appreciate your help in this process. If you find an error, a typographical mistake, or an inaccuracy in NCCER's curricula, please fill out this form (or a photocopy), or complete the online form at **www.nccer.org/olf**. Be sure to include the exact module ID number, page number, a detailed description, and your recommended correction. Your input will be brought to the attention of the Authoring Team. Thank you for your assistance.

Instructors – If you have an idea for improving this textbook, or have found that additional materials were necessary to teach this module effectively, please let us know so that we may present your suggestions to the Authoring Team.

NCCER Product Development and Revision

13614 Progress Blvd., Alachua, FL 32615

Email: curriculum@nccer.org
Online: www.nccer.org/olf

❏ Trainee Guide ❏ Lesson Plans ❏ Exam ❏ PowerPoints Other _____

Craft / Level: _____ Copyright Date: _____

Module ID Number / Title: _____

Section Number(s): _____

Description: _____

_____

_____

_____

Recommended Correction: _____

_____

_____

_____

_____

Your Name: _____

Address: _____

_____

Email: _____ Phone: _____

00103-15

Introduction to Hand Tools

OVERVIEW

Every profession has its tools. A surgeon uses a scalpel, a teacher uses a whiteboard, and an accountant uses a calculator. The construction trades require a broad collection of hand tools, such as hammers, screwdrivers, and pliers, that almost every craftworker uses. Even if you are already familiar with some of these tools, everyone needs to learn how to select, maintain, and use them safely. A quality tool may cost more up front, but if properly maintained, it will last for years and remain safely intact.

Module Three

Trainees with successful module completions may be eligible for credentialing through the NCCER Registry. To learn more, go to **www.nccer.org** or contact us at **1.888.622.3720**. Our website has information on the latest product releases and training, as well as online versions of our *Cornerstone* magazine and Pearson's product catalog.

Your feedback is welcome. You may email your comments to **curriculum@nccer.org,** send general comments and inquiries to **info@nccer.org**, or fill in the User Update form at the back of this module.

This information is general in nature and intended for training purposes only. Actual performance of activities described in this manual requires compliance with all applicable operating, service, maintenance, and safety procedures under the direction of qualified personnel. References in this manual to patented or proprietary devices do not constitute a recommendation of their use.

Objectives

When you have completed this module, you will be able to do the following:

1. Identify and explain how to use various types of hand tools.
   a. Identify and explain how to use various types of hammers and demolition tools.
   b. Identify and expl n how to use various types of chisels and punches.
   c. Identify and explain how to use various types of screwdrivers.
   d. Identify and explain how to use various types of non-adjustable and adjustable wrenches.
   e. Identify and explain how to use various types of socket and torque wrenches.
   f. Identify and explain how to use various types of pliers and wire cutters.
2. Identify and describe how to use various types of measurement and layout tools.
   a. Identify and explain how to use rules and other measuring tools.
   b. Identify and explain how to use various types of levels and layout tools.
3. Identify and explain how to use various types of cutting and shaping tools.
   a. Identify and explain how to use handsaws.
   b. Identify and explain how to use various types of files and utility knives.
4. Identify and explain how to use other common hand tools.
   a. Identify and explain how to use shovels and picks.
   b. Identify and explain how to use chain falls and come-alongs.
   c. Identify and explain how to use various types of clamps.

Performance Tasks

Under the supervision of your instructor, you should be able to do the following:

1. Visually inspect a minimum of five of the following tools to determine if they are safe to use:
   - Hammer or demolition tool
   - Chisel or punch
   - Screwdriver
   - Adjustable or non-adjustable wrench
   - Socket
   - Torque wrench
   - Pliers
   - Wire cutters
   - Measuring too'
   - Layout tool
   - Level
   - Hand saw
   - File
   - Utility knife
   - Shovel or other earth tool
   - Chain fall or hoist
   - Clamps

2. Safely and proper   a minimum of three of the following tools:
   - Hammer or den   ion tool
   - Chisel or punch
   - Screwdriver
   - Adjustable or n   ljustable wrench
   - Socket
   - Torque wrench
   - Pliers
   - Wire cutters
   - Measuring tool
   - Layout tool
   - Level
   - File
   - Utility knife
   - Shovel or other earth tool
   - Chain fall or hoist
   - Clamps

3. Make a straight, squ :e cut in framing lumber using a crosscut saw.

Trade Terms

Adjustable wrench	Foot-pounds	Points
Ball-peen hammer	Hex key wrench	Punch
Bell-faced hammer	Inch-pounds	Rafter angle square
Bevel	Joint	Ripping bar
Box-end wrench	Kerf	Round off
Carpenter's square	Level	Square
Cat's paw	Miter joint	Striking (or slugging) wrench
Chisel	Nail puller	Strip
Chisel bar	Newton-meter	Tang
Claw hammer	Open-end wrench	Tempered
Combination square	Peening	Tenon
Combination wrench	Pipe wrench	Torque
Dowel	Planed	Try square
Fastener	Pliers	Weld
Flats	Plumb	

Industry Recognized Credentials

If you are training through an NCCER-accredited sponsor, you may be eligible for credentials from NCCER's Registry. The ID number for this module is 00103-15. Note that this module may have been used in other NCCER curricula and may apply to other level completions. Contact NCCER's Registry at 888.622.3720 or go to **www.nccer.org** for more information.

Contents

Topics to be presented in this module include:

Contents (continued)

Figures and Tables

1.0.0 COMMON HAND TOOLS

Objective

Identify and explain how to use various types of hand tools.

a. Identify and explain how to use various types of hammers and demolition tools.
b. Identify and explain how to use various types of chisels and punches.
c. Identify and explain how to use various types of screwdrivers.
d. Identify and explain how to use various types of non-adjustable and adjustable wrenches.
e. Identify and explain how to use various types of socket and torque wrenches.
f. Identify and explain how to use various types of pliers and wire cutters.

Performance Tasks

1. Visually inspect the following tools to determine if they are safe to use:
   - Hammer or demolition tool
   - Chisel or punch
   - Screwdriver
   - Adjustable or non-adjustable wrench
   - Socket
   - Torque wrench
   - Pliers
   - Wire cutters
2. Safely and properly use the following tools:
   - Hammer or demolition tool
   - Chisel or punch
   - Screwdriver
   - Adjustable or non-adjustable wrench
   - Socket
   - Torque wrench
   - Pliers
   - Wire cutters

Trade Terms

Adjustable wrench: A smooth-jawed wrench with an adjustable jaw used for turning nuts and bolts. Often referred to as a Crescent® wrench due to brand recognition.

Ball-peen hammer: A hammer with a flat face that is used to strike cold chisels and punches. The rounded end (the peen) is used to bend and shape soft metal.

Bell-faced hammer: A claw hammer with a slightly rounded, or convex, face.

Bevel: To cut on a slant at an angle that is not a right angle (90 degrees). The angle or inclination of a line or surface that meets another at any angle except 90 degrees.

Box-end wrench: A wrench, usually double-ended, that has a closed socket that fits over the head of a bolt.

Cat's paw: A straight steel rod with a curved claw at one end that is used to pull nails that have been driven flush with the surface of the wood or slightly below it.

Chisel: A metal tool with a sharpened, beveled edge used to cut and shape wood, stone, or metal.

Chisel bar: A tool with a claw at each end, commonly used to pull nails.

Claw hammer: A hammer with a flat striking face. The other end of the head is curved and divided into two claws to remove nails.

Combination wrench: A wrench with an open end and a closed end.

Dowel: A pin, usually round, that fits into a corresponding hole to fasten or align two pieces.

Fastener: A device such as a bolt, clasp, hook, or lock used to attach or secure one material to another.

Flats: The straight sides or jaws of a wrench opening; also, the sides on a nut or bolt head.

Foot-pounds: Unit of measure used to describe the amount of pressure exerted (torque) to tighten a large object.

Hex-key wrench: A hexagonal steel bar that is bent to form a right angle. Often referred to as an Allen® wrench.

Inch-pounds: Unit of measure used to describe the amount of pressure exerted (torque) to tighten a small object.

Joint: The point where members or the edges of members are joined. The types of welding joints are butt joint, corner joint, and T-joint.

Level: Perfectly horizontal; completely flat. Also, a tool used to determine if an object is level.

Nail puller: A tool used to remove nails.

Newton-meter: A measure of torque or moment equal to the force of one Newton applied to a lever one meter long.

Open-end wrench: A non-adjustable wrench with a fixed opening at each end that is typically different, allowing it to be used to fit two different nut or bolt sizes.

Peening: The process of bending, shaping, or cutting material by striking it with a tool.

Pipe wrench: A wrench for gripping and turning a pipe or pipe-shaped object; it tightens when turned in one direction.

Pliers: A scissor-shaped type of adjustable wrench equipped with jaws and teeth to grip objects.

Points: Teeth on the gripping part of a wrench. Also refers to the number of teeth per inch on a handsaw.

Punch: A steel tool used to indent metal.

Ripping bar: A tool used for heavy-duty dismantling of woodwork, such as tearing apart building frames or concrete forms.

Round off: To smooth out threads or edges on a screw or nut.

Square: Exactly adjusted; any piece of material sawed or cut to be rectangular with equal dimensions on all sides; a tool used to check angles.

Striking (or slugging) wrench: A non-adjustable wrench with an enclosed, circular opening designed to lock on to the fastener when the wrench is struck.

Strip: To damage the head or threads on a screw, nut, or bolt.

Tempered: Treated with heat to create or restore hardness in steel.

Torque: A rotating or twisting force applied to an object such as a nut, bolt, or screw, using a socket wrench or screwdriver. Torque wrenches allow a specific torque value to be set and applied.

Weld: To heat or fuse two or more pieces of metal so that the finished piece is as strong as the original; a welded joint.

Hand tools are grouped by their purposes. In each group, there are different physical characteristics that determine their individual uses. It is important to know both the purpose and characteristic of each type to complete the task at hand properly and safely. This section presents common hand tools used in the trades, including:

- Hammers and demolition tools
- Chisels and punches
- Screwdrivers
- Wrenches (non-adjustable and adjustable)

- Socket and torque wrenches
- Pliers and wire cutters

In the construction environment, some personal protective equipment (PPE) must be worn consistently. This PPE includes safety glasses, hard-toe safety shoes, hard hats, and gloves. This same PPE should be worn whenever working with hand tools.

Also remember that, when working above the ground, all tools should be controlled so that they cannot fall from your hands or tool belt. Tools should be connected to the tool belt with lanyards to prevent injury to others and damage to the work below.

1.1.0 Hammers and Demolition Tools

Hammers are made in different sizes and weights for specific types of work. The safest hammers are those with heads made of alloys and drop-forged steel. Two of the most common hammers are the **claw hammer** and the **ball-peen hammer**. Demolition tools include various styles of nail pullers and ripping bars. Information on these tools is also presented in this section.

1.1.1 Claw Hammer

The claw hammer (*Figure 1*) has a steel head and a handle made of wood, steel, or fiberglass. This style of hammer is generally associated with carpentry work. The head is used to drive nails, wedges, and **dowels**, and the claw is used to pull nails out of wood. The face of the hammer may be flat or rounded. It is easier to drive nails with the flat face (plain) claw hammer, but the flat face may leave hammer marks when you drive the head of the nail flush (even) with the surface of the work.

A claw hammer with a slightly rounded (or convex) face is called a **bell-faced hammer**. A skilled worker can use it to drive the nail head flush without damaging the surface of the work.

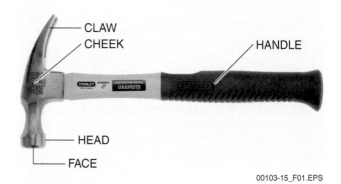

00103-15_F01.EPS

Figure 1 Claw hammer.

Driving a nail with a claw hammer is simple. Using the appropriate PPE, hold the nail and rest the face of the hammer on it to give you a visual starting place. Next, draw the hammer back and lightly tap the nail to start it. Once the nail is started, move your fingers away. Hold the hammer level with the head of the nail, using the motion of the wrist, elbow, and shoulder to strike the head squarely with the center of the hammer face. It requires practice to become proficient with this method.

The hammer is designed to produce a certain amount of force on the object it strikes. If you hold the hammer incorrectly, you cancel out the benefits of its design. Always remember to hold the end of the handle even with the lower edge of your palm. The distance between your hand and the head of the hammer affects the force you apply to drive a nail. The closer you hold the hammer to the head, the harder you will need to swing to achieve the desired force. Make it easier on yourself by holding the hammer properly; it takes less effort to drive the nail.

Pulling a nail with a claw hammer is as easy as driving one. First, slip the claw of the hammer under the nail head. Use the leverage of the hammer to pry the nail up until the hammer's handle is nearly straight up (vertical), partially drawing the nail out of the wood. At this point, pull the nail straight up from the wood. Longer and larger nails require more effort, depending on the material in which they are driven. For longer nails, a small block of wood can be placed under the hammer head to elevate it slightly. This strategy can also be used to protect the material underneath the hammer head from being scarred as the nail is pulled out.

Did You Know?

Hammers

The quality of a hammer is important. The strongest (and safest) hammers have heads made from tough alloy (a mixture of two or more metals) and drop-forged steel (a strong steel formed by pounding and heating). Hammers with cast heads—heads formed by being poured or pressed into a mold—are more brittle. They are not suited for construction work because they tend to chip and break. Hammers with heads made of tough alloy and drop-forged steel tend to be more expensive than hammers with cast heads. When it comes to tools, it pays to invest in quality equipment.

1.1.2 Ball-Peen Hammer

A ball-peen hammer (*Figure 2*) is a type of hammer used in metalworking. It has a flat face for striking and a spherical or hemispherical head for peening (rounding off) metal or rivets. This hammer is used with chisels and punches (discussed later in this module). In welding operations, the ball-peen hammer is used to reduce stress in the weld by peening or striking the joint as it cools.

Ball-peen hammers are also known as engineer's hammers or machinist's hammers. They are classified by weight, which ranges from 4 ounces to 2 pounds (113 g – 0.9 kg).

When using a ball-peen hammer, do not strike the material as though a claw hammer is being used. A ball-peen hammer should strike the material in a controlled manner so that the material is not damaged. Never use a ball-peen hammer to drive nails, because the head of the hammer is made of tough, but milder steel. Continual pounding on nails can deform or damage the hammer's head. A claw hammer is not interchangeable with a ball-peen hammer.

1.1.3 Sledgehammers

A sledgehammer is a heavy-duty tool that is used to drive posts or other large stakes. You can also use it to break up cast iron or concrete. The head of the sledgehammer is made of high-carbon steel and weighs 2 to 20 pounds (0.9 kg to 9 kg). The shape of the head depends on the job the sledgehammer is designed to do. Sledgehammers can be either long-handled or short-handled, depending on the jobs for which they are designed. *Figure 3* shows three styles of sledgehammers.

A sledgehammer can cause injury to you or to anyone working near you. Follow these steps to use a sledgehammer properly and safely:

Step 1 Wear the appropriate PPE, including gloves, eye protection, and hard-toe shoes.

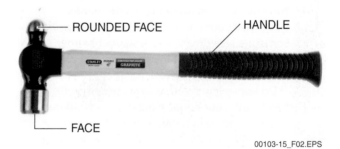

Figure 2 Ball-peen hammer.

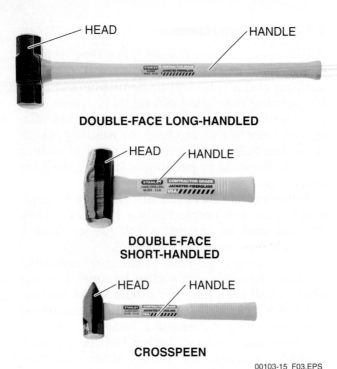

Figure 3 Sledgehammers.

Step 2 Inspect the sledgehammer to ensure that there are no defects.

Step 3 Be sure there is no one nearby that could be struck on the back swing.

Step 4 Hold the sledgehammer with both hands apart (hand over hand).

Step 5 Stand directly in front of the object to be driven.

Step 6 Lift the sledgehammer straight up above the target.

Step 7 Set the head of the sledgehammer on the target.

Step 8 Begin delivering short blows to the target and gradually increase the length and force of the stroke.

> **WARNING!**
>
> Hold the sledgehammer with both hands. Never use your hands to hold an object while someone else drives with a sledgehammer. Doing so could result in serious injury, such as crushed or broken bones.
>
> Do not swing a sledgehammer over your head, with your hands extending beyond the head. Doing so may cause injury to your back and could limit the control you have directing the blow to the target.

1.1.4 Nail Pullers

An existing structure often needs to be torn apart before building can begin. **Nail pullers** and wrecking bars (*Figure 4*) are just the tools for the job.

Three main types of nail-pulling tools are the **cat's paw** (also called nail claws), **chisel bar**, and wrecking bar. There are many names used in the field to describe these tools. The cat's paw is a straight steel rod with a curved claw at one end. It is used to pull nails that have been driven flush with the surface of the wood or slightly below it. The cat's paw is used to pull nails to just above the surface of the wood so they can be pulled completely out with the claw of a hammer or a pry bar.

The chisel bar has a claw at each end and is ground to a chisel-like bevel (slant) on both ends. The chisel bar can be used like a claw hammer to pull a nail. It can also be driven into wood to split and rip apart the pieces.

The wrecking bar (ripping bar, wonder bar, action bar) has a nail slot at the end to pull nails out from tightly enclosed areas. It can also be used as a pry bar to open wooden shipping containers and similar tasks.

Before using a nail puller, it is very important to wear the appropriate PPE to avoid injuries. Once you have donned the PPE, drive the claw into the wood, grabbing the nail head. Once a good grip is achieved, pull the handle of the bar to lift the nail out of the wood.

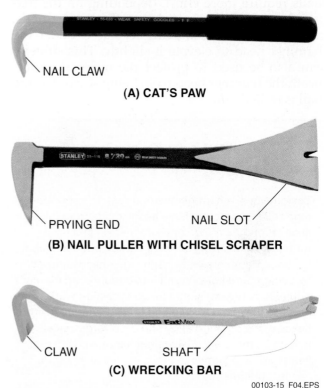

Figure 4 Nail pullers and wrecking bars.

When manipulating the bar, make sure that you never pull the bar toward your face; always pull the bar toward your shoulder. Also, never push in a direction away from your body, because this may cause you to lose your balance.

1.1.5 Ripping Bars

The **ripping bar**, also called a pinch, pry, or wrecking bar, can be 12" to 36" long. Ripping bars are generally those that are at the longer end of the range. This is because the length allows them to develop far more leverage and power. They are used for heavy-duty dismantling of woodwork, such as tearing apart building frames or concrete forms. The ripping bar usually has an octagonal (eight-sided) shaft and two specialized ends. Some may have a deeply curved nail claw at one end for nail pulling; others may not be equipped with a nail puller. An angled, wedge-shaped face at the other end is used as a prying tool to pull apart materials that are nailed together.

When using a ripping bar, make sure that you always wear the appropriate PPE. Eye protection is extremely important due to the risk of flying debris. Use the angled prying end to force apart pieces of wood or use the heavy claw to pull large nails and spikes.

When using a ripping bar or a nail puller, a piece of material can break off and fly through the air. Wear a hard hat, safety glasses, and gloves for protection from flying debris. Make sure others around you are similarly protected.

1.1.6 Safety and Maintenance

No matter what the job is, a safe work environment and safe work practices are extremely important to avoid accidents and injuries. Don the required PPE before beginning any work and always be aware of the surroundings before and during the performance of any task.

Become familiar with the following guidelines for safety and maintenance when using all types of hammers:

- Make sure there are no splinters in the handle of the hammer.
- Make sure the handle is set securely in the head of the hammer.
- Replace cracked or broken handles.
- Make sure the face of the hammer is clean.
- Hold the hammer properly. Grasp the handle firmly near the end and hit the nail squarely.
- Do not hit with the cheek or side of the hammer head.
- Do not use hammers with chipped, mushroomed (overly flattened by use), or otherwise damaged heads.
- Do not use a hammer with a cast head.
- Never strike hammer heads together.

The following are additional considerations for safety and maintenance when working with sledgehammers:

- Replace cracked or broken handles before you use a sledgehammer.
- Use the right amount of force for the job.
- Keep your hands away from the object you are driving.
- Never swing until you have checked behind you to make sure you have enough room and no one is behind you.

The following are safety guidelines to consider for ripping and nail pulling:

- Use two hands when ripping to help ensure that there is even pressure on back muscles when pulling.
- Ensure the material holding the nail is braced securely before pulling to prevent injury.

Demolition Tools

There are many different tools on the market today that are designed for demolition. Small differences in design often make certain tools work far better for a specific task than another style.

00103-15_SA01.EPS

The Stanley Fubar shown here, for example, combines the advantages of a pry bar with those of a hammer. Three different strike-able surfaces allow the tool to be driven under or between parts for disassembly. It can be used to pull nails several different ways, and can be used to chop through softer materials such as drywall. Tougher materials can be chopped by first driving the chisel end through and then using the chopping edge. There are many variations on this type of tool.

Most accidents with prying tools occur when a pry bar slips and the worker falls to the ground. Be sure to keep balanced footing and a firm grip on the tool. This technique also helps reduce damage to materials that must be reused, such as concrete forms.

1.2.0 Chisels and Punches

Chisels are used to cut and shape wood, stone, or metal. Punches are used to indent metal, drive pins, and align holes.

1.2.1 Chisels

A chisel is a metal tool with a sharpened, **beveled** (sloped) edge. Wood chisels and cold chisels (*Figure 5*) are discussed in this section. Both types of chisel are made from steel that is heat-treated to make it harder. A chisel can cut any material that is softer than the steel of the chisel. Both cold and wood chisels are manufactured in a variety of shapes and sizes for the many tasks they are used for.

Cold chisels are designed for hard materials such as steel or stone. The most extensively used is the flat chisel. Common cold chisels have an edge that is ground at a 60-degree angle. The wide cutting edge may have slightly rounded corners that do not dig into the metal, which makes it ideal for cutting out sheet metal and cutting off rods. The cross-cut (cape) chisel is manufactured so the cutting edge is slightly wider than the body. This feature keeps the chisel from binding in the cut when chiseling deep grooves. Cross-cut chisels are used for cutting single grooves as well as cutting a grid pattern of grooves when excessive material needs to be cut from a surface. The round nose (half-round) chisel is generally used for specialized work like forming flutes and channels. The diamond-pointed chisel has a square section at the tip with a single bevel. This chisel is typically used for cleaning square internal angles and chipping through plates.

Before using a cold chisel, it is important to be wearing the appropriate PPE, especially the proper eye protection. The first step to using a cold chisel is to secure the object to be cut in a vise. Place the blade of the chisel at the spot where the material is to be cut. Hit the chisel handle with a ball-peen hammer to force the chisel into and through the material. Repeat as necessary.

There are also several different styles of wood chisel. Bevel-edged chisels are easy to push into corners because of the slightly angled edges. This chisel is commonly used for completing dovetail joints. Firmer chisels have a rectangular

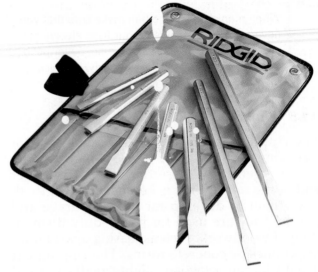

(A) COLD CHISELS

(B) WOOD CHISELS

00103-15_F05.EPS

Figure 5 Cold and wood chisels.

cross-section blade. This chisel is typically used for heavier work requirements due to their strength. Paring chisels are longer and thinner than their counterparts. The feature of this chisel is its ability to reach deeper into long joints to clean them for a precise fit. Note that some chisels are designed for hand-use only; they should not be struck with a hammer or mallet.

Before using a wood chisel, it is important to be wearing the appropriate PPE, especially proper eye protection. First, outline the recessed area to be chiseled out. Next, set the chisel at one

NCCER – *Core Curriculum* 00103-15

end of the outline, with its edge on the cross-grain line and the bevel facing the recess to be made. Lightly strike the chisel head with a mallet. Repeat this process at the other end of the outline, again with the bevel of the chisel blade toward the recess. Then make a series of cuts approximately ¼" apart from one end of the recess to the other. To pare (trim) away the notched wood, hold the chisel bevel-side down to slice inward from the end of the recess (*Figure 6*).

1.2.2 Punches

A punch (*Figure 7*) uses the impact of a hammer to indent metal before you drill a hole, to drive pins, and to align holes in two parts that are being mated. Punches are made of hardened and **tempered** steel, and they come in various sizes.

Three common types of punches are the center punch, the prick punch, and the straight, or tapered, punch. The center and prick punches are used to make small locating points for drilling holes. The straight punch is used to punch holes in thin sheets of metal; its diameter is consistent for much of its length. The tapered punch, pictured in *Figure 7*, has a slight taper along its length. Straight and taper punches typically have blunt ends rather than a point.

To use a punch, hold it straight up and down with one hand and strike it squarely with a hammer.

Stonemasons

Stone used to be a primary building material. Because of its strength, stone was often used for dams, bridges, fortresses, foundations, and monumental buildings. Today, steel and concrete have replaced stone as a basic construction material. Stone is used primarily as sheathing for buildings, for flooring in high-traffic areas, and for decorative uses.

A stonemason's job requires precision. Stones have uneven, rough edges that must be trimmed and finished before each stone can be set. The process of trimming projections and jagged edges is called dressing the stone. This requires skill and experience using specialized hand tools, such as chisels and sledgehammers. Many craftworkers consider stonework to be an art form.

Stonemasons build stone walls, and they also set stone exteriors and floors, working with natural cut and artificial stones including marble, granite, limestone, cast concrete, marble chips, and other masonry materials.

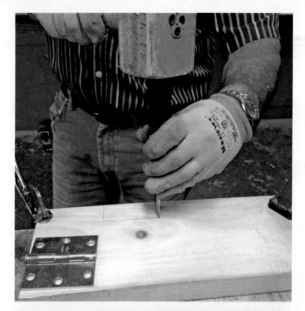

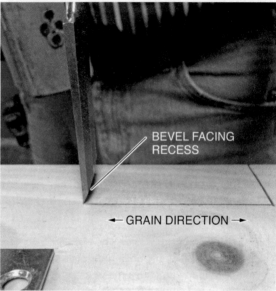

BEVEL FACING RECESS

← GRAIN DIRECTION →

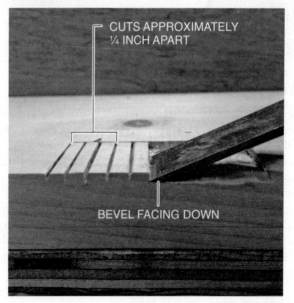

CUTS APPROXIMATELY ¼ INCH APART

BEVEL FACING DOWN

00103-15_F06.EPS

Figure 6 Proper use of a wood chisel.

CENTER PUNCH

PRICK PUNCH

TAPERED PUNCH

00103-15_F07.EPS

Figure 7 Punches.

1.2.3 Safety and Maintenance

Follow these guidelines for safety and proper maintenance of chisels and punches:

- Always wear the appropriate PPE, such as work gloves and safety goggles.
- Make sure the wood chisel blade is beveled at a precise 25-degree angle so it will cut well.
- Ensure the cold chisel blade is beveled at a 60-degree angle so it will cut well.
- Sharpen the cutting edge of a chisel on an oil-stone to produce a keen edge.
- Inspect the point of the punch to ensure it is sharp and not damaged in any way. If necessary, have it sharpened.
- Don't use a chisel, punch, or hammer head that has become mushroomed or flattened (*Figure 8*).
- Ensure the right tool for the job is used.

WARNING!	Striking a chisel or punch that has a mushroom-shaped head can cause metal chips to break off. These flying chips can cause serious injury. If a chisel has a mushroom-shaped head, remove the tool from service until it is repaired or replaced according to company policy.

1.3.0 Screwdrivers

A screwdriver is used to tighten or remove screws. It is identified by the type of screw it fits. The most common screwdrivers are slotted (also known as straight, flat, or standard tip) and Phillips head screwdrivers. Other specialized screwdrivers such as a clutch-drive, Torx®, Robertson®, and Allen® head (hex) may also be needed. The three latter screw heads offer greater grip and **torque** than straight or Phillips screws generally offer. *Figure 9* shows six common types of screw heads.

The screwdrivers designed to fit theses screw heads are described as follows:

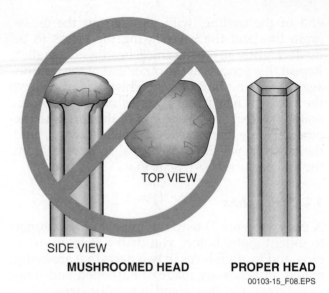

TOP VIEW

SIDE VIEW
MUSHROOMED HEAD   **PROPER HEAD**

00103-15_F08.EPS

Figure 8 Chisel damage.

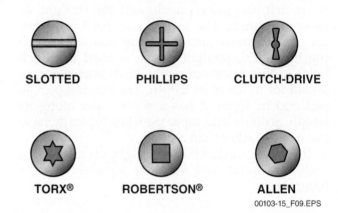

SLOTTED   **PHILLIPS**   **CLUTCH-DRIVE**

TORX®   **ROBERTSON®**   **ALLEN**

00103-15_F09.EPS

Figure 9 Common screw heads.

- *Slotted* – The most common type of standard screwdriver; it fits slotted screws.
- *Phillips* – The most common type of crosshead screwdriver; it fits Phillips head screws.
- *Clutch-drive* – Has an hourglass-shaped tip that is especially useful for extra holding power, such as when working on cars or appliances.
- *Torx®* – Has a star-shaped tip that is useful for replacing parts such as tailgate lenses. It is widely used in automobile repair work. Torx® screws are used in household appliances as well as lawn and garden equipment.
- *Robertson®* (*square*) – Has a square drive that provides high torque power. These screwdrivers may be color-coded according to size.
- *Allen®* (*hex*) – Works with socket-head screws with a female, hex-shaped recess. Hex-key (Allen®) wrenches are L-shaped, hexagonal (six-sided) steel bars. Note that this wrench fits inside the fastener. **Hex-key wrenches** come in an L-shape or in a T configuration with a handle.

To choose the right screwdriver and use it correctly, it helps to know something about screwdriver construction. Each section has a name, as shown in *Figure 10*. The handle is designed to give you a firm grip. The shank is the hardened metal portion between the handle and the blade. The blade is the formed end that fits into the head of a screw. Industrial screwdriver blades are made of tempered steel to resist wear and to prevent bending and breaking. Good-quality screwdrivers have shanks that run through the length of the handle. Some also have a tip that is hardened, resulting in the darker tip color shown in *Figure 10*.

It is important to choose the right screwdriver for the screw. The blade should fit snugly into the screw head and not be too long, too short, loose, or tight (*Figure 11*). If you use the wrong size blade, you might damage the screwdriver or the screw head. Note that there are several different sizes of each screw type. In addition, there are several styles of screws that are commonly called Phillips screws, but with slight differences. A Reed and Prince screw, for example, looks like the Phillips head but comes to a point at the socket base. The base of a Phillips head screw slot is slightly rounded. Using the wrong screwdriver on these two styles will result in slippage and screw damage.

It is very important to use a screwdriver correctly. Using one the wrong way can damage the screwdriver or **strip** the screw head. Wear work gloves to protect your hands from blade slippage. Once the right size and type of blade for the screw head is chosen, position the shank perpendicular (at a right angle) to your work. Apply firm, steady pressure to the screw head and turn clockwise to tighten, or counterclockwise to loosen.

1.3.1 Safety and Maintenance

Properly maintaining a screwdriver not only gives it a longer life, it also helps to prevent injuries. The following guidelines provide for the proper use and maintenance of screwdrivers:

- Keep the screwdriver free of dirt, grease, and grit so the blade will not slip out of the screwhead slot, causing injury or equipment damage.
- Visually inspect your screwdriver before using it. If the handle is worn or damaged, or the tip is not straight and smooth, the screwdriver should be repaired or replaced.
- Never use the screwdriver as a punch, chisel, or pry bar.
- Never use a screwdriver near live wires or as an electrical tester.
- Do not expose a screwdriver to excessive heat.
- Do not use a screwdriver that has a worn or broken handle.
- Never point the screwdriver blade toward yourself or anyone else.

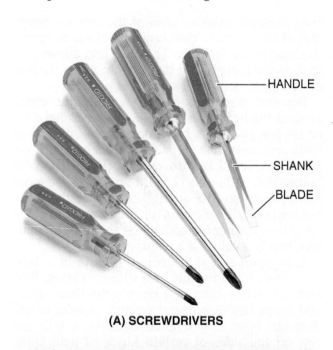

(A) SCREWDRIVERS

HANDLE

SHANK

BLADE

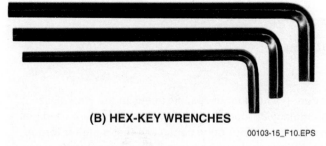

(B) HEX-KEY WRENCHES

00103-15_F10.EPS

Figure 10 Screwdrivers and hex-key wrenches.

Screws

Screws hold better than nails in most situations. The spiral ridges (threads) help hold the screw tightly inside the material, unlike the smooth surface of most nails. Self-tapping screws end in a sharp point and have sharp threads. These types of screws cut their own threads in the material, eliminating the need to drill a starter hole. In woodworking, however, making a small starter hole with a drill helps keep the wood from splitting. This is especially important in cabinetry, trim work, and furniture construction.

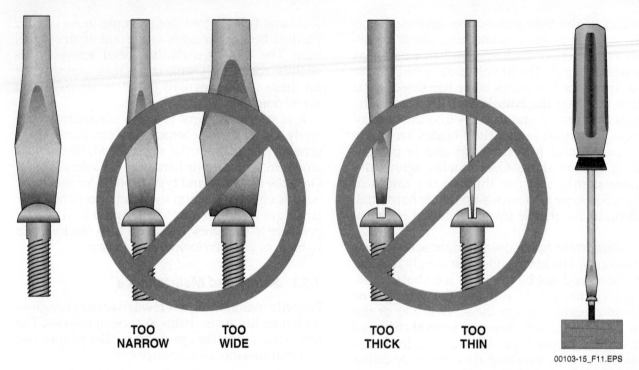

| TOO NARROW | TOO WIDE | | TOO THICK | TOO THIN |

00103-15_F11.EPS

Figure 11 Proper use of a screwdriver.

1.4.0 Wrenches

Wrenches are used to hold and turn screws, nuts, bolts, and pipes. There are many types of wrenches, but they fall into two main categories: non-adjustable and adjustable. Non-adjustable wrenches fit only one size nut or bolt. They come in both Imperial and metric sizes. Adjustable wrenches can be adjusted to fit different-sized nuts and bolts.

1.4.1 Non-Adjustable Wrenches

Non-adjustable wrenches (*Figure 12*) include the open-end wrench, the box-end wrench, the combination wrench, and the striking wrench.

The open-end wrench is one of the easiest wrenches to use. It has an opening at each end that determines the size of the wrench. Often, the wrench has different-sized openings on each end, such as 7/16" and 1/2" (10.0 mm and 12.7 mm). These sizes measure the distance between the flats (straight sides or jaws of the wrench opening) of the wrench, matching the distance across the head of the fastener. The open end allows you to slide the tool around the fastener when there is not enough room to fit a box-end wrench.

Box-end wrenches form a continuous circle around the head of a fastener. The ends have six or twelve points. A hexagonal shape, typical of bolts and nuts, has six sides. Wrenches with six points are less likely to round-off the flats of a bolt or nut. Although 12-point wrenches are more likely to do so, the advantage is that there are twelve possible positions for the wrench to settle on to the fastener instead of just six. In tight quarters, this is a valuable advantage.

Like open-end wrenches, the ends of a box-end wrench are usually two different sizes. Box-end wrenches offer a firmer grip than open-end wrenches. A box-end wrench is safer to use than an open-end wrench because it will not slip off the sides of certain kinds of bolts quite

---

Did You Know?

Types of Screws

The demand for different types of screw heads came about due to a number of specific needs in the workplace. For example, Torx® head screws were developed to be compatible with robotic assembly line equipment used in a number of production applications. Phillips heads were developed with four-point contact so that a higher torque could be applied and the head countersunk into the material. Tamper-resistant (snake eye or pig nose) screws were developed to prevent vandalism to public facilities such as school and park restrooms.

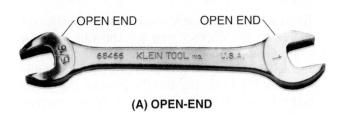

(A) OPEN-END

(B) BOX-END WRENCH

(C) OFFSET RATCHETING BOX WRENCH

(D) COMBINATION

00103-15_F12.EPS

Figure 12 Non-adjustable wrenches.

so easily. Box-end wrenches with a ratcheting feature are popular, but are generally shorter in length, reducing the leverage. Ratcheting box-end wrenches are excellent for assembly work, but are not a good choice for freeing corroded or tight hardware assemblies.

Combination wrenches are, as the name implies, a combination of two types of wrenches. One end of the combination wrench is open and the other is a box-end. Combination wrenches can speed up your work because you don't have to keep changing wrenches. Each combination wrench fits only one size of fastener; each end is the same size.

Striking or slugging wrenches (*Figure 13*) are similar to box-end wrenches in that they have an enclosed circular opening designed to lock onto the fastener. Unlike other wrenches, they are designed to be struck with a hammer to loosen large fasteners. The wrenches have a striking surface to be used with a mallet or handheld sledgehammer. The ends have 6 or 12 points. Striking wrenches are used only in certain situations, such as when a bolt has become stuck to another material because of rust or corrosion. Striking wrenches can damage screw threads and bolt heads. In some cases, it is best to use a power tool, such as an impact wrench, for the task instead. When using a striking wrench, the wrench should be secured with a lanyard to keep it from flying from the fastener out of control. If you are ever in doubt about whether or not to use a striking wrench, ask your instructor or immediate supervisor. Mushroomed hammering surfaces develop just as they do with chisels and punches. Wrenches damaged in such a way should be removed from service until repaired or replaced according to company policy.

When using a non-adjustable wrench, use the correct size wrench for the nut or bolt. Always

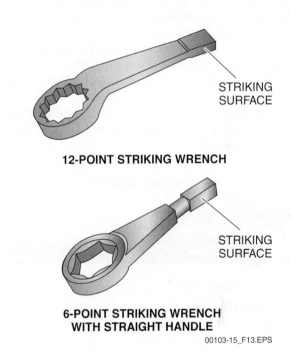

12-POINT STRIKING WRENCH

6-POINT STRIKING WRENCH
WITH STRAIGHT HANDLE

00103-15_F13.EPS

Figure 13 Striking wrenches.

pull the wrench toward you. Pushing the wrench can cause injury.

Be sure that the fit of the wrench is snug and **square** (exactly adjusted) around the nut, bolt, or other fastener. If the fit of the wrench is too loose, it will slip and **round off** or strip the points of the nut or bolt head. Stripped points may make it impossible to remove the fastener.

1.4.2 Adjustable Wrenches

Adjustable wrenches, commonly called Crescent® wrenches due to name recognition, are used to loosen or tighten nuts and bolts like non-adjustable models. Although a fixed-jaw wrench is usually better and safer, an adjustable wrench is a good partner to non-adjustable types. For example, while a nut is being tightened with a box-end wrench, the adjustable wrench might be used to hold the bolt head. Adjustable wrenches have one fixed jaw and one movable jaw. The adjusting nut on the wrench joins the teeth in the body of the wrench and moves the adjustable jaw back and forth. These wrenches typically come in lengths from 4" to 24" (0.10 m to 0.61 m), and open as wide as 2⁷⁄₁₆" (0.05 m). Common types include common adjustable wrenches, **pipe wrenches**, and spud wrenches (*Figure 14*).

Pipe wrenches are used to tighten and loosen all types and sizes of threaded pipe. The upper jaw of the wrench is adjusted by turning the adjusting nut. Both jaws have serrated teeth for gripping power on smooth surfaces. The jaw is spring-loaded and slightly angled so you can release the grip and reposition the wrench without having to readjust the jaw.

The spud wrench operates in the same manner as a pipe wrench. However, the jaws are smooth, so it has no value in gripping round pipe. Instead, the wrench is used on large nuts common to plumbing and piping. These large, thin nuts might be found on the bottom of a sink drain, for example.

To use an adjustable wrench properly, set the jaws to the correct size for the nut, bolt, or threaded pipe. Ensure that tightening pressure is applied to the fixed jaw and not the adjustable jaw (*Figure 15*). Fully tighten the jaws on the piece; it helps to shake the wrench lightly as you use the thumb and forefinger to tighten the adjusting nut. Pull on the wrench to tighten or loosen the component. Pulling towards the body is safer and provides better power than pushing away from the body.

> **WARNING!**
>
> Improperly adjusted jaws could cause the wrench to slip, resulting in loss of balance and serious injury. Whenever possible, pull the wrench toward you. Pushing the wrench can cause injury. If you must push on the wrench, keep your hand open to avoid getting pinched.

1.4.3 Safety and Maintenance

Here are some safety and maintenance guidelines for working with wrenches:

- Focus on your work.
- Pull the wrench toward your shoulder and not toward your face. Leaning into the wrench and pushing could cause serious injury.
- Check the condition of the handle and jaw working surfaces before use.

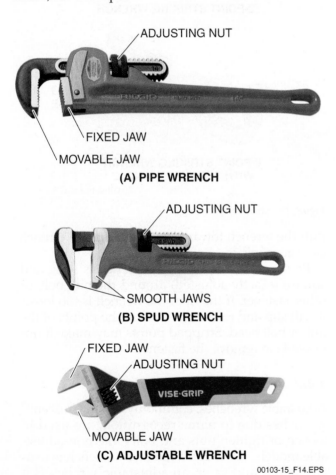

ADJUSTING NUT

FIXED JAW

MOVABLE JAW

(A) PIPE WRENCH

ADJUSTING NUT

SMOOTH JAWS

(B) SPUD WRENCH

FIXED JAW

ADJUSTING NUT

MOVABLE JAW

(C) ADJUSTABLE WRENCH

00103-15_F14.EPS

Figure 14 Adjustable wrenches.

Spud Wrenches

The definition of a spud wrench varies among the trades. Ridgid® and others refer to the example shown in *Figure 14* as a spud wrench. However, another type of spud wrench looks exactly like a common adjustable wrench on one end, while the opposite end comes to a blunt point, shaped like a tapered punch. This wrench was developed to help structural steel workers who need a wrench to tighten large bolts and nuts. However, before the bolt can be inserted through two pieces of steel, another tool is required to align the holes. The spud wrench handles both tasks without changing tools. In some cases, the wrench on the end is not adjustable, but an open-end or box-style instead.

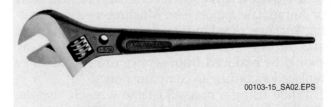

00103-15_SA02.EPS

Figure 15 Loosening a fastener with an adjustable wrench.

- Keep adjustable wrenches clean. Do not allow mud or grease to clog the adjusting screw and slide.
- Don't use the wrench as a hammer.
- Don't use any wrench beyond its capacity. For example, never add an extension to increase its leverage. This could cause serious injury.

1.5.0 Socket and Torque Wrenches

Socket wrenches use a ratcheting mechanism that holds the socket wrench in place when pulled in one direction and releases when pulled in the opposite direction. This enables the user to quickly tighten or loosen a fastener without removing and refitting the wrench after each turn, or when a complete revolution cannot be made because of poor accessibility. Torque wrenches are a wrench and measuring tool, all in one. These wrenches are used to tighten fasteners that require specific amounts of force to be applied.

1.5.1 Socket Wrenches

Socket wrench sets include different combinations of sockets (the part that fits onto and grips

the nut or bolt) and ratcheting handles that are used to turn the sockets. The ratcheting mechanism on the handle attaches to a variety of socket sizes using a square nub. The size of the square nub is referred to as the drive size. Drives are commonly available in ¼", ⅜", and ½" sizes. Much larger sizes are available when required, up to 2½" for industrial work. Ratchet drive sizes are the same globally.

Socket sets are manufactured to fit many Imperial and metric fastener sizes. Most sockets (*Figure 16*) have 6 or 12 gripping points and come in different lengths. Deep sockets, which are longer than standard sockets, allow the socket to accommodate a bolt that protrudes through the nut. A short extension is also shown in *Figure 16*. Extensions are available in a variety of lengths.

Socket sets may contain different types of handles for different uses. The ratchet handle shown in *Figure 16* has a knob used to change the turning direction. Ratchet handles come in a variety of lengths. The longer the ratchet handle, the more leverage can be applied to loosen or tighten the nut or bolt. In addition to ratchets, speed handles and hinge handles are available (*Figure 17*). Speed handles very quickly run a fastener down, but they offer limited torque. However, a speed handle is very handy, especially when a nut is a bit rough and difficult to run down with the fingers. Hinge handles can be used when a ratcheting action is not needed; they are longer and stronger than ratchet handles.

To use sockets and ratchets properly, first select a socket that fits the fastener to be loosened. Place the square end of the socket over the spring-loaded button on the ratchet shaft. Place the socket over the fastener and pull on the handle in the appropriate direction to turn the fastener. Reverse direction by using the button or lever located at the head of the ratchet handle.

Around the World
Metrics and Tools

It is important to know whether the hardware and tools you are working with are metric or Imperial. A proper fit will be unattainable if you try to use an Imperial tool on a metric part. For example, if an Imperial socket is used on a metric bolt, it may tear the points off the bolt head or nut.

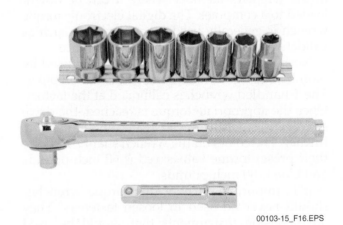

00103-15_F16.EPS

Figure 16 Socket set.

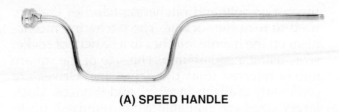

(A) SPEED HANDLE

(B) HINGE HANDLE

00103-15_F17.EPS

Figure 17 Other socket drive tools.

1.5.2 Torque Wrenches

A torque wrench is utilized only when a torque setting is specified for a particular bolt. Torque wrenches measure resistance while applying a twisting force using a common socket. Torque is measured in units of distance × force. In the Imperial system, torque values are usually stated in inch-pounds for small fasteners or foot-pounds for large fasteners. Inches and feet are the units of distance and the pound is the unit of applied force. In the metric system, the unit of torque measure is Newton-meters. Newtons are the measure of force and the meter is the related distance. Users of the metric system often refer to torque as *moment*. Conversion charts allow for easy conversion between the two measurement systems. Torquing procedures will be taught in future modules if the skill is relevant to your chosen craft.

There are three types of commonly used torque wrenches (*Figure 18*). Click-type torque wrenches are preset with the desired torque by adjusting the handle. Once the preset torque is reached, the user feels a slight give in the handle and a clicking sound may be heard. With proper calibration, the benefits of this design include precision and accuracy.

Digital electronic torque wrenches display the desired unit of force needed to tighten a fastener. Measurement details and limit values can be stored and displayed while tightening. The torque values reached on fasteners can even be stored in the wrench's memory where it can be downloaded to a computer. The digital electronic torque wrench is generally used for applications such as validation, in-process, and quality assurance.

No-hub torque wrenches are typically used by plumbers to tighten clamping bands on soil pipes. The T-handled wrench is calibrated at the factory. Once the appropriate torque is reached, the drive socket will stop turning so it does not tighten any further. The colors on the wrench shanks indicate their preset torque values; red is 60 inch-pounds and blue is 80 inch-pounds.

It is important to note that torque wrenches should never be used to loosen fasteners. They are precision instruments that should be used

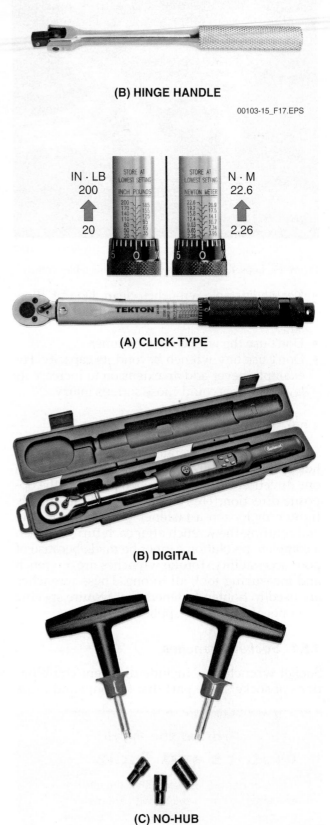

(A) CLICK-TYPE

(B) DIGITAL

(C) NO-HUB

00103-15_F18.EPS

Figure 18 Torque wrenches.

only when precision measurement is required. Use other non-precision hand tools to loosen and remove fasteners.

1.5.3 Safety and Maintenance

Follow these guidelines to maintain and use socket, ratchet, and torque wrenches safely:

- Never use a cheater pipe or bar (a longer piece of pipe slipped over the ratchet handle to provide more leverage). This could snap the tool or break the head off the bolt.
- Thoroughly clean sockets and ratchets after each use. Grease and grit can collect in the socket and ratchet nub, causing it to slip and shear the inside edges.
- Before using a torque wrench of any kind, be sure it is properly calibrated for accuracy and reliability. Check torque settings twice.
- Pay close attention. Over-torquing a nut can cause the bolt head to break.

1.6.0 Pliers and Wire Cutters

Pliers are not generally considered wrenches; they are hinged tools. The jaws are adjustable because the handles move around a hinge point. Pliers are generally used to hold, cut, and bend wire and soft metals. Pliers should not be used for tightening or loosening hardware such as bolts and nuts. If any significant torque is applied, they will round off the corners of the bolt or nut head; as a result, wrenches may no longer fit properly on the damaged fastener.

High-quality pliers are made of hardened steel. Pliers come in many different head styles, depending on their use (*Figure 19*). The following types of pliers are the most commonly used. However, it is important to note that there are many variations on these styles:

- Slip-joint pliers
- Long-nose pliers
- Lineman's pliers
- Tongue-and-groove pliers
- Locking pliers

1.6.1 Slip-Joint Pliers

Slip-joint pliers are used to hold and bend wire and to grip and hold objects during assembly operations. They have adjustable jaws with two possible jaw-opening positions. To use slip-joint pliers properly, place the jaws on the object to be held and squeeze the handles until the pliers grip the object. Adjust the jaw position to improve the grip if necessary.

1.6.2 Long-Nose Pliers

Long-nose (needle-nose) pliers are used to get into tight places where other pliers won't reach, or to grip parts that are too small to hold with your fingers. These pliers are useful for bending angles or circles in wire or narrow metal strips. Most have a set of wire cutting jaws near the pivot. Long-nose pliers, like many other types of pliers, are available with spring openers. This is a spring-like device between the handles that keep the handles apart—and the jaws open—unless you purposely close them. This device makes long-nose pliers easier to use.

Tool Lanyards

Tools are as important when working at heights as they are when working on the ground. However, when working at heights, a dropped tool becomes a serious hazard. If the tool has moving parts, such as a battery-powered drill, the fall may destroy it. Almost any dropped tool can cause a serious injury, especially when the impact comes as a complete surprise.

Tool lanyards specifically designed for work in an elevated environment have been introduced by companies like Snap-on®. Tethering tools to the tool belt or wrist can prevent injuries, protect the tools, and eliminate the time lost to retrieve a dropped tool.

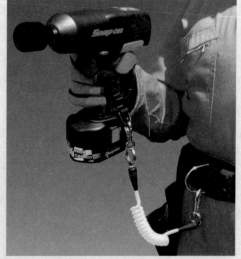

00103-15_SA03.EPS

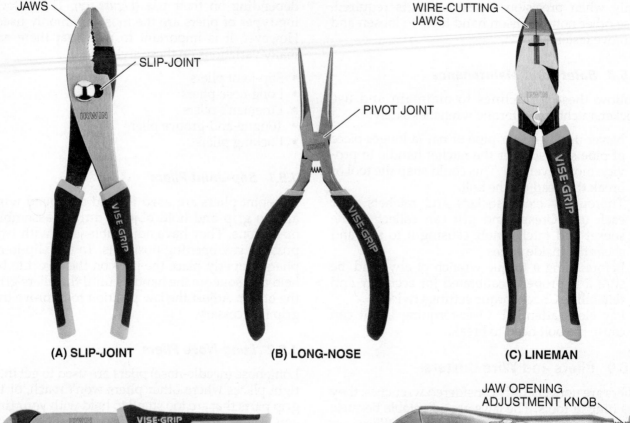

(A) SLIP-JOINT JAWS / SLIP-JOINT

(B) LONG-NOSE PIVOT JOINT

(C) LINEMAN WIRE-CUTTING JAWS

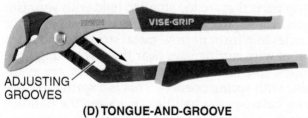

(D) TONGUE-AND-GROOVE ADJUSTING GROOVES

(E) LOCKING JAW OPENING ADJUSTMENT KNOB

00103-15_F19.EPS

Figure 19 Types of pliers.

The following are guidelines for using long-nose pliers properly:

- If the pliers do not have a spring between the handles to keep them open, place your third or little finger inside the handles to keep them open.
- To cut a wire, squeeze the handles to cut at a right angle to the wire.

1.6.3 Lineman Pliers

Lineman pliers, also known as side cutters, have wider, heavier jaws than slip-joint pliers. They are typically used to cut heavy or large-gauge wire and to hold work. The wedged jaws reduce the chance that wires will slip, and the hook bend in both handles allows for a better grip.

To properly cut wire with lineman pliers, always point the loose end of the wire down. Squeeze the handles to cut at a right angle to the wire.

1.6.4 Tongue-and-Groove Pliers

Tongue-and-groove pliers have serrated teeth that grip flat, square, round, or hexagonal objects. You can set the jaws in any of several positions by slipping the curved ridge into the desired groove. Large tongue-and-groove pliers are often used to grip pipe because the longer handles give more leverage. They are available with three basic jaw designs: straight, smooth, and V-shaped or curved. V-shaped or curved jaws work best with round objects. Straight jaws, as shown on the pliers in *Figure 20*, remain parallel to each other and are better for flat material. Smooth jaws are chosen when serrated jaws may damage a delicate surface.

Use the pliers by opening them to the widest position and place the jaws on the object to be held. Determine which groove provides the correct position and adjust, and then squeeze the handles until the pliers grip the object.

00103-15_F20.EPS

Figure 20 Straight-jaw tongue-and-groove pliers.

1.6.5 *Locking Pliers*

Locking pliers clamp firmly onto objects the way a vise does. Due to brand recognition, they are often called Vise-Grips®. They are especially handy for holding metal components together for welding, and for gripping a nut or bolt that has been rounded. A knurled knob on the handle is used to adjust jaw spacing. Simply close the handles to lock the pliers, and release the lock by depressing the lever to open the jaws. There are many different styles of locking pliers; many of these styles have been developed for unique applications. In *Figure 21*, large-jaw locking pliers are being used in a fabrication shop. Note the unusual construction of the locking pliers used to position and hold these materials.

To use locking pliers, first place the jaws on the object to be held. Turn the adjusting screw in the handle until the jaws make contact with the work piece and squeeze the handles together to lock the pliers. Re-adjust the jaws if the locking tension is too loose or too tight. Squeeze the release lever to remove the pliers.

1.6.6 *Safety and Maintenance*

Pliers may not seem to be a dangerous tool, but they can cause injury if misused. Proper safety precautions and maintenance of pliers are very important. Here are some guidelines to remember when using pliers:

- Wear appropriate PPE, especially if you cut wire.
- Hold pliers close to the end of the handles to avoid pinching your fingers in the hinge.
- Do not extend the length of the handles for greater leverage. Use a larger pair of pliers instead.

00103-15_F21.EPS

Figure 21 Using locking pliers on the job.

- Hold the short ends of wires to avoid flying metal bits when you cut.
- Always cut at right angles. Don't rock the pliers from side to side or bend the wire back and forth against the cutting blades. Loose wire can fly up and injure you or someone else.
- Oil pliers regularly to prevent rust and to keep them working smoothly.
- Never use pliers around energized electrical wires. Although the handles may be plastic-coated, they are not generally rated to prevent electrical shock.
- Do not expose pliers to direct heat.
- Do not use pliers to turn nuts or bolts; they are not wrenches.
- Do not use pliers as hammers.

Additional Resources

Easy Ergonomics: A Guide to Selecting Non-Powered Hand Tools. National Institute for Occupational Safety and Health (NIOSH), DHHS Publication No. 2004-164. **www.cdc.gov**

Field Guide to Tools. John Kelsey. 2004. Philadelphia, PA: Quirk Books.

1.0.0 Section Review

1. Ball-peen hammers are classified by ____.
   a. length
   b. the size of the head
   (c.) weight
   d. the material the head is made of

2. The cross-cut chisel is made with the cutting edge wider than the body so it does not bind when chiseling ____.
   (a.) deep grooves   7 page
   b. excess wood
   c. dowels
   d. channels

3. Torx® and Robertson® screws were developed to ____.
   (a.) improve grip and the torque that can be applied to a screw
   b. increase the length of screws
   c. reduce the different types of screws used for a project
   d. improve the holding power of a screw in wood

4. What type of wrench *cannot* be used as a screwdriver?
   a. Robertson®
   b. Torx®
   (c.) Combination wrench
   d. Allen®

5. The direction of a ratchet is reversed by ____.
   (a.) using the button or lever on the ratchet handle
   b. simply reversing direction
   c. removing the socket and turning it around
   d. changing the ratchet handle to a reversible ratchet handle

6. What type of pliers work well for holding fittings together on welding projects?
   a. Needle-nose
   (b.) Locking
   c. Tongue-and-groove
   d. Lineman

2.0.0 MEASUREMENT AND LAYOUT TOOLS

Objective

Identify and describe how to use various types of measurement and layout tools.

a. Identify and explain how to use rules and other measuring tools.
b. Identify and explain how to use various types of levels and layout tools.

Performance Tasks

1. Visually inspect the following tools to determine if they are safe to use:
   - Measuring tool
   - Layout tool
   - Level
2. Safely and properly use the following tools:
   - Measuring tool
   - Layout tool
   - Level

Trade Terms

Carpenter's square: A flat, steel square commonly used in carpentry.

Combination square: An adjustable carpenter's tool consisting of a steel rule that slides through an adjustable head.

Planed: Describing a surface made smooth by using a tool called a plane.

Plumb: Perfectly vertical; the surface is at a right angle (90 degrees) to the horizon or floor and does not bow out at the top or bottom.

Rafter angle square: A type of carpenter's square made of cast aluminum that combines a protractor, try square, and framing square.

Try square: A square whose legs are fixed at a right angle.

K nowing how to achieve precise measurements, straight vertical lines, and accurate leveling are skills that form the foundational knowledge of a skilled craftsperson. An incorrect measurement, crooked vertical line, or improper leveling will have adverse effects as the job progresses. It is important to know how to properly use tools to ensure that the outcome of each measurement is precise.

2.1.0 Rules and Other Measuring Tools

In a previous module, you learned how to read various types of measuring tools. This section presents information on these and other types of measuring tools, and includes guidelines for their use.

Craftworkers use four basic types of measuring tools most often:

- Flat steel rule
- Tape measure
- Wooden folding rule
- Laser measuring tool

Accuracy, ease of use, durability, and readability are a several considerations for choosing a measuring tool.

2.1.1 Steel Rule

The flat steel rule (*Figure 22*) is a simple measuring tool. Flat steel rules are usually 6" or 12" long (15 cm or 30.5 cm), but other lengths are available. Steel rules can be flexible or rigid. While flat steel rules are a very accurate measuring tool, they can also be used as a straightedge for laying out lines and cutting.

2.1.2 Measuring Tape

The tape measure blade (*Figure 23*) is marked in $\frac{1}{16}$" increments or smaller. A tape measure may include both Imperial and metric markings. Tape measures are also available with special markings that apply to residential and light commercial construction wall framing.

The concave (or curve) of a tape measure blade is designed to strengthen the blade when it is extended, helping it to hold its shape without bending. Once the blade tip is secure and the proper measurement is established, rotate the blade edge nearest you slightly until it lies flat on the surface, then mark the material. Using this method makes it easier to read the measurement and mark the material more accurately.

2.1.3 Wooden Folding Rule

A wooden folding rule (*Figure 24*) is usually marked in sixteenths of an inch on both edges of each side. Folding rules come in 6' and 8' lengths, with metric versions being commonly available in 2-meter lengths. Because of its stiffness, a folding rule is better than a cloth or steel tape for

Figure 22 Steel rule.

measuring vertical distance. This is because, unlike tape, it can be held straight up. This makes it easier to measure some distances, such as those that might require someone on a ladder to reach one end. Like a tape measure, the folding rule can also have special marks at the 16", 19.2", and 24" increments to make wall framing easier. These markings may differ on metric models due to differences in construction standards.

2.1.4 Laser Measuring Tools

A laser measuring tool (*Figure 25*) is a battery-powered, electronic version of a tape measure. This hand-held tool works by pointing it at a specific object and then pressing a measurement button on the control panel. When the button is pressed,

a laser shoots out at the object and a reading is sent back to the instrument and displayed on the screen. Laser measuring tools can be designed to register and record in both Imperial and metric measurements. These tools are precision instruments, so ensure they are handled with care and stored appropriately.

The following are some of the advantages a laser measuring tool has over a traditional tape measure:

- Measurements required at higher elevations can be taken from ground level.
- Longer measurements can be taken. Some construction laser measuring tools measure from 1' up to 600' (0.3 m up to 200 m). A target plate may be required for long distances with higher power lasers.
- Some laser measuring tools have buttons on the control panel that make length-related calculations (such as addition, subtraction, and area) easier to perform.
- A number of measurements can be electronically stored on the tool, so that measurements can be written down all at once after the task is completed.

Figure 23 Tape measure.

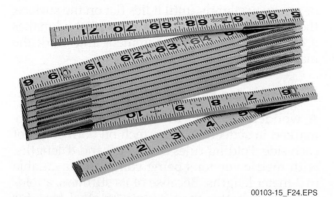

Figure 24 Wooden folding rule.

Figure 25 Laser measuring tool.

- Some laser measuring tools can have built-in electronic levels and spirit levels.

Some disadvantages of a laser measuring tool are that it can have an accuracy range of ¹⁄₁₆" (1.59 mm). Over some distance, this may not represent a problem. It is also difficult to make distant measurements in very hot or smoky environments due to the beam bouncing from particle to particle in the atmosphere, causing the beam to scatter. In addition, some targets may be too reflective, causing the beam to produce inaccurate measurements.

2.1.5 Safety and Maintenance

Follow these guidelines to ensure measuring tools are safely used and maintained properly:

- Occasionally apply a few drops of light oil on the spring joints of a wooden folding rule and steel tape.
- Wipe moisture off steel tape to keep it from rusting.
- Don't kink or twist steel tape, because this could cause it to break.
- Don't use steel tape near exposed electrical parts.
- Don't let laser measuring tools get wet.

2.2.0 Levels and Layout Tools

A level is a tool used to determine how level a horizontal surface is or how **plumb** a vertical surface is. If a surface is described as level, it means it is exactly horizontal. If a surface is described as plumb, it means it is exactly vertical. Levels are used to determine how near to exactly horizontal or exactly vertical a surface is. In most cases, the level allows you to adjust to a level or plumb condition before assembling components.

Layout tools also include carpenter's and combination squares, which are covered here. Note that there are various versions of squares designed specifically for a given craft. For example, although a carpenter's square and a pipefitter's square may appear the same initially, they often have different markings and other information important to their specific type of work. Other layout tools include chalk lines and plumb bobs.

2.2.1 Spirit Levels

The spirit level (*Figure 26*) is the most commonly used leveling instrument in the construction trade. The spirit level got its name because the

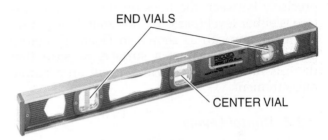

(A) TWO-FOOT LEVEL

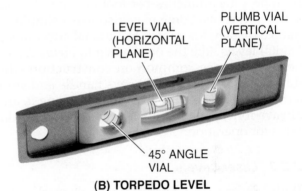

(B) TORPEDO LEVEL

00103-15_F26.EPS

Figure 26 Spirit levels.

Precision Measuring Tools

Laser measuring tools are becoming increasingly common in the construction industry. They allow you to make precise and reasonably accurate measurements. Precision measuring tools, such as micrometers and calipers, make it possible to accurately measure parts that are being machined to one ten-thousandth of an inch (0.0001") or 2.54 microns. However, a standard micrometer's smallest division is 0.001 (25.4 microns). Digital micrometers are available that can read increments as small as 0.00005", or 1.27 microns. Of course, there are metric micrometers as well.

00103-15_SA04.EPS

vials in it are filled with alcohol, which is sometimes called spirits. Alcohol is used because it does not freeze.

Most spirit levels are made of tough, lightweight metals such as magnesium or aluminum, or types of plastic that can reliably maintain their true form. They typically have three vials filled with alcohol. The center vial is used to check for level, and the two end vials are used to check for plumb. Some spirit levels, like the torpedo level shown in *Figure 26*, will also show a true 45-degree angle. Spirit levels come in a variety of sizes; the longer the level, the greater its accuracy.

The amount of liquid in each vial is intentionally not enough to fill it, so there is always a bubble in the vial. When the bubble is centered precisely between the lines on the vial, the surface is either level (perfectly horizontal) or plumb (perfectly vertical) as shown in *Figure 27*. Before placing a level against a surface, be sure that there is no debris that will prevent an accurate measurement.

2.2.2 Digital Levels

Digital levels provide a digital readout of degrees of slope, plus an inches-per-foot of rise to lay out stairs and roofs. Some may also use a simulated bubble display for a more traditional appearance. Digital levels, like the one shown in *Figure 28*, are becoming more common on construction sites due to their versatility. Always handle and store a digital level carefully to avoid damaging it, and always follow the manufacturer's recommendations for operation.

2.2.3 Laser Levels

With a laser level (*Figure 29*), a single worker can accurately and quickly establish plumb, level, or square measurements. Laser levels are used to set foundation levels, establish proper drainage slopes, square framing, and align plumbing and electrical lines. A laser level may be mounted on a tripod, fastened onto pipes or framing studs, or be suspended from ceiling framing. Levels for professional construction jobs are housed in sturdy casings designed to withstand job site conditions. These tools come in a variety of sizes and weights, depending on the application. They are primarily used for leveling over a significant distance. Always handle and store laser levels carefully, and follow the manufacturer's recommendations for use.

> **WARNING!**
>
> Never look directly at the laser beam that is generated by a laser tool. It can impair your vision and damage your eyes.

2.2.4 Safety and Maintenance

All levels are considered to be precision instruments that must be handled with care. Unless the user looks directly at the laser, there is little risk of personal injury when working with this particular tool. Remember these guidelines when working with levels:

- Never look directly at the laser.
- Replace a level if a crack or break appears in any of the vials.
- Keep levels clean and dry.
- Do not bend or apply too much pressure on a level.
- Do not drop or bump a level.

2.2.5 Squares

Squares (*Figure 30*) are used for marking, checking, and measuring. The type of square you use depends on the type of job and your preference.

Bricklayers

Working with your hands to create buildings is a time-honored craft. People have used bricks in construction for more than 10,000 years. Archaeological records prove that bricks were one of the earliest man-made building materials.

In bricklaying, it is important that each course (row) of bricks is level and that the wall is straight. An uneven wall is weak as well as unattractive. To ensure that the work stays true, a bricklayer uses a straight level and a plumb line.

Today, bricklayers build walls, floors, partitions, fireplaces, chimneys, and other structures with brick, precast masonry panels, concrete block, and other materials. They lay brick for houses, schools, sports stadiums, office buildings, and other structures. Some bricklayers specialize in installing heat-resistant firebrick linings inside huge industrial furnaces, called refractory brick.

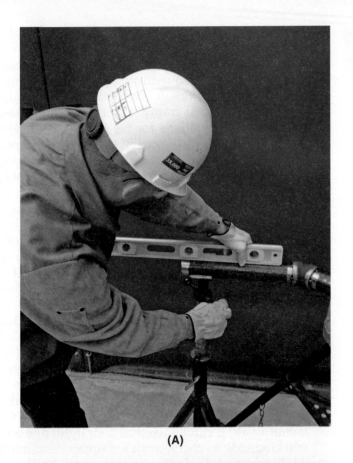

(A)

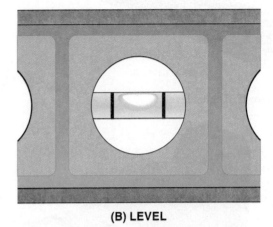

(B) LEVEL

(C)

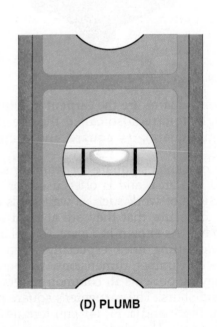

(D) PLUMB

00103-15_F27.EPS

Figure 27 Levels showing level and plumb.

00103-15_F28.EPS

Figure 28 Digital level.

00103-15_F29.EPS

Figure 29 Laser level.

Common squares are the **carpenter's square**, **rafter angle square** (also called the speed square or magic square), **try square**, and **combination square**.

The carpenter's square, or framing square, is shaped like an L and is often used for squaring up sections of work such as wall studs and sole plates to ensure that they are at right angles to each other. As shown in *Figure 31*, it is also used to mark cutting lines, especially when working with wide lumber dimensions. Carpenters use it for laying out cuts in common rafters, hip rafters, and stairs. The carpenter's square has a 24" (61 cm) blade and a 16" (41 cm) tongue, forming a right angle. The blade and tongue are marked with Imperial inches and fractions of an inch, or metric centimeters and millimeters. You can use the blade and the tongue as a rule or a straightedge. Tables and formulas are printed on the blade for making quick calculations such as determining area and volume.

The rafter angle square is another type of carpenter's square, frequently made of cast aluminum or tough plastics. It is a combination protractor, try square, and framing square. It is marked with degree gradations for fast, easy layout. The square is small, so it is easy to store and carry. By clamping the square on a piece of lumber, you can use it as a guide when cutting with a portable circular saw.

The try square has a fixed, 90-degree angle and is used mainly for woodworking. Like other squares, it can be used to lay out cutting lines at 90-degree angles; to check the squareness of adjoining surfaces; to check a joint to ensure it is square; and to check if a **planed** piece of lumber is warped or cupped (bowed).

The combination square has a ruled blade that slides through a head. The position of the ruled blade can be changed in relation to the head. The head is marked with 45-degree and 90-degree angle measures. Some squares also contain a small spirit level and a metal scribe. The combination square is one of the most useful tools for layout work. It can be used as a straightedge and marking tool; to check work for squareness; mark 90-degree and 45-degree angles; check level and plumb surfaces; and measure lengths and widths. Good combination squares have all-metal parts, a blade that slides freely but can be clamped securely in position, and a glass-tube spirit level.

To mark a 90-degree angle using a combination square (*Figure 32*), set the blade at a right angle (90 degrees). Position the square so that the head fits snugly against the edge of the material to be marked. Start at the edge of the material and use the blade as a straightedge to guide the mark.

To mark a 45-degree angle using a combination square (*Figure 33*), set the blade at a 45-degree angle. Position the square so that the head fits snugly against the edge of the material to be marked. Start at the edge of the material and use the blade as a straightedge to guide the mark.

2.2.6 *Use and Maintenance*

Follow these guidelines for using squares:

- Do not use a square for something it wasn't designed for, especially prying or hammering.
- Keep the square dry to prevent it from rusting.
- Use a light coat of oil on the blade, and occasionally clean the blade's grooves and the setscrew.
- Do not bend a square.

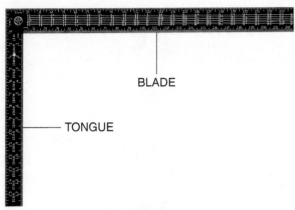

CARPENTER'S SQUARE

RAFTER ANGLE SQUARE
(SPEED SQUARE)

TRY SQUARE

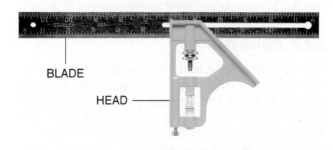

COMBINATION SQUARE

00103-15_F30.EPS

Figure 30 Types of squares.

00103-15_F31.EPS

Figure 31 Marking a cutting line.

- Do not drop or strike the square hard enough to change the angle between the blade and the head.

A square can be checked to ensure that it has not become distorted. Use it against a straight and true surface to scribe a line. Then turn the square over and scribe a line in the same location. The two lines should be identical and true to each other.

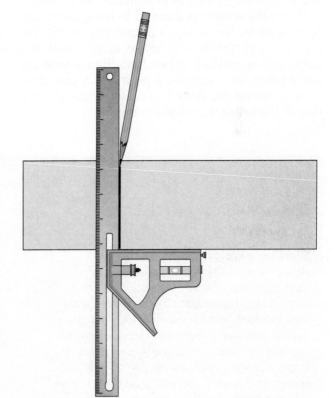

00103-15_F32.EPS

Figure 32 Using a combination square to mark a 90-degree angle.

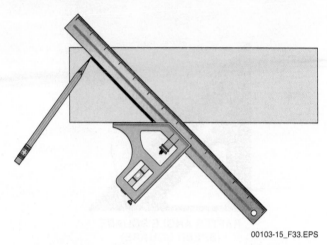

00103-15_F33.EPS

Figure 33 Using a combination square to mark a 45-degree angle.

2.2.7 Plumb Bob

The plumb bob, which is a pointed weight attached to a string (*Figure 34*), uses the force of gravity to make the line hang vertical, or plumb. Plumb bobs come in different weights, with 20-, 16-, 12-, and 8-ounce (≈566-, 454-, 340-, and 227 g) weights being the most common. Although metric equivalents are provided here, metric plumb bobs may be found in even weights such as 570 or 450 grams. The most important criteria is to use one with enough weight to maintain its position.

When the weight is allowed to hang freely and is motionless, the string is plumb (*Figure 35*). You can use a plumb bob to make sure a wall or a doorjamb is vertical. When installing a post under a beam for example, a plumb bob can show what point on the floor is directly under the section of the beam that requires support.

00103-15_F34.EPS

Figure 34 Plumb bob.

Follow these steps to use a plumb bob properly:

Step 1 Make sure the line is attached at the exact top center of the plumb bob.

Step 2 Hang the bob from a horizontal member, such as a doorjamb, joist, or beam. Be careful not to drop a plumb bob on its point; it could be damaged and cause inaccurate readings.

Step 3 When the weight is allowed to hang freely and stops swinging, the string is plumb.

Did You Know?

Geometry and the Square

Some combination squares, like the one shown here, can be set at any angle. The head has a built-in precision protractor. Like other combination squares, the metal rule can slide back and forth in the head, then locked in the desired position. Combination square sets like this one are used to measure and mark any angle, including the common 30-, 45-, 60-, and 90-degree angles. When you use a combination square to measure and mark materials, you are incorporating basic geometric principles to your work.

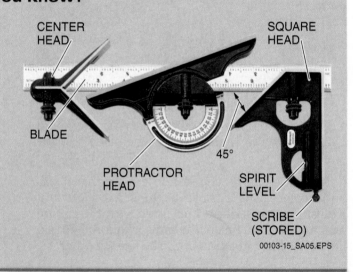

00103-15_SA05.EPS

PLUMB
BOB

00103-15_F35.EPS

Figure 35 Using a plumb bob.

Step 4 When you are using a plumb bob outdoors, be aware that the wind may blow it out of true vertical.

Step 5 Mark the point directly below the tip of the plumb bob. This point is precisely below the point where you attached the bob above.

2.2.8 Chalk Lines

Chalk lines are used to create long, straight lines on smooth surfaces. A chalk line is a tool with a piece of string or cord that is coated with chalk. The line is stretched taut between two points and then snapped to release a chalky line to the surface.

The chalk line contains a line on a reel (*Figure 36*). The case is filled with colored chalk powder. The line is automatically chalked each time you pull it out of the case. Some models can serve as a relatively lightweight plumb bob as well.

Mercury

Years ago, the highest-quality plumb bobs were made with a large cavity in the center, which was then filled with mercury to increase the weight and concentrate the weight directly above the bob's center point. We have since learned about the potential health hazards of mercury and this practice has been discontinued.

00103-15_F36.EPS

Figure 36 Chalk line and chalk.

To snap a chalk line, simply pull the line from the case and secure one end. Stretch the line between the two points to be connected. After the line has been pulled taut, pull the string straight up or away from the work and then release it. This marks the surface underneath with a straight line of chalk (*Figure 37*). Spraying the chalk line with clear lacquer will prevent it from wearing away for a reasonable period of time. Otherwise, the chalk is easily washed away by rain or quickly distorted by foot traffic. Store the chalk line and chalk supply in a dry place, as damp or wet chalk is unusable.

00103-15_F37.EPS

Figure 37 Using a chalk line.

Additional Resources

Easy Ergonomics: A Guide to Selecting Non-Powered Hand Tools. National Institute for Occupational Safety and Health (NIOSH), DHHS Publication No. 2004-164. **www.cdc.gov**

Field Guide to Tools. John Kelsey. 2004. Philadelphia, PA: Quirk Books.

2.0.0 Section Review

1. Why are wooden folding rules good for measuring vertical distances?

    a. They are extra long.
    b. They have hinges.
    c. They are extra wide.
    d. They are stiffer than a measuring tape.

2. What measuring tool is useful when higher elevation measurements need to be taken from ground level?

    a. Laser measuring tool
    b. Wooden folding rule
    c. Tape measure
    d. Steel rule

3. The carpenter's square is also called a ____.

    a. framing square
    b. speed square
    c. try square
    d. magic square

4. What type of square may be used as a guide when using a portable circular saw?

    a. Combination square
    b. Try square
    c. Framing square
    d. Rafter angle square

3.0.0 CUTTING AND SHAPING TOOLS

Objective

Identify and explain how to use various types of cutting and shaping tools.

    a. Identify and explain how to use handsaws.

    b. Identify and explain how to use various types of files and utility knives.

Performance Tasks

1. Visually inspect the following tools to determine if they are safe to use:
   - Hand saw
   - File
   - Utility knife
2. Safely and properly use the following tools:
   - File
   - Utility knife
3. Make a straight, square cut in framing lumber using a crosscut saw.

Trade Terms

Kerf: A cut or channel made by a saw.

Miter joint: A joint made by fastening together usually perpendicular parts with the ends cut at an angle.

Tang: Metal handle-end of a file. The tang fits into a wooden or plastic file handle.

Tenon: A piece that projects out of wood or another material for the purpose of being placed into a hole or groove to form a joint.

There are many different saws on the market. Some have specific purposes, while others can be used for multiple tasks. From logging saws to hacksaws, saws have been used for the past 5,000 years.

3.1.0 Saws

Using the right saw for the job makes cutting easy. The main differences between types of saws are the shape, number, and pitch of their teeth. These differences make it possible to cut straight or curved lines through wood, metal, plastic, or wallboard. As a general rule, the fewer points, or teeth per inch (tpi), on a saw blade, the coarser and faster the cut will be; the more teeth, the slower and smoother the cut will be.

Figure 38 shows several types of saws. The following are brief descriptions of these saws.

> **WARNING!**
> Saw teeth are very sharp. Use gloves and do not handle the saw teeth with bare hands. When cutting with a saw, ensure that your fingers remain clear of the teeth at all times.

- *Backsaw* – The standard blade of this saw is 8" to 14" long (20 cm to 36 cm) with 11 to 14 tpi. A backsaw has a broad, flat blade and a reinforced back edge. It is used for cutting joints, especially **miter joints** and **tenons**. The reinforced back keeps the blade straight and true.
- *Keyhole saw* – The standard blade of this saw is 12" to 14" (30 cm to 36 cm) long with 7 or 8 tpi. This saw cuts curves quickly in wood, plywood, or wallboard. It is also used to cut holes for large-diameter pipes, vents, and plugs or switch boxes.
- *Coping saw* – This saw has a narrow, flexible blade attached to a U-shaped frame. Holders at each end of the frame can be rotated so you can cut at an angle to the frame. They also adjust to maintain proper blade tension. Standard blades range from 10 to 20 tpi. The coping saw is used for making irregular-shaped moldings fit together cleanly. As a general rule, coping saw blades are mounted with the teeth pointing towards the handle, making it cut on the pull stroke. However, not all woodworkers agree, and the blade can be placed in the opposite direction, for cutting on the push stroke, if the situation calls for it.
- *Drywall saw* – A drywall saw is a long, narrow saw used to cut softer building materials, such as drywall. Drywall saws can have a fixed blade or a retractable blade held to either a wood or plastic handle with thumb screws. The blade on a drywall saw has a very sharp point to easily poke a hole to start a cut without drilling a starter hole. The jigsaw and spiral saw are power tools that are sometimes used in the same applications as a drywall saw.
- *Handsaw* – The standard blade of this saw is 26" long with 8 to 14 tpi for a crosscut saw and 5 to 9 tpi for a ripsaw. A crosscut saw, as the name implies, is designed to cut wood across the grain. A ripsaw is designed to cut wood in the same direction as the grain.
- *Hacksaw* – The standard blade of this saw is 8" to 16" (20 cm to 41 cm) long with 14 to 32 tpi. It has a sturdy frame and a pistol-grip handle.

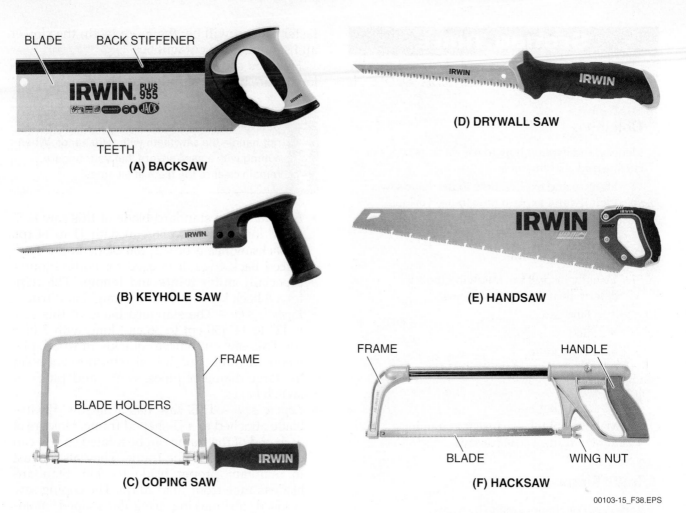

Figure 38 Types of saws.

The blade is tightened using a wing nut and bolt. The hacksaw is used to cut through metal such as nails, bolts, or pipe. When installing a hacksaw blade, be sure that the teeth face away from, not toward, the saw handle. Hacksaws are designed to cut on the push stroke, not on the pull stroke.

3.1.1 Handsaws

The handsaw's blade is made of tempered steel so it will stay sharp and will not bend or buckle.

Handsaws are classified mainly by the number, shape, size, slant, and direction of the teeth. Saw teeth are set or angled alternately in opposite directions to make a cut (or **kerf**) slightly wider than the thickness of the saw blade itself. Two common types of handsaws are the crosscut saw and the ripsaw. However, there are many variations today in blade design that make some versions more universal in their use.

The crosscut saw, which has 8 to 14 tpi, is designed to cut across the grain of wood, cutting slower but smoother than a ripsaw. Blade lengths

Did You Know?

The Romans and Their Tools

Ancient Egyptians and Greeks used a variety of tools. The Romans, however, are known as the toolmakers of the ancient world. The plane, metal-cutting saw, drawknife, frame saw, level, square, and claw hammer were all created by the Romans.

00103-15_SA06.EPS

range from 20" to 28" (51 cm to 71 cm). For most general uses, 24" or 26" (61 cm or 66 cm) is a good length.

Follow these steps to use a crosscut saw properly:

Step 1 Mark the cut to be made with a square or other measuring tool. Also place an X or similar mark on one side of the squared line to show on which side of the line the cut should be made.

Step 2 Make sure the piece to be cut is well-supported on a sawhorse, jack, or other support. Clamp or otherwise secure the material. Support the scrap end as well as the main part of the wood to keep it from splitting as the kerf nears the edge. With short pieces of wood, you can support the scrap end of the piece with your free hand. With longer pieces, you will need additional support or assistance.

Step 3 Don work gloves before beginning. Place the saw teeth on the edge of the wood farthest from you, just at the outside edge of the mark. Note that the edge of the blade should be aligned with the line on the waste side of the lumber.

Step 2 Start the cut with the part of the blade closest to the handle end of the saw, pulling the first stroke toward your body.

Step 4 Use the thumb of the hand that is not sawing to guide the saw so it stays vertical to the work.

Step 5 Place the saw at about a 45-degree angle to the wood and pull the saw to make a small groove.

Step 6 Start sawing slowly, increasing the length of the stroke as the kerf deepens. Remember that the cutting is done on the push stroke. Therefore, the push stroke is often a bit more forceful and faster than the pull stroke.

Step 7 Do not push or ride the saw into the wood. Let the weight of the saw set the cutting rate. This will make it easier to control the saw and is less tiring.

Step 8 Continue to saw with the blade at a 45-degree angle to the wood. If the saw starts to wander from the line, angle the blade back toward it. If the saw blade sticks in the kerf, wedge a thin piece of wood into the cut to hold it open.

The ripsaw has 5 to 9 tpi, and is designed to cut with the grain (parallel to the wood fibers), meeting less resistance than a crosscut saw. Because it has fewer teeth than the crosscut saw, it will make a coarser, but faster cut. To use a ripsaw properly, mark and start a ripping cut the same way you would start cutting with a crosscut saw (*Figure 39*). Once you've started the kerf, saw with the blade at a steeper, 60-degree angle to the wood.

3.1.2 Safety and Maintenance

Saws must be maintained for them to work safely and properly. Also, it is very important to focus on the work when sawing—saws can be dangerous if used incorrectly or if you are not paying attention. Here are the guidelines for working with handsaws:

- Brace yourself when sawing so you are not thrown off balance on the last stroke.
- Clamp or otherwise secure the workpiece to prevent it from moving while sawing.
- Clean your saw blade with a fine emery cloth and apply a coat of silicone lubricant if it starts to rust.
- Always lay a saw down gently.
- Have saws sharpened by an experienced sharpener.
- Do not let saw teeth come in contact with stone, concrete, or metal.

00103-15_F39.EPS

Figure 39 Kerf cut across the wood grain.

3.2.0 Files and Utility Knives

Files and rasps are shaping tools that can be used in areas that chisels cannot reach. They are commonly used to remove wood and metal burrs and smooth rough spots to either finish a surface or prepare it for the next step.

As the name implies, the utility knife is used as a general purpose tool. It is not designed for one particular use. Because of its versatility, utility knives are now quite common among all trades.

> **WARNING!**
>
> All types of files and utility knives have sharp points and edges that can cause serious injury. Handle knives and files with extreme caution and the greatest of care. Never use a file without first installing a handle.

3.2.1 Files and Rasps

Files and rasps are used to cut, smooth, or shape metal and wood parts. A variety of files are shown in *Figure 40*.

Files have slanting rows of teeth, while rasps have individual teeth which are more aggressive (deeper). Rasps are designed primarily for use on wood, and files are used on metal. Files and rasps are usually made from a hardened piece of high-grade steel. They are sized by the length of the body (*Figure 41*). The size does not include the handle because the handle is generally separate from the file or rasp. The sharp metal point at the end of the file, the tang, fits into the handle. Handles can easily be transferred from one file to another. For most sharpening jobs, files and rasps range from 4" to 8" (20 cm to 36 cm) in length.

Choose a file or rasp with a shape that fits the area to be filed. Files are available in round, square, flat, half-round, and triangular shapes (*Figure 42*). For filing large concave (curved inward) or flat surfaces, a half-round shape is used.

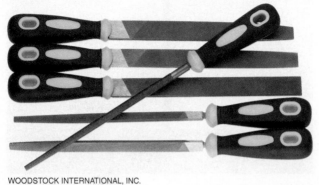

WOODSTOCK INTERNATIONAL, INC.

00103-15_F40.EPS

Figure 40 Files with handles.

For filing small curves or for enlarging and smoothing holes, a round shape with a tapered end is used. For filing angles, a triangular file is used.

There is a specific type of file for each of the common soft metals, hard metals, plastics, and wood. In general, the teeth of files for soft materials are very sharp and widely spaced. Those for hard materials are less sharp and closer together. The shape of the teeth also varies depending on the material to be worked. Using a file designed for soft material on hard material will quickly chip and dull the teeth; whereas using a file designed for hard material on soft material will clog the teeth.

File classifications include single-cut, with all rows of teeth facing the same direction; and double-cut, with rows of teeth crisscrossing each other, forming a diamond pattern. Files are also classified according to their texture and depth of teeth. These include rough, coarse, bastard, second cut or medium, and smooth. Obviously, smoother files apply a smoother finish. Smooth files have more cuts per inch across the face, and the cuts are shallow. Bastard files are relatively coarse models, classified just below coarse files. The unusual name *bastard file* stems from the fact that it is an irregularity in the file line-up. Coarse files have fewer, deeper cuts (teeth) per inch. Rasps are also classified by the nature of their teeth: coarse, medium, and fine. *Table 1* lists types of files and some uses for each.

Trying to use a file the wrong way is inefficient and can be frustrating. Follow these steps to use a file properly:

Step 1 Wear work gloves. Ensure the file handle is securely tightened.

Step 2 Mount the work to be filed in a vise at approximately elbow height.

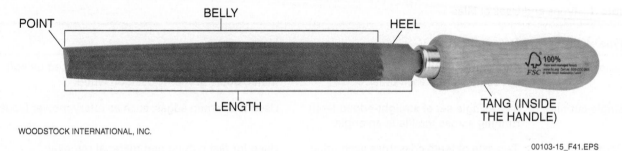

POINT

BELLY

HEEL

LENGTH

TANG (INSIDE THE HANDLE)

00103-15_F41.EPS

Figure 41 Parts of a file.

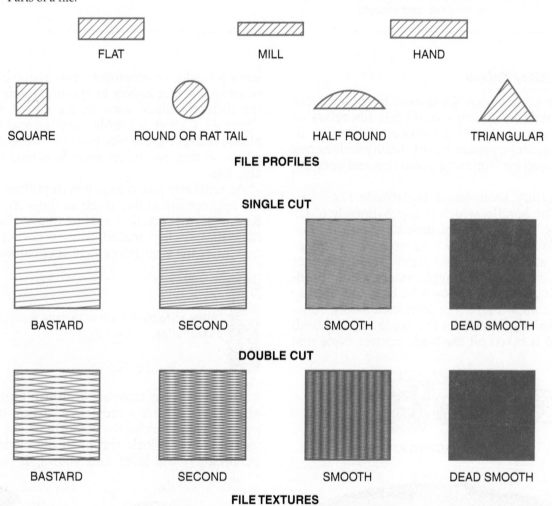

FLAT

MILL

HAND

SQUARE

ROUND OR RAT TAIL

HALF ROUND

TRIANGULAR

FILE PROFILES

SINGLE CUT

BASTARD

SECOND

SMOOTH

DEAD SMOOTH

DOUBLE CUT

BASTARD

SECOND

SMOOTH

DEAD SMOOTH

FILE TEXTURES

00103-15_F42.EPS

Figure 42 File profiles and textures.

Step 3 Do not lean directly over the work. Stand back from the vise with feet approximately 24" (61 cm) apart, with the right foot ahead of the left. (If left-handed, the left foot will be ahead of the right.)

Step 4 Hold the file handle with one hand and steady the tip of the blade with the other hand.

Step 5 For average work, hold the tip with your thumb on top of the blade and the first two fingers under it. For heavy work, use a full-hand grip on the tip.

Step 6 Apply pressure only on the forward stroke. The file cuts on the push stroke. Placing pressure on the file during the back stroke while the file remains in contact with the workpiece only wears out the file.

Step 7 Raise the file from the work on the back stroke to prepare for the next stroke.

Step 8 Keep the file flat on the work.

Table 1 Types and Uses of Files

Type	Description	Uses
Rasp-cut file	The teeth are individually cut; they are not connected to each other.	Leaves a very rough surface. Can be used on soft metal, but is primarily used on wood.
Single-cut file	Has a single set of straight-edged teeth running across the file at an angle.	Used to sharpen edges, such as rotary mower blades.
Double-cut file	Two sets of teeth crisscross each other. Types are bastard (roughest cut), second cut, and smooth.	Used for fast cutting and material removal.

3.2.2 Utility Knives

A utility knife (*Figure 43*) is used to cut a variety of materials including roofing felt, fiberglass or asphalt shingles, vinyl or linoleum floor tiles, fiberboard, and gypsum board. Utility knives can also be used for trimming insulation and opening cartons.

The utility knife has a replaceable razor-like blade. The handle, which is approximately 6" (15 cm) long and is made of die-cast metal or plastic, holds the blade. The least expensive, simpler models are made in two halves, held together with a screw for blade replacement. Other models, such as the ones shown in *Figure 43*, fold to a more compact size. For increased safety, some utility knives self-retract the blade when thumb pressure is taken off the blade control. Note that

some job sites or employers may require the use of self-retracting knives to maximize safety. Note the different blade used in the carpet knife in *Figure 43*. Carpet is highly abrasive and hard on blades. Carpet knife blades have two sharpened edges, so they can be reversed to extend their usable life.

As with any sharp tool, it is important to wear the appropriate gloves (such as those made from Kevlar®) for protection against injury (*Figure 44*). Note that Kevlar® materials are reported to have five times the strength of steel at equal weights.

> **WARNING!**
>
> Utility knife blades are extremely sharp, having a razor edge.

To use a retractable-blade utility knife safely and properly, place a protective barrier, such as a piece of wood, under the object to be cut. Always wear gloves. Unlock and push the blade out by pushing on the lever and sliding it outwards.

SELF-RETRACTING KNIFE

FOLDING UTILITY KNIFE

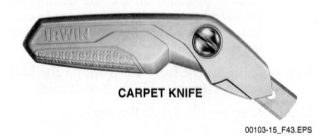

CARPET KNIFE

00103-15_F43.EPS

Figure 43 Utility knives.

00103-15_F44.EPS

Figure 44 Kevlar® gloves.

Release pressure on the lever to lock the blade. Always cut away from the body, or perpendicular to it. Once cutting is complete, unlock and pull the blade in by pushing on the lever and sliding it inwards. Folding models have a blade lock that must be depressed to fold and unfold.

3.2.3 Safety and Maintenance

Files and rasps will become worthless without proper maintenance. Here are some guidelines for the use and maintenance of files and rasps:

- Use the correct file for the material being worked.
- Always put a handle on a file before using it; files and rasps often come without handles when boxed.
- After using a file, brush the filings from between the teeth using a file card or wire brush, pushing in the same direction as the alignment of the teeth. A file card has tightly spaced, short, stiff wires on its face that can dislodge material from between the file teeth.

- Store files in a dry place and keep them separated so that they won't chip or damage each other.
- Do not let the material vibrate in the vise as you file, because it dulls the teeth and produces an irregular surface on the workpiece.

Here are some guidelines for safely using and maintaining a utility knife:

- Replace blades when they stop cutting and start tearing. Sharp blades are safer than dull ones.
- Always keep the blade closed and locked when a utility knife is not in use.
- Be sure to position yourself properly and make the cut in the appropriate direction, keeping your free hand out of the line of the cut. Cut perpendicular to the body (left-to-right or right-to-left) or away from the body.
- Do not apply side loads to the blade, such as trying to pry open a can lid. The brittle blades can easily snap off.

Additional Resources

Easy Ergonomics: A Guide to Selecting Non-Powered Hand Tools. National Institute for Occupational Safety and Health (NIOSH), DHHS Publication No. 2004-164. **www.cdc.gov**

Field Guide to Tools. John Kelsey. 2004. Philadelphis, PA: Quirk Books.

3.0.0 Section Review

1. A coping saw is used for cutting ____.
   a. joints
   b. nails
   c. steel plate
   d. irregularly shaped moldings

2. Which of the following file types would apply the smoothest finish?
   a. Bastard
   b. Coarse
   c. Second-cut
   d. Rasp

4.0.0 OTHER COMMON HAND TOOLS

Objective

Identify and explain how to use other common hand tools.
  a. Identify and explain how to use shovels and picks.
  b. Identify and explain how to use chain falls and come-alongs.
  c. Identify and explain how to use various types of clamps.

Performance Tasks

1. Visually inspect the following tools to determine if they are safe to use:
   • Shovel or other earth tool
   • Chain fall or hoist
   • Clamps
2. Safely and properly use the following tools:
   • Shovel or other earth tool
   • Chain fall or hoist
   • Clamps

The proper selection and safe use of shovels, picks, chain falls, come-alongs, and clamps is discussed in this section. Proper training to use these tools is especially important because improper use could easily cause injuries such as back strains, as well as falling loads and material slippage.

4.1.0 Shovels and Picks

Shovels are used by many different construction trades. An electrician running underground wiring may dig a trench. A concrete mason may dig footers for a foundation. A carpenter may clear dirt from an area for concrete form-building. A plumber may dig a ditch to lay pipe. A welder may use a shovel to clean up scrap metal and slag after a job is finished.

There are three basic shapes of shovel blades: round, square, and spade (*Figure 45*). A round-bladed shovel is used to dig holes or remove large amounts of soil. A square-bladed shovel is used to move gravel or clean up construction debris. A spade is used to move large amounts of soil or dig trenches that need smooth, straight sides.

SPADE ROUND SQUARE

00103-15_F45.EPS

Figure 45 Shapes of shovel blades.

Shovels can have wooden or fiberglass handles. They generally come in two lengths. A long handle is usually 47" to 48" (1.2 m) long; a short handle is usually 27" (66 cm) long. Short handles often have a D-ring on the end to provide a different grip. Generally, the longer handles provide better leverage, but a shovel is not meant to be used as a lever. Roots and other obstructions should be cleared with other tools.

The first step to using any shovel is to select the right type for the job. For a round shovel or spade, place the tip of the shovel blade or spade at the point where digging or soil removal will begin. Balance a foot on the turned step and press down to cut into the soil with the blade.

For a square shovel, place the leading edge of the shovel blade against the gravel or construction debris and push until the shovel is loaded.

A pick is a swinging tool similar to an ax (*Figure 46*). A pick consists of a wooden handle that is 36" to 45" (91 cm to 1.1 m) in length and a forged steel head weighing 2 to 3 pounds (907 g and 1 kg). Depending on the size and strength of the pick, it can be used to break hardened or rocky soil, to level out stones and pavers, to loosen soil, and to break up stones and concrete. Long-handled picks (45" or 1.1 m) are used for tasks that require a normal amount of swing force, such as that used for digging a hole. Short-handled picks are used when a maximum amount of swinging

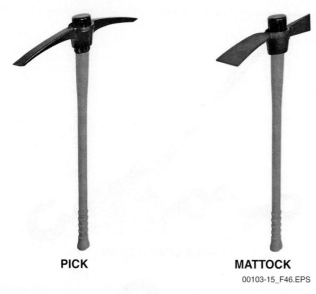

PICK   **MATTOCK**

00103-15_F46.EPS

Figure 46 Pick and mattock.

force is required and when the target is in a depression. The worker may use a short-handled pick while kneeling in these situations.

Very similar to a pick, a mattock is also used for breaking hardened, rocky soil and to dig trenches. Mattocks are much better than picks for clearing tree roots, due to the wide blade. The mattock may have a slightly shorter wooden handle than a pick. One side of the mattock head has a wide cutting blade, as shown in *Figure 46*. The other end may be like a pick, or (as shown) another blade turned 90 degrees from the mattock blade.

The first step to using a pick or mattock safely and correctly is to select the pick appropriate for your height and strength. Gloves should always be worn, as well as steel-toe work boots or shoes. Place one hand at the end of the handle and with the dominant hand about two-thirds of the way up the handle. For tasks using a short-handled pick to strike hard, raise the pick up and over the head like an axe, rapidly bending the knees and back to plunge the tool into the ground. For tasks requiring less forceful strikes, use a long-handled pick. Raise the pick up to chest height and then swing it back toward the ground, using the weight of the tool head and the leverage of the long handle to produce the strike.

4.1.1 Safety and Maintenance

Here are some guidelines for working safely with shovels:

- Always check to ensure that the blade is fixed firmly to the handle and no cracks or splits are present.

- Use the appropriate PPE when digging, trenching, or clearing debris, including safety glasses and gloves. Wear hard-toed boots to protect your feet from dropped materials and tool blades.
- Don't let dirt or debris build up on the blade. Always rinse off the shovel blade after using it.

Here are some guidelines for working safely with picks:

- Always check to ensure that the head is fixed firmly to the handle and no cracks or splits are present in the handle.
- Make sure that there are no other workers in the swing path before beginning the work.
- Always use a pick that is of the appropriate length and weight for your size.
- Only use maximum force swings (over your head) when necessary, because they put more strain on your back and shoulders than do normal swings (chest height).
- Be sure to wear appropriate eye protection, gloves, and steel-toe shoes.

4.2.0 Chain Falls and Come-Alongs

Chain falls and come-alongs are used to move heavy loads safely. A chain fall, also called a chain block or chain hoist, is a tackle device fitted with an endless chain used for hoisting heavy loads by hand. It is usually suspended from an overhead track. A come-along is used to move loads horizontally over the ground for short distances.

4.2.1 Chain Falls

The chain fall (*Figure 47*) has an automatic brake that holds the load after it is lifted. As the load is lifted, a screw forces fiber discs together to keep the load from slipping. The brake pressure increases as the loads get heavier, and the brake holds the load until the lowering chain is pulled. Manual chain falls are operated by hand. Electrical chain falls are operated from a tethered or wireless electrical control box.

The suspension hook is a steel hook used to hang the chain fall. It is often one size larger than the load hook. The gear box contains the gears that provide lifting power. The hand chain is a continuous chain used to operate the gearbox. The load chain is attached to the load hook and used to lift loads. A safety latch prevents the load from slipping off the load hook, which must be securely attached to the load.

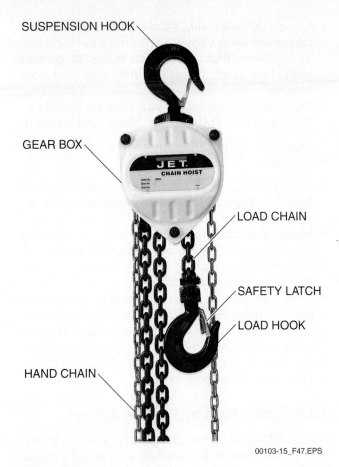

SUSPENSION HOOK

GEAR BOX

LOAD CHAIN

SAFETY LATCH

LOAD HOOK

HAND CHAIN

00103-15_F47.EPS

Figure 47 Parts of a chain fall.

4.2.2 *Come-Alongs and Ratchet Chain Hoists*

Come-alongs and ratchet chain hoists use a ratcheting handle to position or move loads horizontally (*Figure 48*). They can generally move loads from 1 to 6 tons. Ratchet chain hoists may be used for vertical lifting if they are designed and rated for the task. Cable come-alongs should not be used for vertical lifting, as they typically do not have a locking mechanism that is considered safe enough to lift a load vertically.

> **WARNING!**
>
> Never use a cable come-along for vertical overhead lifting. Use this tool only to move loads horizontally for short distances. Cable come-alongs are not equipped with the necessary safety features to ensure the safety of the load and nearby workers. For vertical lifting, use a chain fall or chain come-along designed and rated for lifting purposes.

4.2.3 *Safety and Maintenance*

Guidelines for maintaining and safely using chain falls and come-alongs include the following:

CABLE COME-ALONG

RATCHET CHAIN HOIST

00103-15_F48.EPS

Figure 48 Come-along and ratchet chain hoist.

- Follow the manufacturer's recommendations for lubricating the chain fall or come-along.
- Inspect a chain fall or come-along for wear before each use.
- Try out a chain fall or come-along on a small load first to ensure it is working normally.
- Have a qualified person ensure that the support rigging is strong enough to handle the load.
- Do not get lubricant on the clutches.
- Never stand under a load or allow others to do so.
- Never put hands your near pinch-points on the chain.

4.3.0 Clamps

There are many types and sizes of clamps (*Figure 49*), each designed to satisfy a different holding requirement. Clamps are sized by the maximum opening of the jaw. The depth (or throat) of the clamp determines how far from the edge of the work the clamp can be placed. The following are common types of clamps:

- *C-clamp* – This multipurpose clamp is named for its C-shaped frame. The clamp has a metal shoe at the end of a screw. Using a sliding T-bar handle, the clamp is tightened so it holds material between the metal jaw of the frame and the shoe. C-clamps are strong and durable, providing great holding power.
- *Locking C-clamp pliers* – This clamp works just like locking pliers. A knob in the handle controls the width and tension of the jaws. The handles are closed to lock the clamp and a lever is pressed to unlock and open the jaws.
- *Spring clamp* – Use a hand to open the spring-operated clamp. When the handles are released, the spring holds the clamp tightly shut, applying even pressure to the material. The jaws are usually made of steel, some with plastic coating to protect the material's surface against scarring.

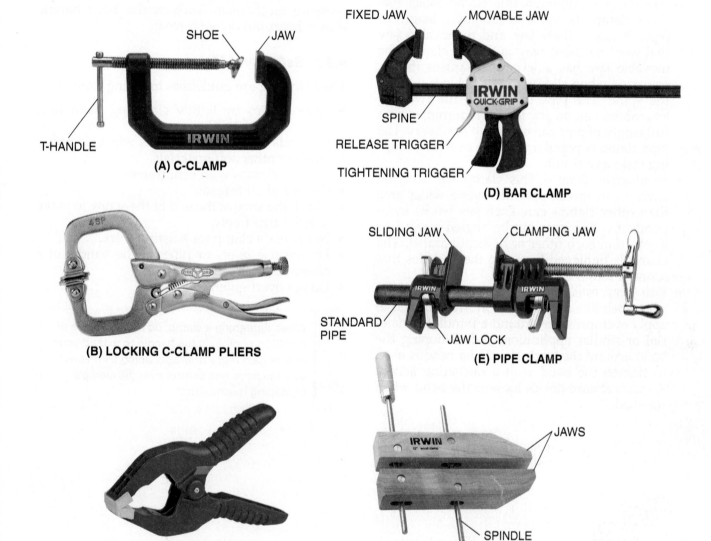

Figure 49 Types of clamps.

- *Bar clamp* – A rectangular piece of steel or aluminum is the spine of the bar clamp. It has a fixed jaw at one end and a sliding jaw (tail slide) with a spring-locking device that moves along the bar. It is equipped with non-marring pads that prevent it from damaging delicate surfaces such as finished woods. It is also designed to release the material quickly and easily without an explosive pressure release simply by squeezing the release trigger. Another feature of this tool is that you can quickly and easily change the direction of the jaw to turn it into a spreader bar. Simpler bar clamps are built more like pipe clamps, as described below.
- *Pipe clamp* – Although this clamp looks like a bar clamp, the spine is actually a length of pipe. It has a fixed jaw and a movable jaw that work the same way as the bar clamp. The movable jaw has a lever mechanism that is squeezed when sliding the movable jaw along the spine. The pipe connecting the two jaw assemblies can be any length required; even a full length of pipe can be used if necessary. The pipe clamp is popular for large or wide clamping tasks as a result.
- *Hand-screw clamp* – This clamp has wooden jaws. It can spread pressure over a wider area than other clamps can. Each jaw works independently. The jaws can be angled toward or away from each other or be kept parallel. The clamp is tightened by using the spindles that screw through the jaws.
- *Web (strap, band) clamp* – This clamp (*Figure 50*) uses a belt-like canvas or nylon strap or band to apply even pressure around a bundle of material or similar applications. After looping the band around the work, the clamp head is used to tighten the band with a ratcheting action. A quick-release device loosens the band when finished.

00103-15_F50.EPS

Figure 50 Web clamp.

When clamping wood or other soft material, place rubber pads or thin blocks of wood between the workpiece and the clamp to protect the work, as shown in *Figure 51*. Tighten the clamp's pressure mechanism (such as the T-bar handle shown here), but do not force it.

4.3.1 *Safety and Maintenance*

The following are guidelines for using clamps:

- Store clamps by lightly clamping them to a rack.
- Use pads or thin wood blocks when clamping wood or other soft materials.
- Discard clamps with bent frames.
- Clean and oil threads.
- Check the shoe at the end of the screw to make sure it turns freely.
- Never use a clamp for hoisting work.
- Do not use pliers or pipe on the handle of a clamp for tightening.
- Do not overtighten clamps.

> **CAUTION**
>
> When tightening a clamp, do not use pliers or a section of pipe on the handle to extend your grip or gain more leverage. Doing so means you will have less control over the clamp's tightening mechanism.

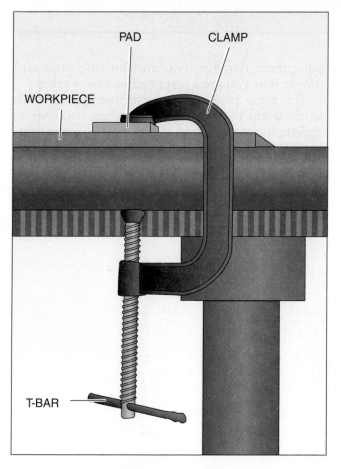
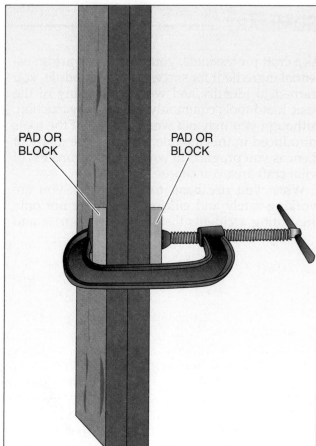

PAD   CLAMP

WORKPIECE

T-BAR

PAD OR BLOCK   PAD OR BLOCK

00103-15_F51.EPS

Figure 51   Placing pads and wood blocks.

Additional Resources

Easy Ergonomics: A Guide to Selecting Non-Powered Hand Tools. National Institute for Occupational Safety and Health (NIOSH), DHHS Publication No. 2004-164. **www.cdc.gov**

Field Guide to Tools. John Kelsey. 2004. Philadelphia, PA: Quirk Books.

4.0.0 Section Review

1. What makes a mattock different from a pick?

   a. There is a wide blade on one end of a mattock.
   b. There is a wide blade on one end of a pick.
   c. A mattock is used only in rock.
   d. A pick has a sledgehammer on one end.

2. Another name for a cable puller is _____.

   a. come-along
   b. ratchet
   c. chain fall
   d. chain block

3. Which type of clamp is best suited for a very wide clamping application, such as 8 feet?

   a. C-clamp
   b. Bar clamp
   c. Spring clamp
   d. Pipe clamp

SUMMARY

As a craft professional, your hand tools are an essential ingredient for success. In this module, you learned to identify and work with many of the basic hand tools commonly used in construction. Although you may not work with all of the tools introduced in this module, you will use many of them as you progress in your career, regardless of what craft area you choose.

When you use hand tools properly, you are working safely and efficiently. You are not only preventing accidents that can cause injuries and equipment damage, you are showing your employer that you are a responsible, safe worker.

The same pride you take in using your tools to do a job well is important when it comes to maintaining your tools. When you maintain your tools properly, they last longer, work better, and function more safely. The simple act of maintaining your tools will also help you perform your job better. Taking the time to learn to use and maintain these tools properly now will help keep you safe and save you time and money down the road.

1. The safest hammers are those with heads that are _____.
   a. welded and alloyed
   b. cast steel and chiseled
   c. chiseled and drop forged
   d. alloy and drop-forged steel

2. A chisel bar can be used to _____.
   a. pry apart steel beams
   b. split and rip apart pieces of wood
   c. break apart concrete
   d. make channels in wood beams

3. Paring chisels are _____.
   a. longer and much wider than other wood chisels
   b. much heavier than other wood chisels
   c. much shorter than other wood chisels
   d. longer and thinner than other wood chisels

4. For safety's sake, industrial screwdriver blades are made of _____.
   a. tempered steel
   b. Torx®
   c. clutch-driven steel
   d. fiberglass

5. An adjustable wrench is a good working partner to _____.
   a. pliers
   b. screwdrivers
   c. non-adjustable wrenches
   d. hammers

6. What type of mechanism do socket wrenches use?
   a. Clutch
   b. Chain
   c. Ratcheting
   d. Universal

7. The longer the ratchet handle, the better the _____.
   a. reach
   b. leverage
   c. gripe
   d. torque

8. The type of torque wrench designed for tightening clamping bands on underground pipe is the _____.
   a. digital type
   b. click type
   c. jointed type
   d. no-hub type

9. Torque wrenches are used to determine how much torque is required to loosen a rusted bolt.
   a. True
   b. False

10. Pliers should *not* be used on a nut or bolt because _____.
    a. they will round off the edges of the hex head
    b. they are not strong enough
    c. they are designed only for tightening
    d. their jaws will not open wide enough

11. What type of pliers could be used for cutting heavy duty wire?
    a. Slip-joint
    b. Locking
    c. Lineman
    d. Tongue-and-groove

12. An advantage of a laser measuring tool is that it _____.
    a. can be stored in your toolbox without breaking
    b. can take longer measurements
    c. has a large accuracy window
    d. is cheap to purchase

13. In order to determine whether a surface is level, check the _____.
    a. vertical surface
    b. spirit
    c. horizontal surface
    d. amount of bubbles

14. When something is plumb, it is _____.
    a. exactly vertical
    b. horizontally level
    c. at a 30-degree angle
    d. bobbed

15. The try square is made at a fixed _____.

    a. 45-degree angle
    b. 90-degree angle
    c. 180-degree angle
    d. 360-degree angle

16. Files have slanting rows of teeth and the teeth on a rasp are _____.

    a. smooth
    b. individual
    c. coarse
    d. wire

17. How is a carpet knife different from a common utility knife?

    a. The blade has a single sharpened edge.
    b. The blade has two sharpened edges.
    c. A carpet knife is much longer.
    d. A carpet knife uses a round, spinning blade.

18. A spade is used to _____.

    a. clear tree roots
    b. move gravel or clean up construction debris
    c. tamp down soil along a building's foundation
    d. move large amounts of soil or dig trenches with straight sides

19. Chain falls are used to _____.

    a. transport light loads safely
    b. supplement come-along pulls
    c. rig light loads safely
    d. safely move heavy loads vertically

20. A hand-screw clamp has _____.

    a. metal jaws
    b. nylon jaws
    c. wooden jaws
    d. fiberglass jaws

Trade Terms Quiz

Fill in the blank with the correct term that you learned from your study of this module.

1. Used mainly for woodworking, the _____ is set at a fixed, 90-degree angle.

2. A(n) _____ is an L-shaped, hexagonal steel bar.

3. The _____ has a flat face for striking and a rounded face for rounding off metal and rivets.

4. Shaped like an L, the _____ is used to make sure wall studs and sole plates are at right angles to each other.

5. A(n) _____ is a metal tool with a sharpened, beveled edge that is used to cut and shape wood, stone, or metal.

6. The _____ is used to drive nails and to pull nails out of wood.

7. To _____ is to cut on a slant at an angle that is not a right angle.

8. The _____ has a 12" blade that moves through a head that is constructed with both 45-degree and 90-degree angles.

9. If you use a screwdriver incorrectly, you can damage the screwdriver or _____ the screw head.

10. A(n) _____ has a moveable jaw that allows it to adjust to different nut or bolt sizes.

11. To fasten or align two pieces or material, you can use a(n) _____, which is a pin that fits into a corresponding hole.

12. A(n) _____ is a device such as a nut or bolt used to attach one material to another.

13. Use a(n) _____ for heavy-duty dismantling of woodwork.

14. The straight sides or jaws of a wrench opening are called the _____.

15. A(n) _____ is a claw hammer with a slightly rounded face.

16. _____ is a unit of measure used to describe the torque needed to tighten a large object.

17. _____ is a unit of measure used to describe the torque needed to tighten a small object.

18. The point at which members or the edges of members are joined is called the _____.

19. The _____ is the cut or channel made by a saw.

20. Using a(n) _____ can speed up your work because it has an open wrench at one end and a box-end at the other.

21. Use a(n) _____ to determine if a surface is exactly horizontal.

22. You make a(n) _____ by fastening together usually perpendicular parts with the ends cut at an angle.

23. A(n) _____ is a tool used to remove nails.

24. A(n) _____ has a fixed opening at each end that allows it to fit more than one size of nut or bolt.

25. To reduce stress in a weld, use a special type of hammer for _____ the joint as it cools.

26. Used for marking, checking, and measuring, a(n) _____ comes in several types: carpenter's, rafter angle, try, and combination.

27. A(n) _____ has serrated teeth on both jaws for gripping power.

28. _____ is the rotating or turning force applied to an object such as a bolt or nut.

29. The _____ is used to pull nails that have been driven flush with the surface of the wood or slightly below it.

30. A box-end wrench has 6 or 12 _____.

31. The _____, a non-adjustable wrench, forms a continuous circle around the head of a fastener.

32. To indent metal before you drill a hole, to drive pins, or to align holes in two parts that are mates, use a(n) _____.

33. Also called a speed square or magic square, the _____ is a combination protractor, try square, and framing square.

34. Using a damaged screwdriver on a screw may _____ the head and make the screw difficult to remove.

35. The _____ is a tool with a claw at each end, commonly used to pull nails.

36. A special type of adjustable wrench, _____ are scissor-shaped tools with jaws.

37. The _____ fits into a wooden file handle.

38. Some tools are made of _____ steel so that they resist wear and do not bend or break.

39. A(n) _____ piece of lumber is one that has had its surface made smooth.

40. If a surface is _____, it is exactly vertical.

41. A(n) _____ is a piece that projects out of wood so it can be placed into a hole or groove to form a joint.

42. A(n) _____ is a joint that has been created by heating pieces of metal.

43. A(n) _____ is a non-adjustable wrench with an enclosed, circular opening designed to lock onto the fastener when the wrench is struck.

44. In the metric system, the unit of measure for torque or moment is the _____.

Trade Terms

Adjustable wrench	Combination wrench	Nail puller	Ripping bar
Ball-peen hammer	Dowel	Newton-meter	Round off
Bell-faced hammer	Fastener	Open-end wrench	Square
Bevel	Flats	Peening	Striking wrench
Box-end wrench	Foot-pounds	Pipe wrench	Strip
Carpenter's square	Hex-key wrench	Planed	Tang
Cat's paw	Inch-pounds	Pliers	Tempered
Chisel	Joint	Plumb	Tenon
Chisel bar	Kerf	Points	Torque
Claw hammer	Level	Punch	Try square
Combination square	Miter joint	Rafter angle square	Weld

Chris Williams
Associated Builders and Contractors, Inc.
Director of Safety

How did you choose a career in the construction industry?
My father owned and operated a light-commercial and residential construction firm, so I was raised in and around construction. I spent many summers on job sites doing every task possible and enjoyed the organization and camaraderie that each crew demonstrated. Continuing in construction after college was a natural progression.

Who inspired you to enter the industry?
My father—witnessing his dedication to not only the industry but also to continuous training and learning to help develop his skills and advance the industry.

What types of training have you completed?
Fall protection competent person, NCCER Master Trainer, and the OSHA 10- and 30-hour courses.

How important is education and training in construction?
Education and training is extremely important, especially from a safety standpoint. Construction safety culture continues to evolve and the concepts—both technical and cultural—need to be reinforced. From a craft standpoint, we can always improve our skills through learning, which enhances our productivity and output.

How important are NCCER credentials to your career?
I would consider NCCER credentials to be of higher value than any others that I have received. Having participated as a Subject Matter Expert and also having gone through the NCCER Master Trainer course, I value the stringent criteria that NCCER holds its students to and the in-depth training materials that are used. In construction, if you're not the very best, you lose business. NCCER helps our members be their very best.

How has training/construction impacted your life?
I grew up in construction so I've witnessed firsthand the value of our industry—and the consequences that a lack of training/understanding can reap. Being involved in construction safety has helped significantly alter my view of how we work—we have evolved from "just get the job done any way you can" to "if you see someone else acting unsafe, it is your job to stop them and help them work safely." The culture of interdependence and the safety training we deliver to our employees is the reason I come to work every day. Construction workers need to know that safety is a core value that affects every other decision made on the job site.

What types of work have you done in your career?
Besides construction safety, I've worked with various construction non-profit associations as both a trainer and consultant.

Tell us about your present job.
I am the Director of Safety for Associated Builders and Contractors, Inc., a merit-shop trade association representing 21,000 chapter members nationwide. I am the point person for our regulatory efforts with OSHA, as well as our safety representative on various bodies. My role over the years has evolved to include the development of new curriculum and training strategies that help advance ABC's safety vision. This vision believes every incident is preventable when leadership is committed to a culture that focuses on safety as a core value, and that every employee is part of the continuing evolution of that culture.

What do you enjoy most about your job?
The interaction with our members and their employees is, by far, the highlight of my job. Being able to visit a job site and see our men and women in action, building and creating—and doing it safely—brings me tremendous pride in being a construction professional.

What factors have contributed most to your success?

A significant contribution is the knowledge and experience of my peers, and their willingness to openly share both their knowledge and safety policies for others to use. Our industry recognizes that the only way to truly protect our employees and reach a zero-incident workplace is by sharing the concepts and ideas with each other. Our industry is truly a team when it comes to construction safety.

Would you suggest construction as a career to others? Why?

Construction is wrongly stigmatized as being the place where people go when they cannot succeed in college. That cannot be further from the truth. In my reality, college is where people go who don't have the creativity, the drive, the passion to succeed in construction! To me, construction is one of the few remaining industries where a man or woman can start a career and, through their own hard work and desire to learn, move up and achieve the American Dream. Construction is the gateway to prosperity if you're willing to seize it.

Interesting career-related fact or accomplishment:

I not only have multiple safety designations, but I'm also a Certified Association Executive (CAE).

How do you define craftsmanship?

Craftsmanship is about the sense of pride and accomplishment in your finished product. It doesn't matter what the project is—a complex electrical system in a refinery or a new light installed in a house—the quality of the finished product is a direct reflection on the quality of your own skills. Satisfaction is found in a sense of pride and in the admiration of your work by others.

Trade Terms Introduced in This Module

Adjustable wrench: A smooth-jawed wrench with an adjustable, moveable jaw used for turning nuts and bolts. Often referred to as a Crescent® wrench due to brand recognition.

Ball-peen hammer: A hammer with a flat face that is used to strike cold chisels and punches. The rounded end—the peen—is used to bend and shape soft metal.

Bell-faced hammer: A claw hammer with a slightly rounded, or convex, face.

Bevel: To cut on a slant at an angle that is not a right angle (90-degree). The angle or inclination of a line or surface that meets another at any angle but 90-degree.

Box-end wrench: A wrench, usually double-ended, that has a closed socket that fits over the head of a bolt.

Carpenter's square: A flat, steel square commonly used in carpentry.

Cat's paw: A straight steel rod with a curved claw at one end that is used to pull nails that have been driven flush with the surface of the wood or slightly below it.

Chisel: A metal tool with a sharpened, beveled edge used to cut and shape wood, stone, or metal.

Chisel bar: A tool with a claw at each end, commonly used to pull nails.

Claw hammer: A hammer with a flat striking face. The other end of the head is curved and divided into two claws to remove nails.

Combination square: An adjustable carpenter's tool consisting of a steel rule that slides through an adjustable head.

Combination wrench: A wrench with an open end and a closed end.

Dowel: A pin, usually round, that fits into a corresponding hole to fasten or align two pieces.

Fastener: A device such as a bolt, clasp, hook, or lock used to attach or secure one material to another.

Flats: The straight sides or jaws of a wrench opening; also, the sides on a nut or bolt head.

Foot-pounds: Unit of measure used to describe the amount of pressure exerted (torque) to tighten a large object.

Hex key wrench: A hexagonal steel bar that is bent to form a right angle. Often referred to as an Allen® wrench.

Inch-pounds: Unit of measure used to describe the amount of pressure exerted (torque) to tighten a small object.

Joint: The point where members or the edges of members are joined. The types of welding joints are butt joint, corner joint, and T-joint.

Kerf: A cut or channel made by a saw.

Level: Perfectly horizontal; completely flat. Also, a tool used to determine if an object is level.

Miter joint: A joint made by fastening together usually perpendicular parts with the ends cut at an angle.

Nail puller: A tool used to remove nails.

Newton-meter: A measure of torque or moment equal to the force of one Newton applied to a lever one meter long.

Open-end wrench: A non-adjustable wrench with a fixed opening at each end that is typically different, allowing it to be used to fit two different nut or bolts sizes.

Peening: The process of bending, shaping, or cutting material by striking it with a tool.

Pipe wrench: A wrench for gripping and turning a pipe or pipe-shaped object; it tightens when turned in one direction.

Planed: Describing a surface made smooth by using a tool called a plane.

Pliers: A scissor-shaped type of adjustable wrench equipped with jaws and teeth to grip objects.

Plumb: Perfectly vertical; the surface is at a right angle (90 degrees) to the horizon or floor and does not bow out at the top or bottom.

Points: Teeth on the gripping part of a wrench. Also refers to the number of teeth per inch on a handsaw.

Punch: A steel tool used to indent metal.

Rafter angle square: A type of carpenter's square made of cast aluminum that combines a protractor, try square, and framing square.

Ripping bar: A tool used for heavy-duty dismantling of woodwork, such as tearing apart building frames or concrete forms.

Round off: To smooth out threads or edges on a screw or nut.

Square: Exactly adjusted; any piece of material sawed or cut to be rectangular with equal dimensions on all sides; a tool used to check angles.

Striking (or slugging) wrench: A non-adjustable wrench with an enclosed, circular opening designed to lock on to the fastener when the wrench is struck.

Strip: To damage the head or threads on a screw, nut, or bolt.

Tang: Metal handle-end of a file. The tang fits into a wooden or plastic file handle.

Tempered: Treated with heat to create or restore hardness in steel.

Tenon: A piece that projects out of wood or another material for the purpose of being placed into a hole or groove to form a joint.

Torque: A rotating or twisting force applied to an object such as a nut, bolt, or screw, using a socket wrench or screwdriver. Torque wrenches allow a specific torque value to be set and applied.

Try square: A square whose legs are fixed at a right angle.

Weld: To heat or fuse two or more pieces of metal so that the finished piece is as strong as the original; a welded joint.

Additional Resources

This module presents thorough resources for task training. The following resource material is suggested for further study.

Easy Ergonomics: A Guide to Selecting Non-Powered Hand Tools. National Institute for Occupational Safety and Health (NIOSH), DHHS Publication No. 2004-164. **www.cdc.gov**

Field Guide to Tools. John Kelsey. 2004. Philadelphia, PA: Quirk Books.

Figure Credits

Fluke Corporation, reproduced with permission, Module opener

The Stanley Works, Figures 1–4, 7, 12D, 16, 26B, 30, 34, SA01

RIDGID®, Figures 5A, 10A, 14A, 14B, 24, 26A, 38F, 45

Courtesy of Irwin Tools, Figures 5B, 14C, 19, 20, 23, 25, 28, 36, 37, 38A–38E, 43, 49

Cianbro Corporation, Figures 6, 27A, 27C, 39

Klein Tools, Inc., Figures 10B, 12A, SA02

Courtesy of SK Professional Tools, Figure 12B

Snap-on Incorporated, Figure 12C

Proto Industrial Tools, Figure 17

TEKTON, Figure 18A

Courtesy of The Eastwood Company, Figure 18B

Lowell Corporation, Figure 18C

The Lincoln Electric Company, Cleveland, OH, USA, Figure 21

Topaz Publications, Inc., Figures 22, SA06

DeWALT Industrial Tool Co., Figure 29

Zachary McNaughton, River Valley Technical Center, Figure 31

Woodstock International, Inc., Figures 40, 41

Youngstown Glove Company, Figure 44

A.M. Leonard, Inc., Figure 46

Walter Meier Manufacturing Americas, Figures 47, 48

Erickson Manufacturing Corp., Figure 50

Courtesy of Snap-on Industrial - Tools at Height Program, SA03

The LS Starrett Company, SA04, SA05

Answer	Section Reference	Objective
Section One		
1. c	1.1.2	1a
2. a	1.2.1	1b
3. a	1.3.0	1c
4. c	1.4.1	1d
5. a	1.5.1	1e
6. b	1.6.5	1f
Section Two		
1. d	2.1.3	2a
2. a	2.1.4	2a
3. a	2.2.5	2b
4. d	2.2.5	2b
Section Three		
1. d	3.1.0	3a
2. c	3.2.1	3b
Section Four		
1. a	4.1.0	4a
2. a	4.2.2	4b
3. d	4.3.0	4c

NCCER CURRICULA — USER UPDATE

NCCER makes every effort to keep its textbooks up-to-date and free of technical errors. We appreciate your help in this process. If you find an error, a typographical mistake, or an inaccuracy in NCCER's curricula, please fill out this form (or a photocopy), or complete the online form at **www.nccer.org/olf**. Be sure to include the exact module ID number, page number, a detailed description, and your recommended correction. Your input will be brought to the attention of the Authoring Team. Thank you for your assistance.

Instructors – If you have an idea for improving this textbook, or have found that additional materials were necessary to teach this module effectively, please let us know so that we may present your suggestions to the Authoring Team.

NCCER Product Development and Revision

13614 Progress Blvd., Alachua, FL 32615

Email: curriculum@nccer.org
Online: www.nccer.org/olf

❏ Trainee Guide ❏ Lesson Plans ❏ Exam ❏ PowerPoints Other _____

Craft / Level: _____ Copyright Date: _____

Module ID Number / Title: _____

Section Number(s): _____

Description: _____

_____

_____

_____

Recommended Correction: _____

_____

_____

_____

Your Name: _____

Address: _____

Email: _____ Phone: _____

00104-15

Introduction to Power Tools

OVERVIEW

Power tools are used in almost every construction trade to make holes; to cut, smooth, and shape materials; and even to demolish pavement. All construction workers are certain to use power tools on the job eventually. This module provides an overview of the common types of power tools and how they work. It also describes the proper techniques required to safely operate these tools.

Module Four

Trainees with successful module completions may be eligible for credentialing through the NCCER Registry. To learn more, go to **www.nccer.org** or contact us at **1.888.622.3720**. Our website has information on the latest product releases and training, as well as online versions of our *Cornerstone* magazine and Pearson's product catalog.

Your feedback is welcome. You may email your comments to **curriculum@nccer.org**, send general comments and inquiries to **info@nccer.org**, or fill in the User Update form at the back of this module.

This information is general in nature and intended for training purposes only. Actual performance of activities described in this manual requires compliance with all applicable operating, service, maintenance, and safety procedures under the direction of qualified personnel. References in this manual to patented or proprietary devices do not constitute a recommendation of their use.

INTRODUCTION TO POWER TOOLS

Objectives

When you have completed this module, you will be able to do the following:

1. Identify and explain how to use various types of power drills and impact wrenches.
   a. Identify and explain how to use common power drills and bits.
   b. Identify and explain how to use a hammer drill.
   c. Identify and explain how to use pneumatic drills and impact wrenches.
2. Identify and explain how to use various types of power saws.
   a. Identify and explain how to use a circular saw.
   b. Identify and explain how to use saber and reciprocating saws.
   c. Identify and explain how to use a portable band saw.
   d. Identify and explain how to use miter and cutoff saws.
3. Identify and explain how to use various grinders and grinder attachments.
   a. Identify and explain how to use various types of grinders.
   b. Identify and explain how to use various grinder accessories and attachments.
4. Identify and explain how to use miscellaneous power tools.
   a. Identify and explain how to use pneumatic and powder-actuated fastening tools.
   b. Identify and explain how to use pavement breakers.
   c. Identify and explain the uses of hydraulic jacks.

Performance Tasks

Under the supervision of your instructor, you should be able to do the following:

1. Safely and properly demonstrate the use of three of the following tools:
   - Electric drill
   - Hammer drill or rotary hammer
   - Circular saw
   - Reciprocating saw
   - Portable band saw
   - Miter or cutoff saw
   - Portable or bench grinder
   - Pneumatic nail gun
   - Pavement breaker

Trade Terms

Abrasive	Chuck key	Ground fault circuit interrupter (GFCI)	Revolutions per minute (rpm)
Alternating current (AC)	Countersink	Ground fault protection	Ring test
Arbor	Direct current (DC)	Kerf	Shank
Auger bit	Forstner bit	Masonry bit	Trigger lock
Carbide	Grit	Reciprocating	
Chuck			

Industry Recognized Credentials

If you are training through an NCCER-accredited sponsor, you may be eligible for credentials from NCCER's Registry. The ID number for this module is 00104-15. Note that this module may have been used in other NCCER curricula and may apply to other level completions. Contact NCCER's Registry at 888.622.3720 or go to **www.nccer.org** for more information.

Contents

Topics to be presented in this module include:

Figures

SECTION ONE

1.0.0 POWER DRILLS

Objective

Identify and explain how to use various types of power drills and impact wrenches.

a. Identify and explain how to use common power drills and bits.
b. Identify and explain how to use a hammer drill.
c. Identify and explain how to use pneumatic drills and impact wrenches.

Performance Task

1. Safely and properly demonstrate the use of the following tool(s):
   - Electric drill
   - Hammer drill or rotary hammer

Trade Terms

Alternating current (AC): The common power supplied to most all wired devices, where the current reverses its direction many times per second. AC power is the type of power generated and distributed throughout settled areas.

Auger bit: A drill bit with a spiral cutting edge for boring holes in wood and other materials.

Carbide: A very hard material made of carbon and one or more heavy metals. Commonly used in one type of saw blade.

Chuck: A clamping device that holds an attachment; for example, the chuck of the drill holds the drill bit.

Chuck key: A small, T-shaped steel piece used to open and close the chuck on power drills.

Countersink: A bit or drill used to set the head of a screw at or below the surface of the material.

Direct current (DC): An electric power supply where the current flows in one direction only. DC power is supplied by batteries and by transformer-rectifiers that change AC power to DC.

Forstner bit: A bit designed for use in wood or similar soft material. The design allows it to drill a flat-bottom blind hole in material.

Ground fault circuit interrupter (GFCI): A circuit breaker designed to protect people from electric shock and to protect equipment from damage by interrupting the flow of electricity if a circuit fault occurs.

Ground fault protection: Protection against short circuits; a safety device cuts power off as soon as it senses any imbalance between incoming and outgoing current.

Masonry bit: A drill bit with a carbide tip designed to penetrate materials such as stone, brick, or concrete.

Revolutions per minute (rpm): The rotational speed of a motor or shaft, based on the number of times it rotates each minute.

Shank: The smooth part of a drill bit that fits into the chuck.

Trigger lock: A small lever, switch, or part that can be used to activate a locking catch or spring to hold a power tool trigger in the operating mode without finger pressure.

This module introduces three kinds of power tools: electric, pneumatic, and hydraulic.

- *Electric tools* – These tools are powered by electricity. They are operated from either an **alternating current (AC)** source (such as a wall receptacle) or a **direct current (DC)** source (such as a battery).
- *Pneumatic tools* – These tools are powered by air. Electric or gasoline-powered compressors produce the air pressure. Air hammers and pneumatic nailers are examples of pneumatic tools.
- *Hydraulic tools* – These tools are powered by fluid pressure. Hand pumps or electric pumps are used to produce the fluid pressure. Pipe benders, jackhammers, and Porta-Powers® are examples of hydraulic tools.

All power tools can be dangerous. Workers should never attempt to operate these tools without proper instruction and supervision. Operating a power tool incorrectly or unsafely can injure the operator as well as other workers. Safety issues for each tool are covered in this module, but general safety issues—such as safety in the work area, safety equipment, and working with electricity—are covered in the *Basic Safety* module. This information is vital for working with power tools, and trainees must complete the *Basic Safety* module before beginning this module.

One of the most important power tool safety rules is to make sure that the tool has been disconnected from its source of energy before parts such as bits, blades, or discs are replaced and before any type of maintenance is performed on the tool. Cords and hoses that provide energy to a power tool should be checked to make sure they are not frayed or damaged. If a power tool is equipped with a trigger lock, do not use the lock. A trigger lock is a small lever, switch, or part that can be used to activate a locking catch or spring to hold a power tool trigger in the operating mode even when the trigger is released. Locking any power tool in the On position can be very dangerous. Although this was a popular new feature when first introduced, it has proven to be an unsafe choice. Many contractors do not allow their use at any time.

Regardless of what power tool is in use, safety glasses are required. Some tools require the use of a face shield in addition to safety glasses. Safety shoes should also be worn, and tight-fitting gloves are required for most power tools. Loose gloves can be a safety hazard of their own.

1.1.0 Types of Power Drills

Various types of power drills are often used in the construction industry. A power drill is most commonly used to make holes by spinning drill bits into wood, metal, plastic, and other materials. However, with different attachments and accessories, the power drill can be also used as a sander, polisher, screwdriver, grinder, or countersink.

Common types of power drills include the following:

- Electric drills
- Cordless drills
- Electromagnetic drills
- Hammer drills
- Pneumatic drills
- Electric screwdrivers

Most of the basic types of power drills are similar and share some common features. For example, most power drills have a pistol grip with a trigger switch for controlling power (*Figure 1*).

The farther back the trigger of a variable-speed drill is pulled, the faster the drill spins. Drills also have reversing switches that enable the drill to spin backwards in order to back the drill bit out if it gets stuck in the material while drilling. Power drills use replaceable bits for different drilling tasks (*Figure 2*). On variable-speed power drills, a screwdriver bit can be used in place of a drill bit so that the drill can be used as a screwdriver. Only screwdriver bits that are designed for use in a power drill should be used.

Twist drill bits are used to drill wood and plastics at high speeds or to drill metal at a lower speed. A Forstner bit is used on wood and is particularly good for boring a flat-bottom hole. A paddle bit or spade bit is also used in wood. The bit size is measured by the paddle's diameter, which generally ranges from ½ to 1½ inches (≈13 to 38 mm). A masonry bit, which has a carbide tip, is used in concrete, stone, slate, and ceramic material. The auger bit is used for drilling wood and other soft materials, but not for drilling metal. The auger bit is designed for use at a very low speed.

00104-15_F01.EPS

Figure 1 Parts of a power drill.

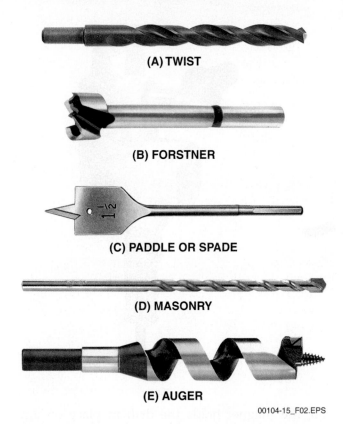

(A) TWIST

(B) FORSTNER

(C) PADDLE OR SPADE

(D) MASONRY

(E) AUGER

00104-15_F02.EPS

Figure 2 Drill bits.

Drill bit sizes can vary a great deal. For most applications, bits are available in both fractional inch sizes and metric sizes. For twist drill bits, common fractional-inch sizes range from ¹⁄₆₄-inch to 1 inch. Common metric sizes for twist drill bits range from 0.5 mm to 25.0 mm. It is important to note that fractional-inch bits and metric bits are not directly interchangeable. In other words, no fractional-inch bit has a corresponding metric bit, and vice versa. However, twist drill bits are also available in sets of numbered and lettered bits. Numbered sets range from #80 up to #1, which is the largest of the set. The next bit size larger is the A bit, with the Z bit being the largest of the lettered set. These sets correspond to decimal fraction and metric sizes. For example, a #8 drill bit is sized at 0.199 inches and 5.055 mm. Number and letter drills are used or specified when the precise size of the hole is important.

All bits are held in a drill by the drill chuck. Keyed chucks are opened and closed using a chuck key (*Figure 3*). Chuck keys are typically interchangeable in design, but there are several different sizes. Keyless chucks are also used, normally found on cordless drills. The size of the bit or tool that a drill can accommodate is limited by the size of the chuck.

Some electric power drills are designed to be used in tight spaces, such as between studs and joists. Drills like those shown in *Figure 4* are

00104-15_F03.EPS

Figure 3 Chuck key.

referred to as right-angle drills. The drill on the right is larger and develops more power for larger holes.

1.1.1 Cordless Drills

Cordless power drills (*Figure 5*) are useful for working in areas where a power source is hard to find.

Cordless drills contain a rechargeable battery pack that runs the motor. The pack can be detached and plugged into a battery charger any time the drill is not in use. Some chargers can recharge the battery pack in short period of time, while others require more time. The quality of charge and the life-cycle of the battery must often be considered when determining how to best charge the battery. Manufacturers provide information regarding the health of the battery in the product literature. Workers who use cordless drills often carry an extra battery pack with them.

Some cordless drills have adjustable clutches so that the drill motor can also serve as a power screwdriver without applying too much power to the screw. Note that most cordless drills are equipped with keyless chucks except the most heavy-duty models.

1.1.2 Electromagnetic Drills

An electromagnetic drill (*Figure 6*) is a portable drill mounted on an electromagnetic base. It is used for drilling thick metal. Once the drill is placed on a metal surface, the power turned on and the magnet energized, the magnetic base will hold the drill in place for drilling. Some drills can also be rotated in place while the base remains stationary.

A switch on the junction box controls the electromagnetic base. When the switch is turned

00104-15_F04.EPS

Figure 4 Right-angle drills.

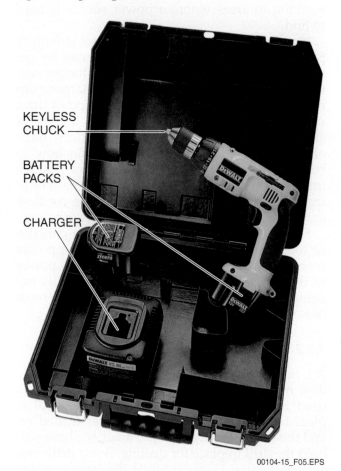

KEYLESS
CHUCK

BATTERY
PACKS

CHARGER

00104-15_F05.EPS

Figure 5 Cordless drill.

on, the magnet holds the drill in place on any surface with magnetic characteristics, such as carbon steel. The drill base must be clean and the surface must be flat. The switch on the top of the drill turns the drill motor on and off. A depth gauge can be used to set the depth of the hole being drilled. It operates like a drill press, with the operator turning a hand wheel to raise and lower the drill against the workpiece. Workers who use electromagnetic drills should be properly trained to safely operate the specific drill being used.

> **CAUTION**
>
> Do not remove power from an electromagnetic drill while it is in use. Interrupting power to the drill will deactivate the electromagnetic base and cause the drill to fall.

1.1.3 How to Use a Power Drill

To prepare a drill for use, first make sure that the drill is disconnected from its power source. Then turn the chuck counterclockwise (to the left) until the chuck opening is large enough to fit the bit shank, which is the smooth part of the bit. Insert the bit shank into the chuck opening (*Figure 7[A]*). Keeping the bit centered in the opening, turn the chuck by hand until the jaws grip the bit shank. Make sure the bit is straight in the chuck and not leaning.

NCCER – *Core Curriculum*   00104-15

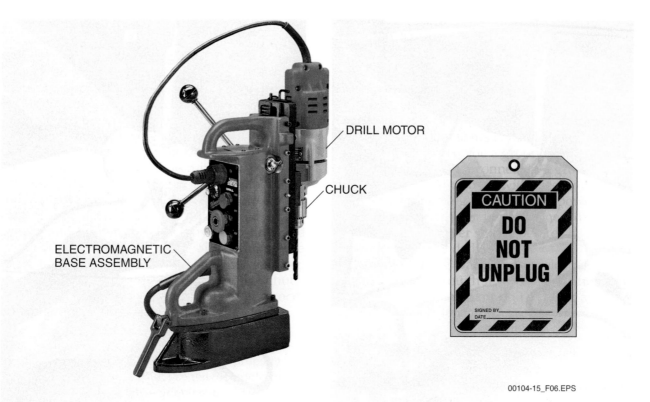

00104-15_F06.EPS

Figure 6 Electromagnetic drill.

For keyed chucks, insert the chuck key (*Figure 7[B]*) for the drill into one of the holes on the side of the chuck. The chuck key has a grooved ring called a gear. Make sure that the proper size chuck key is used so that the key's gear meshes with the matching gears on the end of the chuck. Turn the chuck key clockwise to tighten the grip on the bit. With larger chucks, tighten the bit by inserting the chuck key into each of the holes in the three-jawed chuck. This tightens the individual jaws and ensures that all the jaws close uniformly tight around the bit. Once the bit has been secured, remove the key from the chuck.

> **WARNING!**
>
> Always remember to remove the key from the chuck. Otherwise, when you start the drill, the key could fly out and injure you or a co-worker.

> **CAUTION**
>
> Power drills can be dangerous if you do not use them properly. Always wear the proper PPE, including appropriate eye, head, and hand protection.

To drill a hole with a power drill, start by making a small indent in the material exactly where the hole needs to be drilled. In wood, use a small punch to make the indent; in metal, use a center punch. Firmly clamp or support the work that is being drilled. Then, hold the drill perpendicular (at a right angle) to the material surface and start the drill motor. With a variable-speed drill, start the bit slowly. Be sure the drill is rotating in the right direction (with the bit facing away it should be turning clockwise). Hold the drill with both hands and apply only moderate pressure when drilling. The drill motor should operate at approximately the same **revolutions per minute (rpm)** as it does when it is not loaded. If the sound of the drill indicates it is slowing considerably, use less pressure. *Figure 7(C)* shows the proper way to hold the drill.

Reduce the pressure when the bit is about to emerge from the other side of the work, especially when drilling metal. Be prepared! As the drill bit emerges through the opposite side of the work, it tends to grab, causing the bit to stall. With power still on, the tendency of the drill motor is to rotate the operator, and not the bit! Be sure to maintain firm control. If too much pressure is being applied when the bit comes out the other side, the drill itself will hit the surface of the material. This could damage or dent the metal surface. If the drill bit gets stuck in the material during drilling, release the trigger, use the reversing switch to change the direction of the drill, and back it gently out of the material. After backing out the bit, switch back to the original drilling position.

to block flying objects and safety lines to keep the drill from falling if the power is cut off, are available. In some places, these attachments are required. Instructors or supervisors should be familiar with the requirements for safety attachments in their specific areas. Finally, support the drill before turning it off; otherwise, when the power is turned off, the drill may fall over.

1.2.0 Hammer Drills

A hammer drill (*Figure 10*) has a light pounding action that enables it to drill into concrete, brick, or tile. The bit rotates and hammers at the same time, allowing faster drilling in these materials than a regular drill. The depth gauge on a hammer drill can be set to the depth of the hole to be drilled.

Special bits that can take the pounding are needed for a hammer drill. Hammer drills use carbide-tipped masonry bits whenever they are being using in masonry. Common drilling tasks can be done with other bits when the hammering action is disabled.

The term *hammer drill* is often used for all tools that hammer and drill. However, rotary hammers (*Figure 11*) are for more heavy-duty work. They usually have slower rotational speeds than hammer drills, and hammer harder and less often than a hammer drill. Most rotary hammers require bits that fit into special chuck designs (*Figure 12*). You must select masonry bits that are compatible with the particular type of chuck on the tool. The chucks are not keyed, but rely upon the design of the bit and chuck to hold it in place. Adapters are available to use one bit shank design with another type of chuck when necessary.

DEPTH GAUGE

ADJUSTABLE RING

POWER SWITCH

00104-15_F10.EPS

Figure 10 Hammer drill.

00104-15_F11.EPS

Figure 11 Rotary hammer.

1.2.1 How to Use a Hammer Drill

When using a hammer drill, follow the same safety practices that apply to electric power drills. Always wear proper PPE, including appropriate eye, head, and hand protection. If the drill is being used on concrete or similar materials that produce airborne dust, respiratory protection is needed as well.

Most hammer drills will not hammer until pressure is applied to the drill bit against a surface. An adjustable ring is turned to adjust the number of blows per minute. The hammer action stops when pressure applied to the drill is stopped.

1.3.0 Pneumatic Drills and Impact Wrenches

Pneumatic power tools are powered by compressed air. An air hose transfers the compressed air from an air compressor to the tool. Pneumatic tools tend to have more power for their weight than comparable electric tools. Two common pneumatic power tools used by construction workers are pneumatic drills and impact wrenches.

1.3.1 Pneumatic Drills

Pneumatic drills have many of the same parts, controls, and uses as electric drills. Since there is no motor, they are generally more compact in size. A pneumatic drill is typically used when there is no available source of electricity, or when a high rate of production is desired. Like electric drills, they can also be used as power

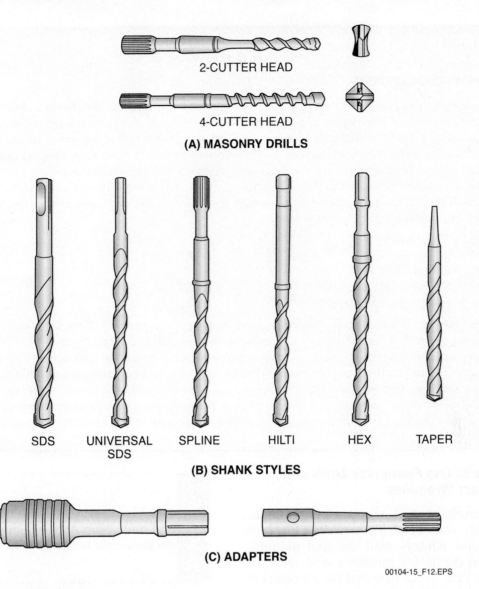

2-CUTTER HEAD

4-CUTTER HEAD

(A) MASONRY DRILLS

SDS UNIVERSAL SDS SPLINE HILTI HEX TAPER

(B) SHANK STYLES

(C) ADAPTERS

00104-15_F12.EPS

Figure 12 Rotary hammer bit designs.

screwdrivers. The pneumatic drill in *Figure 13* has a Phillips-head screwdriver bit in the chuck. They are sometimes equipped with a keyed chuck.

Common sizes of pneumatic drills are ¼-, ⅜-, and ½-inch. The size refers to the diameter of the largest bit shank that can be gripped in the chuck, not the drilling capacity. Some common metric sizes are 8 mm, 10 mm, and 13 mm.

1.3.2 Impact Wrenches

Pneumatic impact wrenches (*Figure 14*) are power tools that are used to fasten, tighten, and loosen nuts and bolts. As with a pneumatic drill, a pneumatic impact wrench must be connected with a hose to an air compressor. The speed and strength (torque) of these wrenches can easily be adjusted depending on the type

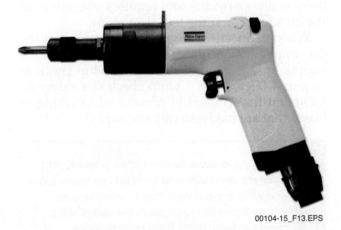

00104-15_F13.EPS

Figure 13 Pneumatic drill.

of job. Note that although pneumatic impact wrenches are more popular, there are electric models as well.

Power Sources

Each country provides power to its residents at a voltage and frequency that they believe best for their own needs. While the voltage often varies slightly from the target voltage, most power tools can accommodate these minor day-to-day differences. However, users must always ensure that the power supply for a given tool is appropriate. Although slight changes in voltage do not represent a problem, changes in power frequency and significant changes in the voltage will damage or destroy a power tool.

Here are the power characteristics for single-phase power from several different countries to show their differences. Note that the frequency, reported in hertz (Hz), refers to the cyclic nature of alternating current and how many cycles are completed per second:

- Portugal: 230 volts @ 50Hz
- Hungary: 220 volts @ 50 Hz
- England: 240 volts @ 50 Hz
- Sweden: 230 or 400 volts @ 50 Hz
- United States: 115 or 230 volts, 60 Hz

Note that the United States is one of very few countries that provide power at 60 Hz. Many devices can operate using power from almost all international power systems. These devices are often battery-charging devices that use a transformer to reduce the voltage and change the voltage to direct current (DC) to charge a battery. Other devices, such as computers and printers, may be equipped with external switches to allow the use of different power sources. However, few if any portable power tools have been designed with such a switch, and corded tools have no transformer. So before packing away your favorite power saw for work in another country, be sure that there is a power supply in place that can support the tool's required voltage and frequency.

1.3.3 How to Use Pneumatic Drills and Impact Wrenches

Using a pneumatic drill or a pneumatic impact wrench safely and effectively requires a few basic considerations. Always read the tool manufacturer's instructions for guidance and wear the appropriate PPE. Make sure that the air compressor is set to the appropriate pressure. Also, ensure there is an oiler if the tool requires one, either at the air source or at the tool (*Figure 15*).

When connecting the tool, make sure that the pneumatic connection between the tool and the supply hose is good, and install a whip check as required (*Figure 16*). A whip check is a safety attachment that is used to prevent whiplashing in hoses that are inadvertently uncoupled.

> **WARNING!**
> The air hose must be connected properly and securely. An unsecured air hose can come loose and whip around violently, causing serious injury. Some fittings require the use of whip checks to keep them from coming loose.

Use only those drill bits and impact sockets that are designed for use with the applicable tool. Operate the tool safely and when the work is completed, disconnect the tool from the air hose.

> **WARNING!**
> Using handheld sockets can damage property and cause injury. Use only impact sockets made for pneumatic impact wrenches.

00104-15_F14.EPS

Figure 14 Pneumatic impact wrench.

1.3.4 Safety and Maintenance

The safety practices that should be followed when using a pneumatic drill or an pneumatic impact wrench are similar to those that apply to power drills. Always wear the appropriate PPE, including eye, hand, and ear protection. Make sure that the workpiece is secure. It is also important to maintain a balanced body stance when operating the tool and keep hands away from the working end of the tool. If hardware being tightened is a bolt-and-nut combination, use a backup wrench to keep the bolt or nut from spinning.

The air supply should be clean, dry, and at the proper pressure. Before changing attachments or performing any maintenance on a pneumatic drill or pneumatic impact wrench, make sure that the air supply is turned off and the tool is physically disconnected from the supply hose.

00104-15_F15.EPS

Figure 15 Inline pneumatic oiler reservoir.

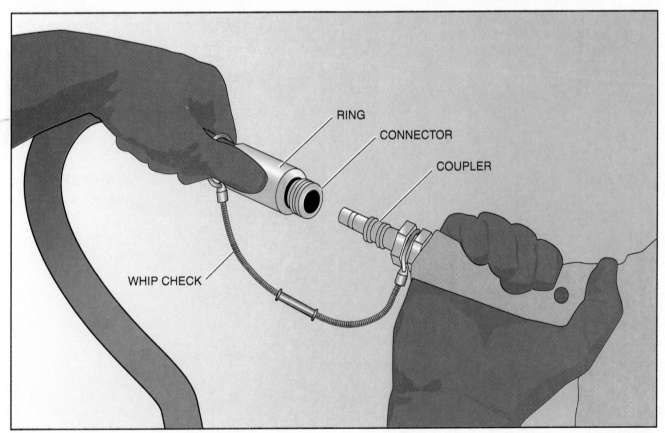

RING

CONNECTOR

COUPLER

WHIP CHECK

00104-15_F16.EPS

Figure 16 Properly connecting a pneumatic tool.

Additional Resources

29 *CFR* 1926, *OSHA Construction Industry Regulations*, Latest Edition. Washington, DC: Occupational Safety and Health Administration, US Department of Labor, US Government Printing Office.

All About Power Tools. Ortho Books; Larry Johnson, ed. 2002. Des Moines, IA: Meredith Books.

Power Tool Institute, Inc. 1300 Sumner Avenue Cleveland, OH 44115-2851. **www.powertoolinstitute.com**.

1.0.0 Section Review

1. A device on a power drill that enables a user to back out a drill bit that is stuck in the work material is called a(n) _____.

   a. chuck key
   b. reversing switch
   c. whip check
   d. auger switch

2. A hammer drill has an adjustable ring that can be turned to adjust the _____.

   a. voltage reaching the drill motor
   b. depth of the hole being drilled
   c. magnetic force applied to the base
   d. number of drill blows per minute

3. A pneumatic power tool that is best suited for fastening, tightening, and loosening nuts and bolts is a _____.

   a. cordless screwdriver
   b. hydraulic ratchet
   c. pneumatic impact wrench
   d. right-angle drill

2.0.0 POWER SAWS

Objective

Identify and explain how to use various types of power saws.

  a. Identify and explain how to use a circular saw.

  b. Identify and explain how to use saber and reciprocating saws.

  c. Identify and explain how to use a portable band saw.

  d. Identify and explain how to use miter and cutoff saws.

Performance Task

1. Safely and properly demonstrate the use of the following tool(s):
   - Circular saw
   - Reciprocating saw
   - Portable band saw
   - Miter or cutoff saw

Trade Terms

Arbor: The end of a circular saw shaft where the blade is mounted.

Kerf: The channel created by a saw blade passing through the material, which is equal to the width of the blade teeth.

Reciprocating: Moving backward and forward on a straight line.

Using the right saw for the job will make your work much easier. Always make sure that the blade is right for the material being cut. This section focuses on the following types of power saws:

- Circular saws
- Saber saws
- Reciprocating saws
- Portable handheld band saws
- Power miter box saws

2.1.0 Circular Saws

Many years ago, a company named Skil® made power tool history by introducing the portable circular saw. Today many different companies make dozens of models, but a lot of people in the United States still call any portable circular saw a Skilsaw. Other names you might hear are utility saw, electric handsaw, and builder's saw. The portable circular saw (*Figure 17*) is designed to cut lumber and boards to size for a project.

The size of a circular saw is based on the diameter of the circular blade. Circular saws used in the United States typically use fractional-inch measurements with blade diameters that range from 3⅜ to 16¼ inches. The most popular blade size for corded saws is 7¼ inches. Many smaller cordless circular saws use a 6½-inch blade. The hole in the center of a circular saw blade fits onto the **arbor**, or shaft, of the saw. The most common arbor size for a circular saw blade is ⅝ inch.

Circular saws are also available in metric sizes. Metric circular saws typically have circular blades of 165 mm, 190 mm, or 235 mm. Most metric circular saws have a 20 mm arbor, although some have a 25.4 mm arbor. It is important to note that standard saws with fractional-inch measurements are not interchangeable with metric saws. For instance, a 7¼-inch blade with a ⅝-inch arbor hole is not the same as a 190 mm blade with a 20 mm arbor hole. Only metric blades can be used on metric saws and only fractional-inch blades can be used on standard saws.

Circular saw weights can vary, but most of them weigh between 7 and 14 pounds (3.175 kg and 6.35 kg). The handle of the circular saw has a trigger switch that starts the saw. The motor is protected by a rigid housing. Blade speed when the blade is not engaged in cutting is stated in rpm. The teeth of the blade point in the direction of rotation.

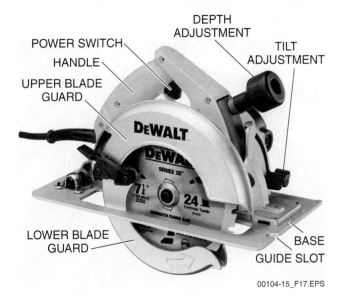

Figure 17 Circular saw.

The blade is protected by two guards. On top, a rigid plastic guard protects workers from flying debris and from accidentally touching the spinning blade. The lower guard is spring-loaded—when the saw is pushed forward, it retracts up and under the top guard to allow the saw to cut.

Saw blades fall into two categories: standard steel, which must be sharpened regularly, and carbide-tipped. Carbide-tipped blades are very common, but you must use the appropriate saw blade for the job. Some common types of saw blades include the following (refer to *Figure 18*):

- *Rip* – These blades are designed to cut with the grain of the wood. The square chisel teeth cut parallel with the grain and are generally larger than other types of blade teeth.
- *Crosscut* – These blades are designed to cut across the grain of the wood; that is, at 90-degree angle. Crosscut teeth cut at an angle and are finer than rip blade teeth.
- *Combination* – These blades are designed to cut hard or soft wood, either with or across the grain. The combination blade features both rip and crosscut teeth with deep troughs (gullets) between the teeth.
- *Nail cutter* – This blade has large carbide-tipped teeth that can make rough cuts through nails that may be embedded in the work.

- *Nonferrous metal cutter* – This blade has carbide-tipped teeth for cutting aluminum, copper, lead, and brass. It should be lubricated with oil or wax before each use.

Always follow the manufacturer's instructions when using saw blades.

2.1.1 How to Use a Circular Saw

Using a circular saw safely and efficiently requires the user to wear the appropriate PPE. The material to be cut must be secured and properly supported. If the work isn't heavy enough to stay in position without moving, it should be weighted or clamped down. The cut to be made should be marked with a pencil or other marking tool.

With the tool unplugged, the blade depth can be adjusted to the thickness of the wood being cut plus ¼-inch (≈6 mm). That way, the blade does not protrude through the material farther than necessary. The front edge of the baseplate should be placed on the work so the guide notch and the blade are in line with the cut mark.

Circular saw blades leave a **kerf** (*Figure 19*) that is roughly ⅛-inch (3.2 mm) wide. Be sure to align the blade with the waste side of the cutting line,

00104-15_F18.EPS

Figure 18 Circular saw blades.

00104-15_F19.EPS

Figure 19 Saw kerf.

or the finished piece will be short. When marking for the cut, mark an X on the waste side of the cut mark as a reminder of which side of the mark to cut. The saw kerf is an unavoidable result of sawing any material, so it must be considered for each and every cut.

After the saw has been started and is up to full speed, it is moved forward to start cutting. The lower blade guard will automatically rotate up and under the top guard when the saw is pushed forward. While cutting with the saw, grip the saw handles firmly with two hands, as shown in *Figure 20*.

If the saw cuts off the line, stop, back up very slightly and restart the cut. Do not force the saw. As the cut nears completion, the guide notch on the baseplate will move off the end of the work. At that point, use the blade as a guide. Once the cut has been completed, release the trigger switch. The blade will stop rotating. Make sure the blade has stopped before setting the saw down.

00104-15_F20.EPS

Figure 20 Proper use of a circular saw.

2.1.2 *Safety and Maintenance*

There are numerous factors to consider in order to use a circular saw safely and effectively. First, always wear the appropriate PPE. Before connecting the saw to its source of power, ensure that the blade is tight and that the blade guard is working correctly. The chosen blade should have a maximum rpm equal to or higher than the speed of the saw. To avoid hitting water lines or electrical wiring, find out what is inside the wall or on the other side of a partition before cutting through a structure.

During operation, keep both hands on the saw grips. Never force the saw through the work. This causes binding and overheating and may cause injury. Never reach underneath the work while operating the saw and never stand directly behind the work. Always stand to one side of it. Use clamps to secure small pieces of material to be cut. Know where the saw's power cord is located at all times. Accidentally cutting through the power cord can cause electrocution.

> **WARNING!**
>
> When using a circular saw, workers should never hold material to be cut with their hands; always use a clamp instead.

The most important maintenance on a circular saw is at the lower blade guard. Sawdust builds up and causes the guard to stick. If the guard sticks and does not move quickly over the blade after it makes a cut, the bare blade may still be turning when the saw is set down and may cause damage. Remove sawdust from the blade guard area. Remember to always disconnect the power source before performing any maintenance.

The Worm-Drive Saw

The worm-drive saw is a heavy-duty type of circular saw. Most circular saws have a direct drive. That is, the blade is mounted on a shaft that is part of the motor. With a worm-drive saw, the motor drives the blade from the rear through two gears. One gear (the worm gear) is cylindrical and threaded like a screw. The worm gear drives a wheel-shaped gear (the worm wheel) that is directly attached to the shaft to which the blade is fastened. This setup delivers much more rotational force (torque), making it easier to cut a double thickness of lumber. The worm-drive saw is almost twice as heavy as a conventional circular saw. This saw should be used only by an experienced craftworker.

00104-15_SA01.EPS

To avoid personal injury and damage to materials, check often to make sure the guard snaps shut quickly and smoothly. To ensure smooth operation of the guard, disconnect the saw from its power source, allow it to cool, and clean foreign material from the track. Be aware of fire hazards when using cleaning liquids such as isopropyl alcohol. Do not lubricate the guard with oil or grease. This could cause sawdust to stick in the mechanism. Always keep blades clean and sharp to reduce friction and kickback. Blades can be cleaned with hot water or mineral spirits. Be careful with mineral spirits; they are very flammable.

2.2.0 Saber and Reciprocating Saws

Two types of saws that are capable of making straight and curved cuts are saber saws and reciprocating saws. Both of these saws have a blade that moves back and forth to enable the cutting action.

2.2.1 Saber Saws

Saber saws, sometimes referred to as jig saws, have very fine blades. This makes the saw an effective tool for doing delicate and intricate work, such as cutting out patterns or irregular shapes from wood or thin, soft metals. They are also some of the best tools for cutting circles.

The saber saw (*Figure 21*) is a very useful portable power tool. It can make straight or curved cuts in wood, metal, plastic, wallboard, and other materials. The saber saw cuts with a blade that moves up and down, unlike the spinning circular saw blade. This means that each cutting stroke (upward) is followed by a return stroke (downward), so the saw is cutting only half the time it is in operation. This is called up-cutting or clean-cutting.

An important part of the saber saw is the baseplate (shoeplate or footplate). Its broad surface helps to keep the blade lined up. It keeps the work from vibrating and allows the blade teeth to bite into the material.

Many models are available with tilting baseplates for cutting beveled edges. Models come with a top handle or a barrel handle. Some cordless models are available.

The saber saw has changeable blades that enable it to cut many different materials, from wood and metal to wallboard and ceramic tile. Fine-toothed blades are used for thin materials and smoother cuts. Coarse blades are used for faster cutting in thicker materials and when smooth cuts are not a concern. Blades are rated by the number of teeth per inch or teeth per centimeter.

Most saber saws can be operated at various blade speeds. Types of saber saws include

(A) TOP HANDLE

(B) BARREL HANDLE

00104-15_F21.EPS

Figure 21 Saber saws.

single-speed, two-speed, and variable-speed. The speed of a variable-speed saber saw is controlled by how far the trigger is depressed. The low-speed range is for cutting hard materials, and the high-speed range is for soft materials.

> **CAUTION**
>
> Do not lift the blade out of the work while the saw is still running. If the blade is lifted out, the tip of the blade may hit the wood surface, marring the work and possibly breaking the blade.

2.2.2 Reciprocating Saws

A reciprocating saw, regardless of the manufacturer, is often referred to as a SawZall® (a trademark of the Milwaukee Electric Tool Company) in the United States because it was the first saw designed to serve as an electric hacksaw. Both the saber saw and the reciprocating saw can make straight and curved cuts. They are used to cut irregular shapes and holes in plaster, plasterboard, plywood, studs, metal, and most other materials that can be cut with a saw.

Both saws have straight blades that move backward and forward along a straight line as they are guided along the cut. The reciprocating saw (*Figure 22*) is designed for more heavy-duty jobs than the saber saw. It uses longer and tougher blades than a saber saw.

The reciprocating saw is used for jobs that require brute strength. It is an excellent choice for general demolition work. It can saw through walls or ceilings and create openings for windows, plumbing lines, and more.

Like the saber saw, reciprocating saws come in single-speed, two-speed, and variable-speed models. The low-speed setting is best for metal work. The high-speed setting is for sawing wood and other soft materials.

The baseplate (shoeplate or footplate) may have a swiveling action, or it may be fixed. Whatever the design, the baseplate is there to provide a brace or support point for the sawing operation.

2.2.3 How to Use Saber and Reciprocating Saws

Many of the steps involved in the safe and efficient use of saber saws and reciprocating saws are the same. Both saws require the user to wear the appropriate PPE. The material being sawed should be clamped to a pair of sawhorses or secured in a vise to reduce vibration. It is also important to check the blade for dulling or damage and to make sure the proper blade is being used for the material. The material should be measured and marked before any cutting takes place.

When cutting from the edge of a board or panel with a saber saw, be sure the front of the baseplate is resting firmly on the surface of the work before starting the saw. The blade should not be touching the work at this stage. If a cut must be made from the middle of a board, drill a hole at the starting point that is large enough to allow the blade to pass through. Once all the preparations have been made, pull the trigger to start the saw and move the blade gently but firmly into the work. Continue feeding the saw into the work at a reasonable pace without forcing it. Do not force the blade into the work. When the cut is finished, release the trigger and let the blade come to a stop before removing it from the work.

Before cutting material with a reciprocating saw, be sure to set the saw to the desired speed. Use lower speeds for sawing metal; use higher speeds for sawing wood and other soft materials. Grip the saw with both hands (*Figure 23*) and place the baseplate firmly against the workpiece. Once the trigger is squeezed to On, the blade moves back and forth, cutting on the backstroke.

> **WARNING!**
> You must use both hands to grip the reciprocating saw firmly. Otherwise, the pull created by the blade's grip might jerk the saw out of your grasp.

2.2.4 Safety and Maintenance

As with any power tool, saber saws and reciprocating saws can be dangerous to operate if certain guidelines are not followed. Always wear appropriate PPE. Secure the material being cut to reduce vibration and ensure safety. Before cutting through a wall or partition, find out what is inside the wall or on the other side of the partition. This will prevent the accidental cutting of water lines or electrical wiring.

Make sure that the saw is disconnected from its power source before installing or changing blades or performing any maintenance on the saw.

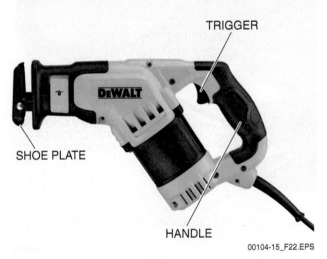

TRIGGER

SHOE PLATE

HANDLE

00104-15_F22.EPS

Figure 22 Reciprocating saw.

00104-15_F23.EPS

Figure 23 Proper use of a reciprocating saw.

When installing a blade in the saw, make sure the blade is in the collar as far as it will go, and tighten the setscrew securely. When replacing a broken blade, look for any pieces of the blade that may be stuck inside the collar. Before using the saw, make sure that the switch is in the Off position before it is plugged into a power source.

Regardless of the type of saw being used, always use a sharp blade and never force the blade through the work. Forcing or leaning into the blade can cause a worker to lose balance, slip or fall forward, and risk being cut by the saw. When cutting metal pieces, use a metal-cutting blade. Lubricate the blade with an agent such as beeswax, to help make tight turns and to reduce the chance of breaking the blade.

Saber and reciprocating blades are available for a wide variety of cutting tasks and materials. To make selection easier, the blades are often labeled for the material or specific use. As a general rule, metal-cutting blades have more teeth per inch, (or per centimeter) than wood-cutting blades. The teeth are also smaller.

When the blade first strikes the surface, it may jump. Keep a steady hand, and the cutting action will eventually allow the blade to enter the workpiece. Plunging the blade into the work with sudden force is a common cause of broken blades. Other causes of broken blades are pushing a saw too fast and mishandling the saw when it is not in use.

2.3.0 Portable Band Saws

The portable band saw (*Figure 24*) can cut pipe, metal, plastics, wood, and irregularly shaped materials. It is especially good for cutting heavy metal, but it will also do fine cutting work. Although it can cut wood, it is used almost exclusively used to cut metal products on the job site.

The band saw has a one-piece blade that runs in one direction around guides at either end of the saw. The blade is a thin, flat piece of steel. The blade must be of the proper length to fit the revolving pulleys that drive and support the blade. The proper blade length is determined from the manufacturer's documentation. Like most blades, their coarseness is rated in teeth per centimeter or teeth per inch. Thicker materials require coarser blades. If the blade is too coarse for the material, the individual teeth will often begin to break off. *Figure 25* shows how the blade is routed around the pulleys and through the blade guides.

Some band saws have multiple speeds, although most do not. The portable band saw generally cuts best at a low speed. Using a high speed will cause the blade's teeth to rub rather than cut. This can create heat through friction, which will cause the blade to wear out quickly.

> **WARNING!**
>
> A portable band saw always cuts in the direction of the user. For that reason, workers must be especially careful to avoid injury when using this type of saw. Always wear appropriate PPE and stay focused on the work.

2.3.1 How to Use a Portable Band Saw

Using a portable band saw safely and efficiently begins with wearing the appropriate PPE. Once prepared, place the stop firmly against the object

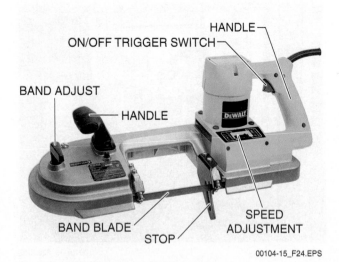

ON/OFF TRIGGER SWITCH
HANDLE
BAND ADJUST
HANDLE
BAND BLADE
STOP
SPEED ADJUSTMENT

00104-15_F24.EPS

Figure 24 Portable handheld band saw.

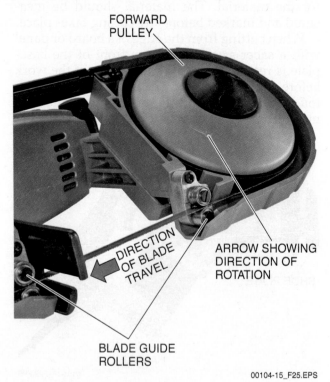

FORWARD PULLEY
DIRECTION OF BLADE TRAVEL
ARROW SHOWING DIRECTION OF ROTATION
BLADE GUIDE ROLLERS

00104-15_F25.EPS

Figure 25 Band saw pulley and blade guide rollers.

to be cut. Note that the direction of blade rotation tends to pull the saw toward the stop. If the stop is not firmly against the workpiece, the saw will jump when started until the stop slams against the material being cut.

Gently pull the trigger. Little or no downward pressure is needed to make a good, clean cut because the weight of the saw provides pressure for cutting.

2.3.2 Safety and Maintenance

There are a few basic guidelines that need to be followed to protect workers and the equipment when a portable band saw is used. First, always wear appropriate PPE. Use only a band saw that has a stop in place. Make sure the saw is disconnected from its power source during any maintenance-related procedure. Before cutting through lines or pipes, always make sure that they do not contain material that could present a hazard; some liquids or gases could be combustible. Keep in mind that the blade of a portable band saw can get stuck or twisted in the work easily. Never force a portable band saw; let the saw do the cutting.

2.4.0 Miter and Cutoff Saws

Miter saws and cutoff saws are similar in that both are used to make straight or miter cuts. Both types of saws can be permanently mounted to be stationary, but portable versions of the saws have the convenience of allowing users to move them from a workshop to a work site.

2.4.1 Power Miter Saws

The power miter saw combines a miter box with a circular saw, allowing it to make straight and miter cuts. There are two types of power miter saws: power miter saws and compound miter saws.

The saw blade of a standard miter saw pivots horizontally from the rear of the table and locks in position to cut angles from 0 to 45 degrees right and left. Stops are pre-set for common angles. The difference between the power miter saw and the compound miter saw (*Figure 26[A]*) is that the blade on the compound miter saw can be tilted vertically, allowing the saw to be used to make a compound cut (combined bevel and miter cut).

Similar to a power miter saw and compound miter saw is the compound-slide miter saw (*Figure 26[B]*). A compound-slide miter saw has a rail in the table that allows the motor and blade assembly to slide forward and backward. This sliding capability enables the tool to cut wider material than a standard miter saw can cut.

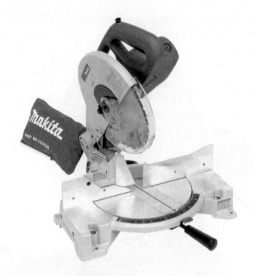

(A) COMPOUND MITER SAW

(B) COMPOUND SLIDING MITER SAW

00104-15_F26.EPS

Figure 26 Miter saws.

2.4.2 Abrasive Cutoff Saws

An **abrasive** cutoff saw, also referred to as a chop saw or cutoff saw (*Figure 27*), can be used to make straight cuts or angular cuts through materials such as angle iron, flat bar, and channel. As with miter saws, cutoff saws can be either stationary or portable.

The abrasive blade on a cutoff saw can be between 10 and 18 inches in diameter. Metric blades are commonly 250 mm to 350 mm in diameter. When the saw is in operation, the blade spins at such a high speed that the resulting friction is hot enough to burn through the material. Like all rotating blades and stones, the maximum rpm of the blade must be equal to or greater than that of the saw.

Figure 27 Abrasive cutoff saw.

00104-15_F27.EPS

2.4.3 *How to Use Miter and Cutoff Saws*

Using a power miter saw and an abrasive cutoff saw safely and efficiently involves many of the same procedures used for other types of power saws. Both types of saws require the user to wear appropriate PPE. It is also important to make sure that the saw is disconnected from its power supply while it is being set up for use. Since miter and cutoff saws have adjustments that can be used to angle and/or tilt the blade, it is important to properly set up the saw according to the cut that is needed. Once the saw has been set up for the proper cut, it can be connected to the power supply and the material can be placed on the saw table and secured.

When the saw is turned on, make sure the saw blade reaches its maximum speed before starting the cut. Hold the workpiece firmly against the fence when making the cut. Once the cut has been made, turn off the saw immediately. Abrasive cutoff saws produce a significant amount of sparks, much like a grinder.

2.4.4 *Safety and Maintenance*

Several basic guidelines should be followed when operating a miter saw or a cutoff saw to ensure worker safety and equipment protection. The appropriate PPE must be worn, which can include long sleeves, gloves, safety goggles, and a face mask to protect against dust and sparks. Never wear a watch or jewelry while operating the saw because they can get caught in the machinery. Do not allow other workers to stand nearby while the saw is being operated. It is also a good idea to make sure the work area is clear of flammable materials such as chemicals and rags that could ignite from sparks. Abrasive cutoff saws in particular produce a brilliant stream of sparks.

Do not check or change the blade or perform any sort of maintenance on a miter saw or a cutoff saw unless the saw is disconnected from its power source. Make sure that the blade is in good condition and secure before using the saw. Verify that the rpm rating of the blade meets or exceeds the saw's spindle speed. Check all saw guards to ensure that they are in place and working properly. Never retract a safety guard to view the material being cut while the saw is in use. Also, make sure that the saw is sitting on a firm base and is properly fastened to the base.

The saw must be properly set up for the cuts being made. It should be securely locked at the correct cutting angle. During operation, keep fingers clear of the blade. Never attempt to adjust the saw while it is running. If long material is being cut, have a helper support the end of the material. Once the cut is complete, stop the saw. Never leave a saw unattended until the blade comes to a stop.

Additional Resources

29 *CFR* 1926, *OSHA Construction Industry Regulations*, Latest Edition. Washington, DC: Occupational Safety and Health Administration, US Department of Labor, US Government Printing Office.

All About Power Tools. Ortho Books; Larry Johnson, ed. 2002. Des Moines, IA: Meredith Books.

Power Tool Institute, Inc. 1300 Sumner Avenue Cleveland, OH 44115-2851. **www.powertoolinstitute.com**.

2.0.0 Section Review

1. The proper way to start cutting material with a circular saw is to _____.

   a. rev the saw to full speed and slowly move it forward into the material
   b. hold the lower blade guard up to position the blade on the cut mark
   c. press the blade against the material being cut and set the saw rpm to Low
   d. tilt the front edge of the baseplate upward and push the saw forward

2. A saber saw is an effective tool for _____.

   a. drilling holes in concrete or pavement
   b. making long straight cuts through thick metal
   c. cutting through walls in demolition jobs
   d. doing delicate work on thin materials

3. The blade on a portable band saw _____.

   a. moves up and down through a shoeplate
   b. spins in a circular path on an arbor
   c. runs in one direction around guides
   d. reciprocates in and out from a guard

4. A type of miter saw in which the blade can be pivoted horizontally and vertically is called a _____.

   a. jig saw
   b. compound miter saw
   c. reciprocating saw
   d. sliding abrasive saw

3.0.0 GRINDERS AND GRINDER ATTACHMENTS

Objective

Identify and explain how to use various grinders and grinder attachments.

a. Identify and explain how to use various types of grinders.
b. Identify and explain how to use various grinder accessories and attachments.

Performance Task

1. Safely and properly demonstrate the use of the following tool:
   • Portable or bench grinder

Trade Terms

Abrasive: A substance, such as sandpaper, that is used to wear away material.

Grit: A granular, sand-like material used to make sandpaper and similar materials abrasive. Grit is graded according to its texture. The grit number indicates the number of abrasive granules in a standard size (per inch or per cm). The higher the grit number, the more particles in a given area, indicating a finer abrasive material.

Ring test: A method of testing the condition of a grinding wheel. The wheel is mounted on a rod and tapped. A clear ring means the wheel is in good condition; a dull thud means the wheel is in poor condition and should be disposed of.

Grinding tools can power all kinds of **abrasive** wheels, brushes, buffs, drums, bits, saws, and discs. These wheels come in a variety of materials and **grits**. They can drill, cut, smooth, and polish; shape or sand wood or metal; mark steel and glass; and sharpen or engrave. They can even be used on plastics.

WARNING!

Always wear safety goggles and a face shield when working with grinders. Make sure that the work area is free of combustible materials such as rags or flammable liquids and that a fire extinguisher is easily accessible. Clothes should be snug and comfortable and free of cuffs at the wrists and ankles. Wearing excessively loose clothing on the worksite can be extremely dangerous.

3.1.0 Grinders

Grinders are available in various configurations. Common handheld grinders include angle grinders (also called side grinders or right-angle grinders), end grinders, and detail grinders. Stationary grinders, called bench grinders, are permanently mounted on a work table or bench.

Angle grinders are used to grind away hard, heavy materials and to grind surfaces such as pipes, plates, or welds (*Figure 28*). The angle grinder has a rotating grinding disc set at a right angle to the motor shaft.

End grinders are sometimes called horizontal grinders or pencil grinders. These smaller grinders are used to smooth the inside of materials, such as pipe (*Figure 29*). The grinding disc on the end grinder rotates in line with the motor shaft. Grinding is also done with the outside of the grinding disc.

Like end grinders, detail grinders (*Figure 30[A]*) have an arbor that extends from the motor shaft onto which small attachments, called points, can be mounted to smooth and polish intricate metallic work. These attachments, a sample of which is shown in *Figure 30*, are commonly made in shaft sizes ranging from 1/16- to 1/4-inch. The shaft of these points is called the spindle. Metric points typically come in spindle sizes of 3 mm and 6 mm. A tremendous variety of point shapes are available to suit the grinding task.

The primary difference between an end grinder and a detail grinder is power; end grinders offer more power than a detail grinder.

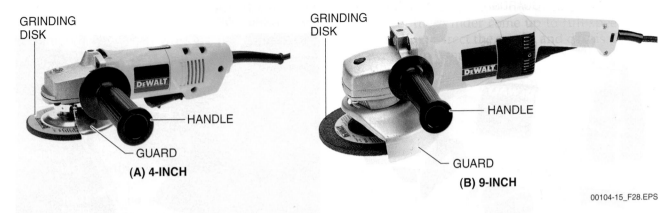

GRINDING DISK

HANDLE

GUARD

(A) 4-INCH

GRINDING DISK

HANDLE

GUARD

(B) 9-INCH

00104-15_F28.EPS

Figure 28 Angle grinders.

00104-15_F29.EPS

Figure 29 End grinder.

Bench grinders (*Figure 31*) are electrically powered stationary grinding machines. They usually have two grinding wheels that are used for grinding, rust removal, and metal buffing. They are also great for renewing worn edges and maintaining the sharp edges of cutting tools. For example, the bench grinder can be used to smooth the mushroomed heads of cold chisels.

Heavy-duty grinder wheels range from 6¾ to 10 inches (150 mm to 250 mm) in diameter. Each wheel's maximum speed is given in rpm. Never use a grinding wheel above its rated maximum speed. Its rated speed must be equal to or faster than the maximum speed of the power tool.

Bench grinders come with an adjustable tool rest. This is the surface on which you position the material you are grinding, such as cold chisel heads. There should be a distance of only ⅛-inch (about 3 mm) between the tool rest and the wheel.

> **WARNING!**
>
> Never change the adjustment of tool rests when the grinder is on or when the grinding wheels are spinning. Doing so may damage the work or cause injuries. Disconnect the power source before making any adjustments.

3.1.1 How to Use Grinders

Using handheld portable grinders safely and efficiently requires workers to follow some basic guidelines. Wear appropriate PPE, especially a face shield. If it is not already secured, secure the material in a vise or clamp it to the workbench. To use an angle grinder, place one hand on the handle of the grinder and one on the trigger. To use an end grinder or detail grinder, grip the grinder

(A) DETAIL GRINDER

(B) STONE CONE

00104-15_F30.EPS

Figure 30 Detail grinder and point.

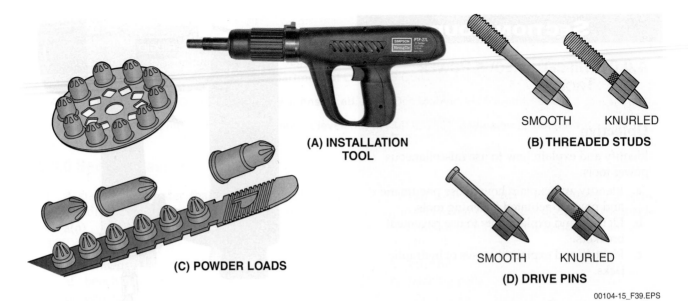

(A) INSTALLATION TOOL

(B) THREADED STUDS
SMOOTH KNURLED

(C) POWDER LOADS

(D) DRIVE PINS
SMOOTH KNURLED

00104-15_F39.EPS

Figure 39 Powder-actuated fastening system.

4.1.1 How to Use Power Fastening Tools

The safe and efficient use of power fastening tools depends on careful preparation, following some common sense guidelines, and being familiar with the tool. Before using any pneumatically powered nailer, always read the manufacturer's instructions. Also, wear the appropriate PPE, including gloves, safety goggles, hearing protection, and a hard hat.

Inspect the nailer for damage and loose connections before operation. Select the proper type of nail for the job and make sure the nails being used match the nailer. Then, load the nails into the nailer.

Verify that the air hose is properly connected to the nailer. Check the air compressor and adjust the pressure to the recommended amount. Most nailers operate at pressures of 70 to 120 pounds per square inch (psi) (≈480 to 830 kPa). Test the nailing ability using a piece of scrap material. If the nail penetration is not correct, follow the manufacturer's instructions for adjusting the air pressure for the particular gun.

When nailing wall materials, locate and mark the wall studs before nailing. Otherwise, it will be difficult to determine when a nail penetrates the wallboard but misses the stud. Hold the nailer firmly against the material to be fastened, then press the trigger (*Figure 40*).

Once the job is finished, turn off the air supply and disconnect the air hose from the nailer. Never leave the nailer connected and unattended.

> **WARNING!**
> A nail gun is not a toy. Playing with a nail gun can cause serious injury. Nails can easily pierce a hand, leg, or eye. Never point a nail gun at anyone or carry one with your finger on the trigger. Use the nail gun only as directed.

Using a powder-actuated fastening tool requires special training and certification. Never try to operate a powder-actuated tool without the proper training. Training and certification to use a powder-actuated tool is typically provided by the manufacturer through an on-site class or similar arrangement.

00104-15_F40.EPS

Figure 40 Proper use of a nailer.

Power Screwdrivers

Power screwdrivers use a power source (this model uses a battery) to speed production in a variety of applications, such as drywall installation, floor sheathing and underlayment, decking, fencing, and cement board installation. A chain of screws feeds automatically into the firing chamber. Most models incorporate a back-out feature to drive out screws as well as a guide that keeps the screw feed aligned and tangle-free. This tool can accept Phillips or square slot screws and weighs an average of six pounds.

00104-15_SA02.EPS

To use a powder-actuated tool, start by wearing the appropriate PPE, including safety goggles, ear protection, gloves, and a hard hat. The pin or stud to be fired is fed into the piston and then the gunpowder cartridge (booster or charge) is placed. The tool is positioned in front of the item to be fastened and pressed firmly against the mounting surface (*Figure 41*). This pressure releases the safety lock, and the trigger fires the charge.

4.1.2 Safety and Maintenance

There are numerous safety and maintenance guidelines associated with pneumatic and powder-actuated fastening tools. Both types of tools require workers to wear appropriate PPE, including ear protection, safety goggles, and a hard hat. It is also important to read and understand the manufacturer's manual for the tool being used.

When using a pneumatic nailer, workers should avoid pointing the nail gun at their own body or towards anyone else. The correct nail gun and nail size and type must be selected for the job. Keep the nailer oiled according to the manufacturer's instructions. Add a few drops to the air inlet before each use, according to the manufacturer's recommendations.

Never load a pneumatic nailer with the compressor hose attached. If the nailer is not firing, disconnect the air hose before attempting any repairs. Remember that there is a chance that air pressure is trapped within the tool if an internal

00104-15_F41.EPS

Figure 41 Using a powder-actuated fastening tool.

failure occurs, although it is unlikely. Check for pipes, electrical wiring, vents, and other materials behind wallboard before nailing. During operation, keep all body parts and co-workers away from the nail path to avoid serious injury. Nails can go through paneling or wallboard and strike someone on the other side if you miss the wood surface behind it. Never leave the nailer connected to the compressor hose when it is not in use.

Powder-actuated fastening tools have several guidelines that apply to them specifically. The most important rule is that workers should never use a powder-actuated tool unless they have been properly trained and certified on the specific model being used. Manufacturer representatives often visit the job site to conduct their training and certification program for a group of workers. The manufacturer's instruction manual also lists safety precautions that must be followed.

Be sure to select the proper size pin for the job and never load the tool until it is time to complete the firing. When loading the driver, put the pin in before loading the charge. Use the correct booster charge according to the manufacturer's instructions for the tool being used. Never hold the end of the barrel against any part of your body or cock the tool against your hand. Never place your hands behind the material being fastened. Do not fire the tool close to the edge of the material, especially concrete. Pieces of concrete may chip off and strike someone, or the projectile could continue past the concrete and strike a co-worker.

4.2.0 Pavement Breakers

Several large-scale demolition tools are frequently used in construction. They include pavement breakers, clay spades, rock drills, and core borers (*Figure 42*). These tools do not rotate like hammer drills; they reciprocate (move back and forth). They can be powered pneumatically or electrically. The name *jackhammer* comes from a trade name, but has come to refer to almost any of the handheld impact tools. While there are slight differences in these tools and their uses, this section will focus on pavement breakers.

The pavement breaker is used for large-scale demolition work, such as tearing down brick and concrete walls and breaking up concrete or pavement. A typical pneumatic pavement breaker weighs between 50 and 90 pounds (≈22 to 40 kilograms). On most pavement breakers, a throttle is located on the T-handle. When the throttle is pushed, compressed air operates a piston inside the tool. The piston drives the steel cutting shank into the material being broken up with a

hammering action. Attachments, such as spades or chisels, can be used on the pavement breaker for different applications.

4.2.1 How to Set Up and Use a Pavement Breaker

To operate a pavement breaker safely and efficiently, start by wearing the appropriate PPE. Prior to connecting the tool to its air supply, make sure that the air pressure is shut off at the main air outlet. Then, hold the coupler at the end of the air supply line, slide the ring back, and slip the coupler on the connector, or nipple, that is attached to the air drill. Verify that the connection is good. (A good coupling cannot be taken apart without first sliding the ring back.) Add a whip check to prevent the hoses from whiplashing if the connector comes uncoupled. Once a good connection has been established, turn on the air supply valve. The pavement breaker is now ready to use.

4.2.2 Safety and Maintenance

Besides always following the tool manufacturer's guidelines for operating a pavement breaker, there are a few other important rules. Always wear the appropriate PPE. This includes gloves, a hard hat, eye protection, appropriate boots, and because some of these tools make a lot of noise, hearing protection (earplugs). Also, know what is under the material being broken up. There could be water, gas, electricity, sewer, and telephone lines below the surface. Always follow the applicable methods to find out what is there and where it is before breaking the pavement.

4.3.0 Hydraulic Jacks

Hydraulic tools are used when an application calls for extreme force to be applied in a controlled manner. These tools do not operate at high speed, but great care should be used when operating them. The forces generated by hydraulic tools can easily damage equipment or cause personal injury if the manufacturer's procedures are not strictly followed.

Hydraulic jacks are portable devices used for a wide variety of purposes. They can be used to move or lift heavy equipment and other heavy material, to position heavy loads precisely, and to straighten or bend frames. Hydraulic jacks have two basic parts: a pump and a cylinder (sometimes called a ram). There are various types of hydraulic jacks including those with internal pumps and those that use a lever-operated pump.

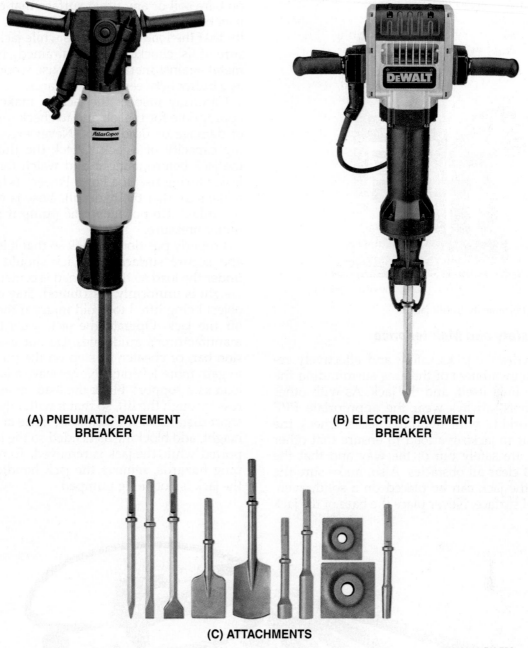

(A) PNEUMATIC PAVEMENT BREAKER

(B) ELECTRIC PAVEMENT BREAKER

(C) ATTACHMENTS

00104-15_F42.EPS

Figure 42 Typical demolition tools and attachments.

The latter type is often referred to as a Porta-Power®, the name of one common brand.

A hydraulic jack with an internal pump is a general-purpose jack that is available in many different capacities (*Figure 43*). The pump inside the jack applies pressure to the hydraulic fluid when the handle is pumped. The pressure on the hydraulic fluid applies pressure to the cylinder, which lifts or moves the load.

A lever-operated pump kit (*Figure 44*) includes a length of hydraulic hose and a cylinder, which are joined by the high-pressure hydraulic hose.

Lever-operated pumps are available in different capacities. Cylinders are available in many sizes; they are rated by the weight (in tons or metric tonnes) they can lift and the distance they can move it. This distance is called stroke and is measured in inches or millimeters. Hydraulic cylinders can lift more than 500 tons (≈454 metric tonnes). Strokes range from ¼-inch (≈6 mm) to more than 48 inches (≈122 cm). Different cylinder sizes and ratings are used for different jobs. Lever-operated pumps are especially useful for horizontal jacking.

Figure 43 Portable hydraulic jack.

4.3.1 Safety and Maintenance

Using hydraulic jacks safely and effectively requires an awareness of the area surrounding the load, the load itself, and the jack. As with other power tools, always wear the appropriate PPE when working with hydraulic jacks. Check the area prior to jacking a load to ensure that other workers are safely out of the way and that the load will clear all obstacles. Also, make sure the base of the jack can be placed on a solid, even, and level surface. Never place the base of the jack on bare soil or any other surface that could compact or shift under the load. If there is a possibility that the load could move while jacking, make sure it is chocked and restrained. Never jack metal against metal. Instead, use wood softeners as a buffer between metal surfaces.

Carefully inspect the jack to make sure it is appropriate for the job and to check for any signs of damage or fluid leaks. Never exceed the lifting capacity of a jack. Check the fluid level in the jack before using it and watch for any fluid leaks during use. If a Porta-Power® is being used, make sure that the hydraulic hose is not twisted or kinked. Do not move the pump if the hose is under pressure.

Properly position the jack so that it is on a level and secure surface. The jack should be placed under the load so that the load is centered and the weight is uniformly distributed. Stay clear of the object being lifted to avoid injury if the load slips off the jack. Operate the jack according to the manufacturer's guidelines. Do not use an extension bar, or cheater, or step on the pump handle to gain more leverage. Never leave a jack under a load as a support. Block the load up as you progress through the lift, so that it will only fall a very short distance if the jack fails. Once at the proper height, add blocking as needed so the load is supported while the jack is removed. To reduce tripping hazards, remove the jack handle any time the jack is not being pumped.

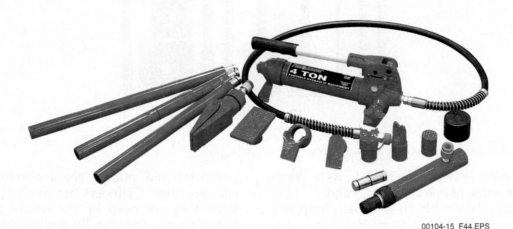

Figure 44 Lever-operated hydraulic pump kit.

Additional Resources

29 *CFR* 1926, *OSHA Construction Industry Regulations*, Latest Edition. Washington, DC: Occupational Safety and Health Administration, US Department of Labor, US Government Printing Office.

All About Power Tools. Ortho Books; Larry Johnson, ed. 2002. Des Moines, IA: Meredith Books.

Power Tool Institute, Inc. 1300 Sumner Avenue Cleveland, OH 44115-2851. **www.powertoolinstitute.com**.

4.0.0 Section Review

1. Pneumatic nailers are designed to fire when the trigger is pressed and the tool is _____.

   a. filled with a charge of compressed air
   b. pressurized with hydraulic fluid
   c. connected to its battery pack
   d. pressed against the material being fastened

2. The cutting shank on a pavement breaker is able to demolish concrete or pavement by _____.

   a. drilling
   b. reciprocating
   c. rotating
   d. expanding

3. When using a portable hydraulic jack, be sure to _____.

   a. avoid twisting or kinking the air lines
   b. leave the jack under the load as a support
   c. use an extension bar on the pump handle
   d. place the base of the jack on a solid, level surface

SUMMARY

Power tools are a necessity in the construction industry, and it is important to understand how they work and what they do. A worker might not use all of the tools covered in this module, but more than likely they will encounter other craft workers on construction sites who will be using them. All workers will be safer if everyone is familiar with the tools being used on the job site.

Power tool maintenance is another subject that workers need to learn and appreciate. The better a tool is maintained, the better it will function. Well-maintained tools operate safer and last longer, which protects workers and saves time and money.

As workers focus on a chosen field within the construction industry, they will learn to use the power tools for that specialized area. Although some of these specific tools might not be covered in this module, the basic safety and usage concepts are always applicable. Always read the manufacturer's manual for any new power tool being used. Never use a tool without the proper training for that tool. Following the basic use and safety guidelines explained in this module, properly maintaining the tools, and becoming educated about any new equipment beforehand will help ensure safe and efficient work habits and promote career advancement.

1. Pneumatic tools get their power from _____.
   a. air pressure
   b. fluid pressure
   c. hand pumps
   d. AC power sources

2. The most common use of the power drill is to _____.
   a. cut wood, metal, and plastic
   b. drive nails into wood, metal, and plastic
   c. make holes in wood, metal, and plastic
   d. carve letters in wood, metal, and plastic

3. A masonry bit is able to drill into concrete and similar material because it has a _____.
   a. countersink shank
   b. ceramic core
   c. whip check
   d. carbide tip

4. An example of an electric power drill that is designed to be used in tight spaces is a(n) _____.
   a. electromagnetic drill
   b. right-angle drill
   c. hammer drill
   d. keyless chuck drill

5. The electromagnetic drill is a _____.
   a. handheld drill used on wood
   b. cordless drill used on masonry and tile
   c. portable drill used on thick metal
   d. pneumatic drill that has a pounding action

6. Hammer drills are designed to drill into _____.
   a. wood, metal, and plastic
   b. concrete, brick, and tile
   c. drywall, fiberglass, and wood
   d. roofing shingles, plastic, and wood

7. A pneumatic impact wrench requires the use of _____.
   a. impact sockets that are designed for the applicable tool
   b. an adapter so that handheld sockets will fit
   c. shear pins between the wrench and the socket
   d. a trigger lock to prevent accidental starting

8. When cutting with a circular saw, grip the saw handles _____.
   a. and pull the saw toward you
   b. firmly with one hand
   c. and engage the trigger lock
   d. firmly with two hands

9. The high speed setting on a reciprocating saw is used for _____.
   a. cutting through drywall
   b. metal work
   c. sawing wood and other soft materials
   d. grinding surfaces

10. When using a saber saw, avoid vibration by _____.
    a. using a low-speed setting
    b. using a clamp or vise to hold the work
    c. setting a heavy object on the workpiece
    d. holding the workpiece down with your free hand

11. Before using a reciprocating saw to cut through a wall or partition, always _____.
    a. find out what is on the other side
    b. remove the lower blade guard
    c. increase the revolutions per minute
    d. lubricate the guard with oil or grease

12. Use only a band saw that has a _____.
    a. breastplate with a broad surface
    b. battery pack
    c. thick, three-piece blade
    d. stop

13. A sliding compound miter saw has a rail that allows the blade to slide forward and backward, which enables the saw to _____.
    a. use much thinner blades than a standard miter saw
    b. cut wider material than a standard miter saw
    c. produce much less dust than a standard miter saw
    d. cut harder material than a standard miter saw

14. The blade of an abrasive cutoff saw spins at such a high speed that _____.
    a. it can only be used for straight cuts
    b. the abrasive particles will melt into some metals
    c. the resulting friction is hot enough to burn through the material
    d. it can never be more than eight inches in diameter

15. The end grinder is used to _____.
    a. polish intricate work
    b. grind surfaces
    c. smooth the work before painting
    d. smooth the inside of materials, such as pipe

16. A detail grinder smoothes and polishes intricate metallic work by using attachments called _____.
    a. points
    b. rollers
    c. pins
    d. studs

17. Powder-actuated fastening systems are used to _____.
    a. penetrate drywall
    b. anchor static loads to steel beams
    c. hammer nails into metal
    d. remove nails

18. Before you begin setting up a pavement breaker for use, make sure that the air pressure is _____.
    a. shut off at the main air outlet
    b. turned on only halfway
    c. turned on full
    d. shut off at the coupler

19. Porta-Power® cylinders are rated by how much weight they can lift and by _____.
    a. their torque
    b. the amount of electromagnetic material they have
    c. the distance they can move the weight
    d. how much they weigh

20. Hydraulic jacks are used when the application calls for _____.
    a. operation at high speed
    b. extreme force to be applied
    c. quiet operation
    d. manually assisted lifting

Trade Terms Quiz

Fill in the blank with the correct term that you learned from your study of this module.

1. Activating the _____ will make the trigger stay in the operating mode even when it is released.

2. _____ reverses its direction at regularly recurring intervals; this type of current is delivered through wall plugs.

3. A(n) _____ saw's straight blades move backward and forward along a straight line.

4. A circular saw blade is attached to the _____ of the saw.

5. Masonry bits and nail-cutter saw blades have a(n) _____ tip.

6. A(n) _____ is a substance, such as sandpaper, that is used to wear away material.

7. A(n) _____ is used to open and close the chuck on a power drill.

8. _____ is the value used to report the rotational speed of a motor or shaft.

9. A(n) _____ is used to set the head of a screw at or below the surface of the material.

10. _____ flows in one direction, from the negative to the positive terminal of the source, such as a battery.

11. Use a(n) _____ to bore holes in wood and other materials.

12. A sand-like material used to make a surface rough, graded by its size, is called _____.

13. The _____ of the drill holds the drill bit.

14. To prevent an electrical shock, do not operate electric power tools without proper _____.

15. If a flat-bottomed hole is needed in a piece of lumber, use a _____.

16. A _____ is used to bore holes in brick, block, and similar materials.

17. A _____ is created by a saw blade as it cuts through the material.

18. Perform a(n) _____ to check the condition of a grinding wheel.

19. The _____ is the smooth part of a drill bit that fits into the chuck.

20. A(n) _____ protects people from electric shock and protects equipment from damage by interrupting the flow of electricity if an electrical fault occurs.

Trade Terms

Abrasive
Alternating current (AC)
Arbor
Auger bit
Carbide
Chuck

Chuck key
Countersink
Direct current (DC)
Forstner bit
Grit

Ground fault circuit
  interrupter (GFCI)
Ground fault protection
Kerf
Masonry bit
Reciprocating

Revolutions per minute
  (rpm)
Ring test
Shank
Trigger lock

Fernando Sanchez
TIC—The Industrial Company
Craft Training Instructor

How did you choose a career in the construction industry?
Honestly, by referral through family. My brother Israel recommended me to TIC. I immediately adapted to it and it grew on me. Helping build something provides a great sense of accomplishment that I really enjoy.

Who inspired you to enter the industry?
My brother Israel and my dad Fernando Sr. They inspired me and helped me get into the industry.

What types of training have you been through?
I have completed a wide variety of training programs, including:
- Real estate school (obtained a license)
- Mortgage training for real estate purchasing
- How to Be an Effective Leader training/seminar
- OSHA 10 hour
- First Aid and CPR
- NCCER *Core*
- NCCER *Pipefitting* Levels 1 through 4 (obtained Certified Plus credential)
- NCCER Master Trainer
- *Welding* Levels 1 and 2 (obtained Structural Pipe Certifications)
- TIC – QA/QC Levels 1 and 2
- TIC – Field Management 1 and 2
- TIC – Bull Rigging Class
- E-learn SkillSoft Training

How important is education and training in construction?
Education and training is extremely important, because an individual needs to be trained in all facets of construction to become professional in their craft.

How important are NCCER credentials to your career?
NCCER is a nationally accredited source of quality craft training and their credentials need to be on the record of all construction workers around the country.

How has training/construction impacted your life?
It has given me the opportunity to grow with my company and experience new and innovative things in the construction industry. Most importantly, it has given me the opportunity to build on my personal skills and allowed me to transfer my knowledge and experience to others in the field. Teaching others is a true passion of mine.

I've also had the opportunity to provide a better way of life for my family, to put my kids through college, and to give them things I never had.

What kinds of work have you done in your career?
I have done a little bit of everything in my career, such as:
- Residential construction
- Environmental services
- Pharmacy technician
- Mortgage loans
- Real estate
- Electrical line clearance (professional tree trimmer)
- Public relations (arborist)
- Restaurant (general manager)
- Industrial construction

Tell us about your present job.
I'm currently in charge of the Pipe Department at the Steamboat Springs TIC training center. I am a craft instructor and my job is to teach our current employed pipe helpers and pipefitters to reach their ultimate goal of becoming journey-level workers and supervisors in their craft.

What do you enjoy most about your job?
The thing I enjoy most about my job is having the ability to teach someone new things and transfer knowledge so that they can become successful in their own careers. I love to teach and train and I truly enjoy seeing others become great.

What factors have contributed most to your success?

Having loving parents has contributed in a big way. They raised me and taught me that being a simple person can take you a long way. My dad showed me the power of friendship and kindness. My mom always told me that if I try my best, anything is possible.

Working for a great company like TIC has also been a major factor. They offer the opportunity to evolve and reach a career goal with their training and support.

I've also have had several mentors in my career, including Paul La Borde, my pipe instructor. Greg Jones and Mike Burris, past foremen of mine, also deserve a place on that list. Above all through, the support of my beautiful wife Gaby has taken me a long way.

Would you suggest construction as a career to others? Why?

Yes. This is a fast-growing field and we need more people in construction. The opportunities are endless, and working for a company like TIC that provides training for their employees opens up opportunities to make a great career out of construction.

What advice would you give to those new to the field?

Knowledge is power; the more you learn, the more you can contribute. Above all, stay safe and keep others safe as well.

Interesting career-related fact or accomplishment:

To me, it is a blessing that I am doing what I love to do for a living. I have been coaching and teaching since the age of 12 and I always knew that's what I wanted to do with my life. Having been at the right place at the right time, I am now living my dream. I can honestly say I do not feel as if I work a day in my life because I love what I do.

How do you define craftsmanship?

Having the ability to perform a particular occupation or trade at your fullest potential, and taking pride in your work.

Abrasive: A substance, such as sandpaper, that is used to wear away material.

Alternating current (AC): The common power supplied to most all wired devices, where the current reverse its direction may times per second. AC power is the type of power generated and distributed throughout settled areas.

Arbor: The end of a circular saw shaft where the blade is mounted.

Auger bit: A drill bit with a spiral cutting edge for boring holes in wood and other materials.

Carbide: A very hard material made of carbon and one or more heavy metals. Commonly used in one type of saw blade.

Chuck: A clamping device that holds an attachment; for example, the chuck of the drill holds the drill bit.

Chuck key: A small, T-shaped steel piece used to open and close the chuck on power drills.

Countersink: A bit or drill used to set the head of a screw at or below the surface of the material.

Direct current (DC): An electric power supply where the current flows in one direction only. DC power is supplied by batteries and by transformer-rectifiers that change AC power to DC.

Forstner bit: A bit designed for use in wood or similar soft material. The design allows it to drill a flat-bottom blind hole in material.

Grit: A granular, sand-like material used to make sandpaper and similar materials abrasive. Grit is graded according to its texture. The grit number indicates the number of abrasive granules in a standard size (per inch or per cm). The higher the grit number, the more particles in a given area, indicating a finer abrasive material.

Ground fault circuit interrupter (GFCI): A circuit breaker designed to protect people from electric shock and to protect equipment from damage by interrupting the flow of electricity if a circuit fault occurs.

Ground fault protection: Protection against short circuits; a safety device cuts power off as soon as it senses any imbalance between incoming and outgoing current.

Kerf: The channel created by a saw blade passing through the material, which is equal to the width of the blade teeth.

Masonry bit: A drill bit with a carbide tip designed to penetrate materials such as stone, brick, or concrete.

Reciprocating: Moving backward and forward on a straight line.

Revolutions per minute (rpm): The rotational speed of a motor or shaft, based on the number of times it rotates each minute.

Ring test: A method of testing the condition of a grinding wheel. The wheel is mounted on a rod and tapped. A clear ring means the wheel is in good condition; a dull thud means the wheel is in poor condition and should be disposed of.

Shank: The smooth part of a drill bit that fits into the chuck.

Trigger lock: A small lever, switch, or part that can be used to activate a locking catch or spring to hold a power tool trigger in the operating mode without finger pressure.

Additional Resources

This module presents thorough resources for task training. The following resource material is suggested for further study.

29 *CFR* 1926, *OSHA Construction Industry Regulations*, Latest Edition. Washington, DC: Occupational Safety and Health Administration, US Department of Labor, US Government Printing Office.

All About Power Tools. Ortho Books; Larry Johnston, ed. 2002. Des Moines, IA: Meredith Books.

Power Tool Institute, Inc. 1300 Sumner Avenue Cleveland, OH 44115-2851. **www.powertoolinstitute.com**.

Figure Credits

Answer	Section Reference	Objective
Section One		
1. b	1.1.0	1a
2. d	1.2.1	1b
3. c	1.3.2	1c
Section Two		
1. a	2.1.1	2a
2. d	2.2.1	2b
3. c	2.3.0	2c
4. b	2.4.1	2d
Section Three		
1. c	3.1.0	3a
2. a	3.2.0	3b
Section Four		
1. d	4.1.0	4a
2. b	4.2.0	4b
3. d	4.3.1	4c

NCCER CURRICULA — USER UPDATE

NCCER makes every effort to keep its textbooks up-to-date and free of technical errors. We appreciate your help in this process. If you find an error, a typographical mistake, or an inaccuracy in NCCER's curricula, please fill out this form (or a photocopy), or complete the online form at **www.nccer.org/olf**. Be sure to include the exact module ID number, page number, a detailed description, and your recommended correction. Your input will be brought to the attention of the Authoring Team. Thank you for your assistance.

Instructors – If you have an idea for improving this textbook, or have found that additional materials were necessary to teach this module effectively, please let us know so that we may present your suggestions to the Authoring Team.

NCCER Product Development and Revision

13614 Progress Blvd., Alachua, FL 32615

Email: curriculum@nccer.org
Online: www.nccer.org/olf

❏ Trainee Guide   ❏ Lesson Plans   ❏ Exam   ❏ PowerPoints   Other _____

Craft / Level: Copyright Date:

Module ID Number / Title:

Section Number(s):

Description:

_____

_____

_____

_____

Recommended Correction:

_____

_____

_____

_____

Your Name:

Address:

_____

Email: Phone:

00105-15

Introduction to Construction Drawings

OVERVIEW

Various types of construction drawings are used to represent actual components of a building project. The drawings provide specific information about the locations of the parts of a structure, the types of materials to be used, and the correct layout of the building. Knowing the purposes of the different types of drawings and interpreting the drawings correctly are important skills for anyone who works in the construction trades. This module introduces common types of construction drawings, their basic components, standard drawing elements, and measurement tools that are typically used when working with construction drawings.

Module Five

Trainees with successful module completions may be eligible for credentialing through the NCCER Registry. To learn more, go to **www.nccer.org** or contact us at **1.888.622.3720**. Our website has information on the latest product releases and training, as well as online versions of our *Cornerstone* magazine and Pearson's product catalog.

Your feedback is welcome. You may email your comments to **curriculum@nccer.org,** send general comments and inquiries to **info@nccer.org**, or fill in the User Update form at the back of this module.

This information is general in nature and intended for training purposes only. Actual performance of activities described in this manual requires compliance with all applicable operating, service, maintenance, and safety procedures under the direction of qualified personnel. References in this manual to patented or proprietary devices do not constitute a recommendation of their use.

Objective

When you have completed this module, you will be able to do the following:

1. Identify and describe various types of construction drawings, including their fundamental components and features.
   a. Identify various types of construction drawings.
   b. Identify and describe the purpose of the five basic construction drawing components.
   c. Identify and explain the significance of various drawing elements, such as lines of construction, symbols, and grid lines.
   d. Identify and explain the use of dimensions and various drawing scales.
   e. Identify and describe how to use engineer's and architect's scales.

Performance Task

Under the supervision of your instructor, you should be able to do the following:

1. Using the floor plan supplied with this module:
   - Locate the wall common to both interview rooms.
   - Determine the overall width of the structure studio.
   - Determine the distance from the outside east wall to the center of the beam in the structure studio.
   - Determine the elevation of the slab.

Trade Terms

Architect
Architect's scale
Architectural plans
Beam
Blueprints
Civil plans
Computer-aided drafting (CAD)
Contour lines
Detail drawings
Dimension line
Electrical plans
Elevation (EL)
Elevation drawing

Engineer
Engineer's scale
Fire protection plan
Floor plan
Foundation plan
Hidden line
HVAC
Leader
Legend
Mechanical plans
Metric scale
Not to scale (NTS)

Piping and instrumentation
  drawings (P&IDs)
Plumbing isometric drawing
Plumbing plans
Roof plan
Scale
Schematic
Section drawing
Specifications
Structural plans
Symbol
Title block

Industry Recognized Credentials

If you are training through an NCCER-accredited sponsor, you may be eligible for credentials from NCCER's Registry. The ID number for this module is 00105-15. Note that this module may have been used in other NCCER curricula and may apply to other level completions. Contact NCCER's Registry at 888.622.3720 or go to **www.nccer.org** for more information.

Contents ──────────────

Topics to be presented in this module include:

Figures

1.0.0 CONSTRUCTION DRAWINGS AND THEIR COMPONENTS

Objective

Identify and describe various types of construction drawings, including their fundamental components and features.

a. Identify various types of construction drawings.
b. Identify and describe the purpose of the five basic construction drawing components.
c. Identify and explain the significance of various drawing elements, such as lines of construction, symbols, and grid lines.
d. Identify and explain the use of dimensions and various drawing scales.
e. Identify and describe how to use engineer's and architect's scales.

Performance Task

1. Using the floor plan supplied with this module:
   - Locate the wall common to both interview rooms.
   - Determine the overall width of the structure studio.
   - Determine the distance from the outside east wall to the center of the beam in the structure studio.
   - Determine the elevation of the slab.

Trade Terms

Architect: A qualified, licensed person who creates and designs drawings for a construction project.

Architect's scale: A specialized ruler used in making or measuring reduced scale drawings. The ruler is marked with a range of calibrated ratios for laying out distances, with scales indicating feet, inches, and fractions of inches. Used on drawings other than site plans.

Architectural plans: Drawings that show the design of the project. Also called architectural drawings.

Beam: A large, horizontal structural member made of concrete, steel, stone, wood, or other structural material to provide support above a large opening.

Blueprints: The traditional name used to describe construction drawings.

Civil plans: Drawings that show the location of the building on the site from an aerial view, including contours, trees, construction features, and dimensions.

Computer-aided drafting (CAD): The making of a set of construction drawings with the aid of a computer.

Contour lines: Solid or dashed lines showing the elevation of the earth on a civil drawing.

Detail drawings: Enlarged views of part of a drawing used to show an area more clearly.

Dimension line: A line on a drawing with a measurement indicating length.

Electrical plans: Engineered drawings that show all electrical supply and distribution.

Elevation (EL): Height above sea level, or other defined surface, usually expressed in feet or meters.

Elevation drawing: Side view of a building or object, showing height and width.

Engineer: A person who applies scientific principles in design and construction.

Engineer's scale: A straightedge measuring device divided uniformly into multiples of 10 divisions per inch so that drawings can be made with decimal values. Used mainly for land measurements on site plans.

Fire protection plan: A drawing that shows the details of the building's sprinkler system.

Floor plan: A drawing that provides an aerial view of the layout of each room.

Foundation plan: A drawing that shows the layout and elevation of the building foundation.

Hidden line: A dashed line showing an object obstructed from view by another object.

HVAC: Heating, ventilating, and air conditioning.

Leader: In drafting, the line on which an arrowhead is placed and used to identify a component.

Legend: A description of the symbols and abbreviations used in a set of drawings.

Mechanical plans: Engineered drawings that show the mechanical systems, such as motors and piping.

Metric scale: A straightedge measuring device divided into centimeters, with each centimeter divided into 10 millimeters. Usually used for architectural drawings and sometimes referred to as a metric architect's scale.

Not to scale (NTS): Describes drawings that show relative positions and sizes only, without scale.

Piping and instrumentation drawings (P&IDs): Schematic diagrams of a complete piping system.

Plumbing isometric drawing: A type of three-dimensional drawing that depicts a plumbing system.

Plumbing plans: Engineered drawings that show the layout for the plumbing system.

Roof plan: A drawing of the view of the roof from above the building.

Scale: The ratio between the size of a drawing of an object and the size of the actual object.

Schematic: A one-line drawing showing the flow path for electrical circuitry or the relationship of all parts of a system.

Section drawing: A cross-sectional view of a specific location, showing the inside of an object or building.

Specifications: Precise written presentation of the details of a plan.

Structural plans: A set of engineered drawings used to support the architectural design.

Symbol: A drawing that represents a material or component on a plan.

Title block: A part of a drawing sheet that includes some general information about the project.

Construction drawings are architectural or working drawings used to represent a structure or system. These were traditionally referred to as blueprints, because years ago the lines on a blueprint were white and the background was blue. Construction drawings are also called prints. Today, most prints are created by computer-aided drafting (CAD), and they have blue or black lines on a white background.

Various kinds of construction drawings, including residential drawings, commercial drawings, landscaping plans, shop drawings, and industrial drawings, are used in construction. In this module, several of the most common types of drawings will be introduced.

Construction drawings, together with the set of specifications (often abbreviated as *specs*), detail what is to be built and what materials are to be used. Specifications are written statements that the architectural and engineering firm provides to the general contractors. The specs define the quality of work to be done and describe the materials to be used.

The set of construction drawings forms the basis of agreement and understanding that a building will be built as detailed in the drawings. Therefore, everyone involved in planning, supplying, and building any structure should be able to read construction drawings. For any building project, also consult the civil engineering plans for that location, including sewer, highway, and water installation plans.

1.1.0 Six Types of Construction Drawings

A set of building construction drawing plans almost always includes six major types of drawings (*Figure 1*). They include the following:

- Civil
- Architectural
- Structural
- Mechanical
- Plumbing
- Electrical
- Fire protection

This section will examine the various characteristics of each type of drawing.

1.1.1 Civil Plans

Civil plans are also called site plans, survey plans, or plot plans. They show the location of the building on the site from an aerial view (*Figure 2*). A civil plan also shows the natural contours of the earth, represented on the plan by contour lines. The civil plans can also include a landscape plan (*Figure 3*) that shows any trees on the property; construction features such as walks, driveways, or utilities; the dimensions of the property; and possibly a legal description of the property.

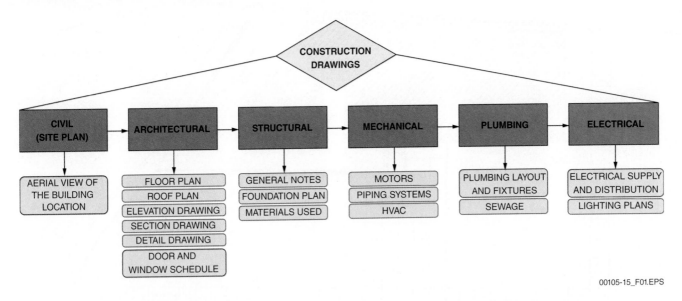

Figure 1 Types of construction drawings.

1.1.2 Architectural Plans

Architectural plans (also called architectural drawings) show the design of the project. One part of an architectural plan is a floor plan, also known as a plan view (refer to *Drawing 1*, First Floor Plan, in the *Appendix*). Any drawing made looking down on an object is commonly called a plan view. The floor plan is an aerial view of the layout of each room. It provides the most information about the project. It shows exterior and interior walls, doors, stairways, and mechanical equipment. The floor plan shows the floor as someone would see it from above if the upper part of the building were removed.

An architectural plan also includes a roof plan (*Figure 4*), which is a view of the roof from above the building. It shows the shape of the roof and the materials that will be used to finish it.

Elevation (EL) is another element of architectural drawings. Elevation drawings are side views that show height. On a building drawing, there are standard names for different elevations. For example, the side of a building that faces south is called the south elevation. Exterior elevations (*Figure 5*) show the size of the building; the style of the building; and the placement of doors, windows, chimneys, and decorative trim.

Building Information Modeling (BIM)

Traditionally, Building Information Modeling (BIM) was a system of paper-based documents and construction drawings designed to monitor and track specific information about a building.

In the last decade, BIM has increasingly moved toward computer technology, which allows the various trades and the building owner to track the lifecycle of the building from the start of planning, through construction, to completion. BIM allows building owners and trades to easily track and identify the various components and assemblies of individual systems (such as electrical, HVAC, and plumbing) and building materials within a specific location or locations of a building.

BIM also has the capabilities of simulating the lifecycles of various systems within the building in order to provide information on determining the wear on, and lifespan of, individual components. Additionally, BIM software allows the trades and the building owner to input and save information on the building as repairs, updates, or renovations are made to the building. Information such as part sizes, manufacturers, part numbers, and related drawings, or any other information ever researched in the past can be stored in a BIM system.

Another advantage of BIM lies in its capabilities of allowing documentation and information on a building to be easily transferred and communicated between all the members of a project.

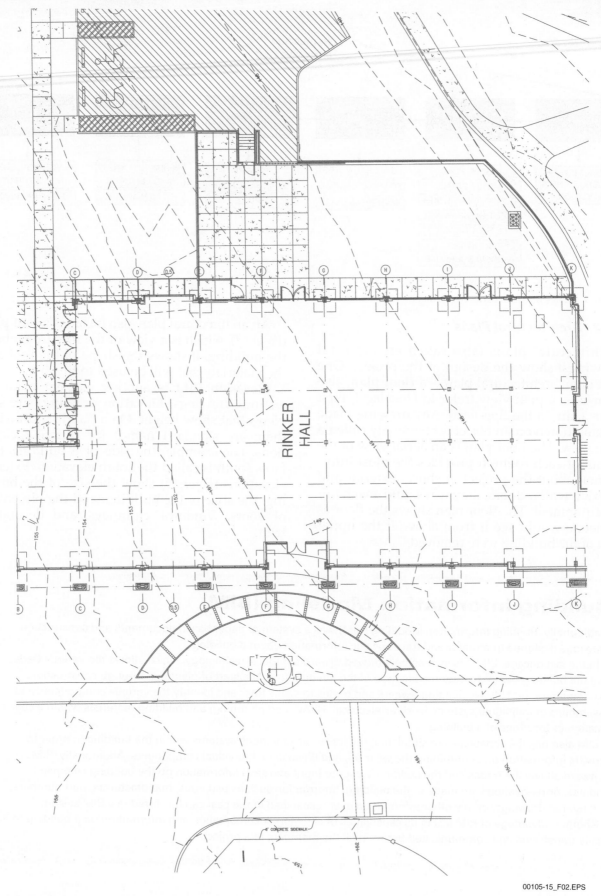

Figure 2 Civil plan, aerial view.

NCCER – *Core Curriculum*   00105-15

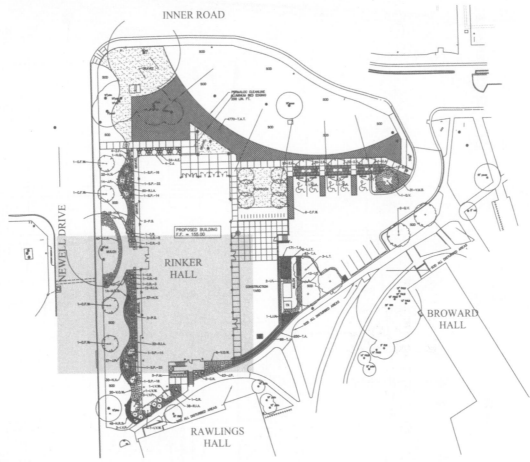

① LANDSCAPE PLAN
1:40

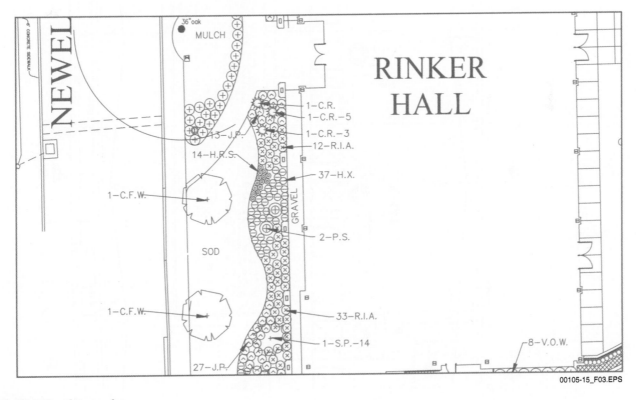

00105-15_F03.EPS

Figure 3 Landscape plan.

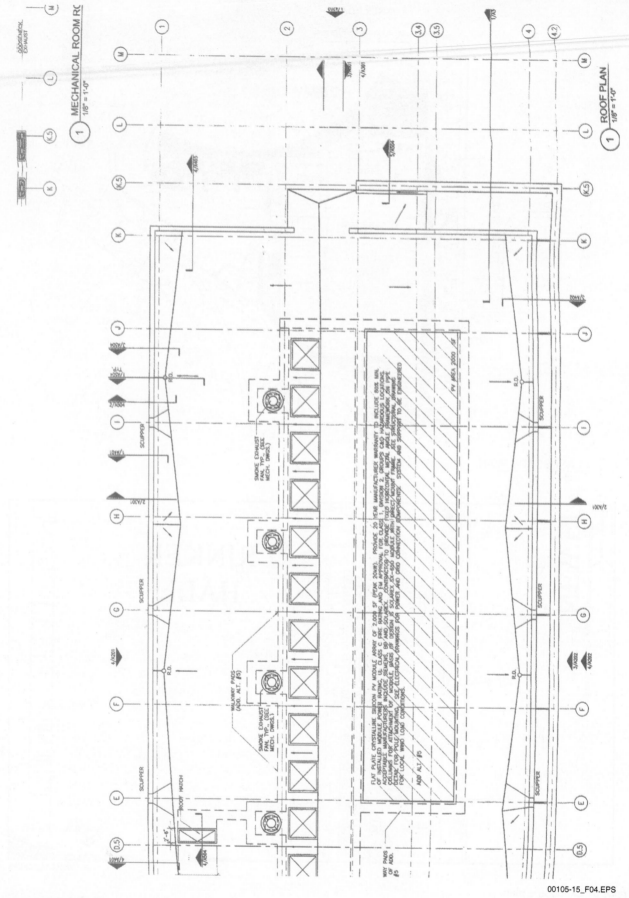

Figure 4 Roof plan.

00105-15_F04.EPS

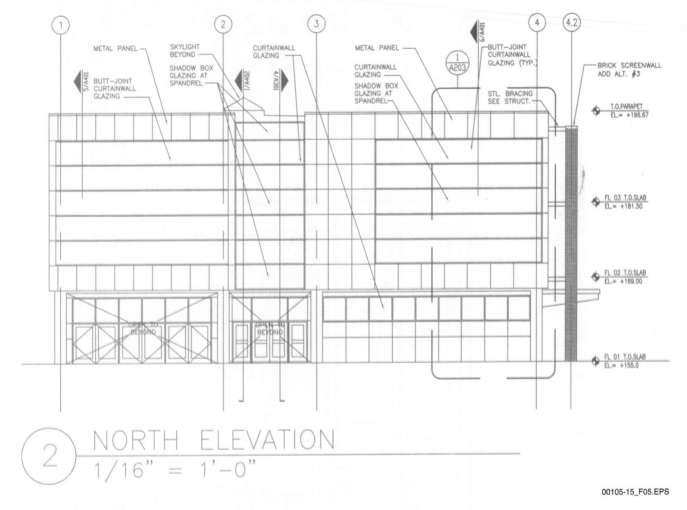

METAL PANEL

BUTT–JOINT
CURTAINWALL
GLAZING

SKYLIGHT
BEYOND

SHADOW BOX
GLAZING AT
SPANDREL

CURTAINWALL
GLAZING

METAL PANEL

CURTAINWALL
GLAZING

SHADOW BOX
GLAZING AT
SPANDREL

BUTT–JOINT
CURTAINWALL
GLAZING (TYP.)

STL. BRACING
SEE STRUCT.

BRICK SCREENWALL
ADD ALT. #3

T.O.PARAPET
EL.= +196.67

FL 03 T.O.SLAB
EL.= +181.50

FL 02 T.O.SLAB
EL.= +169.00

FL 01 T.O.SLAB
EL.= +155.0

OPEN TO BEYOND

OPEN TO BEYOND

2 NORTH ELEVATION
1/16" = 1'–0"

00105-15_F05.EPS

Figure 5 Exterior elevation.

Another element of the architectural plan is section drawings, which show how the structure is to be built. Section drawings (*Figure 6*) are cross-sectional views that show the inside of an object or building. They show what construction materials to use and how the parts of the object or building fit together. They normally show more detail than plan views. Compare the information on the drawing in *Figure 6* with the information on the drawing in *Figure 5*.

Even more detail is shown in detail drawings (*Figure 7*), which are enlarged views of some special features of a building, such as floors and walls. They are enlarged to make the details clearer. Often the detail drawings are placed on the same sheet where the feature appears in the plan, but sometimes they are placed on separate sheets and referred to by a number on the plan view.

The architectural plan also shows the finish schedules to be used for the doors and windows of the building. Door and window schedules, for example, are tables that list the sizes and other information about the various types of doors and windows used in the project. *Figure 8* shows an example of a window schedule and a detail drawing for the Type D1 window. Be aware that in some sets of drawings the window schedule and elevation drawings for each type of window may appear on the same sheet. Schedules may also be included for finish hardware and fixtures. These schedules are not drawings, but they are usually included in a set of working drawings.

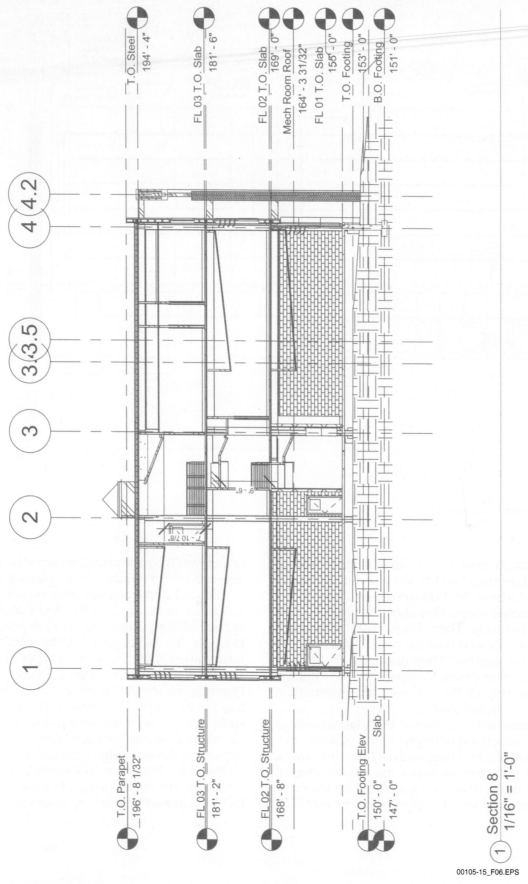

Figure 6 Section drawing (wall section).

NCCER – *Core Curriculum*   00105-15

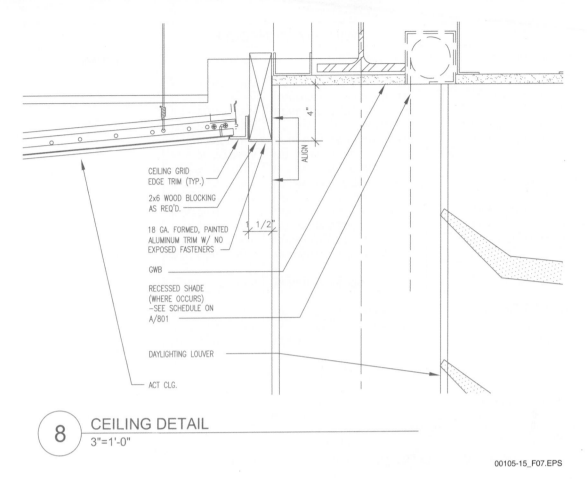

Figure 7 Detail drawing (ceiling detail).

1.1.3 Structural Plans

The **structural plans** are a set of engineered drawings used to support the architectural design. The first part of the structural plans is the general notes (*Figure 9*). These notes give details of the materials to be used and the requirements to be followed in order to build the structure that the architectural plan depicts. The notes, for instance, might specify the type and strength of concrete required for the foundation, the loads that the roof and stairs must be built to accommodate, and codes that contractors must follow. General notes may be on a separate general notes sheet or may be part of individual plan sheets.

The structural plans also include a **foundation plan** (*Figure 10*), which shows the lowest level of the building, including concrete footings, slabs, and foundation walls. They also may show steel girders, columns, or **beams**, as well as detail drawings to show where and how the foundation must be reinforced. Column and spread footing schedules, and foundation notes may be included on the foundation plan. A related element is the structural floor plan, which depicts a wood or metal joist framing and the underlayment of each floor of the structure.

Importance of Architectural Drawings

Look at architectural plans first, because all other drawings follow from them. Architectural plans are the most general; they show how all the parts of the project fit together.

WINDOW SCHEDULE

TYPE	FRAME SIZE	FRAME FINISH	GLAZING	REMARKS
A	4' 2" x 4' 2"	CLEAR ANODIZED ALUMINUM	VIRACON VE-7-2M	
B	24' 2-1/2" x 11' 0"	CLEAR ANODIZED ALUMINUM	VIRACON VE-7-2M	
B-1	12' 10" x 11' 0"	CLEAR ANODIZED ALUMINUM	VIRACON VE-7-2M	
C	15'-11-1/2" x 8' 3-1/2"	CLEAR ANODIZED ALUMINUM	VIRACON VE-7-2M	
C-1	15'-11-1/2" x 4' 1-1/2"	CLEAR ANODIZED ALUMINUM	VIRACON VE-7-2M	
D	15 11-1/2'-0" x 11' 1-1/2"	CLEAR ANODIZED ALUMINUM	VIRACON VE-7-2M	
D-1	15 11-1/2'-0" x 11' 1-1/2"	CLEAR ANODIZED ALUMINUM	VIRACON VE-7-2M	
D-2	31' 11-1/2" x 4' 4"	CLEAR ANODIZED ALUMINUM	VIRACON VE-7-2M	
E	14' 4" x 7' 8"	CLEAR ANODIZED ALUMINUM	VIRACON VE-7-2M	
F	15'-11-1/2" x 4' 4 "	CLEAR ANODIZED ALUMINUM	VIRACON VE-7-2M	
H	11' 11-1/2" x 8' 1/4"	CLEAR ANODIZED ALUMINUM	VIRACON VE-7-2M	
H-1	11' 11-1/2" x 10' 1-3/4"	CLEAR ANODIZED ALUMINUM	VIRACON VE-7-2M	
I	15'-11-1/2" x 4' 1-1/2 "	CLEAR ANODIZED ALUMINUM	VIRACON VE-7-2M	
CW - 1	12' 0" x 27' 11-1/4"	CLEAR ANODIZED ALUMINUM	VIRACON VE-7-2M	
CW - 3	15' 11 1/2" x 37' 3 1/4"	CLEAR ANODIZED ALUMINUM	VIRACON VE-7-2M	
CW - 4	36' 10" x 20' 9 1/2"	CLEAR ANODIZED ALUMINUM	VIRACON VE-7-2M	
CW - 5	28' 11-7/8" x 20' 9-1/2"	CLEAR ANODIZED ALUMINUM	VIRACON VE-7-2M	

NOTES

1. FRAME SIZES HAVE BEEN INDICATED ASSUMING 1/2" JOINT ADJACENT TO METAL PANEL SYSTEM. ALIGNMENT OF JOINTS W/ METAL PANEL MODULE IS REQUIRED. INTERMEDIATE MULLIONS TO BE CENTERED ON ADJACENT METAL PANEL JOINTS △1

2. "SG" INDICATES SAFETY GLAZING (TEMPERED GLASS REQUIRED)

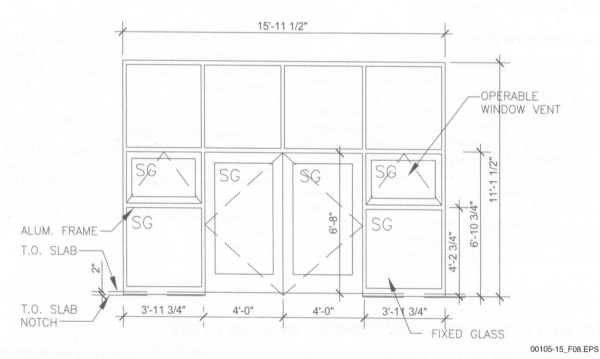

Figure 8 Window schedule and window detail.

00105-15_F08.EPS

GENERAL NOTES

I. DESIGN CRITERIA

A. GENERAL BUILDING CODE
The Contract Documents are based on the requirements of the:
1. Standard Building Code, 1997 edition.

B. DEAD LOADS
1. Partitions. An allowance of 20 PSF has been made for partitions as a uniformly distributed dead load.
2. Hanging Ceiling and Mechanical Loads. An allowance of 10 PSF has been made for hanging ceiling and mechanical equipment loads such as duct work and sprinkler pipes.

C. LIVE LOADS
1. Design live loads are based on the more restrictive of the uniform load listed below or the concentrated load listed acting over an area 2.5 feet square.

CATEGORY	UNIFORM LOAD (PSF)	CONCENTRATED LOAD (LB)
1. Roof	20	N/A
2. Elevated Floors	50	0
3. Terraces, Lobbies	100	0
4. Stairways, Exit Facilities	100	0
5. Elevator Machine Rooms	100	Assumed Eqp. Wt.
6. Mechanical Rooms, typical	150	Assumed Eqp. Wt.

NOTES:
1. Live Load Reduction. Live loads have been reduced on any member supporting more than 150 square feet, including flat slabs, except for floors in places of public assembly and for live loads greater than 100 pounds per square foot in accordance with the following formula:

$$R = r(A-150)$$

The reduction, R, shall not exceed 40 percent for members supporting one level only, 60 percent for other members, or R as calculated in the following formula:

$$R = 23.1 \left(1 + \frac{D}{L}\right)$$

R = Reduction in percent.
r = Rate of reduction equal to .08 percent for floors.
A = Area of floor supported by the member.
D = Total dead load supported by the member.
L = Total, unreduced, live load supported by the member.

2. For storage loads exceeding 100 pounds per square foot, no reduction has been made, except that design live loads on columns have been reduced 20 percent.

D. ELEVATOR LOADS
Machine Beam, Car Buffer, Counterweight Buffer, and Guide Rail Loads. Assumed elevator loads to the supporting structure are shown on the drawings, including machine beam reactions, car buffer reactions, counterweight buffer reactions, and horizontal and vertical guide rail loads. The General Contractor shall submit to the Structural Engineer final elevator shop drawings showing all loads to the structure prior to the installation of the elevators for verification of load carrying capacity.

E. MECHANICAL EQUIPMENT LOADS
The General Contractor shall submit actual weights of equipment to be used in the project to the Structural Engineer for verification of loads used in the design at least three weeks prior to fabrication and construction of the supporting structure.

F. WIND LOADS
1. Wind pressures are based on the American Society of Civil Engineers, Minimum Design Loads for Buildings and Other Structures, ASCE 7-98 with a Wind Speed = 110 MPH (3 sec. gust), Exposure C, Importance Factor 1.15.
2. Wind pressures used in the design of the cladding are shown on these Drawings.

II. FOUNDATION

A. GEOTECHNICAL REPORT
Foundation design is based on the geotechnical investigation report as follows:
1. Reports of Geotechnical exploration, M.E. Rinker Sr. Hall (Revised Location) Near the southeast Corner of Newell Drive and Inner Road, Gainesville, Alachua County, Florida. Law Engineering and Environmental Services, Inc. January 2, 2001.

The geotechnical report is available to the General Contractor upon request to the Owner. The information included therein may be used by the General Contractor for his general information only. The Architect and Engineer will not be responsible for the accuracy or applicability of such data therein.

B. FOUNDATION TYPE
1. Spread Footing.
   a. Design Pressures:
      1. All footings have been designed assuming an allowable bearing pressure of 4000 PSF.
      Allowable pressures are increased 33% for combined gravity and wind loads.

C. SLAB-ON-GRADE
Radon resistant construction guidelines are being followed on this project. The details and specifications for slab-on-grade construction must be adhered to without deviation.
Slab-on-Grade shall be immediately underlain by a 8 mil. vapor barrier. Seams shall be lapped 12 inches and sealed with 2" wide pressure sensitive vinyl tape. All penetrations shall be sealed with tape.

D. CONSTRUCTION DEWATERING
The Contractor shall determine the extent of construction dewatering required for the excavation. The Contractor shall submit to the Geotechnical Engineer for review the proposed plan for construction dewatering, prior to beginning the excavation.

III. REINFORCED CONCRETE

A. CLASSES OF CONCRETE
All concrete shall conform to the requirements as specified in the table below unless noted otherwise on the drawings:

Usage	28 Day Comp. Strength (PSI)	Conc Type	Max Size Agg.	W/C Ratio
1. Elevated Floors	4000	NWT	3/4"	0.48
2. Spread Footings	3000	NWT	1"	0.55
3. Slab-On-Grade	4000	NWT	1"	0.48
4. Fnd. Walls & Plinths	4000	NWT	1"	0.48

All concrete shall be proportioned for a maximum allowable unit shrinkage of 0.03% measured at 28 days after curing in lime water as determined by ASTM C 157 (using air storage).

B. HORIZONTAL CONSTRUCTION JOINTS IN CONCRETE POURS
There shall be no horizontal construction joints in any concrete pours unless shown on the drawings. The Architect/Engineer shall approve all deviations or additional joints in writing.

C. REINFORCING STEEL SPECIFICATION
1. All Reinforcing Steel shall be ASTM A615 Grade 60 unless noted otherwise on the drawings or in these notes.
2. Welded Reinforcing Steel. Provide reinforcing steel conforming to ASTM A706 for all reinforcing steel required to be welded and where noted on the drawings.
3. Galvanized Reinforcing Steel. Provide reinforcing steel galvanized according to ASTM A767 Class II (2.0 oz. zinc PSF where noted on the drawings.
4. Deformed Bar Anchors. ASTM A496 minimum yield strength 70,000 PSI as noted on the drawings. Reinforcing bars shall not be substituted for deformed bar anchors.
5. Welded Wire Fabric. Welded smooth wire fabric, ASTM A 185, yield strength 65,000 PSI where noted on the drawings. Welded deformed wire fabric for, ASTM A 497, yield strength 70,000 PSI where noted on the drawings.

D. PLACEMENT OF WELDED WIRE FABRIC
Wherever welded wire fabric is specified as reinforcement, it shall be continuous across the entire concrete surface and not interrupted by beams or girders and properly lapped one cross wire spacing plus 2".

E. REINFORCEMENT IN TOPPING SLABS
Provide welded smooth wire fabric minimum 6 x 6 W2.9 x W2.9 in all topping slabs unless specified otherwise on the drawings.

F. REINFORCEMENT IN HOUSEKEEPING PADS
Provide welded smooth wire fabric 6 x 6 W2.9 x W2.9 minimum in all housekeeping pads supporting mechanical equipment whether shown on the drawings or not unless heavier reinforcement is called for on the drawings.

G. REINFORCING STEEL COVERAGE
Concrete Cover for reinforcement layer nearest to the surface unless specified otherwise on the drawings.
1. Concrete surfaces cast against and permanently exposed to earth. 3 inches
2. Concrete surfaces exposed to earth or weather or where noted on the drawings 2 inches
3. Concrete surfaces not exposed to weather or in contact with the ground.
   a. #3 to #11 bars 1 inch

H. SPLICES IN REINFORCING STEEL
1. All unscheduled splices shall be Class A tension splice.

IV. STRUCTURAL STEEL

V. STEEL DECKS

VI. CURTAIN WALL

VII. CONCRETE MASONRY

VIII. MISCELLANEOUS

IX. SUBMITTALS

X. DRAWING INTERPRETATION

00105-15_F09.EPS

Figure 9 General notes for structural plans.

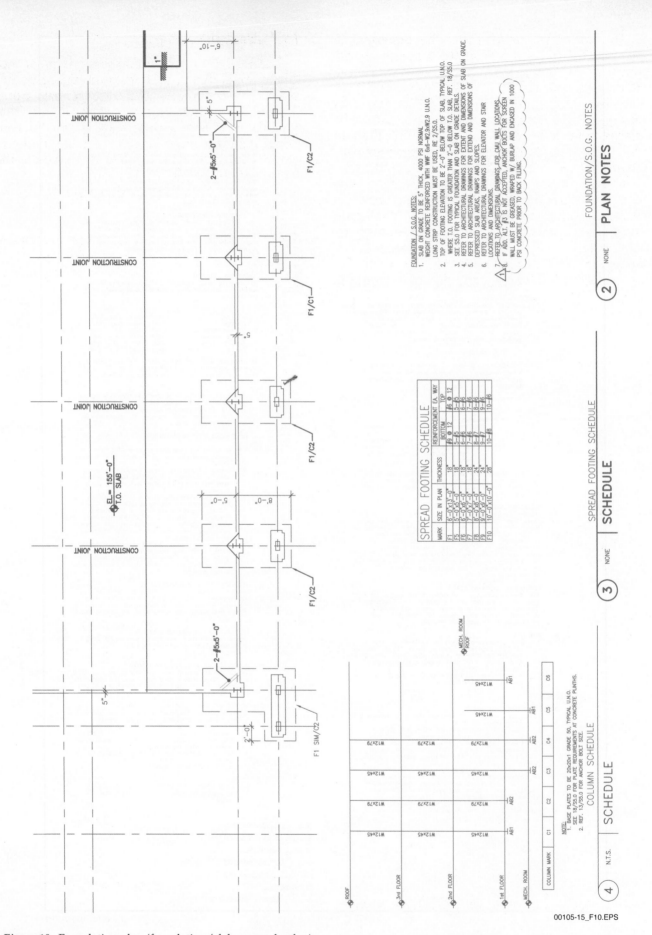

Figure 10 Foundation plan (foundation/slab-on-grade plan).

00105-15_F10.EPS

The structural plans show the materials to be used for the walls, whether concrete or masonry, and whether the framing is wood or steel. Structural plans also include a roof-framing plan, showing what kinds of ceiling joists and roof rafters are to be used and where trusses are to be placed (refer to *Drawing 2, Roof Framing Plan*, in the *Appendix*). Notes for the framing plan are usually found on the same sheet as the drawing.

The structural plans include structural section drawings (*Figure 11*), which are similar to the architectural section drawings but show only the structural requirements. Miscellaneous structural details may also be shown in these sections to provide a better understanding of such things as connections and attachments of accessories.

1.1.4 Mechanical Plans

Mechanical plans are engineered plans for motors, pumps, piping systems, and piping equipment. These plans incorporate general notes (*Figure 12*) containing specifications ranging from what the contractor is to provide to how the contractor determines the location of grilles and registers. A mechanical **legend** (*Figure 13*) defines the **symbols** used on the mechanical plans. A list of abbreviations (*Figure 14*) spells out abbreviations found on the plans.

Piping and instrumentation drawings (P&IDs) (*Figure 15*) are **schematic** diagrams of a complete piping system that show the process flow. They also show all the equipment, pipelines, valves, instruments, and controls needed to operate the system.

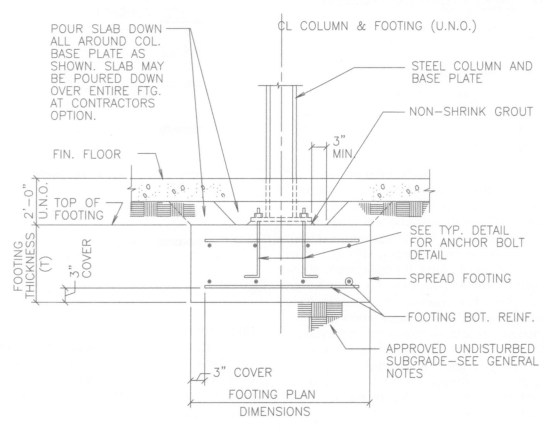

Figure 11 Structural section drawing (foundation details).

```
GENERAL NOTES
(FOR ALL MECHANICAL DRAWINGS)

1. CONTRACTOR IS TO PROVIDE COMPLETE CONNECTIONS TO ALL NEW
   AND RELOCATED OWNER FURNISHED EQUIPMENT.

2. CONTRACTOR TO COORDINATE THE LOCATION OF ALL DUCTWORK
   AND DIFFUSERS WITH REFLECTED CEILING PLAN AND STRUCTURE
   PRIOR TO BEGINNING WORK.

3. DIMENSIONS FOR INSULATED OR NON-INSULATED DUCT ARE OUTSIDE
   SHEET METAL DIMENSIONS.

4. DRAWINGS ARE NOT TO BE SCALED FOR DIMENSIONS.
   TAKE ALL DIMENSIONS FROM ARCHITECTURAL DRAWINGS, CERTIFIED
   EQUIPMENT DRAWINGS AND FROM THE STRUCTURE ITSELF BEFORE
   FABRICATING ANY WORK. VERIFY ALL SPACE REQUIREMENTS
   COORDINATING WITH OTHER TRADES, AND INSTALL THE SYSTEMS IN
   THE SPACE PROVIDED WITHOUT EXTRA CHARGES TO THE OWNER.

5. LOCATION OF ALL GRILLES, REGISTERS, DIFFUSERS AND CEILING
   DEVICES SHALL BE DETERMINED FROM THE ARCHITECTURAL
   REFLECTED CEILING PLANS.

6. THE OWNER AND DESIGN ENGINEER ARE NOT RESPONSIBLE FOR THE
   CONTRACTOR'S SAFETY PRECAUTIONS OR TO MEANS, METHODS,
   TECHNIQUES, CONSTRUCTION SEQUENCES, OR PROCEDURES
   REQUIRED TO PERFORM HIS WORK.

7. ALL WORK SHALL BE INSTALLED IN ACCORDANCE WITH
   PLRC'S SAFETY PLAN AND ALL APPLICABLE STATE
   AND LOCAL CODES.

8. ALL EXTERIOR WALL AND ROOF PENETRATIONS SHALL BE SEALED
   WEATHERPROOF. REFERENCE SPECIFICATION SECTION 15050.

9. ALL MECHANICAL WORK UNDER THIS CONTRACT IS TO FIVE (5)
   FEET OUTSIDE THE BUILDING.
```

00105-15_F12.EPS

Figure 12 Mechanical plan general notes.

P&IDs are not drawn to **scale** because they are meant only to give a representation, or a general idea, of the work to be done. Additionally, P&IDs do not indicate north, south, east, and west directions.

For more complex jobs, a separate heating, ventilating, and air conditioning (**HVAC**) plan is added to the set of plans. Piping system plans for gas, oil, or steam heat may be included in the HVAC plan. The mechanical plans include the layout of the HVAC system, showing specific requirements and elements for that system, including a floor, a reflected ceiling, or a roof. HVAC drawings (*Figure 16*) include an electrical schematic that shows the electrical circuitry for the HVAC system. HVAC plans are both mechanical and electrical drawings in one plan.

Importance of Architectural Symbols

When looking at a section drawing, pay close attention to the way different parts are drawn. Each part of the drawing represents a method of construction or a type of material. These symbols are covered in more detail later in this module.

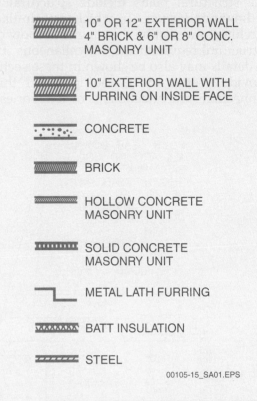

10" OR 12" EXTERIOR WALL 4" BRICK & 6" OR 8" CONC. MASONRY UNIT

10" EXTERIOR WALL WITH FURRING ON INSIDE FACE

CONCRETE

BRICK

HOLLOW CONCRETE MASONRY UNIT

SOLID CONCRETE MASONRY UNIT

METAL LATH FURRING

BATT INSULATION

STEEL

00105-15_SA01.EPS

Be aware that a page with a series of mechanical detail drawings may be included in the mechanical plans. These drawings show specific details of certain components within the mechanical system. *Figure 17* is an example of a mechanical detail drawing.

NCCER – *Core Curriculum*   00105-15

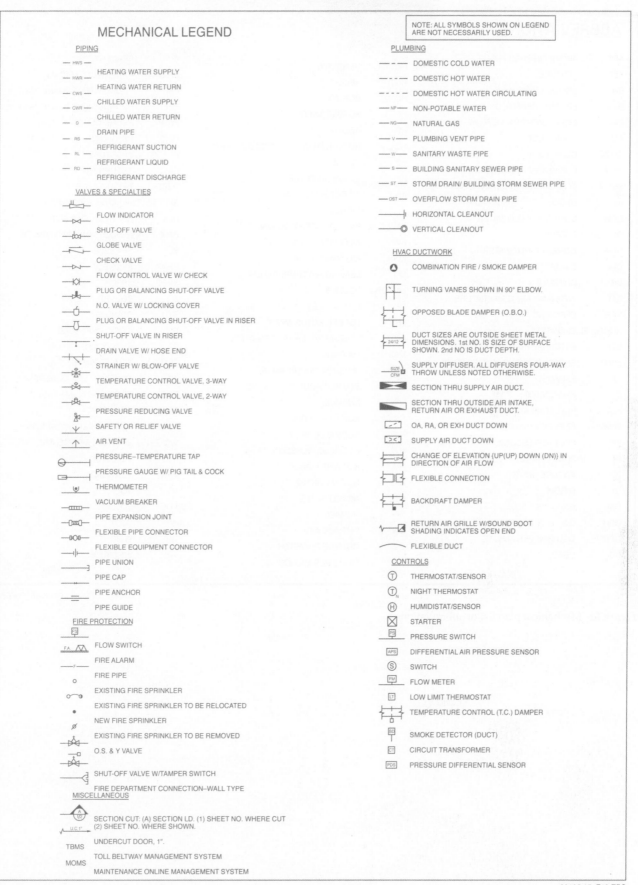

MECHANICAL LEGEND

NOTE: ALL SYMBOLS SHOWN ON LEGEND ARE NOT NECESSARILY USED.

PIPING

— HWS — HEATING WATER SUPPLY
— HWR — HEATING WATER RETURN
— CWS — CHILLED WATER SUPPLY
— CWR — CHILLED WATER RETURN
— D — DRAIN PIPE
— RS — REFRIGERANT SUCTION
— RL — REFRIGERANT LIQUID
— RD — REFRIGERANT DISCHARGE

VALVES & SPECIALTIES

FLOW INDICATOR
SHUT-OFF VALVE
GLOBE VALVE
CHECK VALVE
FLOW CONTROL VALVE W/ CHECK
PLUG OR BALANCING SHUT-OFF VALVE
N.O. VALVE W/ LOCKING COVER
PLUG OR BALANCING SHUT-OFF VALVE IN RISER
SHUT-OFF VALVE IN RISER
DRAIN VALVE W/ HOSE END
STRAINER W/ BLOW-OFF VALVE
TEMPERATURE CONTROL VALVE, 3-WAY
TEMPERATURE CONTROL VALVE, 2-WAY
PRESSURE REDUCING VALVE
SAFETY OR RELIEF VALVE
AIR VENT
PRESSURE–TEMPERATURE TAP
PRESSURE GAUGE W/ PIG TAIL & COCK
THERMOMETER
VACUUM BREAKER
PIPE EXPANSION JOINT
FLEXIBLE PIPE CONNECTOR
FLEXIBLE EQUIPMENT CONNECTOR
PIPE UNION
PIPE CAP
PIPE ANCHOR
PIPE GUIDE

FIRE PROTECTION

FLOW SWITCH
FIRE ALARM
FIRE PIPE
EXISTING FIRE SPRINKLER
EXISTING FIRE SPRINKLER TO BE RELOCATED
NEW FIRE SPRINKLER
EXISTING FIRE SPRINKLER TO BE REMOVED
O.S. & Y VALVE
SHUT-OFF VALVE W/TAMPER SWITCH
FIRE DEPARTMENT CONNECTION–WALL TYPE

MISCELLANEOUS

SECTION CUT: (A) SECTION I.D. (1) SHEET NO. WHERE CUT (2) SHEET NO. WHERE SHOWN.
UNDERCUT DOOR, 1".
TBMS — TOLL BELTWAY MANAGEMENT SYSTEM
MOMS — MAINTENANCE ONLINE MANAGEMENT SYSTEM

PLUMBING

— · · — DOMESTIC COLD WATER
— · · — DOMESTIC HOT WATER
— · · · — DOMESTIC HOT WATER CIRCULATING
— NP — NON-POTABLE WATER
— NG — NATURAL GAS
— V — PLUMBING VENT PIPE
— W — SANITARY WASTE PIPE
— S — BUILDING SANITARY SEWER PIPE
— ST — STORM DRAIN/ BUILDING STORM SEWER PIPE
— OST — OVERFLOW STORM DRAIN PIPE
HORIZONTAL CLEANOUT
VERTICAL CLEANOUT

HVAC DUCTWORK

COMBINATION FIRE / SMOKE DAMPER
TURNING VANES SHOWN IN 90° ELBOW.
OPPOSED BLADE DAMPER (O.B.O.)
DUCT SIZES ARE OUTSIDE SHEET METAL DIMENSIONS. 1st NO. IS SIZE OF SURFACE SHOWN. 2nd NO IS DUCT DEPTH.
SUPPLY DIFFUSER. ALL DIFFUSERS FOUR-WAY THROW UNLESS NOTED OTHERWISE.
SECTION THRU SUPPLY AIR DUCT.
SECTION THRU OUTSIDE AIR INTAKE, RETURN AIR OR EXHAUST DUCT.
OA, RA, OR EXH DUCT DOWN
SUPPLY AIR DUCT DOWN
CHANGE OF ELEVATION {UP(UP) DOWN (DN)) IN DIRECTION OF AIR FLOW
FLEXIBLE CONNECTION
BACKDRAFT DAMPER
RETURN AIR GRILLE W/SOUND BOOT SHADING INDICATES OPEN END
FLEXIBLE DUCT

CONTROLS

THERMOSTAT/SENSOR
NIGHT THERMOSTAT
HUMIDISTAT/SENSOR
STARTER
PRESSURE SWITCH
DIFFERENTIAL AIR PRESSURE SENSOR
SWITCH
FLOW METER
LOW LIMIT THERMOSTAT
TEMPERATURE CONTROL (T.C.) DAMPER
SMOKE DETECTOR (DUCT)
CIRCUIT TRANSFORMER
PRESSURE DIFFERENTIAL SENSOR

00105-15_F13.EPS

Figure 13 Mechanical plan legend.

ABBREVIATIONS

AFF	ABOVE FINISHED FLOOR	HG	MERCURY	%	PERCENT
ALT	ALTITUDE	HGT	HEIGHT	PH OR f	PHASE (ELECTRICAL)
BHP	BRAKE HORSEPOWER	HORZ	HORIZONTAL	PSF	POUNDS PER SQUARE FOOT
BTU	BRITISH THERMAL UNIT	HP	HORSEPOWER	PSI	POUNDS PER SQUARE INCH
Cv	COEFFICIENT, VALVE FLOW	HR	HOUR(S)	PSIA	PSI ABSOLUTE
CU FT	CUBIC FEET	HWC	HOT WATER CIRCULATING (DOMESTIC)	PSIG	PSI GAUGE
CU IN	CUBIC INCH	HZ	HERTZ	PRESS	PRESSURE
CFM	CUBIC FEET PER MINUTE	ID	INSIDE DIAMETER	RA	RETURN AIR
SCFM	CFM, STANDARD CONDITIONS	IE	INVERT ELEVATION	RECIRC	RECIRCULATE
dB	DECIBEL	IN	INCHES	RH	RELATIVE HUMIDITY
DCW	DOMESTIC COLD WATER	IN W.C.	INCHES WATER COLUMN	RLA	RUNNING LOAD AMPS
DEG OR °	DEGREE	KW	KILOWATT	RPM	REVOLUTIONS PER MINUTE
DHW	DOMESTIC HOT WATER	KWH	KILOWATT HOUR	SL	SEA LEVEL
DIA	DIAMETER	LAT	LEAVING AIR TEMPERATURE	SENS	SENSIBLE
DB	DRY-BULB	LBS OR #	POUNDS	SPEC	SPECIFICATION
EAT	ENTERING AIR TEMPERATURE	LF	LINEAR FEET	SQ	SQUARE
EFF	EFFICIENCY	LRA	LOCKED ROTOR AMPS	STD	STANDARD
ELEV or EL	ELEVATION	LWT	LEAVING WATER TEMPERATURE	SP	STATIC PRESSURE
ESP	EXTERNAL STATIC PRESSURE	MAX	MAXIMUM	SA	SUPPLY AIR
EWT	ENTERING WATER TEMPERATURE	MCA	MINIMUM CIRCUIT AMPS	TEMP	TEMPERATURE
EXH	EXHAUST	MBH	BTU PER HOUR (THOUSAND)	TD	TEMPERATURE DIFFERENCE
F	FAHRENHEIT	MIN	MINIMUM	TSP	TOTAL STATIC PRESSURE
FLA	FULL LOAD AMPS	NC	NOISE CRITERIA	TSTAT	THERMOSTAT
FPM	FEET PER MINUTE	N.O.	NORMALLY OPEN	TONS	TONS OF REFRIGERATION
FPS	FEET PER SECOND	N.C.	NORMALLY CLOSED	VAV	VARIABLE AIR VOLUME
FT	FOOT OR FEET	N/A	NOT APPLICABLE	VEL	VELOCITY
FU	FIXTURE UNITS	NIC	NOT IN CONTRACT	VERT	VERTICAL
GA	GAUGE	NTS	NOT TO SCALE	V	VOLT
GAL	GALLONS	NO	NUMBER	VOL	VOLUME
GPH	GALLONS PER HOUR	OA	OUTSIDE AIR	W	WATT
GPM	GALLONS PER MINUTE	OD	OUTSIDE DIAMETER	WT	WEIGHT
HD	HEAD	PPM	PARTS PER MILLION	WB	WET-BULB

00105-15_F14.EPS

Figure 14 Mechanical plan list of abbreviations.

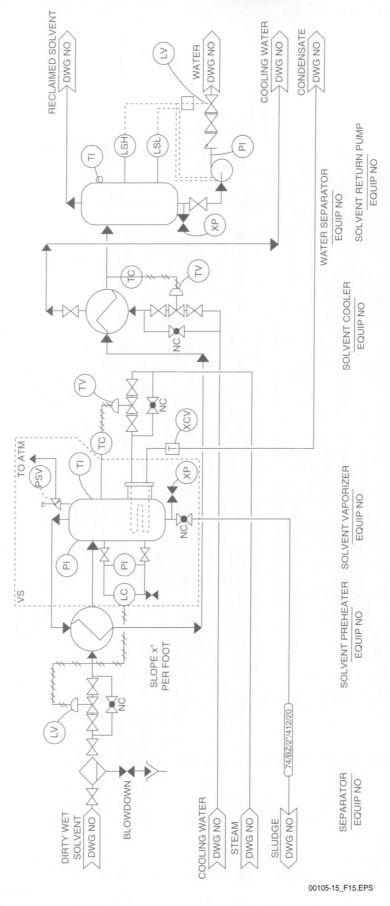

Figure 15 P&ID drawing.

00105-15_F15.EPS

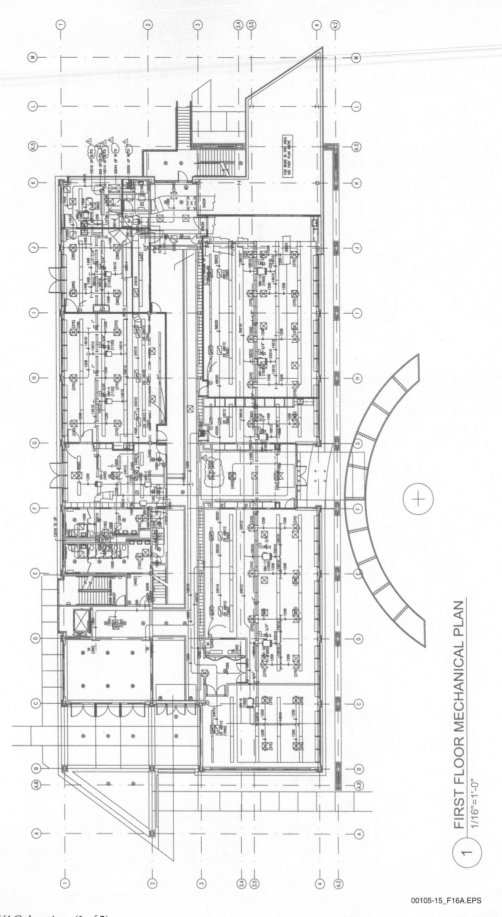

FIRST FLOOR MECHANICAL PLAN
1/16" = 1'-0"

1

00105-15_F16A.EPS

Figure 16A HVAC drawing. (1 of 2)

NCCER – *Core Curriculum* 00105-15

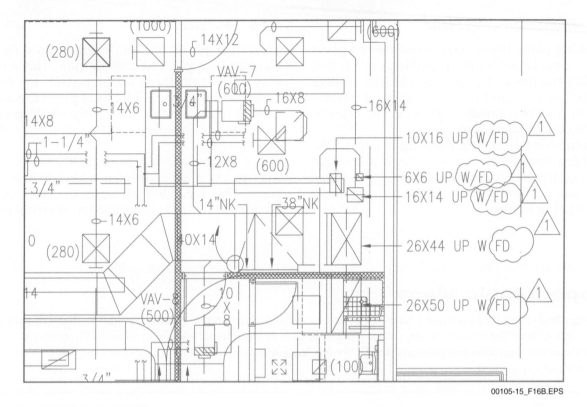

Figure 16B HVAC drawing. (2 of2)

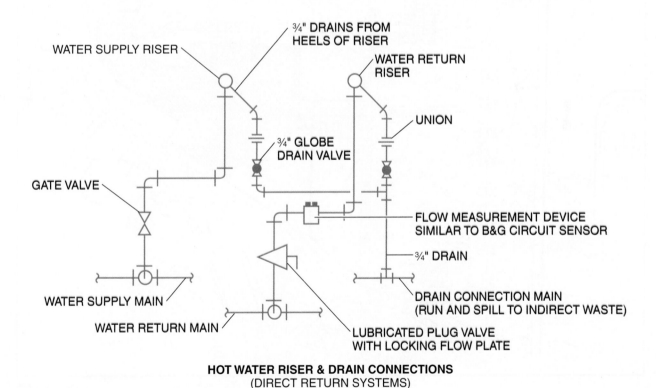

HOT WATER RISER & DRAIN CONNECTIONS
(DIRECT RETURN SYSTEMS)

00105-15_F17.EPS

Figure 17 Mechanical detail drawing for hot water riser and drain connections.

1.1.5 Plumbing/Piping Plans

Plumbing plans (*Figure 18*) are engineered plans showing the layout for the plumbing system that supplies the hot and cold water, for the sewage disposal system, and for the location of plumbing fixtures. For commercial projects, each system may be on a separate plan.

A **plumbing isometric drawing** is part of the plumbing plan. It is a type of three-dimensional drawing that depicts the plumbing system. *Figure 19* shows a plumbing isometric drawing for a sanitary riser system.

1.1.6 Electrical Plans

Electrical plans are engineered drawings for electrical supply and distribution. These plans may appear on the floor plan itself for simple construction projects. Electrical plans include locations of the electric meter, distribution panel, switchgear, convenience outlets, and special outlets.

For more complex projects, the information may be on a separate plan added to the set of plans. This separate plan leaves out construction-related details and shows just the electrical layout.

Topographic Maps

Topographic (topo) maps provide a representation of vertical dimension that gives a feel for the shape or contour of a piece of land. Topo maps identify physical features such as mountains, lakes, and streams. These maps indicate where highways and railroads run. They often include information on drainage and land use such as orchards and woodland.

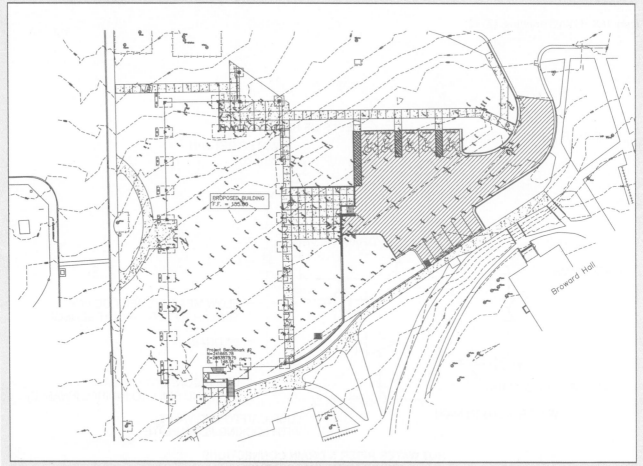

00105-15_SA02.EPS

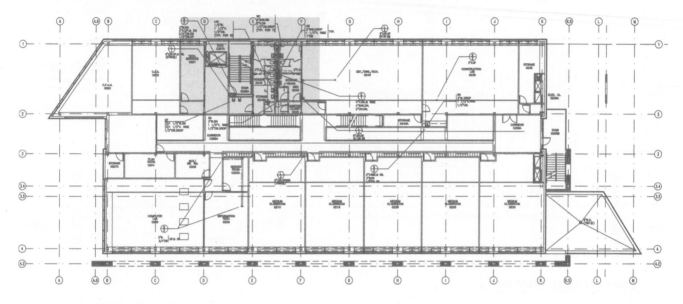

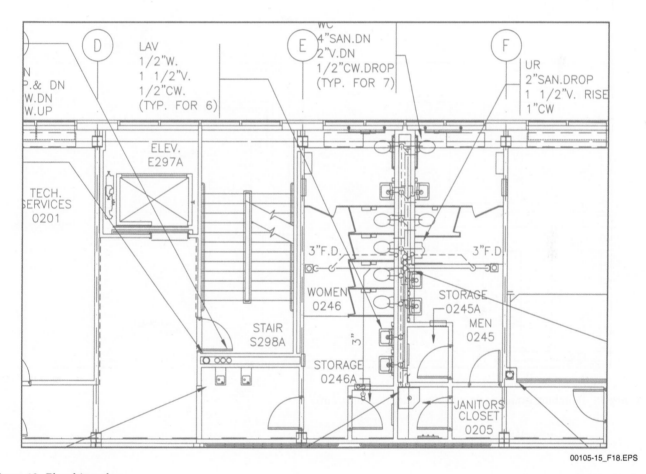

1 FIRST FLOOR PLUMBING PLAN
1/16"=1'-0"

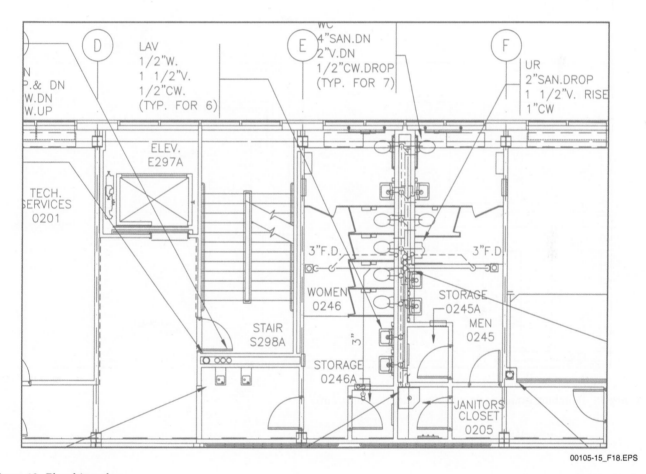

Figure 18 Plumbing plan.

00105-15_F18.EPS

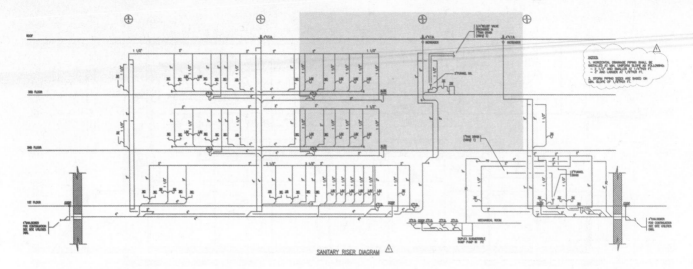

① SANITARY RISER DIAGRAM
N.T.S.

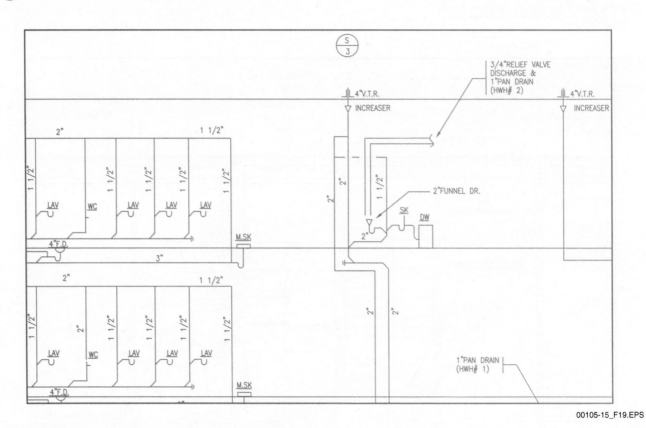

00105-15_F19.EPS

Figure 19 Plumbing isometric drawing (sanitary riser diagram).

More complex electrical plans include locations of switchgear, transformers, main breakers, and motor control centers.

The electrical plans usually start with a set of general notes (*Figure 20*). These notes cover items ranging from main transformers to the coordination of underground penetrations into the building.

The electrical plans can include lighting plans, which show the location of lights and receptacles (refer to *Drawing 3*, First Floor Lighting Plan, in the *Appendix*), power plans (*Figure 21*), and panel schedules (*Figure 22*). Electrical plans have an electrical legend, which defines the symbols (*Figure 23*) used on the plan and a key to the abbreviations (*Figure 24*) used on the plan. Depending on the size and scope of the drawings, the legend is often on a separate drawing sheet of its own, rather than on each individual drawing.

Isometric Drawings

An isometric drawing is a type of three-dimensional drawing also known as a pictorial illustration. Typically in isometric construction drawings, objects are shown at a 30-degree angle in isometric drawings to provide a three-dimensional perspective rather than a flat, two-dimensional view.

BILL OF MATERIAL

P.M.	REQ'D	SIZE	DESCRIPTION
1		1-1/2"	PIPE SCH/40 ASTM-A-120 GR.B
2		3/4"	PIPE SCH/40 ASTM-A-120 GR.B
3	5	1-1/2"	90° ELL ASTM-A-197 BW
4	1	1-1/2"	TEE ASTM-197 STD
5	2	3/4"	45° ELL ASTM-197 STD
6	1	1-1/2" × 3/4"	BELL RED. CONC.
7	2	1-1/2"	GATE VA. BW ASTM-B62
8	1	1-1/2"	CHECK VA. SWING BW 150#

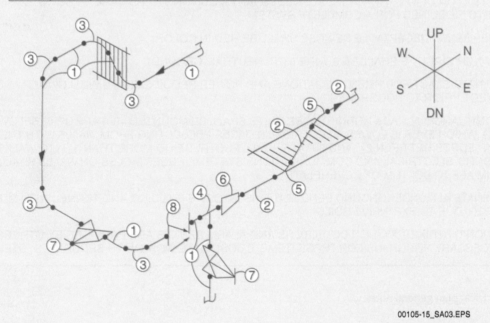

00105-15_SA03.EPS

GENERAL NOTES (FOR ALL ELECTRICAL SHEETS)

1. COORDINATE LOCATION OF LUMINARIES WITH ARCHITECTURAL REFLECTED CEILING PLANS.

2. COORDINATE LOCATION OF ALL OUTLETS WITH ARCHITECTURAL ELEVATIONS, CASEWORK SHOP DRAWINGS AND EQUIPMENT INSTALLATION DRAWINGS.

3. COORDINATE LOCATION OF MECHANICAL EQUIPMENT WITH MECHANICAL PLANS AND MECHANICAL CONTRACTOR PRIOR TO ROUGH-IN.

4. PROVIDE (1) 3/4"C WITH PULL WIRE FROM EACH TELEPHONE, DATA OR COMMUNICATION OUTLET SHOWN, TO ABOVE ACCESSIBLE CEILING, AND CAP.

5. 3-LAMP FIXTURES SHOWN HALF SHADED HAVE INBOARD SINGLE LAMP CONNECTED TO EMERGENCY BATTERY PACK FOR FULL LUMEN OUTPUT. SEE SPECIFICATIONS.

6. SITE PLAN DOES NOT INDICATE ALL OF THE UG UTILITY LINES, RE: CIVIL DRAWINGS FOR ADDITIONAL INFORMATION. CONTRACTOR TO FIELD VERIFY EXACT LOCATION OF ALL EXISTING UNDERGROUND UTILITY LINES OF ALL TRADES PRIOR TO ANY SITE WORK.

7. THE LOCATIONS OF ALL SMOKE DETECTORS SHOWN ARE CONSIDERED TO BE SCHEMATIC ONLY. THE ACTUAL LOCATIONS (SPACING TO ADJACENT DETECTORS, WALLS, ETC.) ARE REQUIRED TO MEET NFPA 72.

8. ANY ITEMS DAMAGED BY THE CONTRACTOR SHALL BE REPLACED BY THE CONTRACTOR.

9. "CLEAN POWER" AND COMMUNICATION/COMPUTER SYSTEM REQUIREMENTS SHALL BE COORDINATED WITH COMMUNICATION/COMPUTER SYSTEMS CONTRACTOR.

10. REFER TO ARCHITECTURAL PLANS, ELEVATIONS AND DIAGRAMS FOR LOCATIONS OF FLOOR DEVICES AND WALL DEVICES. LOCATION WILL INDICATE VERTICAL AND/OR HORIZONTAL MOUNTING. IF DEVICES ARE NOT NOTED OTHERWISE THEY SHALL BE MOUNTED LONG AXIS HORIZONTAL AT +16" TO CENTER.

11. ALL PLUGMOLD SHOWN SHALL BE WIREMOLD SERIES V2000 (IVORY FINISH) WITH SNAPICOIL #V20GB06 (OUTLETS 6" ON CENTER). PROVIDE ALL NECESSARY MOUNTING HARDWARE, ELBOWS, CORNERS, ENDS, ETC. REQUIRED FOR A COMPLETE SYSTEM.

12. ALL EMERGENCY RECEPTACLE DEVICES SHALL BE RED IN COLOR.

13. ALL BRANCH CIRCUITS SHALL BE 3-WIRE (HOT, NEUTRAL, GROUND).

14. COORDINATE EXACT EQUIPMENT LOCATIONS AND POWER REQUIREMENTS WITH OWNER AND ARCHITECT PRIOR TO ROUGH-INS.

15. ADA COMPLIANCE: ALL ADA HORN/STROBE UNITS SHALL BE MOUNTED +90" AFF OR 6" BELOW FINISHED CEILING, WHICH EVER IS LOWER. ELECTRICAL DEVICES PROJECTING FROM WALLS WITH THEIR LEADING EDGES BETWEEN 27" AND 80" AFF SHALL PROTRUDE NO MORE THAN 4" INTO WALKS OR CORRIDORS. ELECTRICAL AND COMMUNICATIONS SYSTEMS RECEPTACLES ON WALLS SHALL BE 15" MINIMUM AFF TO BOTTOM OF COVERPLATE.

16. COORDINATE ALL UNDERGROUND PENETRATIONS INTO THE BUILDING AND TUNNEL WITH STRUCTURAL ENGINEER, DUE TO EXPANSIVE SOILS.

17. ELECTRONIC STRIKES, MOTION DETECTORS AND ALARM SHUNTS ARE PROVIDED BY OTHERS. PROVIDE ALL NECESSARY ROUGH-INS FOR THESE ITEMS. COORDINATE WORK WITH SECURITY SYSTEM PROVIDER.

00105-15_F20.EPS

Figure 20 Electrical plan general notes.

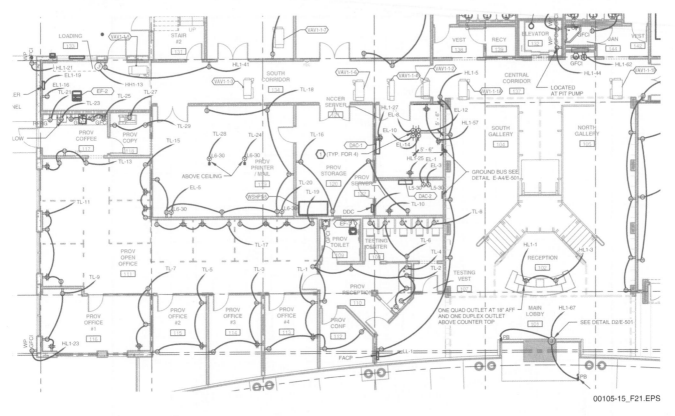

Figure 21 Portion of a power plan.

Visualize Before Building

It is important for a builder to be able to visualize a finished project before starting it. Detail drawings help builders visualize different parts of the structure long before they measure a board or hammer a nail. By visualizing, builders can plan ahead and anticipate potential problems. This saves time and money on the job.

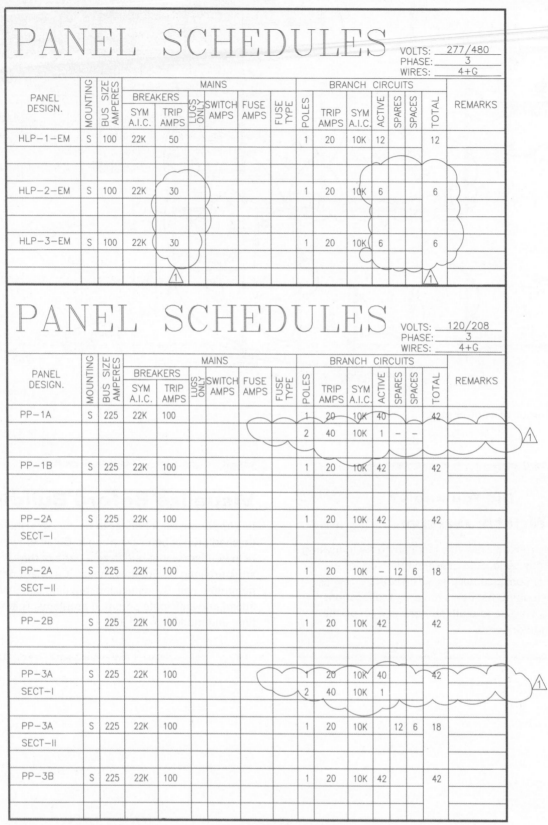

Figure 22 Panel schedules.

00105-15_F22.EPS

ELECTRICAL SYMBOLS LIST

ⓙ / ⑴ / ◔	JUNCTION BOX. CEILING/FLOOR/WALL MOUNTED
▨	PULL BOX SIZED AS REQUIRED
◁▢	BELL
‖▢	CHIME
▷◁ ▷◁◁	HORN (SINGLE/BI– DIRECTIONAL)
∨▢	BUZZER
▣	DOOR PUSHBUTTON
T S	SIGNAL TRANSFORMER
DH	MAGNETIC DOOR HOLDER
DH S	MAGNETIC TYPE DOOR CLOSER, 'S' DENOTES SMOKE DETECTOR TYPE
WF	SPRINKLER WATER FLOW SWITCH
TS	SPRINKLER SUPERVISED VALVE – TAMPER SWITCH
◷ / S-◷ R	INDICATING CLOCK – SINGLE FACE 'S' – DENOTES SKELETON TYPE CLG/WALL MTD 'R' – DENOTES RECESSED TYPE 'D' – DENOTES DOUBLE FACED
Ⓢ / S-◔	SPEAKER; CLG/WALL MTD 'S' – DENOTES SURFACE MOUNTED 'D' – DENOTES DOUBLE FACED
V	VOLUME CONTROL
M / M-	MICROPHONE OUTLET; CLG/WALL MTD
PA	PUBLIC ADDRESS EQUIPMENT RACK
PPP	ELECT. POWER PATCH PANEL WITH 4 RECEPTS.AND R/S SWITCH
A/V	MUTIMEDIA PATCH PANEL
W S	WINDOW ALARM SWITCH
B S	DOOR PUSHBUTTON
▷CCS	CLOSED CIRCUIT SURVEILLANCE CAMERA 'P' DENOTES PAN – P/T TILT
CCSM	COLSED CIRCUIT SURVEILLANCE MONITOR – NUMBERS DENOTE QUANTITY
TV TV-	TELEVISION ANTENNA OUTLET
S 2 TV	TELEVISION SPLITTER NUMBER DENOTES QUANTITY OF SPLITS
▷ TV-	TELEVISION CAMERA OUTLET
TVHE	TELEVISION HEADED EQUIPMENT
▶	TELEPHONE OUTLET WITH BUSHED OPENING LETTERS DENOTE: 'H' – HOUSE PHONE 'J' – JACK TYPE, 'P' – PAY (PUBLIC) PHONE, 'W' – WALL TYPE PHONE
TC	TELEPHONE TERMINAL CABINET
●S	TELEPHONE FLOOR BOX WITH NIPPLE (RISER) EXTENSION. 'S' DENOTES SERVICE FITTING TYPE
▷	DATA OUTLET WITH BUSHED OPENING WALL MOUNTED
▶	COMBINATION TELEPHONE/DATA OUTLET WITH BUSHED OPENING WALL MOUNTED
S	SWITCH
◁ S	BELL
‖ S	CHIME
S S	DOOR STRIKE
SSP	START/STOP WITH PILOT LIGHT
⌁⌁⌁	POWER TRANSFORMER
⊣⊢	GROUND (EARTH)
⊣•⊢	LIGHTNING ARRESTER
⌒	CIRCUIT BREAKER
— G —	GROUND BUS
— N —	NEUTRAL BUS
CR R	CARD READER 'R' – DENOTES RECESSED TYPE
DO	POWER DOOR OPERATOR
▢	INTERCOM
Ⓜ	CEILING MOUNTED OCCUPANCY SENSOR
⊢Ⓜ	SURFACE MOUNTED OCCUPANCY SENSOR
Ⓟ	PHOTOSENSOR

00105-15_F23.EPS

Figure 23 Electrical symbols list.

1.1.7 Fire Protection Plans

Another important drawing that may be included in a set of drawings is the **fire protection plans** (refer to *Drawing 4*, First Floor Fire Protection Plan, in the *Appendix*). This drawing shows the piping, valves, heads, and switches that make up a building's fire sprinkler system. A fire sprinkler symbols list is usually included on a separate sheet along with the fire sprinkler specifications, details and assembly drawings, and riser diagrams.

1.2.0 Basic Components of Construction Drawings

Most construction drawings are laid out in a fairly standardized format. This section describes the following five parts of a construction drawing:

* Title block
* Border
* Drawing area
* Revision block
* Legend

1.2.1 Title Block

The first thing to look at on any drawing is the title block. The title block is normally in the lower right-hand corner of the drawing or across the right edge of the paper (*Figure 25*). The title block has two purposes. First, it gives information about the structure or assembly. Second, it is numbered so the print can be filed easily.

Different companies put different information in the title block. Generally, it contains the following:

* *Company logo* – Usually preprinted on the drawing.
* *Sheet title* – Identifies the project.
* *Date* – Date the drawing was checked and readied for seal, or issued for construction.
* *Drawn by* – Initials of the person who drafted the drawing.
* *Drawing number* – Code numbers assigned to a project
* *Scale* – The ratio of the size of the object as drawn to the object's actual size.

ELECTRICAL ABBREVIATIONS

A	AMPERE(S)	MATV	MASTER ANTENNA TELEVISION SYSTEM
AC	ALTERNATING CURRENT	MC	METAL CLAD CABLE
ACB	AIR CIRCUIT BREAKER	MCC	MOTOR CONTROL CENTER
AFF	ABOVE FINISHED FLOOR	MCM	THOUSAND CIRCULAR MIL(S)
AFG	ABOVE FINISHED GRADE	MCP	MOTOR CONTROL PANEL
AL	ALUMINUM	M.C.	MECHANICAL CONTRACTOR
ALT	ALTERNATE	MH	MANHOLE
ASYM	ASYMMETRICAL	MIC	MICROPHONE
ATS	AUTOMATIC TRANSFER SWITCH	MIN	MINIMUM
AWG	AMERICAN WIRE GAUGE	MS	MAGNETIC STARTER
BC	BOTTOM CONDUIT	MTD	MOUNTED
BD	BUS DUCT	MTG	MOUNTING
BFG	BELOW FINISHED GRADE	MTR	MOTOR
BIL	BASIC IMPULSE LEVEL	MTS	MANUAL TRANSFER SWITCH
BLDG	BUILDING	N	NEUTRAL
BX	ARMORED CABLE	NA	NON-AUTOMATIC
C	CONDUIT	NC	NORMALLY CLOSED
CATV	CABLE ANTENNA TELEVISION SYSTEM	NF	NON-FUSE
CCAB	CONTROL CABINET	N.I.C.	NOT IN CONTRACT
CCTV	CLOSED CIRCUIT TELEVISION	NL	NIGHT LIGHT
CH	CABINET HEATER	NO	NORMALLY OPEN
CKT	CIRCUIT	NP	NETWORK PROTECTOR
CKT BKR/CB	CIRCUIT BREAKER	NTS	NOT TO SCALE
CL	CLOSET	OC	ON CENTER
CLG	CEILING	OL	OVERLOAD ELEMENT
COND	CONDUCTOR	P	POLE
CO	CONDUIT ONLY	PA	PUBLIC ADDRESS
CT	CURRENT TRANSFORMER	PB	PULL BOX
CU	COPPER	P.C.	PLUMBING CONTRACTOR
DB	DUCT BANK	PF	POWER FACTOR
DMB	DIMMER BOARD	Ø	PHASE
DC	DIRECT CURRENT	PL	PILOT (INDICATOR) LIGHT
DH	DUCT HEATER	PNL	PANEL (PANELBOARD)
DIM	DIMMER CONTROL	PP	POWER PANEL
DISC	DISCONNECT	PRI	PRIMARY
DM	DAMPER MOTOR	PT	POTENTIAL TRANSFORMER
DN	DOWN	PVC	POLYVINYL CHLORIDE
DP	DISTRIBUTION POWER PANEL(BOARD)	PWR	POWER
DT	DOUBLE THROW	R	RECESSED
DWG	DRAWING	RC	REMOTE CONTROL
EA	EACH	REC	RECEPTACLE
E.C.	ELECTRICAL CONTRACTOR	S	SURFACE
E.HTR	ELECTRIC HEATER	SC	SEPARATE CIRCUIT
ELEV	ELEVATOR	SDB	SUB-DISTRIBUTION BOARD
EL	ELECTRIC	SEC	SECONDARY
EM	EMERGENCY	SMR	SURFACE METAL RACEWAY
EMT	ELECTRICAL METALLIC TUBING	SP	SINGLE POLE
ENT	ELECTRICAL NON-METALLIC TUBING	SPK	SPEAKER
EWC	ELECTRIC WATER COOLER	ST	SINGLE THROW
EX	EXISTING	SW	SWITCH
F	FUSE	SWBD	SWITCHBOARD
FA	FIRE ALARM	SYM	SYMMETRICAL
FACP	FIRE ALARM CONTROL PANEL	T	THERMOSTAT
FBO	FURNISHED BY OTHERS	TEL	TELEPHONE
FCC	FLAT CONDUCTOR CABLE	TB	TERMINAL BOX
FCU	FAN COIL UNIT	TC	TOP CONDUIT
FDR	FEEDER	TCAB	TELEPHONE CABINET
FL	FLOOR	T.C.C.	TEMPERATURE CONTROL CONTRACTOR
FLUOR	FLUORESCENT	TP	TAMPER PROOF
F.P.C.	FIRE PROTECTION CONTRACTOR	TV	TELEVISION
FS	FUSIBLE SWITCH	TYP	TYPICAL
F.S.C.	FOOD SERVICE CONTRACTOR	UG	UNDERGROUND
FT	FEET OR FOOT	UH	UNIT HEATER
G.C.	GENERAL CONTRACTOR	UNG	UNGROUNDED
GEN	GENERATOR	UON	UNLESS OTHERWISE NOTED
GF	GROUND FAULT	UPS	UNINTERRUPTED POWER SYSTEM
GG	GROUND GRID	V	VOLT(S)
GRD	GROUND	VA	VOLTAMP(S)
HC	HUNG CEILING	VAR	VOLT AMPERES REACTIVE
H.I.D.	HIGH INTENSITY DISCHARGE	VP	VAPORPROOF
HP	HORSEPOWER	W	WATT(S)
H.P.S.	HIGH PRESSURE SODIUM	WP	WEATHERPROOF
HPU	HEAT PUMP UNIT	WT	WATERTIGHT
HT	HEIGHT	XFR	TRANSFORMER
HV	HIGH VOLTAGE	XP	EXPLOSION PROOF
HW	HEAVY WALL RIGID CONDUIT		
HZ	FREQUENCY IN CYCLES PER SECOND		
IC	INTERRUPTING CAPACITY		
IG	ISOLATED GROUND		
IMC	INTERMEDIATE METALLIC CONDUIT		
INC	INCANDESCENT		
JB	JUNCTION BOX		
K	KEY OPERATED		
kVA	KILOVOLT AMPERE(S)		
kVAR	KILOVAR(S)		
kW	KILOWATT(S)		
kWhr	KILOWATT(S) HOUR(S)		
LP	LIGHTING PANEL		
L.P.S.	LOW PRESSURE SODIUM		
LTG	LIGHTING		
LV	LOW VOLTAGE		

Figure 24 Electrical abbreviations.

M.E. RINKER, Sr. HALL
SCHOOL OF BUILDING CONSTRUCTION

UNIVERSITY OF FLORIDA-GAINESVILLE
PROJECT NO. BR-191

ARCHITECT
CROXTON COLLABORATIVE/
GOULD EVANS ASSOCIATES

CROXTON COLLABORATIVE ARCHITECTS

475 FIFTH AVENUE
NEW YORK, NY 10017
TEL 212.683.1998
FAX 212.683.2799

GOULD EVANS ASSOCIATES
5405 WEST CYPRESS STREET
TAMPA, FL 33607
TEL 813.289.0729
FAX 813.288.0231

MECHANICAL ENGINEER
LEHR ASSOCIATES

130 WEST 30TH STREET
NEW YORK, NY 10001-4082
TEL 212.947.8050
FAX 212.967.2059

STRUCTURAL ENGINEER
WALTER P. MOORE

201 EAST KENNEDY BOULEVARD
TAMPA, FL 33602
TEL 813.221.2424
FAX 813.221.2280

CIVIL ENGINEER
BROWN & CULLEN, INC.

3530 NW 43 STREET
GAINESVILLE, FL 32605
TEL 352.375.8900
FAX 352.375.0833

LANDSCAPE ENGINEER
McCLAIN DESIGN GROUP, INC.

1843 NW 39 DRIVE
GAINESVILLE, FL 32605
TEL 352.372.2808
FAX 352.372.6622

CONSULTANT
SEEGY-BISCH

AUSSERE SULZBACHER STR. 118
NURNBERG, GERMANY
TEL 0911/59 90 99
FAX 0911/59 98 05

THIS DRAWING, AS AN INSTRUMENT OF SERVICE, IS AND
SHALL REMAIN, PROPERTY OF CROXTON COLLABORATIVE
ARCHITECTS. THIS DRAWING SHALL NOT BE
REPRODUCED, PUBLISHED, USED ON OTHER PROJECTS,
USED FOR ADDITIONS, OR USED IN COMPLETION OF THIS
PROJECT BY OTHERS, EXCEPT BY AGREEMENT IN WRITING.
© 2001 CROXTON COLLABORATIVE ARCHITECTS/
GOULD EVANS ASSOCIATES.

⚠ REVISION 1   12/19/01

NO.   ISSUE/REVISION   DATE

100% CONSTRUCTION
DOCUMENTS

DRAWING NAME
CEILING & LOUVER
DETAILS

DATE
     11/21/01   DRAWING NUMBER
SCALE
     3"=1'-0"   A704
DRAWN BY/CHECKED BY
          DLH

00105-15_F25.EPS

Figure 25 The title block of a construction drawing.

- *Revision blocks* – Information on revisions, including (at a minimum) the date and the initials of the person making the revision. Other information may include descriptions of the revision and a revision number.

Every company has its own system for such things as project numbers and departments. Every company also has its own placement locations for the title and revision blocks. Your supervisor should explain your company's system to you.

1.2.2 Border

The border is a clear area of approximately half an inch around the edge of the drawing area. It is there so that everything in the drawing area can be printed or reproduced on printing machines with no loss of information.

1.2.3 Drawing Area

The drawing area presents the information for constructing the project, such as the floor plan, elevations of the building, sections, and details.

1.2.4 Revision Block

A revision block is located in the drawing area, usually in the lower right corner inside the title block or near it. Different companies put the revision block in different places. This block is used to record any changes (revisions) to the drawing. It typically contains the revision number, a brief description, the date, and the initials of the person who made the revisions (*Figure 26*). All revisions must be noted in this block and dated and identified by a letter or number.

> **CAUTION**
> It is essential to note the revision designation on a construction drawing and to use only the latest version. Otherwise, costly mistakes may result.

1.2.5 Legend

Each line on a construction drawing has a specific design and thickness that identifies it. Note that some of the lines may be used to identify off-site utilities. The identification of these lines and other symbols is called the legend. Although a legend doesn't automatically appear on every construction drawing, when it does, it explains or defines symbols or special marks used in the drawing (*Figure 27*). Be aware that legends are specific only to the set of drawings in which they are contained.

Importance of Specifications

Specifications clarify information that cannot be shown on the drawings. Specifications are very important to the architect and owner to ensure compliance to the standards set. The figure provided shows one page of the specifications for a building's air handling units.

M.E. RINKER SR. HALL
SCHOOL OF BUILDING CONSTRUCTION
BR-191

AIR HANDLING UNITS
SECTION 15760

2.1   AIR HANDLING UNITS

A.   Units shall be of the type, size and capacity as set forth in the schedule. The fan outlet velocities and coil and filter face velocities shall be within 5% of the values specified in the schedule. Units shall be double wall McQuay, or approved equal.

B.   The units, as assembled, shall be complete with fans, coils, insulated casing, filters, drives and accessories. Each unit, including the fan enclosure, shall have essentially constant cross-sectional dimensions as to width and height. Internal baffles shall be provided as required to prevent bypassing of coils and filters.

C.   The casing shall consist of an independent structural steel frame, properly reinforced and braced for maximum rigidity, having individually removable, flush mounted, insulated panels. The casing shall be of sectionalized construction, consisting basically of individual fan section, coil section, access sections, filter section, and drain pan. Sections shall be joined with continuous gasketing to form an air tight closure. Sections shall be so designed that the method of joining can be performed with relative ease and without damage to the insulation and vapor barrier.

The framework shall be constructed of AISC structural rolled shapes having minimum thickness of 1/8" (3 mm) or die formed sheet steel having the minimum gauges set forth in the following schedules:

Maximum Individual Casing Cross-Section	Minimum Framework Gauge
Up to 30 sq.ft. (2.8 sq.m.)	14
30.5 sq.ft. to 47 sq.ft. (2.81 to 4.4 sq.m.)	12
48 sq.ft. (4.45 sq.m.) and up	10

D.   Framework shall be designed with recesses suitable to receive enclosure panels, providing neat appearance, airtight enclosure, and ease of panel removal.

E.   Enclosure panels 12 sq. ft. in area and larger shall be constructed of not less than 18 gauge die formed sheet steel. Should the sides or top of a casing section exceed 20 sq. ft. in area, the panels shall be fabricated of more than one piece, with the individual panels recessed into intermediate structural members.

Protection for the insulation edges shall be provided around the perimeter of each

GEA 0300-0600

15760 - 2
Revision No. 1
Dec. 19, 2001

00105-15_SA04.EPS

PROJ	NO	REVISION	RVSD	CHKD	APPD	DATE
3483	01	RELEASED FOR CONSTRUCTION		APD	NWS	JULY 92
3483	02	DELETED PART OF LINE 12037		APD	NWS	AUG 92
3483	03	⚠ ADDED WELDING SYMBOL		APD	NWS	AUG 92

00105-15_F26.EPS

Figure 26 The revision block of a construction drawing.

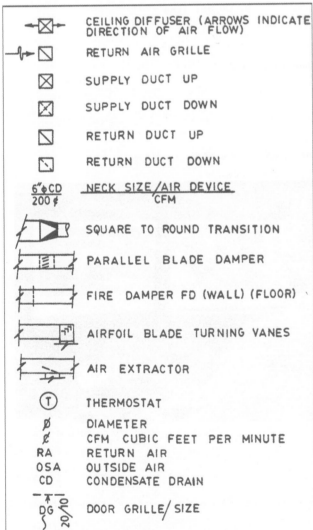

00105-15_F27.EPS

Figure 27 Sample legend.

1.3.0 Drawing Elements

Various drawing elements make it easier and quicker for people to read and understand construction drawings. This section covers several of the most common drawing elements.

1.3.1 Lines of Construction

It is very important to understand the meanings of lines on a drawing. The lines commonly used on a drawing are sometimes called the *alphabet of lines*. Here are some of the more common types of lines (*Figure 28*):

- **Dimension lines** – Establish the dimensions (sizes) of parts of a structure. These lines end with arrows (open or closed), dots, or slashes at a termination line drawn perpendicular to the dimension line.
- **Leaders** *and arrowheads* – Identify the location of a specific part of the drawing. They are used with words, abbreviations, symbols, or keynotes.
- *Property lines* – Indicate land boundaries.
- *Cut lines* – Lines around part of a drawing that is to be shown in a separate cross-sectional view.
- *Section cuts* – Show areas not included in the cutting line view.
- *Break lines* – Show where an object has been broken off to save space on the drawing.
- **Hidden lines** – Identify part of a structure that is not visible on the drawing. You may have to look at another drawing to see the part referred to by the lines.
- *Center lines* – Show the measured center of an object, such as a column or fixture.
- *Object lines* – Identify the object of primary interest or the closest object.

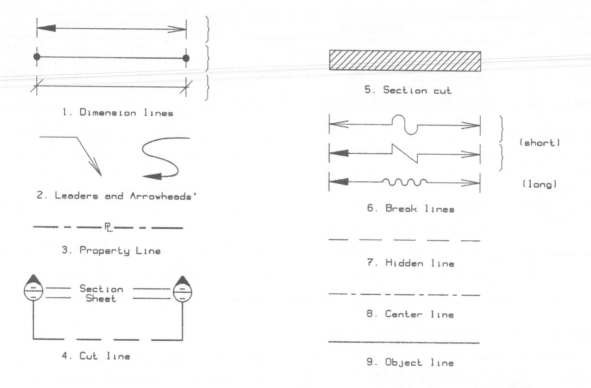

1. Dimension lines

2. Leaders and Arrowheads'

3. Property Line

4. Cut line

5. Section cut

6. Break lines
(short)
(long)

7. Hidden line

8. Center line

9. Object line

'NOTE: Arrowheads may be open or closed

00105-15_F28.EPS

Figure 28 Lines of construction.

What Is Computer-Aided Drafting?

The use of computers is a cost-effective way to increase drafting productivity because the computer program automates much of the repetitive work. A CAD system generates drawings from computer programs. CAD has the following advantages over hand-drawn construction drawings:

- It is automated.
- The computer performs calculations quickly and easily.
- Changes can be made quickly and easily.
- Commonly used symbols can be easily retrieved.
- CAD can include three-dimensional modeling of the structure.

00105-15_SA05.EPS

Regional and Company Differences

Although most symbols are standard, there can be slight variations in different regions of the country. Always check the title sheet or other introductory drawing to verify the symbols you find on the project drawings.

Your company may also use some special symbols or terms. Your instructor or supervisor will tell you about conventions that are unique to your company.

1.3.2 Abbreviations, Symbols, and Keynotes

Architects and engineers use systems of abbreviations, symbols, and keynotes to keep plans uncluttered, making them easier to read and understand. Examples of some of these items appear throughout this module. Following is additional information on these items and some variants that are commonly used.

Each trade has its own symbols, and workers should learn to recognize the symbols used by other trades. For example, an electrician should understand a carpenter's symbols, a carpenter should understand a plumber's symbols, and so on. Then, no matter what symbols are encountered on a project, the workers will understand what the symbols mean.

Abbreviations used in construction drawings are short forms of common construction terms. For example, the term *on-center* is abbreviated O.C. Some common construction abbreviations are listed in *Figure 29*.

Abbreviations should always be written in capital letters. Abbreviations for each project should be noted on the title sheet or other introductory drawing page such as the legend page. Books that list construction abbreviations and their meanings are available. You do not need to memorize these abbreviations, as you will start to remember them as you use them.

Check All the Plans

Always double-check all the plans for a project. Be familiar with other trade work that may affect your job. Determine whether an electrical conduit is close to the plumbing pipes or whether a framing member is placed too close to an HVAC duct. By knowing all the work planned for a particular area, you can prevent errors and save time and money.

If a particular dimension is missing on the plan you will be using, check the other plans to see if it is included there. The part of the building you are working on will also be shown on other plans.

When taking measurements from a set of plans, make sure that the dimensions of each measured section add up to the total measurement. For instance, if you are looking at a drawing of a wall with one window, add up the measurement from the left end of the wall to the left side of the window, the width of the window, and the measurement from the right side of the window to the right end of the wall. Make sure this total matches the measurement of the entire wall.

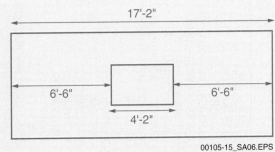

00105-15_SA06.EPS

ABBREVIATIONS

Abbreviation	Meaning
A.B.	ANCHOR BOLT
ADD'L	ADDITIONAL
ADJ.	ADJACENT
A.I.S.C.	AMERICAN INSTITUTE OF STEEL CONSTRUCTION
ALT.	ALTERNATE
ARCH.	ARCHITECTURAL
A.S.T.M.	AMERICAN SOCIETY FOR TESTING & MATERIALS
BLDG.	BUILDING
BM.	BEAM
B.O.	BOTTOM OF
BOT.	BOTTOM
BSMT.	BASEMENT
BTWN.	BETWEEN
CANT.	CANTILEVER
CB.	CARDBOARD
CH.	CHAMFER
C.J.	CONTROL/CONSTRUCTION JOINT
CLR.	CLEAR, CLEARANCE
C.M.U.	CONCRETE MASONRY UNIT
COL.	COLUMN
CONC.	CONCRETE
CONN.	CONNECTION
CONST.	CONSTRUCTION
CONT.	CONTINUOUS
CONTR.	CONTRACTOR
CTRD.	CENTERED
DET.	DETAIL
DIAG.	DIAGONAL
DIAM.	DIAMETER
DIM.	DIMENSION
DISCONT.	DISCONTINUOUS
DWG.	DRAWING
EA.	EACH
E.F.	EACH FACE
EL.	ELEVATION
ELECT.	ELECTRICAL
ELEV.	ELEVATOR
EQ.	EQUAL
E.W.B.	END WALL BARS
E.W.	EACH WAY
EXIST.	EXISTING
EXP. JNT.	EXPANSION JOINT
EXT.	EXTERIOR
F.D.	FLOOR DRAIN
FDN.	FOUNDATION
FIN.	FINISH
FLR.	FLOOR
F.O.B.	FACE OF BRICK
F.O.CONC.	FACE OF CONCRETE
F.O.W.	FACE OF WALL
FS.	FLAT SLAB
FT.	FOOT
FTG.	FOOTING
F.W.	FILLET WELD
GA.	GAUGE
GAL.	GALVANIZED
G.L.	GLU-LAM BEAM
GR.	GRADE
GR. BM.	GRADE BEAM
H.A.S.	HEADED ANCHOR STUD
HORIZ.	HORIZONTAL
H.S.B.	HIGH STRENGTH BOLT
I.D.	INSIDE DIAMETER
IN.	INCH
INT.	INTERIOR
JNT.	JOINT
LB.	POUND
LIN. FT.	LINEAL FEET
L.L.V.	LONG LEG VERTICAL
MAT'L.	MATERIAL
MAX.	MAXIMUM
MECH.	MECHANICAL
MID.	MIDDLE
MIN.	MINIMUM
MISC.	MISCELLANEOUS
MTL.	METAL
N.I.C.	NOT IN CONTRACT
NO.	NUMBER
NOM	NOMINAL
N.T.S.	NOT TO SCALE
O.C.	ON CENTER
O.D.	OUTSIDE DIAMETER
O.H.	OPPOSITE HAND
OPNG.	OPENING
PL	PLATE
P.S.F.	POUND PER SQUARE FOOT
P.S.I.	POUND PER SQUARE INCH
R.	RADIUS
REINF.	REINFORCEMENT
REQ'D.	REQUIRED
RM.	ROOM
SCHED.	SCHEDULE
SECT.	SECTION
SHT.	SHEET
SIM.	SIMILAR
S.L.V.	SHORT LEG VERTICAL
SPC.	SPACE
SPEC.	SPECIFICATION
SQ.	SQUARE
STD.	STANDARD
STIFF.	STIFFENER
STL.	STEEL
STOR.	STORAGE
SYM.	SYMMETRICAL
T.&B.	TOP AND BOTTOM
THK.	THICKNESS
T.O.	TOP OF
TYP.	TYPICAL
U.N.O.	UNLESS NOTED OTHERWISE
VAR.	VARIES
VERT.	VERTICAL
V.I.F.	VERIFY IN FIELD
WT.	WEIGHT

SYMBOLS

Symbol	Meaning
℄	CENTER LINE
⌀	DIAMETER
(filled circle)	ELEVATION
&	AND
W/	WITH
PL	PLATE
X	BY
#	NUMBER
◎	AT
▱	SQUARE
∠	ANGLE

Figure 29 Abbreviations.

Request for Information (RFI)

A request for information (RFI) is used to clarify any discrepancies in the plans. If you notice a discrepancy, you should notify the foreman. The foreman will write up an RFI, explaining the problem as specifically as possible and putting the date and time on it. The RFI is submitted to the superintendent, who passes it to the general contractor, who passes it to the architect or engineer, who then resolves the discrepancy.

Always refer to specifications and the RFI when deciding how to interpret drawings.

DATE 12/07/12   RFI NO. 1

PROJECT NAME __GERMANS FROM RUSSIA__   PROJECT NO. 15-1593

REQUEST: REF D.W.G.NO. M2   REV. DETAIL $\frac{1}{M2}$ OTHER _____

WILL THE 16 X 10 INTAKE AIR DUCTWORK RUNNING THROUGH RM 116
REQUIRE WALL MOUNTED FIRE DAMPERS ON ALL 4 EXIT CORRIDOR
WALL PENETRATIONS AND WILL THE 14 X 10 TRANSFER DUCTWORK
REQUIRE THE INSTALLATION OF A FIRE DAMPER AS WELL?

BY: LARRY MAYRE   REPLY BY (DATE): 12/20/12

REPLY:

ANSWER: ALL DUCTWORK IS ABOVE CEILING, SO OK AS IS.

DATE: 12/19/12

00105-15_SA07.EPS

Symbols are used on a drawing to tell what material is required for that part of the project. A combination of these symbols, expanded and drawn to the same size, makes up the pictorial view of the plan. There are architectural symbols (*Figure 30*); civil and structural engineering symbols (*Figure 31*); mechanical symbols (*Figure 32*); plumbing symbols (*Figure 33*); and electrical symbols. Slightly different symbols may be used in different parts of the country or by different companies. The symbols used for each set of plans should be indicated on the title sheet or other introductory drawing. Many code books, manufacturers' brochures, and specifications include symbols and their meanings.

Some plans use keynotes (*Figure 34*) instead of symbols. A keynote is a number or letter (usually in a square or circle) with a leader and arrowhead that is used to identify a specific object. Part of the drawing sheet (usually on the right-hand side) lists the keynotes with their numbers or letters. The keynote descriptions normally use abbreviations.

1.3.3 Using Gridlines to Identify Plan Locations

Have you ever used a map to find a street? The map may have used a grid to make locating a detailed area easier. For example, the index might have referred you to section B-3, so you located B along the side of the map and 3 along the top. Then you located the intersection of the two to find the street.

The gridline system shown on a plan (*Figure 35*) is used like the grid on a map. On a drawing such as a floor plan, a grid divides the area into small parts called bays.

The numbering and lettering system begins in the upper left-hand corner of the floor plan. The numbers are normally across the top and the letters are along the side. To avoid confusion, certain letters and the symbol for zero are not used. Omitted from the gridline system are the letters I, O, and Q; and numbers 1 and 0.

A gridline system makes it easy to refer to specific locations on a plan. Suppose you want to refer to one outlet, but there are a dozen on a plan. Simply refer to "the outlet in bay C-8".

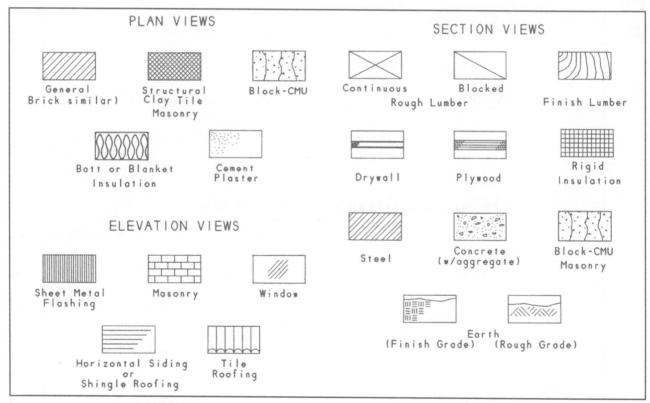

00105-15_F30.EPS

Figure 30 Architectural symbols.

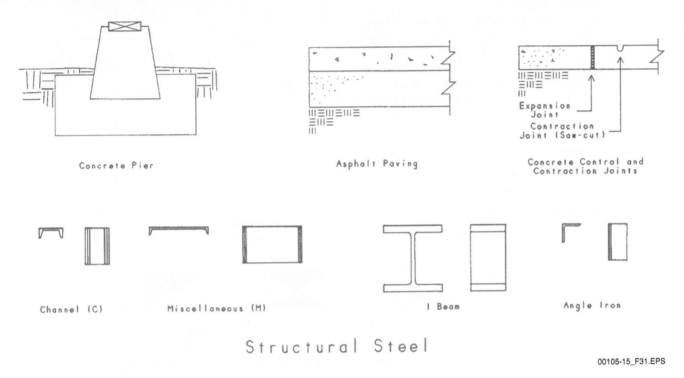

Concrete Pier

Asphalt Paving

Expansion
Joint

Contraction
Joint (Saw-cut)

**Concrete Control and
Contraction Joints**

Channel (C)

Miscellaneous (M)

I Beam

Angle Iron

Structural Steel

00105-15_F31.EPS

Figure 31 Civil and structural engineering symbols.

Green Construction

Green construction refers to a method of designing and building structures using materials and techniques that help minimize the stress on our natural resources and the environment.

With just under 5 percent of the world's population, the United States manages to consume about 19 percent of the world's energy (buildings account for 40 percent of this consumption) and produces 170 million tons (154,221,406 metric tons) of construction and demolition debris a year.

In response to these numbers, and a general increase in environmental awareness, organizations such as the US Green Building Council (USGBC) created environmental assessment systems, such as the LEED (Leadership in Energy and Environmental Design) Green Building Rating System. Systems like LEED provide green standards for the construction industry to follow, and they officially certify structures (nonresidential) that meet USGBC's strict criteria.

The LEED program was launched in 1988. It is a voluntary national standard that awards points for incorporating green strategies into areas such as site planning, safeguarding water quality and efficiency, efficiency in energy use and recycling, conservation of resources and materials, and the design, quality, and efficiency of the indoor environment.

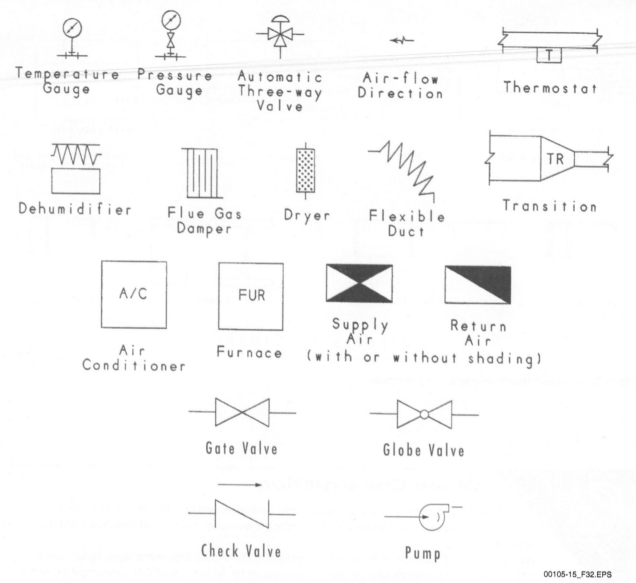

Figure 32 Mechanical symbols.

00105-15_F32.EPS

Care of Construction Drawings

Construction drawings are valuable records and must be cared for. Follow these rules when handling construction drawings:

- Never write on a construction drawing without authorization.
- Keep drawings clean. Dirty drawings are hard to read and can cause errors.
- Fold drawings so that the title block is visible.
- Fold and unfold drawings carefully to avoid tearing.
- Do not lay sharp tools or pointed objects on construction drawings.
- Keep drawings away from moisture.
- Make copies for field use; don't use originals.

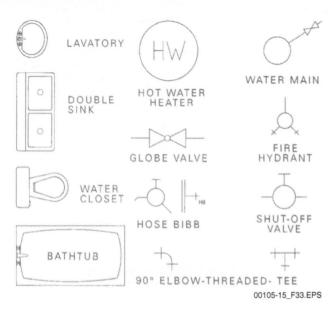

Figure 33 Plumbing symbols.

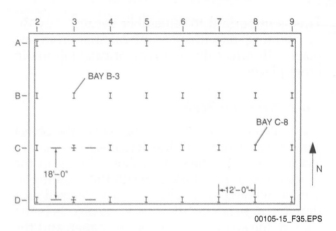

Figure 35 Grid.

1.4.0 Dimensions and Drawing Scale

Dimensions and their associated drawing scales provide information that is essential for correctly translating construction drawings into actual buildings. This section describes the basic features and purpose of dimensions and drawing scales. How to read and use different types of drawing scales will be discussed later.

1.4.1 Dimensions

Dimensions are the parts of the drawings that show the size and the placement of the objects that will be built or installed. Dimension lines can have arrowheads or slashes at both ends, with the dimension itself written near the middle of the line. The dimension is a measurement written as a number, and it may be written in inches with fractions (6½"), in feet with inches (1'-2"), or in inches with decimals (3.2"). When the metric system is used, the dimensions are usually expressed in meters, centimeters, or millimeters (9 mm).

To do accurate work, workers need to know how to read dimensions on construction drawings. This means they need to know whether the dimensions measure to the exterior or the interior of an object. To understand the difference, look at *Figure 36*, which shows a piece of pipe. There are two measurements that could be taken to get the pipe's dimensions.

The first measurement is from the pipe's exterior edge on one side directly across to its exterior edge on the other side. The second measurement is from the pipe's inside (interior) edge on one side directly across to its interior edge on the other side. Even though the difference between these two dimensions may be only a fraction of an inch (the thickness of the pipe), they are still two completely different dimensions.

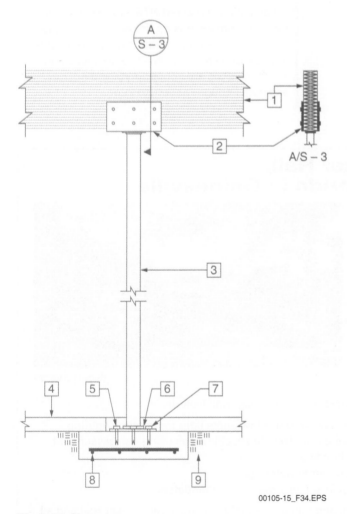

Figure 34 Keynotes.

This is important to remember because any dimensioning inaccuracy or miscalculation in one place will affect the accuracy of calculations in other places.

1.4.2 Drawing Scale

The scale of a drawing tells the size of the object drawn compared with the actual size of the object represented. The scale is shown in one of the spaces in the title block, beneath the drawing itself, or in both places. The type of scale used on a drawing depends on the size of the objects being shown, the space available on the paper, and the type of plan.

On a site plan, the scale may read SCALE: 1" = 20'-0". This means that every 1 inch on the drawing represents 20 feet, 0 inches. The scale used to develop site plans is an **engineer's scale**.

On a floor plan, the scale may read SCALE: ¼" = 1'-0". This means that every ¼ inch on the drawing represents 1 foot, 0 inches. Floor plans are developed using an **architect's scale**. This scale is divided into fractions of an inch. Metric floor plans typically use a ratio of 1:50, indicating that each millimeter on the drawing represents 50 meters.

Some drawings are not drawn to scale. A note on such drawings reads **not to scale (NTS)**.

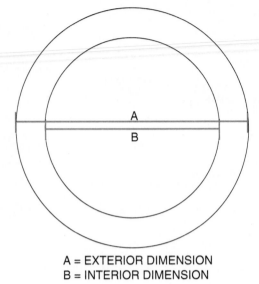

A = EXTERIOR DIMENSION
B = INTERIOR DIMENSION

00105-15_F36.EPS

Figure 36 Exterior and interior dimensions on pipe.

CAUTION	When a plan is marked NTS, workers cannot measure dimensions on the drawing and use those measurements to build the project. Not-to-scale drawings give relative positions and sizes. The sizes are approximate and are not accurate enough for construction.

GOING GREEN

Recycling Rinker Hall, University of Florida in Gainesville

An example of a LEED Gold rated building is Rinker Hall on the campus of the University of Florida in Gainesville, Florida. During the planning phase of the building, special consideration was given to land use and the use of recycled and recyclable materials. Special attention was also given to planning the incorporation of the most recent green technologies in the areas of water consumption and conservation, heating and cooling energies, and other power needs required to run the various systems of a building.

00105-15_SA08.EPS

A few of the green strategies incorporated into the construction of Rinker Hall include:

- Orienting the structure on a pure north-south axis to allow it to use low-angle light for daytime lighting.
- Using rooftop skylights and a central skylight-covered atrium to light specific areas using natural sunlight.
- Using shade walls on the east and west sides of the building.
- Landscaping with native trees and flora that require minimal watering.
- Installing water-free urinals and fixtures that use 20 percent less water than mandated.

With the various green strategies, materials selection, and building techniques employed, Rinker Hall and all of its systems use up to 57 percent less energy than a similar structure designed in minimal compliance with the American Society of Heating, Refrigerating, and Air Conditioning Engineers (ASHRAE).

1.5.0 Measuring Scales

Standard and metric rulers and measuring tapes are usually used in a shop or in the field for a variety of measuring tasks. However, there are other types of rulers called scales used by draftsmen to produce drawings. They are also used by many workers to take scaled measurements on a drawing. These include the architect's scale, the **metric scale** (metric architect's scale), and the engineer's scale. Knowing how these scales work makes it easier to understand the information contained in a set of drawings.

1.5.1 Architect's Scale

The architect's scale is often used to create construction drawings. An architect's scale translates the large measurements of real structures (rooms, walls, doors, windows, duct, etc.) into smaller measurements for drawings. Architect's scales are available in several types, but the most common include the triangular and flat scales. A flat scale is shown in *Figure 37*. The triangular architect's scale is most commonly used because it can combine up to twelve different scales on one tool. Each side of the triangular form has two faces, and two scales are combined on each face. Architect's scales are available in 6- and 12-inch lengths.

Each scale on an architect's scale is designated to a different fraction of an inch that equals a foot. These fraction designations appear on the

Checking Scales on Drawings

Normally the scale of a drawing is marked directly on the drawing itself. If it is not, however, do not guess what the scale is. Rather, compare different scales by identifying a given length on the drawing (for example, an 8-foot, 6-inch pipe), and then find the scale that measures the pipe at 8-feet, 6-inches. Check at least two items on the drawing this way before proceeding.

right and left corners of each scale. You read an architect's scale from left to right or right to left, depending on which scale you are reading.

Look at the point of measurement in *Figure 37*. It represents 57 feet when read from the left on the ⅛-inch scale and equals 18 feet when read from the right on the ¼-inch scale.

Now look at *Figure 38*. Using the ⅜-inch scale and reading from the right, you can determine that the section of duct is 7-feet, 5-inches long. Notice how the 0 point on an architect's scale is not at the extreme end of the measuring line. This is because numbers to the right of the 0 represent fractions of one foot (or inches).

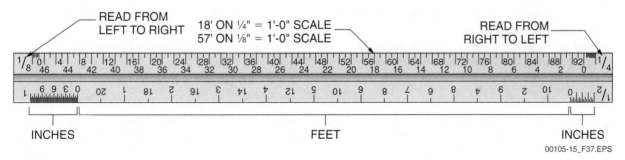

Figure 37 The architect's scale.

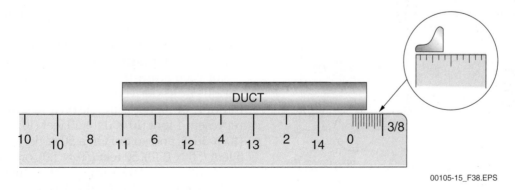

Figure 38 Measuring a section of duct with an architect's scale.

1.5.2 Metric Scale (Metric Architect's Scale)

Similar to the architect's scale is the metric scale, sometimes referred to as a metric architect's scale (*Figure 39*). Common lengths indicated on metric scales are 30 and 60 millimeters. Like the architect's scale, a metric scale can have a number of scales on it and is used to generate drawings. *Figure 39* shows how 10 meters would be shown on a 1:200 scale. The actual length on the plans is 50 mm, equaling 10 meters of real distance on the project.

Metric scales are calibrated in units of 10. Some of the most common metric scales include 1:5, 1:10, 1:20, 1:50, 1:100, 1:200, 1:500, and 1:1,000.

The two common length measurements used with the metric scale on architectural drawings are the meter and millimeter, the millimeter being $\frac{1}{1000}$ of a meter. On drawings drawn to scale between 1:1 and 1:100, the millimeter is typically used. The millimeter symbol (mm) will not be shown, but there should be a note on the drawing indicating that all dimensions are given in millimeters unless otherwise noted.

On drawings with scales between 1:200 (*Figure 39*) and 1:2,000, the meter is generally used. Again, the meter symbol (m) will not be shown,

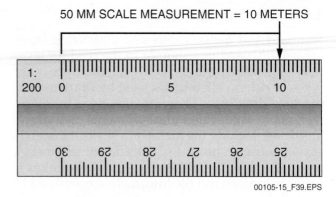

50 MM SCALE MEASUREMENT = 10 METERS

00105-15_F39.EPS

Figure 39 Metric architect's scale.

but the drawing will have a note indicating that all dimensions are in meters unless otherwise noted. Land distances shown on site and plot plans, expressed in metric units, are typically given in meters or kilometers (1,000 meters).

Reading a metric scale is easy once you identify the scale to use and the length that the scale increments represent on the drawing. Here's how it works.

The unit of length in *Figure 39* is the meter and the object on the drawing starts at 0 and extends out to the 10 on the scale. At a 1:200 ratio this object would be 10 meters long. This is because every millimeter on the scale represents 200 millimeters on the object. For example, if there are 50 millimeters from 0 to 10, multiply the 50 by 200, which gives 10,000 mm (50 mm × 200 = 10,000 mm). Because the unit being used is the meter and there are 1000 mm in one meter, divide the 10,000 mm by 1,000, which results in 10 meters. This same process is also used to determine lengths for other ratios on the scale.

1.5.3 Engineer's Scale

The engineer's scale (*Figure 40*) is used mainly for land measurements on site plans, which means the scale must accommodate very large measurements. Each engineer's scale is set up as multiples of 10 and the measurements are taken in decimals. This is different from the architect's scale in that a unit is represented by a portion of an inch. The most common engineer's scales are 10, 20, 30, 40, 50, and 60, which can all be combined on the triangular engineer's scale.

For each scale, the measurements can represent various units derived from that scale number and a multiple of 10. For example on a 10 scale, 1 inch can represent 1 foot (10 × 0.10), 10 feet (10 × 1.0), 100 feet (10 × 10.0), and so on. On a 50 scale, an inch can be 5 feet (50 × 0.10), 50 feet (50 × 1.0), or 500 feet (50 × 10.0), and so on. This same process can also be used to determine lengths on the scales.

Schematic Drawings

Most plumbing and electrical sketches are single-line drawings or schematic drawings. These drawings illustrate the scale and relationship of the project's components. In a single-line or schematic plumbing drawing, the line represents the centerline of the pipe. In a single-line or schematic electrical drawing, the line represents electrical wiring routing or circuit.

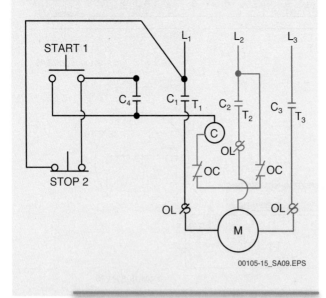

00105-15_SA09.EPS

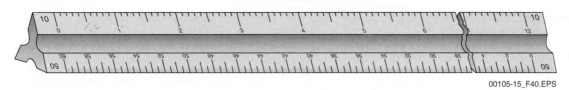

Figure 40 The engineer's scale.

Orthographic Drawings

Orthographic drawings are used for elevation drawings. They show straight-on views of the different sides of an object with dimensions that are proportional to the actual physical dimensions. In orthographic drawings, the designer draws lines that are scaled-down representations of real dimensions. Every 12 inches, for example, may be represented by ¼ inch on the drawing. Similarly, in an example using metric measurements with a ratio of 1:2, every 30 millimeters may be represented by 15 millimeters on the drawing.

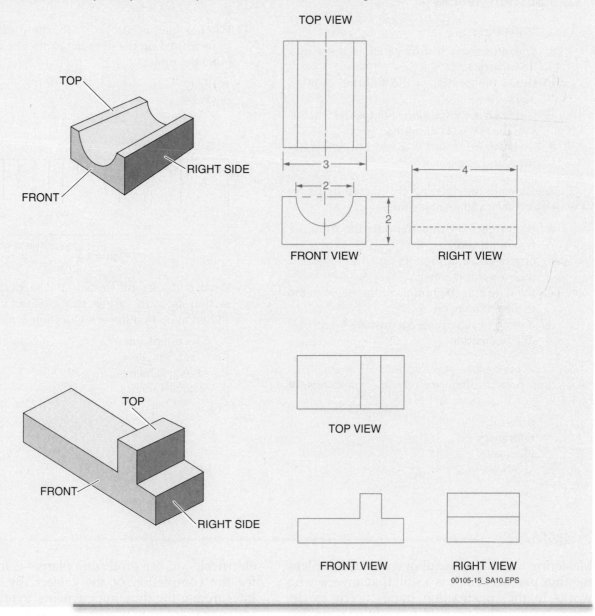

Additional Resources

Blueprint Reading for Construction, James Fatzinger. 2003. Upper Saddle River, NJ: Prentice Hall.

Blueprint Reading for the Construction Trades, Peter A. Mann. 2005. Ontario, Canada: **Micro-press.com**.

Reading Architectural Plans for Residential and Commercial Construction, Ernest R. Weidhaas. 2001. Englewood Cliffs, NJ: Prentice Hall Career & Technology.

Reading Architectural Work Drawings, Edward J. Muller and Phillip A. Grau III. 2003. Upper Saddle River, NJ: Prentice Hall.

Autodesk, 1 Market St, Suite 500, San Francisco, CA 94105, USA; 3D design, engineering and entertainment software and parent company of the AutoCAD software suite. **www.autodesk.com**.

Datacad, P.O. Box 815, Simsbury, CT 06070, USA. Windows-based CADD solutions. **www.datacad.com**.

1.0.0 Section Review

1. A site plan _____.

    a. does not show features such as trees and driveways
    b. shows the location of the building from an aerial view
    c. is drawn after the floor plan is drawn, before the HVAC is designed
    d. includes information about plumbing fixtures and equipment

2. Revisions to the drawing are entered in the revision block and must include _____.

    a. the project tag and the date the drawing was approved
    b. engineering approvals and the intended date of completion
    c. the date and the initials of the person who made the revision
    d. dates and signatures for customer approval documentation

3. Code books and specifications for a construction project often include the meanings of _____.

    a. drawings
    b. references
    c. glossaries
    d. symbols

4. When a plan is marked NTS, the dimensions as measured on the drawing cannot be used to build the project.

    a. True
    b. False

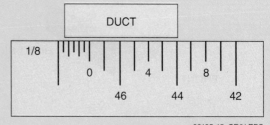

DUCT

1/8   0   4   8
       46   44   42

00105-15_SR01.EPS

Figure 1

5. What is the length (in feet and inches) of the section of duct using the architect's scales shown in Section Review Question *Figure 1*?

    a. 6 feet, ¾ inches
    b. 6 feet, 5 inches
    c. 44 feet, ¾ inches
    d. 48 feet, 4 inches

SUMMARY

Mastering the skill of reading construction plans requires practice, but it is a skill that anyone who works in the construction trades needs to develop. Each of the different types of construction drawings commonly found on a job site—civil, architectural, structural, mechanical, plumbing, electrical, and fire protection plans—is important for the completion of the project. By correctly interpreting the drawing elements, symbols, and scales that are used workers can visualize the entire project, detect inconsistencies or errors early, and possibly avoid costly mistakes or rework.

Review Questions

1. Which type of plan shows the layout of the HVAC system?

   a. Plumbing
   b. Structural
   c. Mechanical
   d. Foundation

2. Electrical plans can include _____.

   a. exterior elevations
   b. section drawings
   c. plumbing isometrics
   d. lighting plans

3. The title block generally contains _____.

   a. revision blocks
   b. special marks used in the drawing
   c. the mechanical plans
   d. the legend

4. The latest revision date on a set of construction drawings can be found _____.

   a. inside the detail
   b. in the schedule
   c. inside the title block
   d. in the legend

5. The Alphabet of Lines consists of _____.

   a. the line types used on a construction drawing
   b. lines that are indicated using letters of the alphabet
   c. lines that match up detail and section drawings
   d. lines that indicate land boundaries on the site plan

6. On a gridline system, a grid divides the area into small parts called _____.

   a. segments
   b. bays
   c. sections
   d. PODS

7. When the metric system is used, dimensions are written in _____.

   a. inches, feet, and yards
   b. degrees, radians, and gradians
   c. CADS, RADS, and KRADS
   d. meters, centimeters, and millimeters

8. If the scale on a site plan reads SCALE: 1" = 20'-0", then every _____.

   a. ¹⁄₂₀th of an inch on the drawing represents 20 feet, 0 inches
   b. 20 inches on the drawing represents 1 foot, 0 inches
   c. inch on the drawing represents 20 feet, 0 inches
   d. 20 inches on the drawing represents 20 feet, 0 inches

1:100 0 1 2 3 4 5

09 55

00105-15_RQ01.EPS

Figure 1

9. What is the length in meters at the point indicated by the arrow on the metric scale in Review Question *Figure 1*? The unit of length used on the scale is the meter.

   a. 3500
   b. 350
   c. 35
   d. 3.5

10. An engineer's scale is set up in multiples of _____.

   a. 5
   b. 10
   c. 12
   d. 39.37

Trade Terms Quiz

Fill in the blank with the correct term that you learned from your study of this module.

1. A(n) _____ is a side view that shows height.

2. A(n) _____ usually has an arrowhead at both ends, with the measurement written near the middle of the line.

3. A(n) _____ is a qualified, licensed person who creates and designs drawings for a construction project.

4. _____, which show the design of the project, include many parts, such as the floor plan, roof plan, elevation drawings, and section drawings.

5. _____ support the architectural design of a building to show how the building is supported. A foundation plan would be part of this category.

6. Also called site plans or survey plans, _____ show the location of the building from an aerial view, as well as the natural contours of the earth.

7. Almost all construction drawings today are made by _____.

8. _____ are solid or dashed lines showing the elevation of the earth on a civil drawing.

9. _____ are enlarged views of some special features of a building, such as floors and walls.

10. A(n) _____ is a person who applies scientific principles in design and construction.

11. _____, or engineered drawings for electrical supply and distribution, include locations of the electric meter, switchgear, and convenience outlets.

12. A large, horizontal support made of concrete, steel, stone, or wood that may be shown in structural plans is a _____.

13. Schematic drawings called _____ show all the equipment, pipelines, valves, instruments, and controls needed to operate a piping system.

14. An element of architectural drawings, _____ refers to the height above sea level or other defined surface.

15. Architectural, civil and structural engineering, mechanical, and plumbing _____ may be used on a drawing to tell what material is required for that part of the project.

16. Also called a plan view, a(n) _____ is an aerial view of the layout of each room.

17. A(n) _____ is a one-line drawing showing the flow path for electrical circuitry or the relationship of all parts of a system.

18. Part of the structural plans, the _____ shows the lowest level of the building.

19. Plans for gas, oil, or steam heat piping may be included in the _____ plan.

20. When a plan is marked _____, it means that the drawing gives approximate positions and sizes only.

21. A(n) _____ is a dashed line on a plan showing an object obstructed from view by another object.

22. A(n) _____ is a type of three-dimensional drawing that is included in a plumbing plan.

23. The meter and millimeter are two common length measurements used with the _____ on architectural drawings.

24. In drafting, an arrowhead is placed on a(n) _____ to identify a component.

25. _____ are written statements provided by the architectural and engineering firm to define the quality of work to be done and to describe the materials to be used.

26. The _____ defines the symbols used in architectural plans.

27. _____ are engineered plans for motors, pumps, piping systems, and piping equipment.

28. A(n) _____ is a cross-sectional view that shows the inside of an object or building.

29. The triangular and flat scales are the two most common types of _____.

30. A(n) _____ shows the shape of the roof and the materials that will be used to finish it.

31. The _____ of a drawing tells the size of the object drawn compared with the actual size of the object represented.

32. _____ show the layout for the plumbing system that supplies hot and cold water, for the sewage disposal system, and for the location of plumbing fixtures.

33. Part of the construction drawing, the _____ gives information about the structure and is numbered for easy filing.

34. _____ are the traditional name for construction drawings.

35. The drawing that makes up a building's piping, valves, and switches is a(n) _____.

36. A(n) _____ is used mainly for land measurements on site plans.

Trade Terms

Architect
Architect's scale
Architectural plans
Beam
Blueprints
Civil plans
Computer-aided drafting (CAD)
Contour lines
Detail drawings
Dimension line
Electrical plans
Elevation (EL)
Elevation drawing

Engineer
Engineer's scale
Fire protection plan
Floor plan
Foundation plan
Hidden line
HVAC
Leader
Legend
Mechanical plans
Metric scale
Not to scale (NTS)

Piping and instrumentation drawings (P&IDs)
Plumbing isometric drawing
Plumbing plans
Roof plan
Scale
Schematic
Section drawing
Specifications
Structural plans
Symbol
Title block

Trade Terms Introduced in This Module

Architect: A qualified, licensed person who creates and designs drawings for a construction project.

Architect's scale: A specialized ruler used in making or measuring reduced scale drawings. The ruler is marked with a range of calibrated ratios for laying out distances, with scales indicating feet, inches, and fractions of inches. Used on drawings other than site plans.

Architectural plans: Drawings that show the design of the project. Also called architectural drawings.

Beam: A large, horizontal structural member made of concrete, steel, stone, wood, or other structural material to provide support above a large opening.

Blueprints: The traditional name used to describe construction drawings.

Civil plans: Drawings that show the location of the building on the site from an aerial view, including contours, trees, construction features, and dimensions.

Computer-aided drafting (CAD): The making of a set of construction drawings with the aid of a computer.

Contour lines: Solid or dashed lines showing the elevation of the earth on a civil drawing.

Detail drawings: Enlarged views of part of a drawing used to show an area more clearly.

Dimension line: A line on a drawing with a measurement indicating length.

Electrical plans: Engineered drawings that show all electrical supply and distribution.

Elevation (EL): Height above sea level, or other defined surface, usually expressed in feet or meters.

Elevation drawing: Side view of a building or object, showing height and width.

Engineer: A person who applies scientific principles in design and construction.

Engineer's scale: A straightedge measuring device divided uniformly into multiples of 10 divisions per inch so that drawings can be made with decimal values. Used mainly for land measurements on site plans.

Fire protection plan: A drawing that shows the details of the building's sprinkler system.

Floor plan: A drawing that provides an aerial view of the layout of each room.

Foundation plan: A drawing that shows the layout and elevation of the building foundation.

Hidden line: A dashed line showing an object obstructed from view by another object.

HVAC: Heating, ventilating, and air conditioning.

Leader: In drafting, the line on which an arrowhead is placed and used to identify a component.

Legend: A description of the symbols and abbreviations used in a set of drawings.

Mechanical plans: Engineered drawings that show the mechanical systems, such as motors and piping.

Metric scale: A straightedge measuring device divided into centimeters, with each centimeter divided into 10 millimeters. Usually used for architectural drawings and sometimes referred to as a metric architect's scale.

Not to scale (NTS): Describes drawings that show relative positions and sizes only, without scale.

Piping and instrumentation drawings (P&IDs): Schematic diagrams of a complete piping system.

Plumbing isometric drawing: A type of three-dimensional drawing that depicts a plumbing system.

Plumbing plans: Engineered drawings that show the layout for the plumbing system.

Roof plan: A drawing of the view of the roof from above the building.

Scale: The ratio between the size of a drawing of an object and the size of the actual object.

Schematic: A one-line drawing showing the flow path for electrical circuitry or the relationship of all parts of a system.

Section drawing: A cross-sectional view of a specific location, showing the inside of an object or building.

Specifications: Precise written presentation of the details of a plan.

Structural plans: A set of engineered drawings used to support the architectural design.

Symbol: A drawing that represents a material or component on a plan.

Title block: A part of a drawing sheet that includes some general information about the project.

Additional Resources

This module presents thorough resources for task training. The following resource material is suggested for further study.

Blueprint Reading for Construction, James Fatzinger. 2003. Upper Saddle River, NJ: Prentice Hall.

Blueprint Reading for the Construction Trades, Peter A. Mann. 2005. Ontario, Canada: **Micro-press.com**.

Reading Architectural Plans for Residential and Commercial Construction, Ernest R. Weidhaas. 2001. Englewood Cliffs, NJ: Prentice Hall Career & Technology.

Reading Architectural Work Drawings, Edward J. Muller and Phillip A. Grau III. 2003. Upper Saddle River, NJ: Prentice Hall.

Autodesk, 1 Market St, Suite 500, San Francisco, CA 94105, USA; 3D design, engineering and entertainment software and parent company of the AutoCAD software suite. **www.autodesk.com**.

Datacad, P.O. Box 815, Simsbury, CT 06070, USA. Windows-based CADD solutions. **www.datacad.com**.

Figure Credits

Answer	Section Reference	Objective
Section One		
1. b	1.1.1	1a
2. c	1.2.4	1b
3. d	1.3.2	1c
4. a	1.4.2	1d
5. b	1.5.1	1e

NCCER CURRICULA — USER UPDATE

NCCER makes every effort to keep its textbooks up-to-date and free of technical errors. We appreciate your help in this process. If you find an error, a typographical mistake, or an inaccuracy in NCCER's curricula, please fill out this form (or a photocopy), or complete the online form at **www.nccer.org/olf**. Be sure to include the exact module ID number, page number, a detailed description, and your recommended correction. Your input will be brought to the attention of the Authoring Team. Thank you for your assistance.

Instructors – If you have an idea for improving this textbook, or have found that additional materials were necessary to teach this module effectively, please let us know so that we may present your suggestions to the Authoring Team.

NCCER Product Development and Revision

13614 Progress Blvd., Alachua, FL 32615

Email: curriculum@nccer.org
Online: www.nccer.org/olf

❏ Trainee Guide   ❏ Lesson Plans   ❏ Exam   ❏ PowerPoints   Other _____

Craft / Level: _____   Copyright Date: _____

Module ID Number / Title: _____

Section Number(s): _____

Description: _____

_____

_____

_____

Recommended Correction: _____

_____

_____

_____

Your Name: _____

Address: _____

_____

Email: _____   Phone: _____

Drawing 1,

00106-15

Introduction to Basic Rigging

Overview

A common activity at nearly every construction site is the movement of material and equipment from one place to another using various types of lifting gear. The procedures involved in performing this task are known as rigging. Not every worker will participate in rigging operations, but nearly all will be exposed to it at one time or another. This module provides an overview of the various types of rigging equipment, common hitches used during a rigging operation, and the related Emergency Stop hand signal.

Module Six

Trainees with successful module completions may be eligible for credentialing through the NCCER Registry. To learn more, go to www.nccer.org or contact us at **1.888.622.3720**. Our website has information on the latest product releases and training, as well as online versions of our *Cornerstone* magazine and Pearson's product catalog.

Your feedback is welcome. You may email your comments to **curriculum@nccer.org,** send general comments and inquiries to **info@nccer.org**, or fill in the User Update form at the back of this module.

This information is general in nature and intended for training purposes only. Actual performance of activities described in this manual requires compliance with all applicable operating, service, maintenance, and safety procedures under the direction of qualified personnel. References in this manual to patented or proprietary devices do not constitute a recommendation of their use.

Objective

When you have completed this module, you will be able to do the following:

1. Identify and describe various types of rigging slings, hardware, and equipment.
   a. Identify and describe various types of slings.
   b. Describe how to inspect various types of slings.
   c. Identify and describe how to inspect common rigging hardware.
   d. Identify and describe various types of hoists.
   e. Identify and describe basic rigging hitches and the related Emergency Stop hand signal.

Performance Task

Under the supervision of your instructor, you should be able to do the following:

1. Demonstrate the proper ASME Emergency Stop hand signal.

Trade Terms

Block and tackle	Master link	Sling reach
Bridle	One-rope lay	Splice
Bull ring	Plane	Strand
Competent person	Qualified person	Tag line
Core	Rated capacity	Tattle-tail
Hitch	Rejection criteria	Threaded shank
Hoist	Rigging hook	Unstranding
Lifting clamp	Shackle	Warning yarn
Load	Sheave	Wire rope
Load control	Sling	
Load stress	Sling legs	

Industry Recognized Credentials

If you are training through an NCCER-accredited sponsor, you may be eligible for credentials from NCCER's Registry. The ID number for this module is 00106-15. Note that this module may have been used in other NCCER curricula and may apply to other level completions. Contact NCCER's Registry at 888.622.3720 or go to **www.nccer.org** for more information.

Note

This module is an elective. It is not required for successful completion of the *Core Curriculum*.

Contents

Topics to be presented in this module include:

Figures

SECTION ONE

1.0.0 BASIC RIGGING EQUIPMENT

Objective

Identify and describe various types of rigging slings, hardware, and equipment.

- a. Identify and describe various types of slings.
- b. Describe how to inspect various types of slings.
- c. Identify and describe how to inspect common rigging hardware.
- d. Identify and describe various types of hoists.
- e. Identify and describe basic rigging hitches and the related Emergency Stop hand signal.

Performance Task

1. Demonstrate the proper ASME Emergency Stop hand signal.

Trade Terms

Block and tackle: A simple rope-and-pulley system used to lift loads.

Bridle: A configuration using two or more slings to connect a load to a single hoist hook.

Bull ring: A single ring used to attach multiple slings to a hoist hook.

Competent person: An individual capable of identifying existing and predictable hazards in the surroundings and working conditions which are unsanitary, hazardous, or dangerous to employees; an individual who is authorized to take prompt corrective measures to eliminate these issues.

Core: Center support member of a wire rope around which the strands are laid.

Hitch: The rigging configuration by which a sling connects the load to the hoist hook. The three basic types of hitches are vertical, choker, and basket.

Hoist: A device that applies a mechanical force for lifting or lowering a load.

Lifting clamp: A device used to move loads such as steel plates or concrete panels without the use of slings.

Load control: The safe and efficient practice of load manipulation, using proper communication and handling techniques.

Load stress: The strain or tension applied on the rigging by the weight of the suspended load.

Master link: The main connection fitting for chain slings.

One-rope lay: The lengthwise distance it takes for one strand of a wire rope to make one complete turn around the core.

Plane: A surface in which a straight line joining two points lies wholly within that surface.

Qualified person: A person who, through the possession of a recognized degree, certificate, or professional standing, or one who has gained extensive knowledge, training, and experience, has successfully demonstrated his or her ability to solve problems relating to the subject matter, the work, or the project.

Rated capacity: The maximum load weight a sling or piece of hardware or equipment can hold or lift. Also referred to as the working load limit (WLL).

Rejection criteria: Standards, rules, or tests on which a decision can be based to remove an object or device from service because it is no longer safe.

Rigging hook: An item of rigging hardware used to attach a sling to a load.

Shackle: Coupling device used in an appropriate lifting apparatus to connect the rope to eye fittings, hooks, or other connectors.

Sheave: A grooved pulley-wheel for changing the direction of a rope's pull; often found on a crane.

Sling: Wire rope, alloy steel chain, metal mesh fabric, synthetic rope, synthetic webbing, or jacketed synthetic continuous loop fibers made into forms, with or without end fittings, used to handle loads.

Sling legs: The parts of the sling that reach from the attachment device around the object being lifted.

Sling reach: A measure taken from the master link of the sling, where it bears weight, to either the end fitting of the sling or the lowest point on the basket.

Splice: To join together.

Strand: A group of wires wound, or laid, around a center wire, or core. Strands are laid around a supporting core to form a rope.

Tag line: Rope that runs from the load to the ground. Riggers hold on to tag lines to keep a load from swinging or spinning during the lift.

Tattle-tail: Cord attached to the strands of an endless loop sling. It protrudes from the jacket. A tattle-tail is used to determine if an endless sling has been stretched or overloaded.

Threaded shank: A connecting end of a fastener, such as a bolt, with a series of spiral grooves cut into it. The grooves are designed to mate with grooves cut into another object in order to join them together.

Unstranding: Describes wire rope strands that have become untwisted. This weakens the rope and makes it easier to break.

Warning yarn: A component of the sling that shows the rigger whether the sling has suffered too much damage to be used.

Wire rope: A rope made from steel wires that are formed into strands and then laid around a supporting core to form a complete rope; sometimes called cable.

Rigging is the planned movement of material and equipment from one location to another, using slings, hoists, or other types of equipment. Some rigging operations use a loader to move materials around a job site. Other operations require cranes to lift such loads. Two common types of cranes are overhead cranes (*Figure 1*) and mobile cranes (*Figure 2*). As with other types of equipment, cranes are available in many different configurations and capacities.

Rigging operations can be extremely complicated and dangerous. Do not experiment with rigging operations, and never attempt a lift without the supervision of an officially recognized, qualified person. A lift may appear simple while it is in progress. That is because the people performing the lift know exactly what they are doing; there is no room for guesswork. No matter whether rigging operations involve simple or complicated equipment, only qualified persons may perform rigging operations without supervision.

1.1.0 Slings

During a rigging operation, the load being lifted or moved must be connected to the apparatus, such as a crane, that will provide the power for movement. The connector—the link between the load and the apparatus—is often a sling made of synthetic, chain, or wire rope materials. This section focuses on three types of slings:

- Synthetic slings
- Alloy steel chain slings
- Wire rope slings

00106-15_F01.EPS

Figure 1 Overhead crane.

NCCER – *Core Curriculum* 00106-15

1.1.1 Sling Tagging Requirements

All slings are required to have identification tags (*Figure 3*). An identification tag must be securely attached to each sling and clearly marked with the information required for that type of sling. For all three types of slings, that information will include the manufacturer's name or trademark and the rated capacity of the type of hitch used with that sling. The rated capacity is the maximum load weight that the sling is designed to carry. Rated capacity is technically referred to as working load limit (WLL). The rated capacity, or WLL, of a sling must never be exceeded. Overloading a sling can result in catastrophic failure.

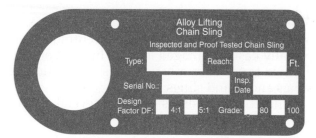

00106-15_F03.EPS

Figure 3 Identification tag.

> **WARNING!**
>
> All slings and hardware have a rated capacity, or working load limit (WLL). Rated capacity, or WLL, is defined as the maximum load weight that a sling or piece of hardware or equipment can hold or lift. Under no circumstances should the rated capacity ever be exceeded. Overloading a piece of rigging equipment can result in catastrophic failure.

The following are the tagging requirements for synthetic slings:

- Manufacturer's name or trademark
- Manufacturer's code or stock number (unique for each sling)
- Rated capacities for the types of hitches used
- Type of synthetic material used in the manufacture of the sling

The following are the tagging requirements for alloy steel chain slings:

- Manufacturer's name or trademark
- Manufactured grade of steel
- Link size (diameter)
- Rated load and the angle on which the rating is based
- Sling reach
- Number of sling legs

The following are the tagging requirements for wire rope slings:

- Manufacturer's name or trademark
- Rated capacity in a vertical hitch (other hitches optional)
- Diameter of wire rope (optional)
- Manufacturer's code or stock number of sling (optional)

Any sling without an identification tag must be removed from service immediately, since its characteristics and rating can no longer be identified. A qualified person may be able to install a new tag in some cases.

00106-15_F02.EPS

Figure 2 Mobile cranes.

1.1.2 Synthetic Slings

Synthetic slings are widely used to lift loads, especially easily damaged loads. This section covers two types of synthetic slings: synthetic web slings and round slings.

Synthetic web slings provide several advantages over other types of slings:

- They are softer and wider than chain or wire rope slings. Therefore, they do not scratch or damage machined or delicate surfaces (*Figure 4*).
- They do not rust or corrode and therefore will not stain the loads they are lifting.
- They are lightweight, making them easier to handle than wire rope or chain slings. Most synthetic slings weigh less than half as much as a wire rope that has the same rated capacity. Some new synthetic fiber slings weigh one-tenth as much as wire rope.
- They are flexible. They mold themselves to the shape of the load (*Figure 5*).
- They are very elastic, and they stretch under a load much more than wire rope. This stretching allows synthetic slings to absorb shocks and to cushion the load.
- Loads suspended in synthetic web slings are less likely to twist than those in wire rope or chain slings.

For all the advantages that synthetic web slings provide, there are some concerns that must be kept in mind when using them. For example, synthetic web slings should not be exposed to temperatures above 180°F (82°C). They are also susceptible to cuts, abrasions, and other wear-and-tear damage. To prevent damage to synthetic web slings, riggers use protective pads (*Figure 6*).

> **CAUTION**
>
> If the sling does not come with protective pads, use other kinds of softeners of sufficient strength or thickness to protect the sling where it makes contact with the load. Pieces of old sling, fire hose, canvas, or rubber can be used. Manufactured softeners can also be purchased.

Most synthetic web slings are manufactured with red-core **warning yarns**. These are used to let the rigger know whether the sling has suffered too much wear or damage to be used. When the red yarns are exposed, the synthetic web sling should not be used (*Figure 7*).

00106-15_F04.EPS

Figure 4 Web slings provide surface protection.

NCCER – *Core Curriculum*   00106-15

00106-15_F05.EPS

Figure 5 Synthetic web sling shaping.

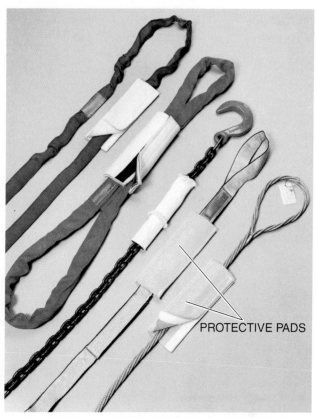

PROTECTIVE PADS

00106-15_F06.EPS

Figure 6 Protective pads.

Synthetic web slings are available in several designs. The most common are the following (*Figure 8*):

- Endless web slings, which are also called grommet slings.
- Synthetic web eye-and-eye slings, which are made by sewing an end of the sling directly to the sling body. Standard eye-and-eye slings (*Figure 9*) have eyes on the same **plane** as the sling material; twisted eye-and-eye slings have eyes at right angles to the main portion of the sling and are primarily used for choker hitches (which are discussed later).
- Round slings, which are endless and made in a continuous circle out of polyester filament yarn. The yarn is then covered by a woven sleeve.

Synthetic web eye-and-eye slings are also available with hardware end fittings instead of fabric eyes. The standard end fittings are made of either aluminum or steel. They come in male and female configurations (*Figure 10*).

Round slings are made by wrapping a synthetic yarn around a set of spindles to form an endless loop. A protective jacket encases the **core** yarn (*Figure 11*). One type of round sling is a

Twin-Path® sling. Twin-Path® is a trademarked sling made by Slingmax®, Inc., although the term is sometimes used generically.

Twin-Path® slings are made of a synthetic fiber, such as polyester. The material used to make up the **strands** and the number of wraps in a loop determines the rated capacity of the sling, or how much weight it can handle. Aramid round slings, for example, have a greater rated capacity for their size than the web-type slings do.

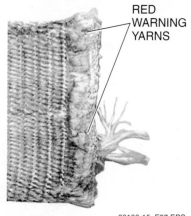

RED
WARNING
YARNS

00106-15_F07.EPS

Figure 7 Synthetic web sling warning yarns.

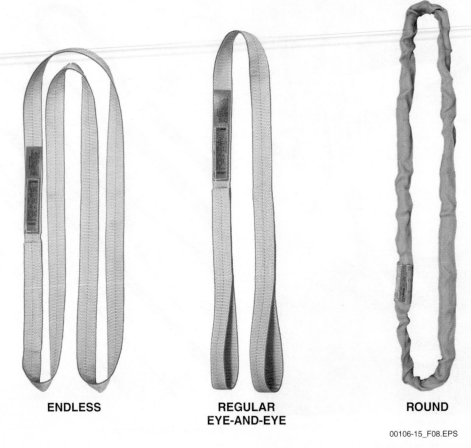

ENDLESS

REGULAR EYE-AND-EYE

ROUND

00106-15_F08.EPS

Figure 8 Synthetic web slings.

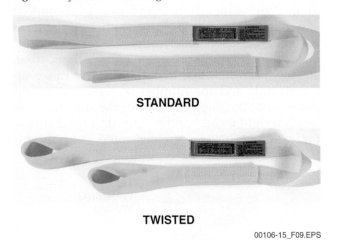

STANDARD

TWISTED

00106-15_F09.EPS

Figure 9 Eye-and-eye synthetic web slings.

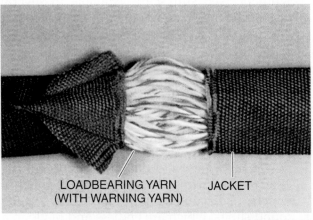

LOADBEARING YARN (WITH WARNING YARN)   JACKET

00106-15_F11.EPS

Figure 11 Synthetic endless-strand jacketed sling.

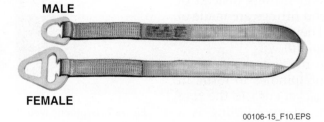

MALE

FEMALE

00106-15_F10.EPS

Figure 10 Synthetic web sling hardware end fittings.

Twin-Path® slings are also available in a design with two separate wound loops of strand jacketed together side-by-side (*Figure 12*). This design greatly increases the lifting capacity of the sling.

The jackets of these slings are available in several materials for various purposes, including heat-resistant Nomex®, polyester, and bulked nylon (Covermax™) (*Figure 13*). Twin-Path® slings featuring K-Spec® yarn weigh at least 50 percent less than a polyester round sling of the same size and capacity.

Twin-Path® slings are equipped with **tattle-tail** yarns to help riggers determine whether the sling has become overloaded or stretched beyond a safe limit (*Figure 14*). Note that these devices are different from red-core yarn, which is incorporated directly into the fabric of the sling. These slings are also available with a fiber-optic inspection cable running through the strand (*Figure 15*).

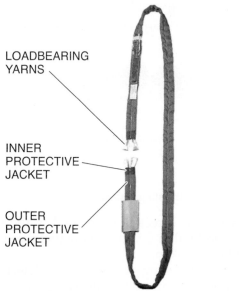

LOADBEARING YARNS

INNER PROTECTIVE JACKET

OUTER PROTECTIVE JACKET

00106-15_F12.EPS

Figure 12 Twin-Path® sling.

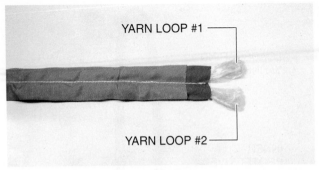

Figure 13 Twin-Path® sling makeup.

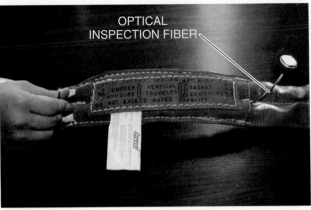

00106-15_F15.EPS

Figure 15 Fiber-optic inspection cable.

When a light is directed at one end of the fiber, the other end will light up to show that the strand has not been broken.

The way the sling will be used determines what material is used for the jackets of round slings. Polyester is normally used for light- to medium-duty sling jackets. Covermax™, a high-strength material with much greater resistance to cutting and abrasion, is used to make a sturdier jacket for heavy-duty uses.

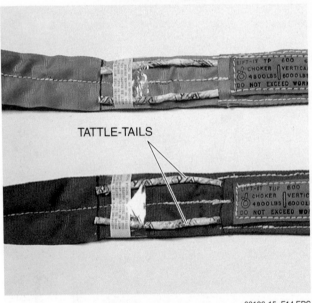

00106-15_F14.EPS

Figure 14 Tattle-tails.

1.1.3 Alloy Steel Chain Slings

Alloy steel chain, like wire rope (which is covered later), can be used in many different rigging operations. Chain slings are often used for lifts in high heat or rugged conditions. Chain slings can be adjusted over a center of gravity, making it a versatile sling. They are also the most durable. However, a chain sling weighs much more than a wire rope sling, and it also may be harder to inspect. Workers will encounter both types of slings in the field and must decide which type of sling is best for each situation.

Steel chain slings used for overhead lifting must be made of alloy steel. The higher the alloy grade, the safer and more durable the chain is. Alloy steel chains commonly used for most overhead lifting are marked with the number 8, the number 80, the number 800, or the letter A (see *Figure 16*).

Steel chain slings have two basic designs with many variations:

- Single- and double-basket slings (*Figure 17*) do not require end-fitting hardware. The chain is attached to the **master link** in a permanent basket hitch or hitches.
- Chain **bridle** slings are available with two to four legs (*Figure 18*).

Regulations and Site Procedures

The information in this module is intended as a general guide. The techniques shown here are not the only methods that can be used to perform a lift. Many techniques can be safely used to rig and lift different loads.

Some of the techniques for certain kinds of rigging and lifting are spelled out in requirements issued by federal government agencies. Some will be provided at the job site, where written site procedures that address any special conditions that affect lifting procedures on that site can be found. Questions about any of these procedures should be directed to the supervisor at the site.

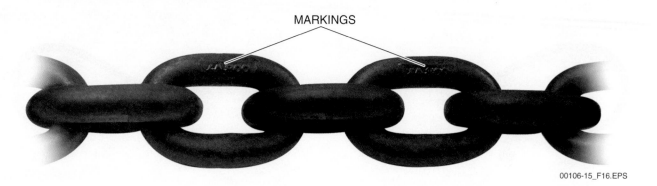

MARKINGS

00106-15_F16.EPS

Figure 16 Markings on alloy steel chain slings.

Alloy steel chain slings are used for lifts when temperatures are high or where the slings will be subjected to steady and severe abuse. Most alloy steel chain slings can be used in temperatures up to 500°F (260°C) with little loss in rated capacity.

Even though alloy steel chain slings can withstand extreme temperature ranges and abusive working conditions, these slings can be damaged if loads are dropped on them; if they are wrapped around loads with sharp corners (unless softeners are used); or if they are exposed to intense temperatures.

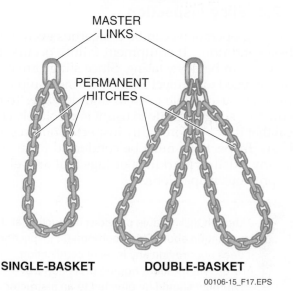

MASTER LINKS

PERMANENT HITCHES

SINGLE-BASKET   **DOUBLE-BASKET**

00106-15_F17.EPS

Figure 17 Chain slings.

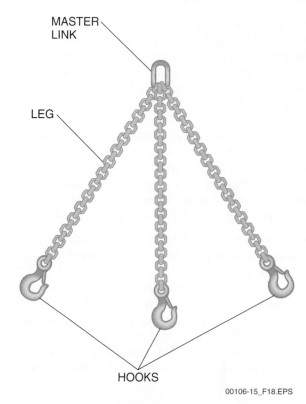

MASTER LINK

LEG

HOOKS

00106-15_F18.EPS

Figure 18 Three-leg chain bridle sling.

1.1.4 *Wire Rope Slings*

Wire rope slings (*Figure 19*) are made of high-strength steel wires that are formed into strands and wrapped around a supporting core. While there are many types of wire rope, they all share this common design. As a general rule, wire rope slings are lighter than chain, can withstand substantial abuse, and are easier to handle than chain slings. They can also withstand relatively high temperatures.

The wire rope that makes up a wire rope sling is designed so that the strand wires and the supporting core interact with one another by sliding and adjusting. This movement compensates for the ever-changing stresses placed upon the rope.

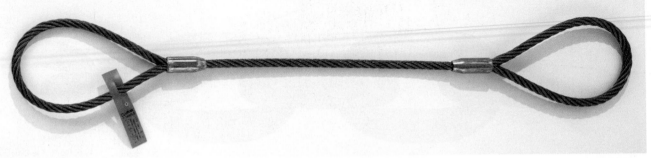

00106-15_F19.EPS

Figure 19 Wire rope sling.

It makes it less likely that the wires and the core will be damaged when the rope is bent around **sheaves** or loads, or when it is placed at an angle during rigging. The wires must be able to move the way they were designed to move, so a wire rope's rated capacity depends on it being in good condition.

Figure 20 shows the three basic components of a wire rope: the supporting core, high-grade steel wires, and multiple center wires.

The supporting core enables the strands to keep their original shapes. There are three basic types of supporting cores for wire rope (*Figure 21*):

- *Fiber cores* – Usually made of synthetic fibers, but can also be made from natural vegetable fibers, such as sisal.
- *Independent wire rope cores* – Made of a separate wire rope with its own core and strands; the core rope wires are much smaller and more delicate than the strand wires in the outer rope.
- *Strand cores* – Made by using one strand of the same size and type as the rest of the strands of rope.

The materials used for the supporting core have both desirable characteristics and drawbacks, depending on how they are to be used. Fiber core ropes, for example, may be damaged by heat at relatively low temperatures (180°F to 200°F [82°C to 93°C]), as well as by exposure to caustic chemicals.

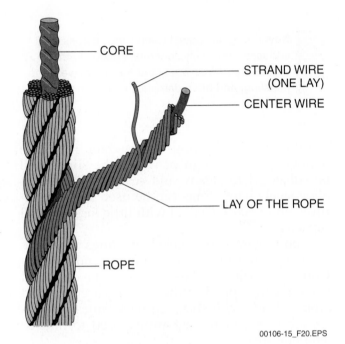

00106-15_F20.EPS

Figure 20 Wire rope components.

1.2.0 Sling Inspection

Rigging operations can be dangerous even in the best conditions. If equipment failure occurs, the results can be devastating. Since slings are commonly used to connect loads to lifting equipment, they are susceptible to wear and damage. To ensure that slings are safe to use, it is absolutely critical that they be thoroughly inspected on a regular basis. Inspections must be conducted regardless of any existing markings or tags that are related to previous inspections.

Did You Know?

Sling Cut Protection

Cuts are the primary cause of synthetic sling failure. Often these cuts come from making contact with sharp edges on the load. Besides sling jackets, or wear pads, devices called edge protectors are often placed on the edges of a load to help reduce the likelihood of cutting the sling.

WARNING!

Although this module provides instruction and information about lifting component inspection, it does not provide any level of certification. Any questions about rigging procedures or equipment should be directed to an instructor or supervisor. Always refer to the manufacturer's instructions for any type of rigging equipment.

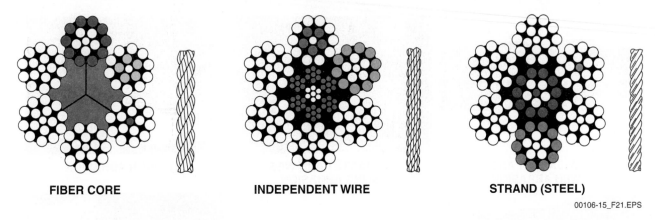

FIBER CORE　　　　**INDEPENDENT WIRE**　　　　**STRAND (STEEL)**

00106-15_F21.EPS

Figure 21 Wire rope supporting cores.

1.2.1 Synthetic Sling Inspection

Like all slings, synthetic slings must be inspected before each use to determine whether they are in good condition and can be used. They must be inspected along the entire length of the sling, both visually (looking at them) and manually (feeling them). If any rejection criteria are met, the sling must be removed from service.

> **WARNING!**
> A competent person must inspect synthetic slings before they are used, every time. The inspection must involve a visual and a touch examination. Sometimes defects and damage can be seen; other times they can be felt.

If any synthetic sling meets any of the rejection criteria presented in this section, it must be removed from service immediately. In addition, the rigger has to exercise sound judgment. Along with looking for any single major problem, the rigger must also watch for combinations of relatively minor defects in the sling. Combinations of minor damage may make the sling unsafe to use, even though the specific defects found may not be listed in the rejection criteria.

Workers who help inspect synthetic slings must alert a qualified person if any defects are suspected, especially if any of the following synthetic sling damage rejection criteria are found (*Figure 22*):

- A missing identification tag or a tag that cannot be read. Any synthetic sling without an identification tag must be removed from service immediately.
- Abrasion that has worn through the outer jacket or has exposed the loadbearing yarn of the sling, or abrasion that has exposed the warning yarn of a web sling.
- A cut that has severed the outer jacket or exposed the loadbearing yarn (single-layer jacket) of a round sling, or has exposed the warning yarn of a web sling.

- A tear that has exposed the inner jacket or the loadbearing yarns (single-layer jacket) of a round sling, or has exposed the warning yarn of a web sling.
- A puncture.
- Broken or worn stitching in the splice or stitching of a web sling.
- A knot that cannot be removed by hand in either a web or round sling.
- A snag in the sling that reveals the warning yarn of a web sling or tears through the outer jacket, or exposes the loadbearing yarns of a round sling.
- Crushing of either a web sling or a round sling. Crushing in a web sling feels like a hollow pocket or depression in the sling. Crushing in a round sling feels like a hard, flat spot underneath the jacket.
- Damage from overload (often called tensile damage or overstretching) in a web or round sling. Tensile damage in a web sling is evident when the weave pattern of the fabric begins to pull apart. Twin-Path® slings have tattle-tail features. When the tail has been pulled into the jacket, it indicates that the sling may have been overloaded.
- Chemical damage, including discoloration, burns, and melting of the fabric or jacket.
- Heat damage, ranging from friction burns to melting of the sling material, the loadbearing strands, or the jacket. Friction burns give the webbing material a crusty or slick texture. Heat damage to the jacket of a round sling looks like glazing or charring. In round slings with heat-resistant jackets, you may not be able to see heat damage to the outside of the sling; however, the internal yarns may have been damaged. You can detect that damage by carefully handling the sling and feeling for brittle or fused fibers inside the sling, or by flexing and folding the sling and listening for the sound of fused yarn fibers cracking or breaking inside.

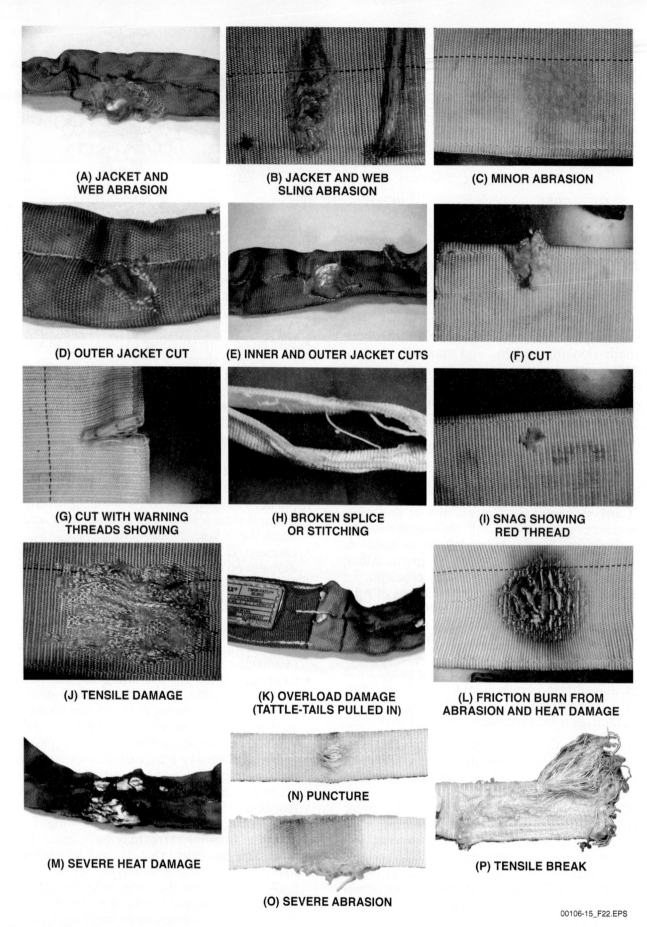

(A) JACKET AND WEB ABRASION

(B) JACKET AND WEB SLING ABRASION

(C) MINOR ABRASION

(D) OUTER JACKET CUT

(E) INNER AND OUTER JACKET CUTS

(F) CUT

(G) CUT WITH WARNING THREADS SHOWING

(H) BROKEN SPLICE OR STITCHING

(I) SNAG SHOWING RED THREAD

(J) TENSILE DAMAGE

(K) OVERLOAD DAMAGE (TATTLE-TAILS PULLED IN)

(L) FRICTION BURN FROM ABRASION AND HEAT DAMAGE

(M) SEVERE HEAT DAMAGE

(N) PUNCTURE

(O) SEVERE ABRASION

(P) TENSILE BREAK

00106-15_F22.EPS

Figure 22 Sling damage rejection criteria.

- Ultraviolet (UV) damage. The evidence of UV damage is a bleaching-out of the sling material, which breaks down the synthetic fibers. Web slings with UV damage will release a powder-like substance when they are flexed and folded. Round slings, especially those made of Cordura® nylon and other specially treated synthetic fibers, have a much greater resistance to UV damage. In round slings, UV damage shows up as a roughening of the fabric texture where no other sign of damage, such as abrasion, can be found.
- Loss of flexibility caused by the presence of dirt or other abrasives. A sling that has lost its flexibility becomes stiff. Abrasive particles embedded in the sling material act like tiny blades that cut apart the internal fibers of the sling every time it is stretched, flexed, or wrapped around a load.

1.2.2 Alloy Steel Chain Sling Inspection

Alloy steel chain slings must be carefully inspected before each use to determine if they are safe to use. There are numerous rejection criteria that indicate when an alloy steel chain sling must be removed from service. A chain sling may also need to be removed from service if something shows up that does not exactly match the rejection criteria. Not all riggers are qualified to make decisions about the condition of the equipment being used. If questions arise about whether a sling is defective or unsafe, a qualified person should be consulted. Note that the chain links in a chain sling cannot be repaired by replacement.

An alloy steel chain sling must be removed from service for any of the following defects (*Figure 23*):

- Missing or illegible identification tag
- Cracks
- Heat damage
- Stretched links; damage is evident when the link grows long and when the barrels—the long sides of the links—start to close up
- Bent links
- Twisted links
- Evidence of replacement links that have been used to repair the chain
- Excessive rust or corrosion, meaning rust or corrosion that cannot be easily removed with a wire brush

- Cuts, chips, or gouges resulting from impact on the chain
- Damaged end fittings, such as hooks, clamps, and other hardware
- Excessive wear at the link-bearing surfaces
- Scraping or abrasion

1.2.3 Wire Rope Sling Inspection

Like other slings, wire rope slings must be inspected before each use by a **competent person**. Remember that a competent person is defined by OSHA as a person who is capable of identifying existing and predictable hazards in the surroundings or working conditions that are unsanitary, hazardous, or dangerous to employees, and who has authorization to take prompt corrective measures to eliminate those conditions.

If the wire rope is damaged, it must be removed from service. Only a competent person can decide whether to use a wire rope in a rigging operation or discard it if it is damaged. Any worker who suspects that a wire rope sling may be damaged must bring it to the attention of a competent person, stopping an active lift if necessary to do so. If a wire rope sling is simply missing its tag or the tag has become unreadable due to wear, it can be retagged by a qualified person. Remember that a qualified person is defined by OSHA as a person who, by possession of a recognized degree, certificate, or professional standing, or by extensive knowledge, training, and experience, has demonstrated the ability to solve or prevent problems relating to a certain subject, work, or project.

Sling Jackets

If only the outer jacket of a sling is damaged, under the rejection criteria, that sling can be sent back to the manufacturer for a new jacket. It must be tested and certified by the manufacturer before it is used again. It is much less costly to do this than to replace the sling. However, slings that are removed because of heat- or tension-related jacket damage cannot be returned for repair and must be disposed of by a qualified person.

Following an inspection, a wire rope sling may be rejected based on several common types of damage (*Figure 24*), including broken wires, kinks, birdcaging, crushing, corrosion and rust, and heat damage.

- *Broken wires* – Broken wires in the strands of a wire rope weaken the material strength of the rope and interfere with the interaction among the rope's moving parts. External broken wires usually mean normal fatigue, but internal or severe external breaks should be investigated closely. Internal or severe breaks in a wire rope mean it has been used improperly. Rejection criteria for broken wires consider how many wires are broken in one lay length of rope, or **one-rope lay**. One-rope lay

(identifying a single rope lay and not a rope lay made with one rope) is a term that defines the lengthwise distance it takes for one strand of wire to make one complete turn around the core (*Figure 25*). Different wire ropes have different one-rope lays, so it is important for a competent person to inspect each wire rope closely when looking for broken wires.

WARNING!

Broken wires in a sling are extremely sharp and can easily cut or puncture the skin. Never use a bare hand to handle wire rope slings or to inspect them by running a hand along their length.

HEAT DAMAGE AND CRACK

IMPACT DAMAGE BENT LINKS

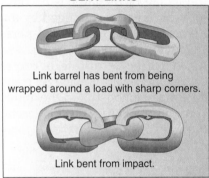

Link barrel has bent from being wrapped around a load with sharp corners.

Link bent from impact.

EXCESSIVE WEAR

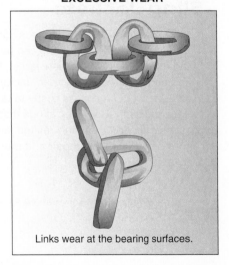

Links wear at the bearing surfaces.

OVERLOAD DAMAGE

As the link stretches the barrels will close up.

TWISTED LINKS

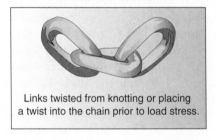

Links twisted from knotting or placing a twist into the chain prior to load stress.

CUTS, CHIPS, AND GOUGES

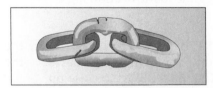

RUST AND CORROSION

00106-15_F23.EPS

Figure 23 Damage to chains.

Figure 25 One-rope lay.

BROKEN WIRES

KINKING

BIRDCAGING

CRUSHING

CORROSION

00106-15_F24.EPS

Figure 24 Common types of wire rope damage.

- *Kinks* – Kinking, or distortion of the rope, is a very common type of damage. Kinking can result in serious accidents. Sharp kinks restrict or prevent the movement of wires in the strands at the area of the kink. This means the rope is damaged and must not be used. Ropes with kinks in the form of large, gradual loops in a corkscrew configuration must be removed from service.
- *Birdcaging* – This damage occurs when a load is released too quickly and the strands are pulled or bounced away from the supporting core. The wires in the strands cannot compensate

for the change in the stress level by adjusting inside the strands. The built-up stress then finds its own release out through the strands. Birdcaging usually occurs in an area where already-existing damage prevents the wires from moving to compensate for changes in stress, position, and bending of the rope. Any sign of birdcaging is cause to remove the rope from service immediately.

- *Unstranding* – Unstranding describes wire rope strands that have become untwisted. This weakens the rope and makes it easier to break.
- *Signs of core failure* – Failure or breakage within the core is usually spotted by the protrusion of core strands through or into the outer jacket of strands.
- *Crushing* – This results from setting a load down on a sling or from hammering or pounding a sling into place. Crushing of the sling prevents the wires from adjusting to changes in stress, changes in position, and bends. A crushed sling usually results in the crushing or breaking of the core wires directly beneath the damaged strands. If crushing occurs, the sling must be removed from service immediately.
- *Corrosion and rust* – Corrosion and rust of wire rope are the result of improper or insufficient lubrication. Corrosion and rust are considered excessive if there is surface scaling or rust that cannot easily be removed with a wire brush, or if they occur inside the rope. If corrosion and rust are excessive, the rope must be removed from service.
- *Heat damage* – Heat damage makes wires in the strands and core in the affected area become brittle. Heat damage appears as discoloration and sometimes the actual melting of the wire rope. A wire rope that has been damaged by heat must be removed from service.
- *Integrity of end connections* – Any end connections that have been applied to the wire rope must be inspected for any signs of damage or failure, such as marks indicating the connection has slipped or moved.

1.3.0 Rigging Hardware

Rigging hardware is as crucial as the crane, the slings, or any specially designed lifting frame or hoisting device. If the hardware that connects the slings to either the load or the master link were to fail, the load would drop just as it would if the crane, hoist, or slings were to fail. Hardware failure related to improper attachment, selection, or inspection contributes to a great number of the deaths, serious injuries, and property damage events in rigging accidents. The importance of hardware selection, maintenance, inspection, and proper use cannot be stressed enough. The requirements for rigging hardware are as stringent as those governing cranes and slings.

1.3.1 Shackles

A shackle is an item of rigging hardware used to attach an item to a load or to couple slings together. For example, a shackle can be used to couple the end of a wire rope to eye fittings or hooks. It consists of a U-shaped body and a removable pin.

Shackles used for overhead lifting should be made from forged steel, not cast steel. Quenched and tempered steel is the preferred material because of its increased toughness, but at a minimum, shackles must be made of drop- or hammer-forged steel.

All shackles must have a stamp that is clearly visible, showing the manufacturer's trademark, the size of the shackle (determined by the diameter of the shackle's body, not by the diameter of the pin), and the rated capacity of the shackle.

Shackles are available in two basic classes, identified by their shapes: anchor shackles and chain shackles. Both anchor and chain shackles have three basic types of pin designs, each one unique, as shown in *Figure 26*. The screw pin shackle design is the most widely used type in general industry. Shackle pins can be threaded into the shackle (screw pin); be threaded and use a nut to secure them; use a locking pin through the end; or be both threaded and pinned to secure them to the shackle.

Specialty shackles are available for specific applications where a standard shackle would not work well. For example, wide-body shackles (*Figure 27*) are for heavy-lifting applications.

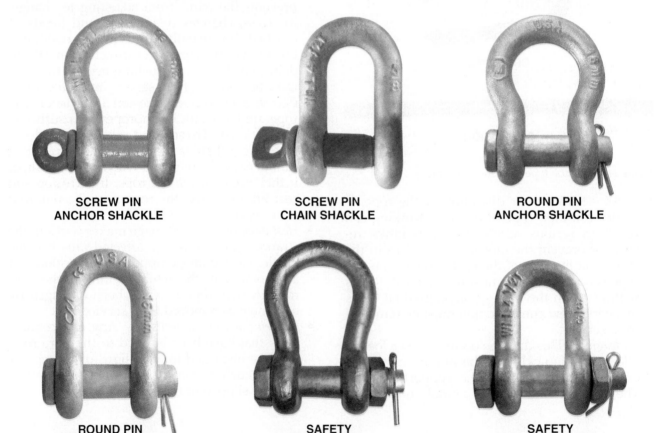

**SCREW PIN
ANCHOR SHACKLE**

**SCREW PIN
CHAIN SHACKLE**

**ROUND PIN
ANCHOR SHACKLE**

**ROUND PIN
CHAIN SHACKLE**

**SAFETY
ANCHOR SHACKLE**

**SAFETY
CHAIN SHACKLE**

00106-15_F26.EPS

Figure 26 Shackles.

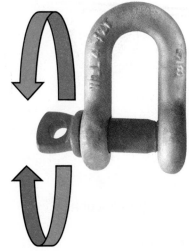

Figure 27 Wide-body shackle.

Synthetic web sling shackles (*Figure 28*) are designed with a wide throat opening and a wide bow that is contoured to provide a larger, nonslip surface area to accommodate the wider body of synthetic web slings.

When a screw pin shackle is used, it is important to tighten the pin the proper amount. The pin should be screwed in until it is fully engaged and the shoulder of the pin is in contact with the shackle body (*Figure 29*). If the pin is left loose, the shackle can stretch under load. If the pin is over-tightened, the load stress from the sling can torque the pin even more. This can stretch and jam the pin's threads.

When any type of shackle is used in a rigging operation, it must be positioned so that the load is centered in the bow of the shackle and not on the shackle pin (*Figure 30*). If the shackle uses a screw pin and the load is positioned on the pin, the shifting of the load could cause the pin to twist and unscrew.

Shackles, like any other type of hardware, must be inspected by the rigger before each use to make sure there are no defects that would make the shackle unsafe. Each lift may cause some degree of damage or may further reveal existing damage.

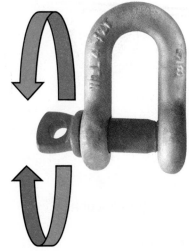

Figure 29 Don't over-tighten the shackle pin.

If any of the following conditions exists, a shackle must be removed from use:

- Bends, cracks, or other damage to the shackle body
- Incorrect shackle pin or improperly substituted pin
- Bent, broken, or loose shackle pin
- Damaged threads on threaded shackle pin
- Missing or illegible capacity and size markings

1.3.2 Eyebolts

An eyebolt is an item of rigging hardware with a threaded shank. The eyebolt's shank end is attached directly to the load, and the eyebolt's eye end is used to attach a sling to the load.

Eyebolts for overhead lifting should be made of drop- or hammer-forged steel. Eyebolts are available in three basic designs with several variations (*Figure 31*).

Figure 28 Synthetic web sling shackle.

Figure 30 Shackle correctly positioned.

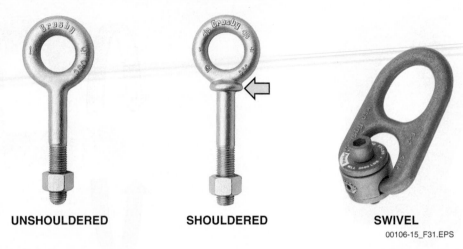

UNSHOULDERED   **SHOULDERED**   **SWIVEL**

00106-15_F31.EPS

Figure 31 Eyebolt variations.

Unshouldered eyebolts are designed for straight vertical pulls only. Shouldered eyebolts have a shoulder that is used to help support the eyebolt during pulls that are slightly angular. Swivel eyebolts, also called hoist rings, are designed for angular pulls from 0 to 90 degrees from the horizontal plane of the load.

In a few cases, eyebolts must be torqued to a specific value when they are installed for lifting use.

1.3.3 Lifting Clamps

Lifting clamps are used to move loads such as steel plates, sheet piles, large pipe, or concrete panels without the use of slings. These rigging devices are designed to bite down on a load and use the jaw tension to secure the load (*Figure 32*).

All lifting clamps must be made of forged steel, and they must be stamped with their rated capacity. Some clamps, such as the one in *Figure 33*, use the weight of the load to produce and sustain the clamping pressure; the grip tightens as the load increases. Others use an adjustable cam that is set and tightened to maintain a secure grip on the load.

Lifting clamps are designed to carry one item at a time, regardless of the capacity or jaw dimensions of the clamp or the thickness or weight of the item being lifted. In order for the clamp to hold an item securely, the cam and the jaw must bite or grip both sides of a single item. Placing more than one plate or sheet into the clamp prevents both the cam and the jaw from securing both sides of the plate or sheet. The clamps must be placed to ensure that the load remains balanced. For larger loads, two lifting clamps are typically required.

Did You Know?

Shackles and Pins

Pins should never be swapped in shackles. The thread forms of different brands of shackles and different brands of pins may not engage properly with one another. In addition, there is no easy way to tell how much reserve strength is left in a pin or a shackle. Further, the rated capacities of different shackles and pins may not be compatible. The shackle pin should be placed and secured into a single shackle between uses.

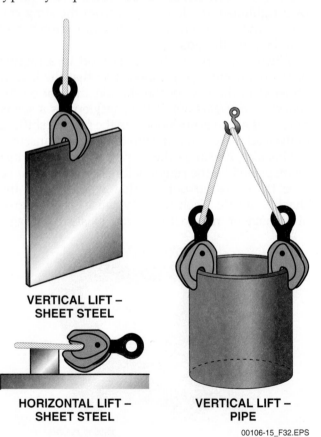

VERTICAL LIFT –
SHEET STEEL

HORIZONTAL LIFT –
SHEET STEEL

VERTICAL LIFT –
PIPE

00106-15_F32.EPS

Figure 32 Using lifting clamps.

NCCER – *Core Curriculum*   00106-15

Figure 33 Basic nonlocking clamp.

Lifting clamps are available in a wide variety of designs, so it is important to match the type of clamp with the intended application. *Figure 34* shows some lifting clamps designed for specific uses.

Lifting clamps, like any other type of rigging hardware, must be inspected by the rigger before every use to make sure there are no defects that would make the clamp unsafe. Each lift may cause some degree of damage or may further reveal existing damage.

If any of the following conditions exists, a clamp must be removed from use (*Figure 35*):

- Cracks
- Abrasion, wear, or scraping
- Any deformation or other impact damage to the shape that is detectable during a visual examination

- Excessive rust or corrosion, meaning rust or corrosion that cannot be removed easily with a wire brush
- Excessive wear of the teeth
- Heat damage
- Loose or damaged screws or rivets
- Worn springs

1.3.4 *Rigging Hooks*

A **rigging hook** is an item of rigging hardware used to attach a sling to a load. There are many classes of rigging hooks used, but most rigging hooks fall into a few basic categories (*Figure 36*). Most rigging hooks have safety latches or gates to prevent slings or other connectors from accidentally coming out of the hook during use. Hooks that do not have safety latches are usually designed so that a latch can be added.

- Eye hooks are the most common type of end fitting hook. A rigging eye hook has a large eye that can accommodate large couplers.
- Reverse eye hooks position the point of the hook perpendicular to the eye.
- Sliding choker hooks are installed onto the sling when it is made. The hooks, which can be positioned anywhere along the sling body, are used to secure the sling eye in a choker hitch. Sliding choker hooks are available for steel chain slings.
- Grab hooks are used on steel chain slings. These hooks fit securely in the chain link, so that choker hitches can be made and chains can be shortened.
- Shortening clutches, a more efficient version of the grab hooks, provide a secure grab of the shortened sling leg with no reduction in the capacity of the chain because the clutch fully supports the links.

LOCKING CLAMP SCREW-ADJUSTED CLAMP NON-MARRING

00106-15_F34.EPS

Figure 34 Lifting clamps.

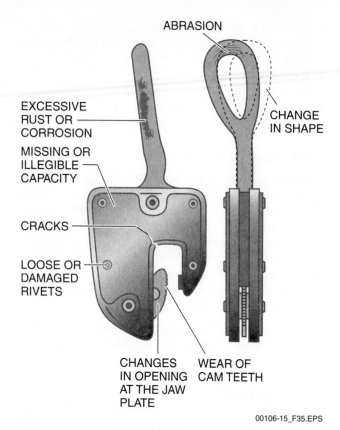

ABRASION

CHANGE
IN SHAPE

EXCESSIVE
RUST OR
CORROSION

MISSING OR
ILLEGIBLE
CAPACITY

CRACKS

LOOSE OR
DAMAGED
RIVETS

CHANGES
IN OPENING
AT THE JAW
PLATE

WEAR OF
CAM TEETH

00106-15_F35.EPS

Figure 35 Rejection criteria for lifting clamps.

Hooks used for rigging must be made of drop- or hammer-forged steel. Although most rigging hooks have safety latches or gates, some hooks used for special applications may not have them. If a safety latch is installed in a rigging hook, the latch must be in good working condition. Damaged safety latches can be easily replaced. Any damage to a safety latch must be reported to an instructor or supervisor.

When hooks are installed as end fittings, they must be inspected along with the rest of the sling before each use. Slings with hook-type end fittings need to be removed from service for any of the following defects (*Figure 37*):

- Wear, scraping, or abrasion
- Cracks
- Cuts, gouges, nicks, or chips
- Excessive rust or corrosion, meaning rust or corrosion that cannot be easily removed with a wire brush
- An increase in the throat opening of the hook—easy to detect if the hook is equipped with a safety latch, because the latch will no longer bridge the throat opening
- A twist in the hook
- An elongation of the hook
- A broken or missing safety latch

1.4.0 Hoists

A hoist is a device that uses pulleys or gears to provide a mechanical advantage for lifting a load, allowing objects to be lifted that cannot be lifted manually. Some hoists are mounted on trolleys and use electricity or compressed air for power. In this section, a simple hoisting mechanism called a **block and tackle**, and a more complex hoisting mechanism called a chain hoist, (*Figure 38*) are discussed.

A block and tackle is a simple rope-and-pulley system used to lift loads. By using fixed pulleys and a wire rope attached to a load, a rigger can raise and lower the load by pulling the rope or by using a winch to lift the load. A block and tackle assembly multiplies the power applied, so that a worker can lift a much heavier load than could be lifted without it.

Chain hoists may be operated manually or mechanically. There are three types of chain hoists (*Figure 39*): manual, electric, and pneumatic. Because electric and pneumatic chain hoists use mechanical power, they are known as powered chain hoists. All chain hoists use a gear system to lift heavy loads (*Figure 40*). The gearing is coupled to a sprocket that has a chain with a hook attached to it. The load is hooked onto a chain and the gearing turns the sprocket, causing the chain to travel over the sprocket and move the load. The hoist can be suspended by a hook connected to an appropriate anchorage point, or it can be suspended from a trolley system (*Figure 41*).

1.4.1 Operation of Chain Hoists

Chain hoists are operated by hand or by electric or pneumatic power. Some of the fundamental operating procedures for hand chain hoists and powered chain hoists are as follows:

- *Hand chain hoist* – To use a hand chain hoist, the rigger suspends the hoist above the load to be lifted, using either the suspension hook or the trolley mount. The rigger then attaches the hook to the load and pulls the hand chain drop to raise the load. The load will either rise or fall, depending on which side of the chain drop is pulled.
- *Powered chain hoist* – To use an electric or a pneumatic powered chain hoist, the rigger positions the chain hoist on the trolley above the load to be lifted, attaches the hook to the load, and uses the control pad to operate the hoist. Only qualified persons may use powered chain hoists.

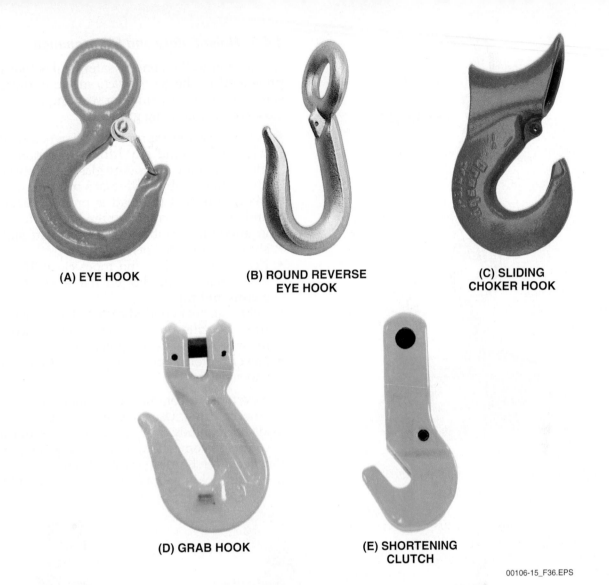

Figure 36 Rigging hooks.

(A) EYE HOOK

(B) ROUND REVERSE EYE HOOK

(C) SLIDING CHOKER HOOK

(D) GRAB HOOK

(E) SHORTENING CLUTCH

00106-15_F36.EPS

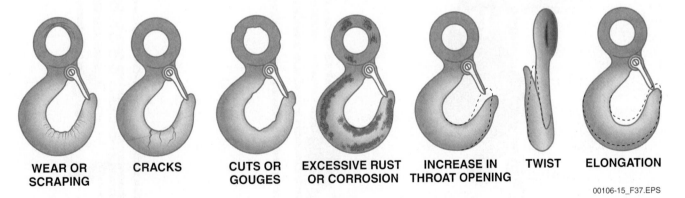

WEAR OR SCRAPING

CRACKS

CUTS OR GOUGES

EXCESSIVE RUST OR CORROSION

INCREASE IN THROAT OPENING

TWIST

ELONGATION

00106-15_F37.EPS

Figure 37 Common rigging hook defects.

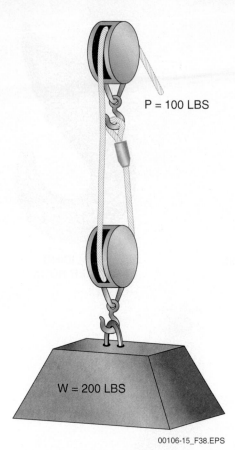

$P = 100 \text{ LBS}$

$W = 200 \text{ LBS}$

00106-15_F38.EPS

Figure 38 Block and tackle hoist system.

1.4.2 Hoist Safety and Maintenance

In addition to the core general safety rules that were presented in the *Basic Safety* module, there are some specific safety rules for working with hoists. Observe the following guidelines:

- Always use the appropriate PPE when working with and around any lifting operations. This includes a hard hat, safety glasses, safety shoes, and a high-visibility vest (at a minimum).
- Make sure that the load is properly balanced and attached correctly to the hoist before a lift is attempted. Unbalanced loads can slide or shift, causing the hoist to fail.
- Keep gears, chains, and ropes clean. Improper maintenance can shorten the working life of chains and ropes.
- Lubricate gears periodically to keep the wheels from freezing up. All such maintenance work must be done by a competent person.
- Never perform a lift of any size without proper supervision.

Never use a common come-along for vertical overhead lifting. Use a come-along only to move loads horizontally over the ground. Be careful not to confuse a come-along with a ratchet lever hoist. Ratchet lever hoists (*Figure 42*) have both a friction-type holding brake and a ratchet-and-pawl

ELECTRIC

PNEUMATIC

MANUAL

00106-15_F39.EPS

Figure 39 Types of chain hoists.

NCCER – *Core Curriculum* 00106-15

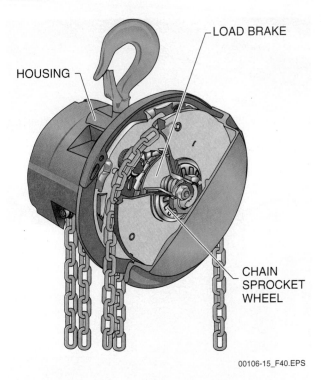

Figure 40 Chain hoist gear system.

load control brake. Cable come-alongs have only a spring-load ratchet that holds the pawl in place to secure the cable. If the ratchet and pawl fail, an overhead load falls. Be certain that any device used to lift is designed and specified for that task. Among manufacturers, the terms *come-along* and *ratchet lever hoist* are often used in the description of a single item, creating confusion in the marketplace and on the job. In most cases, cable come-alongs are not designed for lifting loads.

1.5.0 Hitches

In a rigging operation, the link between the load and the lifting device is often a sling made of synthetic, alloy steel chain, or wire rope material. The way the sling is arranged to hold the load is called the rigging configuration, or hitch. Hitches can be made using just the sling or by using connecting hardware, as well. There are three basic types of hitches:

- Vertical
- Choker
- Basket

One of the most important parts of the rigger's job is making sure that the load is held securely. The type of hitch the rigger uses depends on the type of load to be lifted. Different hitches are used to secure, for example, a load of pipes, a load of concrete slabs, or a load of heavy machinery.

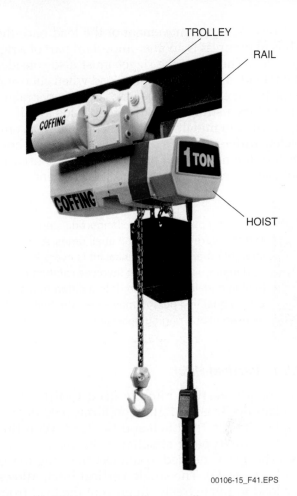

Figure 41 Hoist suspended from a trolley system.

RATCHET LEVER HOIST

Figure 42 Use ratchet lever hoists for vertical lifting.

Controlling the movement of the load once the lift is in progress is another important part of a rigger's job. Therefore, the rigger must also consider the intended movement of the load when choosing a hitch. For example, some loads are lifted straight up and then straight down. Other loads are lifted up, turned in midair, and then set down in a completely different place. This section examines how each of the three basic types of hitches is used to both secure a load and control its movement.

> **WARNING!**
>
> All rigging operations are dangerous, and extreme care must be used at all times. A straight up-and-down vertical lift is every bit as dangerous as a lift that involves rotating a load in midair and moving it to a different place. Only a qualified person may select the hitch to be used in any rigging operation.

1.5.1 Vertical Hitch

The single vertical hitch is used to lift a load straight up. This configuration forms a 90-degree angle between the hitch and the load. With this hitch, some type of attachment hardware, such as a shackle, is needed to connect the sling to the load (*Figure 43*). The single vertical hitch allows the load to rotate freely. To prevent the load from rotating, some method of load control, such as a tag line, must be used.

Another classification of hitch is the bridle hitch (*Figure 44*). The bridle hitch consists of two or more vertical hitches attached to the same hook, master link, or **bull ring**. The bridle hitch allows the slings to be connected to the same load without the use of such devices as a spreader beam, which is a stiff bar used when lifting large objects with a crane hook.

The multiple-leg bridle hitch (*Figure 45*) consists of three or four single hitches attached to the same hook, master link, or bull ring. Multiple-leg bridle hitches provide increased stability for the load being lifted. A multiple-leg bridle hitch is always considered to have only two of the legs supporting the majority of the load and the rest of the legs balancing it.

1.5.2 Choker Hitch

The choker hitch is used when a load has no attachment points or when the attachment points are not practical for lifting. The choker hitch is made by wrapping the sling around the load and passing it through one eye to form a constricting

00106-15_F43.EPS

Figure 43 Single vertical hitch.

loop around the load. In many configurations, the sling is wrapped around the load and passed through a shackle to form the constricting loop. In those situations, it is important that the shackle used in the choker hitch be oriented properly, as shown in *Figure 46*. Remember that the pressure of the load should never be placed on the pin of the shackle.

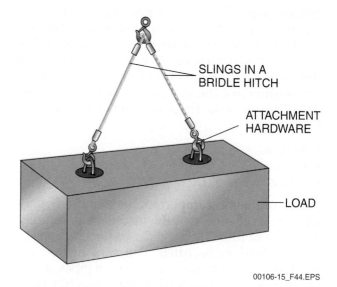

00106-15_F44.EPS

Figure 44 Bridle hitch.

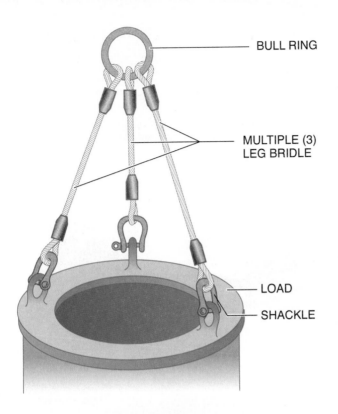

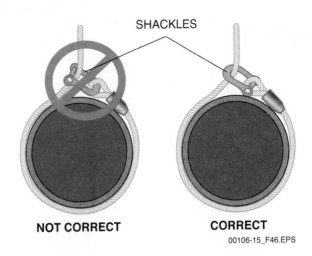

SHACKLES

NOT CORRECT   CORRECT

00106-15_F46.EPS

Figure 46 Choker hitches.

The choker hitch affects the capacity of the sling, reducing it by a minimum of 25 percent. This capacity reduction must be considered when choosing the proper sling. Further, the choker hitch does not grip the load securely. It is not recommended for loose bundles of materials because it tends to push loose items up and out of the choker. Many riggers use the choker hitch for bundles, mistakenly believing that forcing the choke down provides a tight grip. In fact, it serves only to drastically increase the stress on the choke leg (*Figure 47*).

When an item more than 12 feet (≈3.5 m) long is being rigged, the general rule is to use two choker hitches spaced far enough apart to provide the stability needed to transport the load (*Figure 48*).

To lift a bundle of loose items such as pipes and structural steel, or to maintain the load in a certain position during transport, double-wrap choker hitches (*Figure 49*) may be useful. A double-wrap choker hitch is made by wrapping the sling completely around the load, and then wrapping the choke end around again and connecting to the running end like a conventional choker hitch. This enables the load weight to produce a constricting action that binds the load into the middle of the hitch, holding it firmly in place throughout the lift.

Forcing the choke down will drastically increase the stress placed on the sling at the choke point. The double-wrap choker uses the load weight to provide the constricting force, so there is no need to force the sling down into a tighter choke; it serves no purpose. As with a single choker hitch, lifting a load longer than 12 feet (≈3.5 m) requires two double-wrap choker hitches.

BULL RING

MULTIPLE (3) LEG BRIDLE

LOAD

SHACKLE

00106-15_F45.EPS

Figure 45 Multiple-leg bridle hitch.

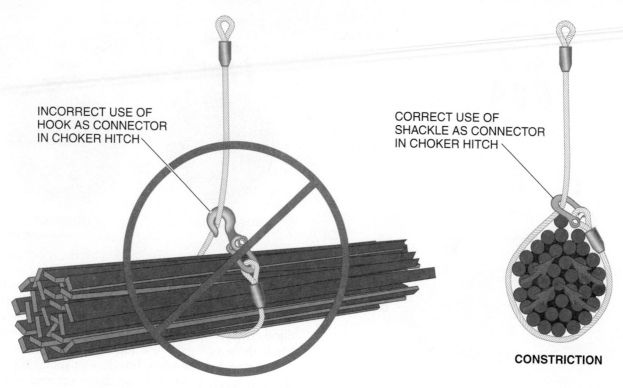

INCORRECT USE OF
HOOK AS CONNECTOR
IN CHOKER HITCH

CORRECT USE OF
SHACKLE AS CONNECTOR
IN CHOKER HITCH

CONSTRICTION

00106-15_F47.EPS

Figure 47 Choker hitch constriction.

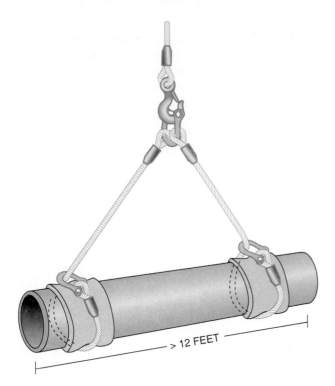

1.5.3 Basket Hitch

Basket hitches are very versatile and can be used to lift a variety of loads. A basket hitch is formed by passing the sling around the load (or, in some cases, through the load) and placing both eyes in the hook (*Figure 50*). Placing a sling into a basket hitch has the effect of doubling the capacity of the sling. This is because the basket hitch creates two sling legs from one sling.

A double-wrap basket hitch combines the constricting power of a double-wrap choker hitch with the capacity advantages of a basket hitch. This means it is able to hold a larger load more tightly. A double-wrap basket hitch requires a considerably longer sling length than a double-wrap choker hitch. If it is necessary to join two or more slings together, the load must be in contact with the sling body only, not with the hardware used to join the slings. The double-wrap basket hitch provides support around the load. As is true of the double-wrap choker hitch, the load weight provides the constricting force for the hitch.

> **CAUTION**
>
> A basket hitch should not be used to lift loose materials. Loads placed in a basket hitch should be balanced.

1.5.4 The Emergency Stop Signal

A major aspect of rigging safety involves maintaining clear communication between the crane operator and the designated signal person on the ground. This communication is normally accomplished using common signals—either verbal signals given by radio or hand signals. Hand signals used in rigging operations in the United States have been developed and standardized by the American Society of Mechanical Engineers (ASME) for all cranes. With the exception of the Emergency Stop signal, hand signals can only be given by the designated signal person on the ground.

00106-15_F48.EPS

Figure 48 Double choker hitches.

The Emergency Stop signal used in rigging operations is shown in *Figure 51*. In the event of an emergency, this signal can be given by anyone on the ground within sight of the crane operator. The Emergency Stop signal is made by extending both arms horizontally with palms down, and then quickly moving the arms back and forth by repeatedly extending and retracting them.

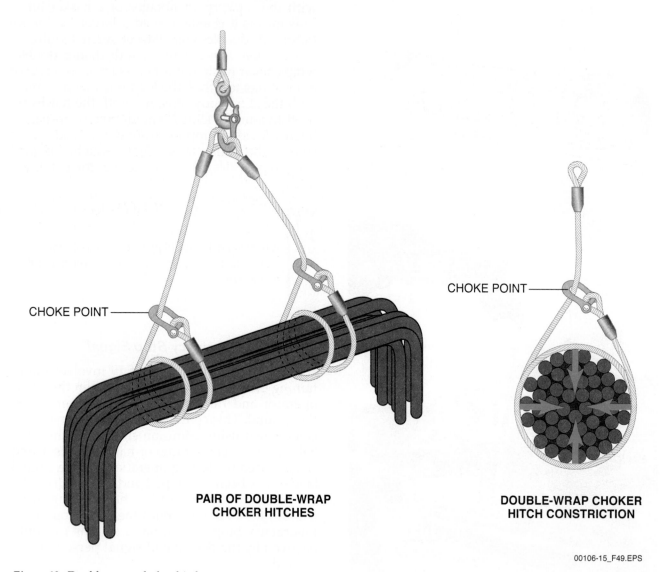

CHOKE POINT

PAIR OF DOUBLE-WRAP CHOKER HITCHES

CHOKE POINT

DOUBLE-WRAP CHOKER HITCH CONSTRICTION

00106-15_F49.EPS

Figure 49 Double-wrap choker hitches.

Figure 50 Basket hitch.

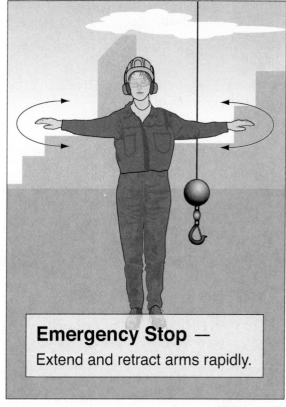

Emergency Stop —
Extend and retract arms rapidly.

00106-15_F51.EPS

Figure 51 Emergency Stop hand signal.

Additional Resources

Bob's Rigging and Crane Handbook, Bob De Benedictis. 2006. Leawood, KS: Pellow Engineering Services, Inc.

Mobile Crane Manual, Donald E. Dickie, D. H. Campbell. 1999. Toronto, Ontario, Canada: Construction Safety Association of Ontario.

Rigging Handbook, Jerry A. Klinke. 2012. Stevensville, MI: ACRA Enterprises, Inc.

Rigging Manual, 2005. Toronto, Ontario, Canada: Construction Safety Association of Ontario.

Rigging, James Headley. 2012. Sanford, FL: Crane Institute of America, Inc.

1.0.0 Section Review

1. Slings are commonly made out of _____
   a. synthetic fibers, cast-iron links, and wire rope
   b. fiberglass strands, exotic metals, and synthetic rope
   c. synthetic material, alloy steel, and wire rope
   d. carbon fiber, stainless steel, and twisted pair cable

2. If the identification tag on a synthetic sling is missing or illegible, the sling must be _____.
   a. re-tagged by a qualified rigger on site
   b. removed from service immediately
   c. used only for minimal loads
   d. tested on a load to determine its capacity

3. The size of a shackle is determined by the diameter of its _____.
   a. body
   b. shank
   c. eye
   d. pin

4. A type of rigging equipment that uses a pulley system to provide a mechanical advantage for lifting a load is a _____.
   a. fulcrum
   b. hitch
   c. hoist
   d. come-along

5. When a load has no attachment points or when the attachment points are not practical for lifting, it is best to use a _____.
   a. bridle hitch
   b. vertical hitch
   c. multiple-leg hitch
   d. choker hitch

SUMMARY

Although rigging operations are complex procedures that can present many dangers, a lift executed by fully trained and qualified rigging professionals can be a rewarding operation to watch or participate in. In order to accomplish a successful rigging operation, workers need to have a comprehensive understanding of the equipment used in rigging as well as some basic procedures for connecting and lifting loads. This module provided an overview of various types of rigging equipment and some common hitch configurations used for moving material and equipment from one location to another. Understanding these basic rigging standards will help provide the groundwork for a safe, productive, and rewarding construction career.

1. Identification tags for slings must include the _____.

   a. type of protective pads to use
   b. type of damage sustained during use
   c. color of the tattle-tail
   d. manufacturer's name or trademark

2. The type of wire rope core that is susceptible to heat damage at relatively low temperatures is the _____.

   a. fiber core
   b. strand core
   c. independent wire rope core
   d. metallic link supporting core

3. Synthetic slings must be inspected _____.

   a. once every month
   b. visually at the start of each work week
   c. before every use
   d. once wear or damage becomes apparent

4. An alloy steel chain sling must be removed from service if there is evidence that _____.

   a. the sling has been used in different hitch configurations
   b. replacement links have been used to repair the chain
   c. the sling has been used for more than one year
   d. strands in the supporting core have weakened

5. A piece of rigging hardware used to couple the end of a wire rope to eye fittings, hooks, or other connections is a(n) _____.

   a. eyebolt
   b. hitch
   c. shackle
   d. U-bolt

6. A lifting clamp is most likely to be used to move loads such as _____.

   a. steel plates
   b. piping bundles
   c. concrete blocks
   d. plastic tubing

7. Chain hoists are able to lift heavy loads by utilizing a _____.

   a. rope and pulley system
   b. rigger's strength
   c. stationary counterweight
   d. gear system

8. Before attempting to lift a load with a chain hoist, make sure that the _____.

   a. hoist is secured to a come-along
   b. load is properly balanced
   c. tag lines are properly anchored
   d. tackle is connected to its power source

9. A hitch configuration that allows slings to be connected to the same load without using a spreader beam is a _____.

   a. double-wrap hitch
   b. choker hitch
   c. bridle hitch
   d. basket hitch

10. To make the Emergency Stop signal that is used by riggers, extend both arms _____.

   a. horizontally with palms down and quickly move both arms back and forth
   b. directly in front and then move both arms up and down repeatedly
   c. vertically above the head and wave both arms back and forth
   d. horizontally with clenched fists and move both arms up and down

Trade Terms Quiz

Fill in the blank with the correct term that you learned from your study of this module.

1. A simple rope-and-pulley system called a(n) _____ is used to lift loads.

2. Use a single ring called a(n) _____ to attach multiple slings to a hoist hook.

3. The _____ is the distance between the master link of the sling to either the end fitting of the sling or the lowest point on the basket.

4. The way the sling is arranged to hold the load is called the rigging configuration, or _____.

5. A(n) _____ hitch uses two or more slings to connect a load to a single hoist hook.

6. A(n) _____ uses a pulley system to give you a mechanical advantage for lifting a load.

7. A(n) _____ is used to move loads such as steel plates or concrete panels without the use of slings.

8. The _____ is the total amount of what is being lifted.

9. The tension applied on the rigging by the weight of the suspended load is called the _____.

10. The _____ is the main connection fitting for chain slings.

11. To form an endless-loop web sling, _____ the ends together.

12. _____ equals the lengthwise distance it takes for one strand of wire to make one complete turn around the core.

13. Standard eye-and-eye slings have eyes on the same _____, whereas twisted eye-and-eye slings have eyes at right angles to each other.

14. The _____ is the link between the load and the lifting device.

15. The maximum load weight that a sling is designed to carry is called its _____.

16. Examples of _____ for synthetic slings include a missing identification tag, a puncture, and crushing.

17. Use a(n) _____ to attach a sling to a load.

18. Often found on a crane, a grooved pulley-wheel for changing the direction of a rope's pull is called a(n) _____.

19. The parts of the sling that reach from the attachment device around the object being lifted are called the _____.

20. A(n) _____ is a group of wires wound around a center core.

21. Riggers use a(n) _____ to limit the unwanted movement of the load when the crane begins moving.

22. If the _____ is showing, the sling is not safe for use.

23. A wire rope sling consists of high-strength steel wires formed into strands wrapped around a supporting _____.

24. An eyebolt is a piece of rigging hardware with a(n) _____, which means it has a series of spiral grooves cut into it.

25. An individual that has a college degree in Safety Technology would be considered a _____ in the field of safety.

26. To prevent the load from rotating freely, you must use some method of _____.

27. _____ slings are made of high-strength steel wires formed into strands wrapped around a core.

28. A(n) _____ is used to determine if an endless sling has been overloaded.

29. If a cable sling is found with the strands become untwisted in a section, the problem is referred to as _____.

30. A worker capable of identifying existing or potential hazards in the work area and authorized to take steps to eliminate such hazards is referred to as a(n) _____.

Trade Terms

Block and tackle
Bridle
Bull ring
Competent person
Core
Hitch
Hoist
Lifting clamp

Load
Load control
Load stress
Master link
One-rope lay
Plane
Qualified person
Rated capacity

Rejection criteria
Rigging hook
Shackle
Sheave
Sling
Sling legs
Sling reach
Splice

Strand
Tag line
Tattle-tail
Threaded shank
Unstranding
Warning yarn
Wire rope

John Stronkowski
Industrial Management & Training Institute
Director of Education

How did you get started in the construction industry?
I worked in the trades to pay my way through college. At that time, it was a means to an end.

Who or what inspired you to enter the industry? Why?
My family members worked in various trades. My grandfather was a master carpenter and that served as an inspiration to me.

What do you enjoy most about your career?
I enjoy the teaching aspect the most. It allows me to share my knowledge and experience, and hopefully impact the lives of my students.

Why do you think training and education are important in construction?
Properly trained people in the trades are crucial to the safety and welfare of the general public. Mistakes and shortcuts have significant costs beyond money.

Why do you think credentials are important in construction?
Documented credentials represent a properly trained tradesperson who is up-to-date with today's technology.

How has training/construction impacted your life and career?
In my dual role as a multiple-licensed tradesperson and a Roman Catholic priest, the construction field has allowed me to be in charge of all major construction projects for the Diocese of Bridgeport, CT. With my vast knowledge of different yet related systems, I am able to provide added benefits to the church without extraordinary costs to our parishioners.

Would you recommend construction as a career to others? Why?
Yes, definitely. The trades offer highly rewarding careers. When you start a construction job and see it through to the end, it gives you a sense of accomplishment.

What does craftsmanship mean to you?
It is professionalism at its highest level and being proud of the work accomplished.

Trade Terms Introduced in This Module

Block and tackle: A simple rope-and-pulley system used to lift loads.

Bridle: A configuration using two or more slings to connect a load to a single hoist hook.

Bull ring: A single ring used to attach multiple slings to a hoist hook.

Competent person: An individual capable of identifying existing and predictable hazards in the surroundings and working conditions which are unsanitary, hazardous, or dangerous to employees; an individual who is authorized to take prompt corrective measures to eliminate these issues.

Core: Center support member of a wire rope around which the strands are laid.

Hitch: The rigging configuration by which a sling connects the load to the hoist hook. The three basic types of hitches are vertical, choker, and basket.

Hoist: A device that applies a mechanical force for lifting or lowering a load.

Lifting clamp: A device used to move loads such as steel plates or concrete panels without the use of slings.

Load: The total amount of what is being lifted, including all slings, hitches, and hardware.

Load control: The safe and efficient practice of load manipulation, using proper communication and handling techniques.

Load stress: The strain or tension applied on the rigging by the weight of the suspended load.

Master link: The main connection fitting for chain slings.

One-rope lay: The lengthwise distance it takes for one strand of a wire rope to make one complete turn around the core.

Plane: A surface in which a straight line joining two points lies wholly within that surface.

Qualified person: A person who, through the possession of a recognized degree, certificate, or professional standing, or one who has gained extensive knowledge, training, and experience, has successfully demonstrated his or her ability to solve problems relating to the subject matter, the work, or the project.

Rated capacity: The maximum load weight a sling or piece of hardware or equipment can hold or lift; also referred to as the working load limit (WLL).

Rejection criteria: Standards, rules, or tests on which a decision can be based to remove an object or device from service because it is no longer safe.

Rigging hook: An item of rigging hardware used to attach a sling to a load.

Shackle: Coupling device used in an appropriate lifting apparatus to connect the rope to eye fittings, hooks, or other connectors.

Sheave: A grooved pulley-wheel for changing the direction of a rope's pull; often found on a crane.

Sling: Wire rope, alloy steel chain, metal mesh fabric, synthetic rope, synthetic webbing, or jacketed synthetic continuous loop fibers made into forms, with or without end fittings, used to handle loads.

Sling legs: The parts of the sling that reach from the attachment device around the object being lifted.

Sling reach: A measure taken from the master link of the sling, where it bears weight, to either the end fitting of the sling or the lowest point on the basket.

Splice: To join together.

Strand: A group of wires wound, or laid, around a center wire, or core. Strands are laid around a supporting core to form a rope.

Tag line: Rope that runs from the load to the ground. Riggers hold on to tag lines to keep a load from swinging or spinning during the lift.

Tattle-tail: Cord attached to the strands of an endless loop sling. It protrudes from the jacket. A tattle-tail is used to determine if an endless sling has been stretched or overloaded.

Threaded shank: A connecting end of a fastener, such as a bolt, with a series of spiral grooves cut into it. The grooves are designed to mate with grooves cut into another object in order to join them together.

Unstranding: Describes wire rope strands that have become untwisted. This weakens the rope and makes it easier to break.

Warning yarn: A component of the sling that shows the rigger whether the sling has suffered too much damage to be used.

Wire rope: A rope made from steel wires that are formed into strands and then laid around a supporting core to form a complete rope; sometimes called cable.

Additional Resources

This module presents thorough resources for task training. The following resource material is suggested for further study.

Bob's Rigging and Crane Handbook, Bob De Benedictis. 2006. Leawood, KS: Pellow Engineering Services, Inc.

Mobile Crane Manual, Donald E. Dickie; D. H. Campbell. 1999. Toronto, Ontario, Canada: Construction Safety Association of Ontario.

Rigging Handbook, Jerry A. Klinke. 2012. Stevensville, MI: ACRA Enterprises, Inc.

Rigging Manual, 2005. Toronto, Ontario, Canada: Construction Safety Association of Ontario.

Rigging, James Headley. 2012. Sanford, FL: Crane Institute of America, Inc.

Figure Credits

Section Review Answer Key

Answer	Section Reference	Objective
Section One		
1. c	1.1.0	1a
2. b	1.2.1	1b
3. a	1.3.1	1c
4. c	1.4.0	1d
5. d	1.5.2	1e

NCCER CURRICULA — USER UPDATE

NCCER makes every effort to keep its textbooks up-to-date and free of technical errors. We appreciate your help in this process. If you find an error, a typographical mistake, or an inaccuracy in NCCER's curricula, please fill out this form (or a photocopy), or complete the online form at **www.nccer.org/olf**. Be sure to include the exact module ID number, page number, a detailed description, and your recommended correction. Your input will be brought to the attention of the Authoring Team. Thank you for your assistance.

Instructors – If you have an idea for improving this textbook, or have found that additional materials were necessary to teach this module effectively, please let us know so that we may present your suggestions to the Authoring Team.

NCCER Product Development and Revision

13614 Progress Blvd., Alachua, FL 32615

Email: curriculum@nccer.org
Online: www.nccer.org/olf

❑ Trainee Guide   ❑ Lesson Plans   ❑ Exam   ❑ PowerPoints   Other _____

Craft / Level: _____ Copyright Date: _____

Module ID Number / Title: _____

Section Number(s): _____

Description: _____

_____

_____

_____

_____

Recommended Correction: _____

_____

_____

_____

_____

Your Name: _____

Address: _____

_____

Email: _____ Phone: _____

00107-15

Basic Communication Skills

The construction professional communicates constantly. The ability to communicate skillfully will help to make you a better worker and a more effective leader. This module provides guidance in listening to understand, and speaking with clarity. It explains how to use and understand written materials, and it also provides techniques and guidelines that will help you to improve your writing skills.

Module Seven

Trainees with successful module completions may be eligible for credentialing through the NCCER Registry. To learn more, go to **www.nccer.org** or contact us at **1.888.622.3720**. Our website has information on the latest product releases and training, as well as online versions of our *Cornerstone* magazine and Pearson's product catalog.

Your feedback is welcome. You may email your comments to **curriculum@nccer.org,** send general comments and inquiries to **info@nccer.org**, or fill in the User Update form at the back of this module.

This information is general in nature and intended for training purposes only. Actual performance of activities described in this manual requires compliance with all applicable operating, service, maintenance, and safety procedures under the direction of qualified personnel. References in this manual to patented or proprietary devices do not constitute a recommendation of their use.

Objectives

When you have completed this module, you will be able to do the following:

1. Describe the communication, listening, and speaking processes and their relationship to job performance.
   a. Describe the communication process and the importance of listening and speaking skills.
   b. Describe the listening process and identify good listening skills.
   c. Describe the speaking process and identify good speaking skills.
2. Describe good reading and writing skills and their relationship to job performance.
   a. Describe the importance of good reading and writing skills.
   b. Describe job-related reading requirements and identify good reading skills.
   c. Describe job-related writing requirements and identify good writing skills.

Performance Tasks

Under the supervision of your instructor, you should be able to do the following:

1. Perform a given task after listening to oral instructions.
2. Fill out a work-related form provided by your instructor.
3. Read and interpret a set of instructions for properly donning a safety harness and then orally instruct another person on how to don the harness.

Trade Terms

Active listening
Appendix
Body language
Bullets
Change order
Electronic signature
Font
Glossary
Graph
Index

Italics
Jargon
Memo
Nonverbal communication
Paraphrase
Permit
Punch list
Table
Table of contents

Industry Recognized Credentials

If you are training through an NCCER-accredited sponsor, you may be eligible for credentials from NCCER's Registry. The ID number for this module is 00107-15. Note that this module may have been used in other NCCER curricula and may apply to other level completions. Contact NCCER's Registry at 888.622.3720 or go to **www.nccer.org** for more information.

Contents

Topics to be presented in this module include:

Figures

SECTION ONE

1.0.0 COMMUNICATION

Objective

Describe the communication, listening, and speaking processes and their relationship to job performance.

   a. Describe the communication process and the importance of listening and speaking skills.

   b. Describe the listening process and identify good listening skills.

   c. Describe the speaking process and identify good speaking skills.

Performance Task

   1. Perform a given task after listening to oral instructions.

Trade Terms

Active listening: A process that involves respecting others, listening to what is being said, and understanding what is being said.

Body language: A person's facial expression, physical posture, gestures, and use of space, all of which communicate feelings and ideas.

Jargon: Specialized terms used in a specific industry.

Nonverbal communication: All communication that does not use words. This includes appearance, personal environment, use of time, and body language.

Paraphrase: Express something heard or read using different words.

Every construction professional learns how to use tools. Depending on your trade, the tools you use could include welding machines and cutting torches, press brakes and plasma cutters, or surveyor's levels and pipe threaders. However, some of the most important tools you will use on the job are not tools you can hold in your hand or put in a toolbox. These tools are your abilities to read, write, listen, and speak.

At first, you might think that these are not really construction tools. They are things you already learned how to do in school, so why do you have to learn them all over again? The types of communication that take place in the construction workplace are very specialized and technical, just like the communications between pilots and air traffic control. Good communications result in a job done safely—a pilot hears and understands the message to change course to avoid a storm, and a construction worker hears and understands the message to install a water heater according to the local code requirements. In a way, you are learning another language, a special language that only trained professionals know how to use. Even though you will use a professional language that other people may not understand, the same communication skills apply to all professions, whether doctors, builders, managers, or mechanics.

The following are some specific examples of why these skills are so important in the construction industry:

- *Listening* – Your supervisor tells you where to set up safety barriers, but because you did not listen carefully, you missed a spot. As a result, your co-worker falls and is injured.
- *Speaking* – You must train two co-workers to do a new task, but you mumble, use words they don't understand, and don't answer their questions clearly. Your co-workers do the task incorrectly, and all of you must work overtime to fix the mistakes.
- *Reading* – Your supervisor tells you to read the manufacturer's basic operating and safety instructions for the new drill press before you use it. You don't really understand the instructions, but you don't want to ask him. You go ahead with what you think is correct and damage the drill press.
- *Writing* – Your supervisor asks you to write up a material takeoff (supply list) for a project. You rush through the list and don't check what you've written. The supplier delivers 250 feet of PVC piping cut to your specified sizes instead of 25 feet.

As you can see, good communication on the work site has a direct effect on safety, schedules, and budgets. A good communications toolbox is a badge of honor; it lets everyone know that you have important skills and knowledge. And like a physical toolbox, the ability to communicate well verbally and in writing is something that you can take with you to any job. You will find that good communications skills can help you advance your career. This module introduces you to the techniques you will need to read, write, listen, and speak effectively on the job.

1.1.0 The Communication Process

There are two basic steps to clear communication (*Figure 1*). First, a sender sends a spoken or written message through a communication channel to a receiver (examples of communication channels include meetings, phones, two-way radios, and email). When the receiver gets the message, he or she figures out what it means by listening or reading carefully. If anything is not clear, the receiver gives the sender feedback by asking the sender for more information.

This process is called two-way communication, and it is the most effective way to make sure that everyone understands what's going on. It sounds simple, doesn't it? So why is good communication so hard to achieve? When we try to communicate, a lot of things—called noise—can get in the way. Following are some examples of communication noise:

- The sender uses work-related words, or jargon, that the receiver does not understand.
- The sender does not speak clearly.
- The sender's written message is disorganized or contains mistakes.
- The sender is not specific.
- The sender does not get to the point.
- The receiver is tired or distracted or just not paying attention.
- The receiver has poor listening or reading skills.
- Actual noise on the construction site makes it physically hard to hear a message.
- There is a mechanical problem with the equipment used to communicate, such as static on a phone or radio line.

1.1.1 *Nonverbal Communication*

It is obvious that humans communicate with their words, known technically as verbal communication. If a mechanic asks for a ratchet set, he or she expects to get it because the words have a shared meaning between the listener and speaker. However, did you know that people communicate constantly without using any words at all? This kind of communication is called nonverbal communication. It is hard to communicate some kinds of information with nonverbal communication. For example, try asking for that ratchet set without any words. For expressing attitudes and emotion, nonverbal communication is a very powerful form of communication.

Imagine that you are talking to a doctor in a clinic. The whole time you are talking, he is yawning, checking his cell phone messages, and cleaning his fingernails. After you finish, he writes out a prescription. Do you trust the diagnosis and the value of the prescription? His nonverbal communication told you that he was bored and inattentive. Perhaps the doctor missed the most important thing you said about your condition, including an allergy to certain medications.

Similarly, you can express feelings and attitudes in a variety of ways without intention. You may show you are nervous by fidgeting in your chair or fiddling with your hands. You may show you are angry by raising your voice, folding your arms, and furrowing your eyebrows. You may show you are happy to see someone by smiling widely and giving him or her a warm handshake. The ways for a person to communicate nonverbally are limitless.

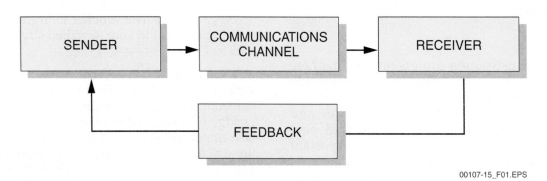

00107-15_F01.EPS

Figure 1 The communication process.

Did You Know?
Following Instructions Prevents Accidents

Many accidents are the result of not listening to or not understanding instructions. For example, according to a study by the US Occupational Safety and Health Administration (OSHA), over a 10-year period, 39 percent of crane operator deaths in the United States resulted from electrocution caused by contact with electrical power lines. This was the single largest cause of death in the study. How many of those accidents could have been prevented if the operator had heard, understood, and followed instructions? The answer is unknown, but it may have prevented all of them.

There are several basic categories of nonverbal communication:

- *Grooming* – Generally, people who maintain an attractive appearance have more successful careers. A groomed appearance communicates self-discipline and awareness. Shaving or keeping a trimmed beard could make the difference between getting promoted at work or being passed up by newcomers. Likewise, messy hair can communicate that someone is incompetent and unable to take on new responsibilities.
- *Dress* – Dress appropriately and neatly. Appropriate dress doesn't automatically mean formal dress. (Imagine working on a hot rooftop in a suit.) The best way to know how to dress in your work environment is to observe the people around you who are most successful, particularly the people in the position that you hope to obtain next. Even on occasions in which casual dress is allowed, such as a company outing to a baseball game, you must make sure that your casual clothing is clean and attractive. Casual does not mean messy. Many workplaces require a uniform. Of course, people working in these environments should always arrive at work with a neat and clean uniform.
- *Condition of one's personal environment* – People make their spaces their own by the way they arrange things even if the space does not belong to them. An office worker can show a sloppy work ethic by having binders, pens, paper clips, and candy wrappers strewn about his or her cubicle. A technician also might show a lack of responsibility by having gloves, coupons, magazines, and bags from fast food restaurants on the dashboard of his or her work vehicle. Whoever sees those environments unconsciously makes judgments about the character of the people who work there.
- *Use of time* – Clearly, people show respect and care by arriving on time or early to their scheduled events. Workers who show up late may be considered lazy or inconsiderate by those around them. Further, they may hinder the productivity of the entire organization by making others wait for them. Not only should you arrive or start on time, but when you hold meetings, you should also try to end on time. Doing so shows a humility on your part that others will appreciate.
- *Facial expressions* – One of the main ways that humans express and read emotions is through facial expressions. A smile can express interest and excitement. A frown could express displeasure or pain. An expressionless face expresses an emotion too, perhaps boredom or the desire to get away from whoever is talking. Of course, eyes are known as the window of the soul. Direct eye contact can express interest, understanding, intelligence, and confidence. On the other hand, a lack of eye contact could show inattention, a lack of confidence, or deceit. A person should not look into the eyes of another person the entire time while speaking, however. Too much eye contact could give the impression of initiating a challenge or trying to dominate. There is no solid rule for what amount of eye contact is appropriate. Different cultures have different rules. For example, in the West, authority figures expect subordinates to look at them when they are reprimanding them. In parts of Asia, however, etiquette requires the one being scolded to look down at the ground as a sign of respect. In the United States, people look at each other's eyes 50 to 60 percent of the time as they communicate.

- *Posture and gestures* – Slouching may feel comfortable, but it may also make people think of you as a sloppy person. Likewise, folding your arms in front of you may make you feel warmer on a cold day, but it will make you seem distant from others and unwilling to talk. Having a confident and powerful posture shows other people that you are confident and powerful. What if you feel timid and weak? Assume a powerful posture anyway. Research shows that people who kept a certain powerful posture for a few minutes—such as leaning slightly over a desk with their hands wide apart on the desk—experienced less stress and actually behaved more boldly when doing a task later. Those who had practiced low-power poses for a few minutes—such as folding their hands in their lap or touching their neck—showed a higher level of stress and behaved more timidly.
- *Physical distance* – Various cultures have different accepted personal distances. People of some countries stand close together while speaking, whereas others stand further apart. Also, different social situations call for a different personal range. People stand closer to their close friends and family members than they do to their business associates. Here is one rule of thumb to follow: let the more powerful person choose the distance. A manager or company owner may politely come and stand close to an employee, but an employee would be rude to do the same to his or her manager.

When it comes to safety and the accuracy of information communicated on the job, there is no doubt that what you say plays a greater role than the way you say it. However, beyond that, your nonverbal communication habits have a major effect on the quality of what is said and how it is received.

1.1.2 Listening and Speaking Skills

Every day on the job can be a learning experience. The more you learn, the more you will be able to help others learn, too (*Figure 2*). An effective method of learning and teaching is through verbal communication—that is, through speaking and listening. As a construction professional, you need to be able to state your ideas clearly. You also need to be able to listen to and understand ideas that other people express. The following are some of the ways that verbal teaching and learning take place on the job:

- *Giving and taking instructions* – One worker may read the steps in a calibration process while a second worker accomplishes the task.
- *Offering and listening to presentations* – Equipment manufacturers may visit the job site or offices to provide operating instruction for a new piece of equipment.
- *Participating in team discussions* – Safety is often discussed among teams; offer your input.

00107-15_F02.EPS

Figure 2 Teaching and learning are often accomplished by speaking and listening.

A Sense of Humor

A special tip: Maintain a sense of humor! A good sense of humor will get you through many situations. As a construction professional, you should always take yourself seriously. This means speaking well and conveying the proper professional attitude. But your listeners will always appreciate you more if you show them that you have a sense of humor and have a light side to you. Humor can diffuse tension and relieve frustration over things that aren't working properly. Remember, though, never to tell off-color or offensive jokes or make jokes at another person's expense. Also, never play practical jokes, as they can lead to accidents and injuries on the job.

Listening in the Classroom

When you are in the classroom, be aware of things that affect your ability to listen well. Take action to correct these problems. Is someone on the other side of the room speaking too softly? Ask other classmates to face the class when they speak and to speak loudly. Is there noise out in the hall? Ask permission to close the door to shut out noise from outside. Did your instructor say something you did not understand? Ask your instructor to explain things you don't understand.

- *Talking with your co-workers and your supervisor* – Listen carefully as they speak, without distracting yourself by thinking of a response too quickly. A slight pause in the conversation to prepare a response after a speaker is finished actually encourages others to listen more carefully to you.
- *Talking with clients* – Again, listening skills are critical.

Before we discuss some of the ways to become a more effective listener and speaker, evaluate your current speaking and listening skills by completing the self-assessment quizzes in *Figures 3* and *4*.

At this stage in your career, you will probably do more listening than speaking. You may be wondering why it is so important to be a good listener. The answer is simple: experience. People learn by listening, not by speaking. You are only beginning to learn how the construction industry works, and there is a lot to learn! Teachers, supervisors, and experienced workers can guide you to make sure you are learning what you need to know (*Figure 5*).

Are You a Good Listener?

Do you have good listening habits? Take the following self–assessment quiz to find out.
Be sure to answer each question honestly.

	Always	Sometimes	Rarely
1. I maintain eye contact when someone is talking to me.	☐	☐	☐
2. I pay attention when someone is talking to me.	☐	☐	☐
3. I ask questions when I don't understand something I hear.	☐	☐	☐
4. I take notes when receiving instructions.	☐	☐	☐
5. I repeat instructions my supervisor has given me to make sure I understand them.	☐	☐	☐
6. I nod my head or say I understand to show others I am listening to them.	☐	☐	☐
7. I let others speak without interrupting.	☐	☐	☐
8. I move to a quieter spot or ask someone to speak up if I am in a noisy location.	☐	☐	☐
9. I put aside what I am doing when someone is speaking to me.	☐	☐	☐
10. I listen with an open mind.	☐	☐	☐

00107-15_F03.EPS

Figure 3 Listening skills self-assessment.

Are You a Good Speaker?

How good are your speaking skills? This self-assessment quiz will help you understand your speaking strengths and weaknesses.

	Always	Sometimes	Rarely
1. When giving instructions to co-workers, I explain words they might not understand.	☐	☐	☐
2. When giving instructions for a task with several steps, I organize my thoughts first, then give the instructions.	☐	☐	☐
3. I give more details when explaining a task to inexperienced co-workers.	☐	☐	☐
4. When giving instructions to others, I try to keep from sounding like a know-it-all.	☐	☐	☐
5. When giving instructions to others, I encourage them to ask questions about anything they don't understand.	☐	☐	☐
6. I am patient and will explain instructions more than once if necessary.	☐	☐	☐
7. I try to speak more carefully when giving instructions to a co-worker for whom English is a second language.	☐	☐	☐
8. When speaking on the phone or over a two-way radio, I repeat instructions and spell out words when necessary.	☐	☐	☐
9. If someone asks me a question I don't know the answer to, I admit it and then try to find the answer.	☐	☐	☐
10. When I give instructions to others, I am confident, upbeat, and encouraging.	☐	☐	☐

00107-15_F04.EPS

Figure 4 Speaking skills self-assessment.

00107-15_F05.EPS

Figure 5 Your supervisor can help you learn what you need to know.

1.2.0 Active Listening on the Job

You might think that listening just happens automatically, that someone says something and someone else hears it. However, real listening, the process not only of hearing, but of understanding what is said, is an active process. You have to be involved and pay attention to really listen. In other words, understanding begins with active listening. You must develop good listening skills to be able to listen actively. This section presents some tips and suggestions that you can use to develop good listening skills.

First of all, you should understand the possible consequences of not listening. Poor listening skills can cause mistakes that waste time and money (*Figure 6*), and may even result in injury or loss of life. Recognize that other people may have

Figure 6 Listen carefully and ask questions to make sure you understand.

Figure 7 Body language shows whether you are paying attention.

important things to say. Even if what they are currently saying is boring, they may happen to mention something vitally important, like what to do when an alarm goes off.

Refuse to allow yourself to be distracted. If a work-related conversation or presentation is taking place, be there in the room listening, not darting around from place to place in your mind. If it helps you to stay focused, keep a pad of paper with you. When you are sidetracked by thoughts of something you need to do later, jot down a couple words about it. Then put the matter out of your mind.

One way to stay focused is to show that you are listening with your **body language**, that is your facial expressions, your posture, and your gestures (*Figure 7*). Not only does body language communicate something to other people, it affects you as well. It was mentioned earlier that behaving confidently can actually make you feel more confident. Similarly, acting like you are listening can actually help you listen better. While you listen, maintain eye contact and acknowledge what is being said by nodding or making appropriate sounds to show that you are paying attention.

While the other person is speaking, do not interrupt or criticize. Instead, wait until he or she has finished, and then ask plenty of questions, especially if he or she gave incomplete or unclear instructions. Asking questions can minimize misunderstandings and maximize your memory of the content. Of course, your mind will never remember perfectly, so take notes on what is said, especially if the content is technical in nature—dealing with numbers, times, amounts, temperatures, or other such data.

Finally, end a conversation by **paraphrasing** what you heard back to the conversation partner. Paraphrasing is the act of repeating what you heard in your own words. By using your own words, you can check the accuracy of your understanding. Therefore, it is important not to simply repeat what the person said exactly. The original speaker must listen carefully to ensure the paraphrased information is accurate.

Imagine how many hours of wasted work you can avoid by just confirming what you think you hear. For example, your supervisor may tell you, "Recalibrate the machines according to the specifications on this paper." You can respond by saying, "So you want me to recalibrate these three machines in this building according to the chart posted here." Then he or she replies, "Oh, no! I mean, recalibrate the machines like these in the building next door, and follow the chart on the back of this paper, since they are slightly different models." Just providing a simple summary in your own words of what was said could save time, money, or even someone's life. Paraphrase instructions whenever you can. It completes the act of listening just as installing a roof completes the construction of a building.

1.2.1 Barriers to Listening

As you have learned, listening well takes some work on your part. However, even when you have mastered effective listening skills, you will still have to overcome some barriers that will keep the message from getting through. Just like the warning signs on a construction site that indicate hazards to be avoided, there are signs that indicate problems in the process of listening and understanding (*Figure 8*). Some of those barriers are listed below, along with tips to overcome them:

00107-15_F08.EPS

Figure 8 Learn to read the warning signs of listening problems.

- *Emotion* – When angry or upset, humans tend to stop listening actively. Try counting to 10 or asking the speaker to excuse you for a minute. Go get a drink of water and calm down. When you are ready, come back and focus your mind on the task to be done.

- *Boredom* – Maybe the speaker is dull or over-bearing. Maybe you think you know it all already. There is no easy tip for overcoming this barrier. You just have to force yourself to stay focused. Keep in mind that the speaker has important information you need to hear.
- *Distractions* – Anything from too much noise and activity on the site to problems at home can steal your attention. If the problem is noise, ask the speaker to move away from it. If a personal problem is keeping you from listening, concentrate harder on staying focused. In some cases it may help if you explain to your supervisor why you are having trouble concentrating.
- *Your ego* – Do you finish people's sentences for them? Do you interrupt others a lot? Do you think about the things you are going to say next instead of listening? That is your ego putting itself squarely between you and effective listening. Be aware of your ego and try to tone it down a bit so you can get the information you need.

Ten Tips for Dealing with Conflict

Conflict can cause emotions to heat up, it can ruin relationships, and it can decrease productivity. Maybe you don't care to be friends with the person who bothers you at work. However, you still have to get along in order to accomplish your work. Allowing conflicts to get worse can only make your work less pleasurable and less effective. Here are some tips for dealing with conflicts at work:

- *Deal with it quickly.* While it is still a small problem, it will be easier to solve. Further, your frustration will not have increased to the breaking point.
- *Set aside a time to discuss the problem.* Dedicating a time later will help emotions cool and provide a neutral emotional atmosphere. You will also have more time to discuss the issue fully.
- *Discuss the problem with civility.* As the old saying goes, "A soft answer turns away wrath." Getting angry will only make the other person angry as well, escalating the situation.
- *Ask questions first.* If a co-worker had a reason for doing what made you angry, you should know what it is. Maybe just knowing why the person behaved that way will solve the problem.
- *Never say "never."* Do not say "always" either. "You never help me" is probably not true, and it will make the other person feel defensive. "You always criticize me" will also make the other person angry.
- *Always focus on specifics, not generalities.* Notice the difference between, "You are a careless person who hurts the whole team," and, "On the last two jobs we had, you forgot to bring the key for the building. We all had to wait while you went back for it. Is there anything you can do or I can do to make sure this doesn't happen again?"
- *Do not bring up things from the past that do not relate to the current situation.* Bringing up something that the co-worker did last year will only cause his or her temper to flare up.
- *Put yourself in the other person's shoes.* Understand what the other person needs in order to come out of the conflict happy. One of the main things people need is an intact sense of honor. If you publicly insult that person or force him or her with an overwhelming use of authority or even logic, you will implant a grudge that will inevitably hurt you later. Beyond giving the person the honor he or she needs, see if you can also compromise on some of the issues you disagree on.
- *Praise the person's good points.* This will show that you are not attacking his or her character, and it will help focus attention on the issue rather than the interpersonal tension.
- *Take the blame for your part of the problem.* Sincerely accepting some blame for the issue can make a huge difference. However, do not take the blame in a self-righteous or manipulative manner in order to shame the other person into submission.

1.3.0 Speaking on the Job

Although you can use the skills presented in this module to give a speech if necessary, the term *speaking skills* does not refer merely to your ability to give a speech or make a presentation to a group of people. It also refers to your ability to communicate effectively, one-on-one, to others on the job every day.

Effective listening depends on effective speaking. After all, you cannot be expected to understand what has not been made clear to you. Look at the following examples of sentences spoken by one worker to another. Which one is the clearest and most effective? Which one would you like to hear if you were the listener?

"Hand me that tool there."

"Hand me the grinder on that bench."

"Hand me the 4-inch angle grinder that's on the bench behind you."

The third example has enough information for you to identify the correct tool and its location. You do not have to ask the speaker, "Which tool? Where is it?" You will not accidentally give the other worker the wrong tool. As a result, time is not wasted trying to clear up confusion. The time it takes for someone to stop what he or she is doing and explain something again because it was not clear the first time is time that the job is not getting done. Time lost this way can add up quickly.

One of the best ways to learn to speak effectively is to listen to someone who speaks well. Think about what makes that person such an effective speaker. Is it the person's choice of words? Or perhaps their body language? Or their ability to make something complex sound simple? Keep the following things in mind when you speak, and they will make a difference for you:

- Think about what you are going to say before you say it, but not at the expense of listening actively.
- Follow the old advice about giving speeches, "Tell them what you are going to tell them. Tell them. Then tell them what you told them." This means you should introduce your topic with a brief summary, then share all of the detailed information, and finally end with a paraphrased summary of what you said.
- As with writing, take time to organize your ideas logically.
- Choose an appropriate place and time. For example, if you need to give detailed assembly instructions to your team, pick a quiet place, and do not hold the meeting just before lunch or the end of the shift.

- Encourage your listeners to take notes if necessary.
- Do not over-explain if people are already familiar with the topic.
- Always speak clearly, and maintain eye contact with the person or people you are speaking to.
- Do not talk on the phone, send text messages, or listen to music while communicating with the work crew.
- When using jargon be sure that everyone knows what the term means.
- Give your listeners enough time to ask questions, and take the time to answer questions thoroughly.
- When you are finished, make sure that everyone understands what you were saying.

Keep these things in mind when you speak, and they will make a difference for you.

1.3.1 Placing Telephone Calls

You may remember when telephones were anchored to walls and desks. To make or receive a phone call, you had to stop what you were doing and go to the telephone. Today, cell phones allow you to make and receive calls from just about anywhere. A cell phone can be a useful tool on the job site, but keep in mind the following guidelines:

- Cell phones can distract you from your job, so never make or receive personal calls while working.
- Wait until a designated break time to make or receive calls.
- Do not operate cell phones where they would pose a safety hazard, such as while operating a piece of machinery, a power tool, or driving a vehicle.

Let It Wait

Never make or receive phone calls while driving or operating heavy equipment. Distraction by cell phones is a leading cause of accidents. In 2013, a Spanish train derailed due to excess speed, killing 79 people just moments after the driver finished a work-related conversation on a cell phone. Subsequently, Spain banned cell phone use by train drivers except for emergency communication. Many areas of the United States likewise have laws banning cell phone use while operating heavy equipment, and even while behind the wheel of any motor vehicle.

- Be aware of the regulations regarding cell phone use at your workplace. While some companies allow cell phone use, many others do not even allow a phone to be in your possession. Companies make these rules to avoid accidents and wasted time.
- Recognize that cell phone cameras can be a potential threat to intellectual property. Workers may photograph secret equipment or data and sell the photographs, causing the company to lose its competitive edge. This is especially worth noting on projects for the government or military. Obviously, the military fiercely protects its secrets for good reason, and may ban employees and contractors from taking cell phones into certain areas.

When you speak to people face-to-face, you can see them and judge how they react to what you say. When you are on the telephone, you don't have these clues. Effective speaking is all the more important in such cases.

When making a call, keep the following points in mind:

- Start by identifying yourself and ask who you are speaking to.
- Speak clearly and explain the purpose of your call.
- Take notes to help you remember the conversation later (*Figure 9*).

00107-15_F09.EPS

Figure 9 Take notes to help you remember important details.

If you leave a message for someone, remember the following:

- Keep it brief.
- Prepare your message ahead of time so you will know what to say.
- Be sure to leave a number where you can be reached and the best time to reach you.

1.3.2 Receiving Telephone Calls

How you answer a phone call is just as important as how you place a phone call. Remember to be professional and courteous when answering your phone because you don't know who is going to be on the other end of the line. When you receive a phone call, remember the following guidelines:

- Don't just say "Hello." Identify yourself immediately by giving your name and the company name.
- Don't keep people on hold for long. People generally resent it. Instead, ask the caller if you can call back at a later time.
- Transfer calls courteously and introduce the caller to the recipient.
- Keep your calls brief.
- If the call is of a personal nature, continue working and do not answer. There are other opportunities throughout the day when you can return the call without interfering with the job or creating a safety hazard.
- Finally, never talk on the phone in front of co-workers, supervisors, or customers. This is usually considered rude and unprofessional.

Did You Know?

Cultural Interpretation

Communication is culturally diverse. Our experiences and surroundings, along with context, individual personality, and mood, help to form the ways we learn to speak and give nonverbal messages. For instance, in Europe the correct form for waving hello or goodbye is to keep your hand and arm stationary, palm out, with fingers moving up and down. In America, the common wave is the whole hand in motion moving back and forth. In Europe, the common American hand gesture for a wave means no, and in Greece it is considered an insult. It is important to remember that not all people communicate in the same way, so be aware of your surroundings when choosing your words and actions.

Recycling Used Cell Phones

Recycling your mobile devices (cell phones, pagers, PDAs) is a great way to save energy and protect the environment. These mobile devices contain toxic materials. Recycling keeps them out of landfills, cutting down on air and water pollution and greenhouse gas emissions. These devices contain precious metals, copper, and plastics, which require energy to manufacture and mine.

Over 100 million cell phones are retired each year. The EPA estimates that the energy saved from recycling cell phones alone could power more than 194,000 households for one year.

Your community benefits from cell phone donations. Cell phones can be reconditioned for reuse. They are often donated to charities and law enforcement, and for emergency use by individuals. Recycling is free and some organizations will even pay for your old cell phone.

Additional Resources

Tools for Success: Critical Skills for the Construction Industry, NCCER. 2009. Upper Saddle River, NJ: Pearson Education, Inc.

How to Win Friends and Influence People. Dale Carnegie. 2013. New York, NY: Simon & Schuster.

Listen Up: How to Improve Relationships, Reduce Stress, and Be More Productive by Using the Power of Listening. Larry Barker; Kittie Watson. 2000. New York: St. Martin's Press.

1.0.0 Section Review

1. Making meetings last longer than they were scheduled is a good way to show that you are a diligent person that wants to encourage others to work hard, too.

   a. True
   b. False

2. If the content of a speech or conversation is technical in nature, including lots of data such as numbers, be sure to _____.

   a. take notes
   b. ask others to summarize
   c. check to see if the information is true
   d. have the speaker repeat the key information

3. When is the best place and time to give assembly instructions to the team?

   a. In a quiet place when they are fresh and not too tired
   b. In the lunchroom while they are eating
   c. In the work area right at the end of their shift before they go home
   d. In the break room while people are getting snacks and watching TV

4. What of the following is banned in some workplaces because it can cause wasted time, accidents, and the theft of intellectual property?

   a. Pagers
   b. Watches
   c. Hearing aids
   d. Cellphones

5. When making a telephone call, start by _____.

   a. making small talk
   b. explaining the purpose of your call
   c. identifying yourself
   d. asking who you are talking to

SECTION TWO

2.0.0 READING AND WRITING

Objective

Describe good reading and writing skills and their relationship to job performance.

a. Describe the importance of good reading and writing skills.
b. Describe job-related reading requirements and identify good reading skills.
c. Describe job-related writing requirements and identify good writing skills.

Performance Tasks

2. Fill out a work-related form provided by your instructor.
3. Read and interpret a set of instructions for properly donning a safety harness and then orally instruct another person on how to don the harness.

Trade Terms

Appendix: A source of detailed or specific information placed at the end of a section, a chapter, or a book.

Bullets: Large, vertically aligned dots that highlight items in a list.

Change order: A written order by the owner of a project for the contractor to make a change in time, amount, or specifications.

Electronic signature: A signature that is used to sign electronic documents by capturing handwritten signatures through computer technology and attaching them to the document or file.

Font: The type style used for letters and numbers.

Glossary: An alphabetical list of terms and definitions.

Graph: Information shown as a picture or chart. Graphs may be represented in various forms, including line graphs and bar charts.

Index: An alphabetical list of topics, along with the page numbers where each topic appears.

Italics: Letters and numbers that lean to the right rather than stand straight up.

Memo: Informal written correspondence. Another term for memorandum (plural: *memoranda*).

Permit: A legal document that allows a task to be undertaken.

Punch list: A written list that identifies deficiencies requiring correction at completion.

Table: A way to present important text and numbers so they can be read and understood at a glance.

Table of contents: A list of book chapters or sections, usually located at the front of the book.

Imagine for a moment that you wake up one morning and head to work just like any other day, except for one difference: writing has not yet been invented. You grab a diagram (which of course has no notes on it) for installing a heating system, but you do not understand one part of the drawing. You head back to the office to ask the engineer what he meant. He starts to explain, but then he cannot clearly remember what he had discussed with the client. The discussion with the client was so long, and of course, there is no formal written contract or even an informal list of what the client wanted. Therefore, you and the engineer go to the client's office together to resolve the issue.

An existence without reading or writing may have worked for the Aztecs and Mayans, but clearly in our complex modern society, the written word is indispensable. Reading and writing is not just for academics. Almost all Americans read, if not write, dozens of times per day regardless of whether they are plumbers, professors, statisticians, or stay-at-home parents.

2.1.0 The Importance of Reading and Writing Skills

The construction industry, just like any other, depends on written materials of all kinds to carry on business, from routine office paperwork to construction drawings and building codes (*Figure 10*). Written documents allow workers to follow instructions accurately, help project managers ensure that work is on schedule, and enable the company to meet its legal obligations. As a construction professional, you need to be able to read and understand the written documents that apply to your work, whether they are printed in a book or letter, or on a computer. Further, as you gain professional experience, you will eventually take on responsibilities for writing documents.

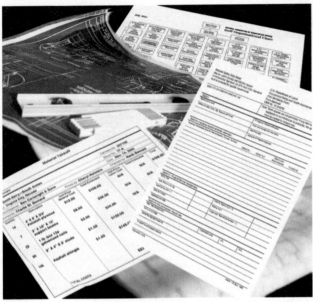

Figure 10 The written word is extremely important in the construction trade.

Other forms of written construction documents that you are very familiar with are textbooks and codebooks (*Figure 11*). A textbook is an instructional document that contains information that a reader needs to know to carry out a task or a series of tasks. A codebook is a guide that provides the codes and standards for certain areas of construction, such as electrical and building codes. Instruction manuals for tools, and installation manuals from manufacturers, are other common examples of documents used on the job. It is crucial for workers to be able to read and understand these important documents. When a person's reading skills are poor, it typically causes them to avoid reading technical data whenever possible. This can lead to errors and accidents. This section reviews some techniques that you may find helpful in reading and writing construction documents.

2.2.0 Reading on the Job

When your company issues new safety guidelines, or the latest version of the local building code contains new standards that affect how you do your job, you need to be able to understand these documents to perform your job safely and effectively. You should not always rely on someone else to tell you how to do your job. Reading is an essential skill for construction professionals.

You may think that you do not have to read on the job, but, in fact, the written word is at the center of the construction trade. The following are some typical examples of things construction workers read on the job:

- Safety instructions
- Construction drawings
- Manufacturer's installation instructions
- Materials lists
- Signs and labels
- Work orders and schedules
- Permits
- Specifications
- Change orders
- Industry magazines and company newsletters
- Emails

This section reviews some simple tips and techniques that will help you read faster and more efficiently on the job. Further, some of these tips and techniques will help you become a better writer as well. Because you have been reading for some time now, it is probably something you do without even thinking. With practice, these guidelines will become second nature, too.

You should always have a purpose in mind when you read. This will help you find the information faster. For example, say you are looking for a specific installation procedure in a manual. Do not waste time reading other parts of the manual; go straight to the section that deals with that procedure. Read slowly enough to be able to concentrate on what you are reading. Often, technical publications are packed with detailed information; you do not want to misunderstand it because you rushed.

Most books have special features that can help you locate information, including the following:

- Table of contents
- Index
- Glossary
- Appendixes
- Tables and graphs

Tables of contents (*Figure 12*) are lists of chapters or sections in a book. They are usually at the front of a book. This is the first place experienced readers look when they want to know about the book in general or when they want to find some specific information in the book.

Indexes are alphabetical listings of topics with page numbers to show where those topics appear. Indexes are also used to list information in documents other than books—for example, in construction drawings (*Figure 13*). Say you want to look up information about transistors. Whereas the table of contents will tell you where to find the chapter that focuses on transistors, the index will tell you every page that mentions transistors even if the page is about another topic and mentions transistors only briefly.

00107-15_F11.EPS

Figure 11 Codebooks are critical resources for your job.

Glossaries are alphabetical lists of terms used in a book, along with definitions for each term. A glossary is usually located at the back of a publication.

Appendixes are sources of additional information placed at the end of a section, chapter, or book. Appendixes are separate because the information in them is more detailed or specific than the information in the rest of the book. If the information were placed in the main part of the book, it could distract readers or slow them down too much. Some of the NCCER *Core Curriculum* trainee guide modules contain an Appendix.

Tables and graphs summarize important facts and figures in a way that lets the reader understand them at a glance. Tables usually contain text or numbers, and graphs use images or symbols to convey their meaning.

Given the importance of written language to your job, make sure to pay close attention when reading. This section details some techniques you can use to maximize your comprehension of written material.

The noise and activity on a construction site or in a shop can easily distract you. While you are reading, try to avoid distractions. Radios, cellphones, televisions, conversations, and nearby machinery and power tools can all affect your concentration.

Take notes or use a highlighter to help you remember and find important text. Your notes on the blank pages at the beginning of a book can be an invaluable resource. If you find the information useful now, you may want to look it up again later, so make a note of the topic and record the page number. You can find it in an instant later. This old-fashioned technique is faster than pulling out your smartphone and doing a web search.

Also, highlighting a main sentence of an especially useful section will help you find the section quickly when you flip through the book.

When reading an unfamiliar book, skim or scan the chapter titles and section headings before you start to read. This will help you organize the material in your mind. Look for visual clues that indicate important material. For example, a bold **font** or *italics* indicate important words or information. Bold fonts are letters and numbers that are heavier and darker than the surrounding text. Italics are letters and numbers that lean to the right (like *this*), rather than stand straight up.

When reading instructions or a series of steps for performing a task, such as turning on a welding machine, imagine yourself performing the task; you may find the steps easier to remember that way. Be sure to read all the directions through before you begin to follow them. You will then be able to understand how all of the steps work together.

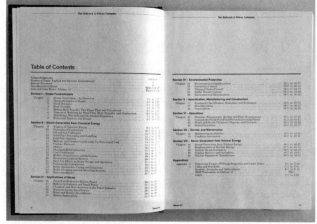

00107-15_F12.EPS

Figure 12 A table of contents lists the chapters or sections in a book.

GERMANS FROM RUSSIA HERITAGE
SOCIETY OFFICE BUILDING
BISMARCK, NORTH DAKOTA

DESIGN TEAM

RITTERBUSH – ELLIG – HULSING P.C.
ARCHITECTS – PLANNERS
209 NORTH 7TH STREET
BISMARCK, NORTH DAKOTA 58501
TELE: 701-223-7780   FAX: 701-258-6564

TRAEHOLT ASSOCIATES – STRUCTURAL
204 WEST THAYER AVENUE
BISMARCK, NORTH DAKOTA 58501
TELE: 701-255-2545   FAX: 701-255-2644

PRAIRIE ENGINEERING – MECHANICAL & ELECTRICAL
909 BASIN AVENUE
BISMARCK, NORTH DAKOTA 58504
TELE: 701-258-3493   FAX: 701-258-6857

SWENSON, HAGEN & Co.- CIVIL
909 BASIN AVENUE
BISMARCK, NORTH DAKOTA 58504
TELE: 701-223-2600   FAX: 701-223-2606

SHEET INDEX

CIVIL
C1   SITE SURVEY
C2   SITE PLAN

ARCHITECTURAL
A0.1   TITLE SHEET
A2.1   FLOOR PLAN, SCHEDULES
       & DOOR DETAILS
A3.1   BUILDING ELEVATIONS
A3.2   BUILDING SECTIONS
A3.3   WALL SECTIONS, DETAILS
A5.1   INTERIOR ELEVATIONS,
       CASEWORK & DETAILS
A6.1   REFLECTED CEILING PLAN

STRUCTURAL
S1   FOUNDATION & FRAMING PLANS & NOTES
S2   FOUNDATION & FRAMING DETAILS

MECHANICAL
M1   PIPING PLAN
M2   HVAC PLAN

ELECTRICAL
E1.1   ELECTRICAL SITE PLAN
E2.1   ELECTRICAL PLAN
E3.1   ELECTRICAL DETAILS
E4.1   ELECTRICAL SCHEDULES

RITTERBUSH – ELLIG – HULSING P.C.
ARCHITECTS – PLANNERS
309 NORTH SEVENTH STREET
BISMARCK, NORTH DAKOTA 58501-4496
TELE: 701-223-7780   FAX: 701-258-6564

GERMANS FROM RUSSIA HERITAGE SOCIETY
BISMARCK, NORTH DAKOTA

CODE INFORMATION
CODE USED: 1997 UBC
OCCUPANCY GROUP: B
TYPE OF CONSTRUCTION: TYPE V N
BUILDING AREA:
MAIN FLOOR: 5,890   SQ. FT.
MECHANICAL ROOM: 324   SQ. FT.
TOTAL: 6,214   SQ. FT.

Contren™ Learning Series
Sheet Metal Level 2
Trainee Module 04205-02
Blueprints and Specifications
Blueprint A0.1
Print 1 of 17 prints
Not to Scale   2002 Revision

TITLE SHEET

A
0.1

I HEREBY CERTIFY THAT THIS PLAN, SPECIFICATION, OR
REPORT WAS PREPARED BY ME OR UNDER MY DIRECT
SUPERVISION AND THAT I AM A DULY REGISTERED
ARCHITECT UNDER THE LAWS OF THE STATE OF NORTH
DAKOTA.

DATE: 5-30-2000   REG. NO. 578

00107-15_F13.EPS

Figure 13 Indexes are often used in construction drawings to identify key information.

Always re-read what you have just read to make sure you understand it. This is especially important when the reading material is complex or very long. And finally, take a break every now and then. If you read slowly or you find that you are getting tired or frustrated, put the material down and do something else to give your mind a break. Reading to understand is hard work and does require effort.

2.3.0 Writing on the Job

At this stage in your career, you will probably do more reading on the job than writing. But no matter what job you have in the construction industry, you'll eventually have to write something. Writing skills are essential if you want to succeed. Construction workers write work orders, health and safety reports, **punch lists** (written lists that identify deficiencies requiring correction at completion), memoranda (informal office correspondence, or **memos** for short), emails, work orders or change orders, and a whole range of other documents as part of their job (see *Figure 14*).

Writing skills will allow you to perform these tasks, too, as you gain experience on the job. They will help you move into positions of greater responsibility. It is never too early to start brushing up on your writing skills, so that when the opportunity comes, you will be ready. This section reviews the writing process and then explains the characteristics of good writing.

First, before you begin to write, do some prewriting. Organize in your mind what you want or need to write. One prewriting technique is brainstorming, which is simply writing down a list of ideas. Do not pressure yourself to make a good list of ideas; just get ideas down on paper. Next, cross out rejected ideas and arrange the better ideas in a logical order to create an outline. After you have built a foundation of basic concepts and have a general sense of what you want to write about, do the necessary research. It may be necessary to ensure that the statements are appropriate for the situation and not something you may regret writing later.

After you have researched, write a rough draft. A rough draft is called this because it is rough, not polished. Do not attempt to write a masterpiece at this time; doing so will merely make you feel discouraged and overwhelmed. Just follow your outline, incorporate your ideas, and include your research. You can improve it at a later point.

The third step is to simply rest or do something else for a while. When you come back to the project later, your mind will be fresh and critical. Maybe you thought something was understandable when you were writing it, but when you read it a day later, you may see that it is unclear. On the other hand, you could also discover that some important information is missing.

Burning - Welding - Hot Work Permit

Valid from _____ to _____, _____   Master Card No._____
     (am/pm) (am/pm)   DATE

1. Work Description
Equipment Location or Area _____

Work to be done:

2. Gas Test

☐ None Required				
☐ Instrument Check	Test Results	Other Tests		Test Results
☐ Oxygen 20.8% Min				
☐ Combustible % LFL				

Gas Tester Signature   Date   Time

3. Special Instructions   ☐ None   ☐ Check with issuer before beginning work

4. Hazardous Materials   ☐ None   What did the line / equipment last contain?

5. Personal Protection   ☐ Standard Equipment: welder's hood with long sleeves; cutting goggles

☐ Goggles or Face Shield   ☐ Respirator   ☐ Forced Air Ventilation

☐ Standby Man   ☐ Other, specify: _____

6. Fire Protection   ☐ None Required   ☐ Portable Fire Extinguisher

☐ Fire Watch   ☐ Fire Blanket   ☐ Other, specify: _____

7. Condition of Area and Equipment

Required

Yes	No		THESE KEY POINTS MUST BE CHECKED
		a.	Lines disconnected & blanked or if disconnecting is not possible, blinds installed?
		b.	Lines steamed, purged, or otherwise properly cleared of combustibles?
		c.	Area and equipment satisfactorily clean of oil or combustibles?
		d.	Trenches, catch basins & sewer connections properly covered or sealed?
		e.	Immediate area and/or area under the work barricaded or roped off?
		f.	Adjoining equip. & operations checked to have any effect on the job?
		g.	Area fire suppression (fire water and sprinkler system) in service?

Comments

00107-15_F14A.EPS

Figure 14A A hot work permit is a typical written product in a construction project. (1 of 2)

Burning - Welding - Hot Work Permit

8. Approval

	Permit Authorization			Permit Acceptance		
	Area Supv.	Date	Time	Maint. Supv./Engineer Contractor Supv.	Date	Time
Issued by						
Endorsed by						
Endorsed by						

9. Individual Review

I have been instructed in the proper Hot Work Procedures

	Signed	Signed
Persons Authorized to Perform Hot Work	______________	______________
	______________	______________
	______________	______________
	______________	______________
	______________	______________
Fire Watch	______________	______________

10. Job Completion

☐ Yes ☐ No Is the work on the equipment completed?

☐ Yes ☐ No Has the worksite been cleaned and made safe?

Worker answering above questions ______________________________

Issuer's Acceptance ______________________________

Forward to Production Superintendent within 7 days of job completion

00107-15_F14B.EPS

Figure 14B A hot work permit is a typical written product in a construction project. (2 of 2)

Next, revise the paper. Come back to your writing and make major changes. You may need to rearrange some sections, add information or try to simplify them, or delete unnecessary sections. Look for wordy phrases.

Finally, proofread. Now that you have fixed the major problems in the paper, look for the smaller errors, like grammar or spelling mistakes. Watch for the errors that spell-check cannot find. If you write a different word than the one you meant to (like *four* instead of *for*) spell-check will not notice.

After you have finished writing, it may be helpful to ask a friend or a co-worker to read it. Another reader can often spot mistakes or problems that you have missed. For example, you might ask a friend to review your job application or resume. You can ask a co-worker to check over a work permit or material takeoff to make sure you haven't left out any information or made any math mistakes. Don't be embarrassed to ask for this type of help, and don't be upset if your friend or co-worker spots a mistake. Professionals are always glad to catch mistakes before those mistakes create bigger problems. Note that how you view the material may affect the accuracy of proofreading, too. Writers often find errors when they read a piece printed on paper rather than on a computer display.

Now that you understand the process of writing, we can examine the characteristics of good writing. Being a good writer is not as hard as you might think. You do not need to be a master of English composition and grammar, or have a college degree, to write well. You just need to remember these five characteristics of good writing: clear, concise, correct, complete, and considerate of the reader. Let's examine them one by one.

- *Be clear* – Make your main point easy to find, and include all the necessary details so the reader can understand what you are saying.

One way to highlight important information is to list it. Lists often use **bullets**, like this one does, or numbers. Several different bullet formats can be used. When you finish writing, reread what you have written to make sure it is clear and accurate. Check for mistakes and take out words you don't need.

- *Be concise* – More words do not mean better writing. Instead of writing "In order to prevent unwanted electrical shocks, workers should make sure they disconnect the main power supply before work commences," write "To avoid being shocked, disconnect the main power supply before starting." Instead of writing "at the present time," simply write "now." It is also important to think about the reading level of the intended audience.

- *Be correct* – Accuracy is vital in the construction industry. For example, if you are filling out a form to order supplies from a manufacturer's catalog, make sure you use the correct terms and quantities. When writing material takeoffs or other documents that will be used to purchase equipment and assign workers, you need to get it right. Get in the habit of looking over your writing, whether it's a handwritten note, a supply list, a permit, or an email. Errors cost money and time.

- *Be complete* – When writing, ask yourself the following questions:

  - Have I identified myself to the reader?
  - Have I said why I'm writing this?
  - Will the reader know what to do and, if necessary, how to do it?
  - Will the reader know when to do it?
  - Will the reader know where to do it?
  - Will the reader know whom to call with questions?

You should always ask these questions when you are writing something for others to read, even if you are writing something short and simple.

The Journalist's Questions

Make sure that you cover all of the necessary information in your writing by asking yourself the following journalist's questions:

- **Who?** Who told you that? Who wants it done? Who are you?
- **What?** What is the problem? What do you need? What do you want to do?
- **When?** When did it happen? When do you want it finished?
- **Where?** Where is it taking place? Where are you going? Where can you be found?
- **Why?** Why do you want to do that? Why is this problem happening? Why did you decide to write about it now?
- **How?** How will you fix it? How can you be reached? How can they help?

- *Be considerate of the reader* – Not everyone who reads what you write will be a professional in the same trade that you are in, and not everyone will understand the same words and concepts that you do. For example, you may need to write a report to a client to describe a technical problem. In that report you must use words that intelligent, yet uninformed, people would understand. Similarly, not every professional in your field will know all of the inside information about your company or the history of the problem you are writing about.

2.3.1 Emails

Email has become the standard way to send written information, files, and pictures as long as the files are not too large. Like printed letters and memos, email is used to ask and answer questions, and to provide information quickly. However, remember that a direct telephone call may be in order if email is not providing a timely result.

There are advantages and disadvantages to email. Email delivery is much faster than paper-based delivery, and emails can be stored permanently on a computer. They can be replicated quickly and distributed to many people at the same time. However, because of their ability to be copied and sent so easily, emails are not as private as paper-based documents. Business emails should not include private, sensitive, or confidential information. They can easily be sent by accident to people who are not authorized to read them. Information such as this should be sent in a more secure fashion. Remember that emails may be stored permanently, and can represent a binding agreement or contract. Negative, aggressive emails can lead to dismissal, just like a verbal attack on someone.

Another advantage to transmitting information electronically is that it reduces the amount of paper, printing materials, and shipping or processing fees associated with written documents. In some cases, however, an email is not considered legally binding. Written documents with handwritten signatures remain the standard for contracts and other formal agreements. However, this situation is changing with the advent of **electronic signatures**.

Because email is now used daily in business to communicate with associates, clients, supervisors, and co-workers, it is important to know the proper way to write an email (see *Figures 15* and *16*). When writing an email, the nonverbal part of communication (facial expressions, body language, tone of voice) is lost, so people may misinterpret the message you are trying to convey if you are not careful with words. Many of the rules that apply to writing paper-based documents still apply to emails, but there is a certain etiquette involved. Business emailing standards exist so that emails are composed in a professional, efficient, and responsible manner.

The following are some general rules for writing a proper business email. Keep in mind that some rules will vary according to the nature of the business and company culture.

- Write business email the same way you would write a formal business letter or memo.
- Make sure you are sending the email to the correct individual(s) to maintain confidentiality.
- Always start with a clear subject line that indicates the purpose of the message.
- Begin the email by addressing the recipient.
- Try to keep the body of the email brief and to the point, typically no longer than one screen length, so that the recipient does not have to scroll when reading.

From:	JQSmith@smithcontracting.com	Sent: Mon 3/17/2014 1:47 PM
To:	WJones@paintersplus.com	
Cc:		
Subject:	Quick Note	
Attachments:		

View As Web Page

Here are some paint colors and faucets available for your bathroom: Color #1415, Soft Jade; Color #1416, Garden Moss; and Color #1417, Forest Glen. All are available in semi-gloss or eggshell finish. There are also three faucet sets: the Meridian (single handle) $109.88; the Mermaid (dual handles) $83.50; and the Monitor (dual handles) $95.75. All are available in polished brass or polished chrome. I've included paint samples and photos of the faucets. Please tell me your choices by Friday. If you have any questions, call me at 703-555-1212.

00107-15_F15.EPS

Figure 15 Functional but unattractive email example.

From: JQSmith@smithcontracting.com   Sent: Mon 3/17/2014 1:47 PM

To: WJones@paintersplus.com

Cc:

Subject: Bathroom Paint and Faucet Options
Attachments:

View As Web Page

Dear Mr. Jones,

The paint colors and faucets available for your bathroom are listed below (photos of faucets and paint colors are attached to this email). Please let me know what you decide by 5:00 pm on Friday, March 21st. If you have any questions, please do not hesitate to contact me at 703-555-1212.

Paint Colors (Available in semi-gloss or eggshell finish)
- #1415 – Soft Jade
- #1416 – Garden Moss
- #1417 – Forest Glen

Faucet Sets (Available in polished brass or polished chrome)

Model	Price	Handle Style
Meridian	$109.88	Single
Mermaid	$83.50	Dual
Monitor	$95.75	Dual

Regards,
John Q. Smith
Smith Contracting

00107-15_F16.EPS

Figure 16   Better example for an email when a sale is at stake.

- Use a concise format, such as numbers or bulleted points, to clarify what the email is about and what response or action is required of the recipient.
- Write in a positive tone and avoid using negative or blaming statements.

Reducing Your Carbon Footprint

GOING GREEN

Many companies are taking part in the paperless movement. They reduce their environmental impact by reducing the amount of paper they use. Using email helps to reduce the amount of paper used, and there are even postscripts on emails that asking you to reconsider printing the email unless necessary.

- Do not type in all capital letters, as it gives the impression of shouting.
- Use italicized, bold, or underlined text if you need to stress a point.
- Avoid sarcasm, as it can be easily misunderstood.
- Do not forward junk mail.
- Do not address private issues and concerns.
- When sending an attachment, state the name of the file, and its format.
- Double-check the email for spelling and grammatical mistakes before sending. Remember, once an email is sent it cannot be retrieved.
- Include additional contact information other than your email address, such as a phone number or a mailing address.
- Email is not a substitute for face-to-face interaction. Know when to pick up the phone or schedule a meeting if you cannot express your thoughts or concerns through an email. Never send bad news through email, and don't use email to avoid your responsibilities.

2.3.2 Texting

Like email, texting has become nearly universal. According to a 2013 Pew Research Center study, 91 percent of American adults own a cell phone. Tablet computers are also common and used for both texting and email. Texting is also an immediate form of communication. Although an email may take several hours or more to be seen, a text message is more likely to be seen right away. This depends on whether or not an individual's cell phone is set up to receive emails as well as text messages. Texting has become an important part of business and industry today. Therefore, the construction professional must be able to text effectively as well. However, how you write a text to your friend is likely quite different than a text you may write to your supervisor.

Remember that both texting and email can create a safety hazard. Safety must always be given a higher priority than a text message. Be familiar with and follow company policy on the use of phones and tablets on the job. If an immediate response is not vital, focus on the job at hand first.

You must consistently ask yourself if sending a text message is the best way to communicate. Text messages should not be used when the message is long or complex. Entering text into a cell phone is much slower than discussing the same information over the phone or even typing it into an email. The average speaking speed is 130 words per minute. Therefore, before you start a long text-message conversation, consider calling. Text messages should not be used to convey highly emotional messages, apologies, or criticism. Emotion cannot be effectively communicated this way.

Under what circumstances should text messages be used? Extremely brief messages that should be viewed immediately are best communicated with a text message. That's why text messages are also called short messages. Any snippet of information that does not require much interaction can be texted: for example, "I'm leaving now. I'll be at the factory in 10 minutes." Texts are also great for confirming information after a telephone or face-to-face conversation. For example, "Looking forward to meeting you at 3:30 Friday afternoon."

When sending a text message, be sure that it is accurate. Texting poses a special problem in the form of auto-corrections. Most smartphone systems correct typos automatically; for example, "sprll" instantly becomes "spell" after hitting the space key. This feature saves time, but it also frequently causes embarrassment. Websites commonly post autocorrect incidents that completely changed the meanings of the messages or added offensive words, but these mistakes are avoidable. Always take a moment and proofread before sending a text.

Not only should your text message be accurate, but it must be clear. Many people use shorthand in their texts in order to save time. The use of abbreviations such as LOL (laugh out loud), TTYL (talk to you later), and IDK (I don't know) is controversial. There are a few factors to consider when deciding whether to use shorthand. If the message is informal, shorthand is usually acceptable. This is especially true if your work group has its own abbreviations for certain phrases. If the person you are sending the message to is a friend or close colleague of the same level in the

He Said *What?*

Being careful about the words you use when writing is especially important when dealing with clients. Always make sure that the person you are communicating with understands the jargon that you are using. To appreciate the humor in the following story, you have to know a couple of terms used in the electrical trade. Two types of electrical wires enter a building: drops (overhead wires) and laterals (underground wires).

A power company customer called to report a power outage. The service technician looked at the problem, wrote up the work order, gave the customer a copy, and left. The customer took one look at the work order, got angry, and called the company to complain that its service technician was rude. What made the customer angry? Here is what the service technician wrote on the work order:

INVESTIGATED POWER OUTAGE – DROP DEAD

The technician described the problem accurately, but the customer took it as an insult!

Source: Adapted from *Tools for Success: Soft Skills for the Construction Industry*, Steven A. Rigolosi. 2009. Upper Saddle River, NJ: Pearson Learning.

company, you can use shorthand. If the message is to a stranger or someone in upper management, it may be better to send messages with correct spelling and grammar in order to leave a good impression. If you are sure that the person you are texting knows the shorthand, it is fine to use it. Otherwise, using abbreviations that your friend or co-worker does not understand could cause frustration.

The warnings in the previous section about making phone calls while operating machinery apply to sending texts as well. According to research by the Federal Motor Carrier Safety Administration, typing or sending a text takes a driver's eyes off the road for an average of 4.6 seconds. At 55 miles per hour, the car will travel the length of a football field in that time. Never send texts while driving or operating machinery.

Additional Resources

Successful Writing, Maxine Hairston and Michael Keene. 2003. New York, NY: W. W. Norton & Company.

The College Writer's Reference, Alan R. HayaKawa and Toby Fulwiler. 1998. Upper Saddle River, NJ: Prentice Hall.

The Elements of Style, Willian Strunk, Jr. 2015. Grammar, Inc.

2.0.0 Section Review

1. When a person's reading skills are poor, it typically causes him or her to _____.

   a. rewrite the material themselves
   b. avoid reading technical information
   c. have another worker read to them
   d. file for disability

2. Where can a reader look to discover if a book has a chapter on a particular topic?

   a. Table of contents
   b. Index
   c. Glossary
   d. Appendix

3. Bold fonts can be used to indicate that words _____.

   a. have recently been changed
   b. need to be checked for accuracy
   c. are important
   d. are illustrated below

4. After you finish writing a rough draft, you should _____.

   a. ask a colleague to proofread it for you
   b. do some research
   c. check for spelling or grammar errors
   d. have a rest and then review and revise

5. Which of the following is the best advice for sending an email?

   a. Treat a business email the same way you would treat a formal business letter.
   b. Type in all capital letters in order to emphasize the importance of your email and set it apart from others.
   c. Send bad news or emotional information via email so that the recipient can read the message in privacy.
   d. Spelling and grammar do not matter in email communication since it is informal.

SUMMARY

Communications skills—the ability to read, write, listen, and speak effectively—are essential for success in the construction workplace. Effective communication ensures that work is done correctly, safely, and on time. The construction industry relies on a wide variety of written documents. As your responsibilities increase, you will be called on to write some of these documents yourself. This module introduced you to simple tips and techniques that you can use every day to read and write effectively.

Construction professionals state their ideas clearly, and they also listen to the ideas of others. Real listening is an active process entailing feedback, interaction, and retention of details. Not listening causes mistakes, which waste time and money. Of course, effective listening depends on effective speaking. No one can understand what is not expressed clearly. Once you master these basic communication skills and add them to your toolbox of skills, you will find that you can succeed at more things than you ever thought possible.

1. Good communication on the job site _____.

   a. affects safety, schedules, and budgets
   b. will make you popular
   c. takes too much time
   d. cannot be learned

2. Nonverbal communication is best for communicating _____.

   a. feelings and attitudes
   b. complex ideas
   c. what a person desires
   d. instructions for installing new equipment

3. Which of the following are examples of positive nonverbal communication?

   a. Sitting up straight, keeping a clean workspace, and looking at people who are talking to you.
   b. Giving compliments to others and encouraging them.
   c. Speaking with an even tone and looking away most of the time during a conversation.
   d. Telling people all of the information they need in order to do a task properly and answering questions.

4. Which of the following is an example of positive verbal communication on the job?

   a. Talking over or interrupting another person.
   b. Giving and taking instructions.
   c. Tuning out when someone is speaking.
   d. Ducking out of team discussions.

5. Real listening is _____.

   a. tedious
   b. an active process
   c. unnecessary
   d. an art

6. A co-worker asks you to hand her a tool, but does not specify which tool she needs. The most appropriate response would be: _____

   a. "Get it yourself."
   b. "How am I supposed to figure out which tool you need?"
   c. "I'm sorry; which tool do you need?"
   d. "I am too busy; ask someone else."

7. A supervisor wants an apprentice to insert a plug into a water supply line for a pressure test. The clearest way to instruct the apprentice would be for the supervisor to say, _____.

   a. "Put the plug in"
   b. "Insert that one plug into the pipe"
   c. "Get the water supply line ready for the pressure test"
   d. "Insert the plug into the end of the water supply line so I can do a pressure test"

8. If you are in the middle of a task and you receive a personal phone call, _____.

   a. put the caller on hold until you complete the task
   b. continue working and do not answer
   c. tell the caller you don't have time to talk
   d. hand the phone to a co-worker

9. In the construction industry, a codebook provides _____.

   a. the keypad entry codes for locked work areas
   b. codes and standards, such as building codes or electrical codes
   c. passwords for company web services, such as databases and training software
   d. keys for understanding blueprint symbols

10. A reader that wants to find any mention of a certain topic in a book can look in the _____.

    a. table of contents
    b. glossary
    c. appendixes
    d. index

11. Which is a typical example of an item a construction worker might read for work on a regular basis?

    a. Newspaper
    b. Roadmap
    c. Materials list
    d. Summons

12. When you are looking for a specific installation procedure in a manual, the best way to find the information is to _____.

    a. identify the correct section and go straight there
    b. start reading the book from the beginning
    c. study the terms in the glossary
    d. skim the book and take notes on all important information

13. Which of the following is one disadvantage to using email versus paper-based delivery?

    a. Email is an expensive way to communicate.
    b. The spell-checker is unreliable.
    c. Emails cannot easily be replicated.
    d. Emails are not as private as paper-based documents.

14. Many of the rules that apply to writing paper-based documents also apply to _____.

    a. writing emails
    b. leaving voice messages
    c. having a face-to-face conversation
    d. taking orders

15. Text messages are best used for _____.

    a. long conversations since a record of what was said will be preserved
    b. telling people how one feels
    c. detailed or complex information
    d. brief messages that should be viewed immediately

Trade Terms Quiz

Fill in the blank with the correct term that you learned from your study of this module.

1. Often, books will include additional detailed information in a(n) _____.

2. In a list, a large dot is called a(n) _____.

3. Words and numbers can be printed using many different _____, or type styles.

4. A(n) _____ includes the definition of terms used in a book.

5. Information can be presented in picture form using a(n) _____.

6. To find an alphabetical listing of topics in a book, consult the _____.

7. A type style that uses slanted print to emphasize text is called _____.

8. Construction workers write _____ to list deficiencies requiring correction at completion.

9. You can use a(n) _____ to send an informal written message to someone in your company.

10. You can find a book's chapters and headings listed in a(n) _____.

11. Numerical or written information can be presented for quick visual scanning in a(n) _____.

12. A(n) _____ is another way to sign documents.

13. Providing feedback is an essential part of the _____ process.

14. Your _____ silently communicates whether or not you are paying attention to a speaker.

15. As you gain experience, you will learn more of the special _____ that you can use to communicate with other workers in your trade.

16. A legal document that allows a task, such as constructing a building, to be undertaken is called a(n) _____.

17. The owner of a project may change the expected completion date by writing a _____.

18. After you receive instructions from people, be sure to _____ what they say back to them to make sure that you understand correctly.

19. No matter what you are doing, other people can get clues about your attitude and character from your _____.

Trade Terms

Active listening	Electronic signature	Italics	Paraphrase
Appendix	Font	Jargon	Permit
Body language	Glossary	Memo	Punch list
Bullets	Graph	Nonverbal	Table
Change order	Index	communication	Table of contents

Dr. Michael John Sandroussi

Craft Training Center of the Coastal Bend (CTCCB)
President

How did you choose a career in the construction industry?
As I was growing up, my career paths took me through many different construction worlds, including the petrochemical and oil field environment. Although my path ultimately led me to public education, I always believed that students that took career and technology education courses were better prepared to function in the real world.

Who inspired you to enter the industry?
I was actually inspired by economics and the need to survive in a complex world. I started working with my grandfather when I was 8 years old and developed a strong work ethic. I learned how to shake hands with adults, look them in the eye, and communicate with others in a professional manner. In working with him and other individuals, I also learned the value of ambition and learned how to set and achieve goals. I realized that a quality education was a neutralizing factor in the US.

What types of training have you been through?
In my career, I have learned plumbing, electrical, carpentry, roofing, foundation work, HVAC, flooring, heavy equipment operations, grain elevator operations, farming, pipefitting, instrumentation, oilfield flow production, wireline operations, and many others.

How important is education and training in construction?
Knowledge is power, and without a trained workforce, an organization will not be fully productive. A lack of training also results in accidents and fatalities.

How important are NCCER credentials to your career?
My NCCER credentials have empowered me to assume the responsibilities as President of a large training facility that trains adults, high school students, junior high students, and veterans in many different crafts.

How has training/construction impacted your life?
Working in construction allowed me to feed my family. Receiving training in construction kept me safe and healthy.

What kinds of work have you done in your career?
Grafting orange trees; plant production; agricultural field technician; plumbing; electrical; carpentry; roofing; masonry work; HVAC; flooring; heavy equipment operator; grain elevator/gin operations; farming; pipefitting; instrumentation; oilfield flow production; and wireline operations.

Tell us about your present job.
I am the President of the Craft Training Center of the Coastal Bend (CTCCB) and an NCCER Certified Primary Administrator. This center trains adults, high school students, junior high students and veterans. Over 1,700-plus students per year are trained in the areas of welding, pipefitting, instrumentation, scaffold-building, industrial painting, electrical, plumbing, mobile crane operations, rigging, field safety, and safety technology. The CTCCB also has an on-line assessment center that has tested over 2,500 journey-level experienced craftsmen.

What do you enjoy most about your job?
NCCER training is changing the quality of life for our students. In the real world, quality of life is what you wear, what you eat, where you live, and what you drive. To my knowledge, the CTCCB is the only training center that catapults high school students immediately into the workforce upon high school graduation. These high school graduates secured NCCER training credentials and entered the workplace as trained, educated, drug-free craftsmen and women.

What factors have contributed most to your success?
My family structure is very strong. My grandmother told me at a very young age that the only goal in her life was for her children to do better, and the only way that was going to happen was to work hard, learn as much as possible, and seek out a quality education.

Would you suggest construction as a career to others? Why?

A career in construction is very rewarding because you are actually building or constructing something that will be around for a long time to come. Your children, grandchildren, great-grandchildren and many others will be able to see what you did during your life. The "backbone" of America is the craftsman; when they are working, our nation is strong!

What advice would you give to those new to the field?

Study, learn, and strive to keep track of any changes in the industry.

Interesting career-related fact or accomplishment:
The Craft Training Center of the Coastal Bend was featured in the *Wall Street Journal* in 2014. The article focused on technical training, and the welder featured is one of our students that attended during his high school career.

How do you define craftsmanship?
Craftsmanship is skill and/or an expertise in a craft.

Trade Terms Introduced in This Module

Active listening: A process that involves respecting others, listening to what is being said, and understanding what is being said.

Appendix: A source of detailed or specific information placed at the end of a section, a chapter, or a book.

Body language: A person's facial expression, physical posture, gestures, and use of space, which communicate feelings and ideas.

Bullets: Large, vertically aligned dots that highlight items in a list.

Change order: A written order by the owner of a project for the contractor to make a change in time, amount, or specifications.

Electronic signature: A signature that is used to sign electronic documents by capturing handwritten signatures through computer technology and attaching them to the document or file.

Font: The type style used for letters and numbers.

Glossary: An alphabetical list of terms and definitions.

Graph: Information shown as a picture or chart. Graphs may be represented in various forms, including line graphs and bar charts.

Index: An alphabetical list of topics, along with the page numbers where each topic appears.

Italics: Letters and numbers that lean to the right rather than stand straight up.

Jargon: Specialized terms used in a specific industry.

Memo: Informal written correspondence. Another term for memorandum (plural: memoranda).

Nonverbal communication: All communication that does not use words. This includes tone of voice, appearance, personal environment, use of time, and body language.

Paraphrase: Express something heard or read using different words.

Permit: A legal document that allows a task to be undertaken.

Punch list: A written list that identifies deficiencies requiring correction at completion.

Table: A way to present important text and numbers so they can be read and understood at a glance.

Table of contents: A list of book chapters or sections, usually located at the front of the book.

Additional Resources

This module presents thorough resources for task training. The following resource material is suggested for further study.

How to Win Friends and Influence People. Dale Carnegie. 2013. New York, NY: Simon & Schuster.

Listen Up: How to Improve Relationships, Reduce Stress, and Be More Productive By Using the Power of Listening. Larry Barker; Kittie Watson. 2000. New York, NY: St. Martin's Press.

Successful Writing, Maxine Hairston; Michael Keene. 2003. New York, NY: W. W. Norton & Company.

The College Writer's Reference, Alan R. Hayakawa; Toby Fulwiler. 1998. Upper Saddle River, NJ: Prentice Hall.

The Elements of Style, WIlliam Strunk, Jr. 2015. Grammar, Inc.

Tools for Success: Critical Skills for the Construction Industry, NCCER. 2009. Upper Saddle River, NJ: Pearson Education, Inc.

Figure Credits

Courtesy of Oak Ridge National Laboratory, Figures 2, 8

M.C. Dean, Inc., Figure 5

Ed Gloninger, Figures 7, 10

Topaz Publications, Inc., Figures 11, 12

Ritterbush-Ellig-Hulsing PC, Figure 13

Section Review Answer Key

Answer	Section Reference	Objective
Section One		
1. b	1.1.1	1a
2. a	1.2.0	1b
3. a	1.3.0	1c
4. d	1.3.1	1c
5. c	1.3.1	1c
Section Two		
1. b	2.1.0	2a
2. a	2.2.0	2b
3. c	2.2.0	2b
4. d	2.3.0	2c
5. a	2.3.1	2c

NCCER CURRICULA — USER UPDATE

NCCER makes every effort to keep its textbooks up-to-date and free of technical errors. We appreciate your help in this process. If you find an error, a typographical mistake, or an inaccuracy in NCCER's curricula, please fill out this form (or a photocopy), or complete the online form at **www.nccer.org/olf**. Be sure to include the exact module ID number, page number, a detailed description, and your recommended correction. Your input will be brought to the attention of the Authoring Team. Thank you for your assistance.

Instructors – If you have an idea for improving this textbook, or have found that additional materials were necessary to teach this module effectively, please let us know so that we may present your suggestions to the Authoring Team.

NCCER Product Development and Revision

13614 Progress Blvd., Alachua, FL 32615

Email: curriculum@nccer.org
Online: www.nccer.org/olf

❏ Trainee Guide   ❏ Lesson Plans   ❏ Exam   ❏ PowerPoints   Other _____

Craft / Level: _____ Copyright Date: _____

Module ID Number / Title: _____

Section Number(s): _____

Description: _____

_____

_____

_____

Recommended Correction: _____

_____

_____

_____

Your Name: _____

Address: _____

_____

Email: _____ Phone: _____

00108-15

Basic Employability Skills

Overview

Becoming gainfully employed in the construction industry takes more preparation than simply filling out a job application. It is essential to understand how the construction industry and potential employers operate. Your trade skills are extremely important, but all employers are also looking for those who are eager to advance and demonstrate positive personal characteristics. This module discusses the skills needed to pursue employment successfully.

Module Eight

Trainees with successful module completions may be eligible for credentialing through the NCCER Registry. To learn more, go to **www.nccer.org** or contact us at **1.888.622.3720**. Our website has information on the latest product releases and training, as well as online versions of our *Cornerstone* magazine and Pearson's product catalog.

Your feedback is welcome. You may email your comments to **curriculum@nccer.org,** send general comments and inquiries to **info@nccer.org**, or fill in the User Update form at the back of this module.

This information is general in nature and intended for training purposes only. Actual performance of activities described in this manual requires compliance with all applicable operating, service, maintenance, and safety procedures under the direction of qualified personnel. References in this manual to patented or proprietary devices do not constitute a recommendation of their use.

Objectives

When you have completed this module, you will be able to do the following:

1. Describe the opportunities in the construction businesses and how to enter the construction workforce.
   a. Describe the construction business and the opportunities offered by the trades.
   b. Explain how workers can enter the construction workforce.
2. Explain the importance of critical thinking and how to solve problems.
   a. Describe critical thinking and barriers to solving problems.
   b. Describe how to solve problems using critical thinking.
   c. Describe problems related to planning and scheduling.
3. Explain the importance of social skills and identify ways good social skills are applied in the construction trade.
   a. Identify good personal and social skills.
   b. Explain how to resolve conflicts with co-workers and supervisors.
   c. Explain how to give and receive constructive criticism.
   d. Identify and describe various social issues of concern in the workplace.
   e. Describe how to work in a team environment and how to be an effective leader.

Performance Tasks

This is a knowledge-based module; there are no Performance Tasks.

Trade Terms

Absenteeism	Hallucinogen	Reference
Amphetamine	Harassment	Self-presentation
Barbiturate	Initiative	Sexual harassment
Bullying	Leadership	Synthetic drugs
Cannabinoids	Methamphetamine	Tactful
Compromise	Mission statement	Tardiness
Confidentiality	Opiates	Work ethic
Constructive criticism	Professionalism	Zero tolerance

Industry Recognized Credentials

If you are training through an NCCER-accredited sponsor, you may be eligible for credentials from NCCER's Registry. The ID number for this module is 00108-15. Note that this module may have been used in other NCCER curricula and may apply to other level completions. Contact NCCER's Registry at 888.622.3720 or go to **www.nccer.org** for more information.

Contents

Topics to be presented in this module include:

Figures

1.0.0 OPPORTUNITIES IN THE CONSTRUCTION INDUSTRY

Objective

Describe the opportunities in the construction business and how to enter the construction workforce.

a. Describe the construction business and the opportunities offered by the trades.
b. Explain how workers can enter the construction workforce.

Trade Terms

Mission statement: A statement of how a company does business.

Reference: A person who can confirm to a potential employer that you have the skills, experience, and work habits that are listed in your resume.

Whether a laborer (blue collar) or professional (white collar), the construction business offers an abundance of opportunities. It is up to each individual to take the necessary steps to advance their own career.

To establish a career in your trade of choice, it is important to know where to look for the opportunities and be prepared to act when opportunities arise.

1.1.0 The Construction Business

The construction industry is made up of a wide variety of specialized skills (*Figure 1*). This diverse industry is made up of both laborers and professionals working together. Industry opportunities have a global presence; skilled tradespeople are needed in areas such as war zones and major construction sites around the world. The trades are not limited solely to building construction. There are also global opportunities for many craftworkers such as wind turbine technicians, welders, powerline workers, crane operators, and instrumentation technicians that are not directly related to new construction.

The construction industry consists of independent companies of all sizes that specialize in one or more types of work. For example, a smaller company might install HVAC systems in residences, whereas a larger mechanical contractor might be responsible for the HVAC and piping systems associated with a hospital.

The following are some examples of the workforce needed to build a large-scale project:

- Architects to create and design the shape, color, and spaces of the structure.
- Civil engineers to analyze the architects' drawings to determine how to make the construction design possible and suggest modifications if it is not. They are also responsible for determining suitable building materials.
- Project managers to plan, coordinate, and control a project's progress or a portion thereof.
- Heavy equipment operators to excavate the site.
- Concrete pourers to pour the foundation.
- Crane operators to lift the steel beams.
- Ironworkers to cut and fit the steel beams.
- Welders to secure the framework of buildings and other metal structures.
- Electricians to install an array of electrical devices and wiring.
- Plumbers to properly install fixtures and other plumbing equipment.
- Heating, ventilation, and air conditioning (HVAC) technicians to install building comfort systems.
- Pipefitters to install chilled and hot water systems, as well as process piping.
- Carpenters to frame the walls.
- Instrumentation technicians to install measuring and control instruments.
- Landscape designers to decide on the aesthetics of the project.

(A) WIND TURBINE TECHNICIAN

(B) IRONWORKER

(C) WELDER

(D) HVAC TECHNICIAN

(E) SURVEYOR

(F) CREW LEADER/SUPERVISOR

00108-15_F01.EPS

Figure 1 The construction industry offers many career options.

Many companies have a **mission statement**, which explains how a company does business. The company may describe its philosophy in an employee handbook or other materials during a new-hire orientation. However a company describes its role in the construction industry, it is very important to become familiar with your company's policies and procedures to know what is expected of you. In fact, the time to learn about a potential employer is before an application is ever submitted. The more you know about a company and what is important to them, the better you can plan your responses for an interview.

1.2.0 Entering the Construction Workforce

Upon completion of your training, it will be time to find a job where your newly aquired skills can be put to use. The first step of a job search is writing a resume that will get an employer's attention.

A good resume can make the task of finding a job much easier. The following are some guidelines for resume writing:

- Use a readable font, such as Times New Roman or Arial, for the text of the resume.
- Write your objective using key words found in the job description to create a relationship between the two.
- Format your resume chronologically, putting your current or most recent job first.
- Use short sentences and bulleted lists to describe your skills, again using key words found in the job description.
- Include certifications and training credentials in specific trade areas. Credentials could include such items as NCCER training certificates or wallet cards, or an automated external defibrillator (AED) certification.
- Make sure that all contact information is current, accurate, and easy to find on the resume.
- Personal information, such as hobbies or volunteer work and commitments can be listed. Employers often appreciate employees who have such interests, as they indicate a well-rounded individual. However, such information should be kept brief.
- Check your resume for inaccurate or missing dates. The timeline of your career should be easy to follow. Large gaps of undocumented time on a resume may require a reasonable explanation.
- Have someone proofread your resume to make sure it is free of spelling errors, typos, and poor grammar.
- Provide clear personal references or indicate that such a list can be provided upon request.

> **Did You Know?**
> # The High Cost of Benefits
> While the pay rates for craftworkers vary widely, the actual cost of an employee to a company averages 35 percent more than the employee's wage. For example, if a worker's hourly wage is $25.00, the company will spend another $8.75 per hour to pay for benefits such as healthcare and vacation time. Even higher percentages are not uncommon.

Look for a job that will match your skill level. Typically, a job posting will include a summary of the position, the required qualifications, and an overview of benefits. Applying for positions well above your skill level usually leads to disappointment.

Most employers advertise openings on the Internet and require online applications. To perform an Internet search, use a website dedicated to job searching or go directly to a company's website. If a computer is not available, libraries and unemployment offices generally have computer stations for public use. Job openings can also be found in local newspapers and trade magazines.

Today, one of the most important and productive methods to identify job opportunities is through networking. Studies indicate that as many as 70 percent of available positions are not publicly advertised. At the same time, job seekers are spending roughly 70 percent of their time looking through public job postings. Obviously, there are many more opportunities available than job seekers are able to identify through postings. There are many ways, including social media, to share your interest in finding a specific type of job. Identify companies that you would like to work with, regardless of job postings, and then look for contacts within the company that could help you identify opportunities. Express your interest in finding a job to family and friends. Networking is quite often the most productive method of finding and landing a job.

Ask your current supervisor, your co-workers, and friends that know you well if they would act as **references** for you. Current and accurate contact information for references must be provided. Both personal and professional references are recommended.

Ensure that your resume is up-to-date, well organized, and easy to read (*Figure 2*). Apply the effective writing techniques that you learned in *Basic Communication Skills* when writing a resume.

John Q. Smith
123 Main Street
Anytown, MD 67890
(555) 111-2233 – Home phone
(555) 333-4455 – Cell phone
johnqsmith@email.com

EMPLOYMENT OBJECTIVE

- To obtain a position within a construction company where I can use my carpentry and custom woodworking skills.

SKILLS AND QUALIFICATIONS

- Able to work independently and with a crew.
- Practical knowledge of carpentry hand and power tools, measuring tools, and woodworking tools.
- Trained and certified to use powder-actuated tools.
- Qualified to read residential and commercial construction drawings.
- Experience with energy efficient and conservation building techniques.
- Knowledge of LEED standards.

WORK EXPERIENCE

Journeyman Carpenter   *May 2011 – present*
LJL Construction, 123 Hammer Heights Road, Fairfax, VA
800-222-2222

- Built custom cabinets and closet systems
- Framed single family homes
- Trained apprentice carpenters

Carpenter   *June 2005 – May 2011*
Hammer Company, 456 Lathe Lane, Philadelphia, PA
866-555-1234

- Built closets and storage systems for commercial storage units

Carpenter   *February 2002 – June 2005*
Blue Ridge Carpenters, 789 Mountain Road, Harrisburg, PA
888-123-4567

- Framed condo units
- Built fences
- Raise wood walkways

CONSTRUCTION EDUCATION

- Carpentry apprenticeship. MK Builders Inc., Gettysburg, PA. Certificate, 2002.
- Professional carpentry certification program. ABC Training Corp., Philadelphia, PA. Certificate, 2000.
- Powder-actuated tools certification program. All States Community College, Marietta, PA. Certificate, 2000.

Figure 2 Resume.

By carefully selecting jobs that you are qualified for and by submitting an accurate, well-written resume, you improve your chances of being called for interviews (*Figure 3*). The effective communication skills that you learned in *Basic Communication Skills* will help you present yourself well.

Before accepting an offer, ask yourself the following questions to ensure the position will be the right fit:

- Is the salary enough to meet my needs?
- Does the company offer a benefits package that covers what I need?
- Will the work be interesting and challenging, but not more than I can handle?
- Does the company have a good reputation in the industry?
- Do the people appear to be nice to work with?
- Is travel required and am I able to agree to that?
- Does the company offer training?
- Are advancement opportunities within the organization evident?
- What is the company's safety record?

Before an offer is made, some trades require credentials or qualification testing to assess the applicant's skill level. To ensure workplace safety, many companies also require drug testing as part of the pre-employment process. Appropriate identification, such as a driver license, US Social Security card, national identification card, and/or a valid passport must be given to the employer. If a position outside of your country of citizenship is applied for, check the hiring and documentation requirements for that specific country and what may be required of non-citizens.

00108-15_F03.EPS

Figure 3 Job interviews are an important part of the hiring process.

After accepting a position, and even before the interview process begins, it is wise to obtain an organizational chart (*Figure 4*). Organizational charts are used to help all employees understand the structure of the company and what position and/or individual has responsibility for various areas. Understanding the organizational structure and being able to recall some important details about a prospective employer during an interview can make a strong impression.

Green-Collar Jobs

GOING GREEN

In recent years, there has been a big shift to environmentally friendly construction. This has created many new sectors in the US economy, including everything from recycling to energy retrofits; green building; solar installation/maintenance; water retrofits; and whole house performance (HVAC). These jobs are in green businesses and all involve work that directly affects environmental quality. Any job that involves the design, manufacture, installation, operation, and/or maintenance of renewable energy and energy-efficient technologies are considered green-collar jobs. Many of these jobs have opened new doors or removed barriers to employment. In the construction industry, opportunities exist in the following areas:

- Installation, construction, maintenance, and repair of energy retrofits, HVAC, solar panels, and whole home performance
- Installation, construction, maintenance, and repair of water conservation and adaptive grey water reuse
- Construction, carpentry, and demolition in green building and recycling

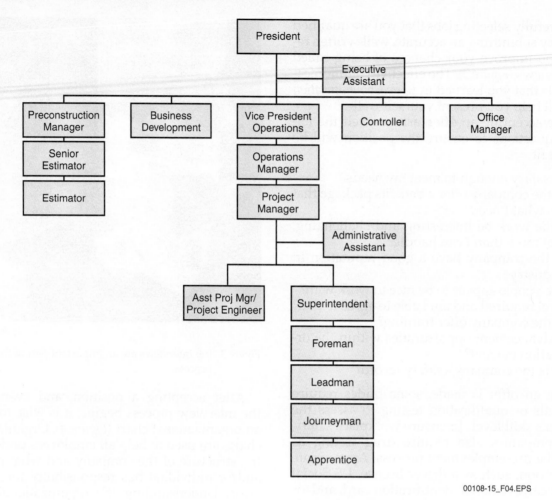

Figure 4 An organizational chart.

00108-15_F04.EPS

TWIC

In the United States, the Transportation Security Administration (TSA) and the US Coast Guard have worked together to create the guidelines for the Transportation Worker Identification Credential (TWIC). The tamper-resistant, biometric-based card is required for all workers who require unescorted access into secure areas of port facilities, outer continental shelf facilities, and all vessels regulated under the Maritime Transportation Security Act of 2002, as well as for all US Coast Guard-credentialed merchant mariners. An integrated circuit chip on the card stores the holder's information and biometric data for positive identification. This credential will likely expand to other forms of transportation in the future.

Completing an Application: Then and Now

Before computers and the Internet became part of daily life, job applicants approached the job market in a very different way. Instead of searching websites and sending their resumes off with the click of a mouse, more footwork was involved. Reading through the classifieds of the local newspaper or going door-to-door to fill out applications were arduous, time-consuming tasks.

Additional Resources

Knock 'em Dead Resumes: A Killer Resume Gets More Job Interviews! Martin Yate. 2014. Avon, MA: Adams Media.

Knock 'em Dead: The Ultimate Job Search, Martin Yate. 2014. Avon, MA: Adams Media.

1.0.0 Section Review

1. Who plans, coordinates, and controls a project's progress, or a portion thereof?

   a. Civil engineer
   b. Architect
   c. Project manager
   d. The funding bank

2. Which of the following is a valid statement about a resume?

   a. Always use a highly creative and unusual font to set your resume apart from all others.
   b. Format your resume chronologically, putting your current or last job first.
   c. Do not allow others to proofread your resume, as they will likely be critical of it.
   d. Resumes should not include certifications; present them only if called for an interview.

2.0.0 CRITICAL THINKING AND PROBLEM SOLVING

Objective

Explain the importance of critical thinking and how to solve problems.

   a. Describe critical thinking and barriers to solving problems.
   b. Describe how to solve problems using critical thinking.
   c. Describe problems related to planning and scheduling.

Trade Terms

Absenteeism: A consistent failure to show up for work.

Tardiness: Arriving late for work.

Having the ability to solve problems using critical thinking skills is a valuable skill in the workplace. There will be situations in which timely job completion is at risk because a problem suddenly arises. Using the tools of critical thinking skills can get the project moving forward and will also demonstrate your problem-solving capabilities.

2.1.0 Critical Thinking and Barriers

Throughout your construction career, you will encounter a variety of problems that must be solved. Critical thinking skills allow you to solve such problems effectively. Critical thinking means evaluating information and then using it to reach a conclusion or to make a decision. Critical thinking allows you to draw sound conclusions and make good decisions when you use the following approach:

- Do not let personal feelings get in the way of fairly evaluating information; put them aside and remain objective.
- Determine the cause (why it happened) and effect (what happened as a result).

- Think about the expertise or experience of people who are sources of information. Ask experts and people you trust for their advice. There is no reason to think through a problem in isolation.
- Compare new information with what you already know. If it does not fit, question it and try to separate fact from fiction.
- Weigh the merits of each option and alternative, and consider if one or more can be justified.

2.1.1 Barriers to Problem Solving

When searching for the solution to a problem, it is easy to fall into a trap that prevents you from making the best possible decision. The following are the most common barriers to effective problem solving:

- Closed-mindedness
- Personality conflicts
- General fear of change

To be closed-minded is to distrust any new ideas. Closed-minded people may also resist any suggested changes. Effective problem solving, however, requires you to be open to new ideas. Sometimes the best solution is one that you would have never considered on your own. Remember that other people have good ideas, too. You must be willing to listen and evaluate their input fairly.

Sometimes you may fail to appreciate the value of information or advice simply because you do not get along with the person offering it. One of the most important skills one can master is the ability to separate the message from the messenger. Weigh the value of the information separately from your feelings about the individual. This ability will show people that you are a true professional.

People often fear change when they believe it threatens them somehow, while it is often the lack of change that is the problem. In addition, it is often difficult to clearly see the path to implementing change or to envision the final result. When the path toward, or results of, a proposed change is difficult to see, the fear of change is combined with a fear of the unknown. If you embrace the concept of change, you will never stop finding new ways to solve problems. An ancient Greek philosopher named Heraclitus is credited for first speaking the words "change is the only constant in life."

2.2.0 Solving Problems Using Critical Thinking

Problems arise when there is a difference between the way something is and the way it should (or needs to) be. You might feel frustrated or intimidated by a problem, or you might feel that you do not have enough time to solve it. Your reaction might be to simply ignore the problem. Instead, try to look at it as an opportunity to demonstrate your skills. By actively seeking solutions to problems, you will demonstrate to your colleagues and supervisors that you are responsible and capable. Encountering and conquering a challenging problem can have significant effect on your reputation, in addition to building your confidence and level of experience.

To solve problems that arise on the job, use the following five-step process (*Figure 5*). Reviewing each of these steps in detail will help you understand how to apply them.

Step 1 – Define the problem. Before you can solve a problem, you need to know exactly what it is. This step might seem obvious, but people often find themselves trying to solve the wrong problem or one that doesn't even exist.

Step 2 – Analyze and explore alternatives. Once you have defined the problem, consider different ways to solve it. Collect information from a wide range of sources, and ask people for their opinions. Perhaps you have a solution in mind, or maybe just one step towards the solution. It can be very productive to seek opinions on solving small portions of a problem rather than seeking opinions on the big picture.

Teamwork is an important part of this step. The more suggestions you get from experts and co-workers, the better your chances are of finding the right answer. Identify and compare the alternatives. Look for solutions that will be the most cost-effective, take the least time, and ensure the highest-quality result. Try to identify the short- and long-term consequences, both good and bad, of each possible solution. Eventually, you will have created several possible options and given each one some thought.

Step 3 – Choose a solution and plan its implementation. Move forward by choosing the solution that you think will work best. Next, develop a plan to put the solution into action. The plan should include all necessary tools and materials. It should also specify all the tasks involved and who is responsible for them, and estimate how long each task will take.

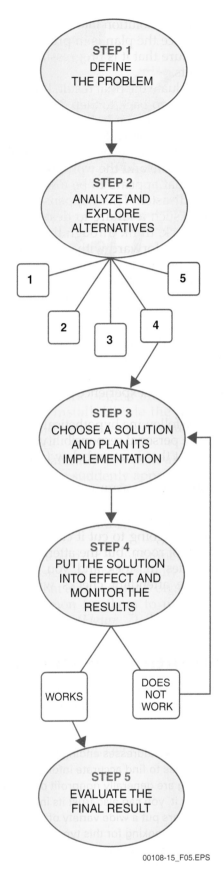

00108-15_F05.EPS

Figure 5 The five-step problem-solving process.

2.3.1 Materials

The materials required for a job are identified during the estimating stages of a project. Once the contract has been completed, they are then ordered from suppliers and delivered according to a prearranged schedule. Many job sites do not have sufficient room to store all the needed materials for a structure at one time. Problems with materials can include errors of quantity or type, delays in delivery, and unavailability due to backorders or even to abrupt business closure. A shortage of materials due to waste, theft, or vandalism is another source of project delays.

In many cases, material problems lead to changes in the material to be used. If the new material was not listed as an option in the contract specifications, a significant amount of time and effort may be required for the change request to make its way through the approval process.

2.3.2 Equipment

Major construction equipment is usually selected and scheduled well before the need arises. If a site planner is a member of the construction team, he or she may be responsible for ensuring that equipment is on site at the right time. Otherwise, project managers and other supervisors may be assigned this responsibility. This generally depends on the scope of the project.

Often, multiple pieces of equipment must be scheduled at the same time: backhoes to excavate a foundation and dump trucks to haul away the excavated dirt, for example. Problems with equipment can include unavailability on the scheduled day(s), lack of qualified operators, mechanical breakdown, and extended maintenance.

The lack of major equipment can create significant delays. In many cases, multiple crafts and subcontractors are relying on the equipment to move forward with their responsibilities. For example, if a crane is required to lift steel but it is not available for any reason, ironworkers, welders, and possibly other crafts find themselves at a standstill.

2.3.3 Tools

Workers and supervisors alike are responsible for ensuring that the appropriate tools are available. Workers are often expected to have common but specific tools when they are employed. Special tools that are expensive or unique are usually the responsibility of the employer. Both parties must ensure that they do their part to provide and maintain proper and functional tools.

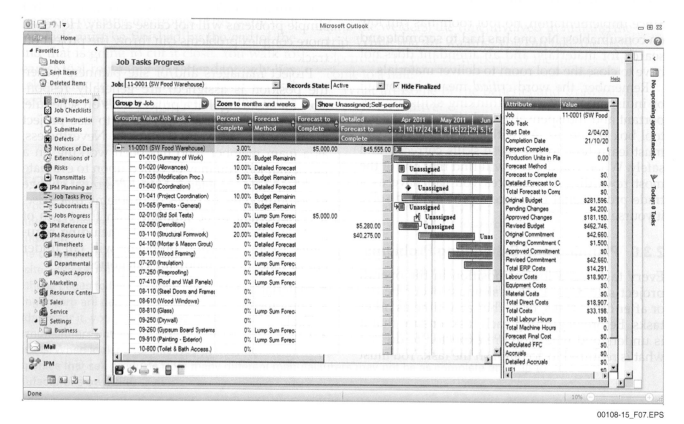

00108-15_F07.EPS

Figure 7 Project management software.

Imagine how a house project would stall if all the carpenters arrived for a day's work without hammers. On the other hand, if pneumatic nail guns are being used extensively, but the supervisor for the contractor does not ensure an air compressor is on site, the project comes just as quickly to a standstill. Other common problems related to tools include breakage or damage, loss or theft, or a lack of skilled users.

2.3.4 Labor

Labor—the men and women who perform the work on the job site—is the most important component of a project. Labor is often one of the most expensive project components as well, depending upon the country, project, and type of workforce needed. Project planners estimate the number of people needed for each day and identify the range of skills required of those workers. Foremen ensure that the right people are available and working on the right tasks. Common problems related

to labor include **tardiness** and **absenteeism**, lack of experience or qualifications to perform a given task, and not enough available workers. Idle time due to a lack of guidance or laziness can also contribute to delays.

2.3.5 Handling Delays

Always be aware of possible delays as you carry out your assigned tasks. Look as far forward into the portion of a project that you are responsible for, and consider the types of delays that could happen. If you see a potential cause of delay, notify your supervisor immediately. Having a plan in place for delays, especially delays that occur regularly or happen often, enables a solution to be applied promptly. Always know clearly what your responsibilities are in such situations. Be prepared to solve the problem yourself if that is part of your assignment. Even then, always keep your supervisors updated on delicate situations involving delays.

GOING GREEN

Banner Bank Building in Boise

What makes the Banner Bank Building one of the greenest buildings in the US? It was designed to do more with less. It uses many modular components that can be reused over and over again as changes are made. Modular wall systems, along with modular power and data components, are easy to relocate without waste. The pressurized raised floor system delivers air at a higher temperature and a lower velocity than common overhead systems. It also eliminates a great deal of ductwork, and allows the location of supply air grilles to be changed at will. The structural components were designed to use less steel and concrete than most buildings in its class. Materials with a high level of recycled components were used where available and when quality would not be compromised.

00108-15_SA01.EPS

The building was intentionally constructed close to public transportation nodes to encourage their use by employees. It also has inside storage for bicycles and showers to accommodate those who want to take green living a step further. Water is reclaimed and a geothermal heating system was installed. The building is even equipped with backup generators that run on biodiesel derived from used vegetable oil.

Additional Resources

The Re-Discovery of Common Sense – A Guide to the Lost Art of Critical Thinking. Chuck Clayton. 2007. Lincoln, NE: iUniverse, Inc.

2.0.0 Section Review

1. In critical thinking, if new information you receive about an issue does not fit with what you already know, _____.

   a. question it and try to separate fact from fiction
   b. change what you previous held as factual in your mind
   c. berate the person and let them know you don't appreciate that kind of input
   d. thank them and forget about the information they offered

2. Why is it important to follow a plan through once it is implemented?

   a. To be sure there are no other options
   b. To be sure no one asks questions
   c. To look for other opinions
   d. To be sure the results are what you are looking for

3. If material that was not specified in contract documents must be substituted for another type, the change will likely need to _____.

   a. pass through an approval process
   b. pass through a testing procedure
   c. be agreeable to all other trades
   d. be considered after the project is done

3.0.0 RELATIONSHIP AND SOCIAL SKILLS

Objective

Explain the importance of social skills and identify ways good social skills are applied in the construction trade.

- a. Identify good personal and social skills.
- b. Explain how to resolve conflicts with co-workers and supervisors.
- c. Explain how to give and receive constructive criticism.
- d. Identify and describe various social issues of concern in the workplace.
- e. Describe how to work in a team environment and how to be an effective leader.

Trade Terms

Amphetamine: A class of drugs that causes mental stimulation and feelings of euphoria.

Barbiturate: A class of drugs that induces relaxation, slowing the body's ability to react.

Bullying: Unwanted, aggressive behavior that involves a real or perceived power imbalance. This form of harassment may include offensive, persistent, insulting, or physically threatening behavior directed at an individual.

Cannabinoids: A diverse category of chemical substances that repress neurotransmitter releases in the brain. Cannabinoids have a variety of sources; some are created naturally by the human body, while others come from cannabis (marijuana). Still others are synthetic.

Compromise: When people involved in a disagreement make concessions to reach a solution that everyone agrees on.

Confidentiality: Privacy of information.

Constructive criticism: A positive offer of advice intended to help someone correct mistakes or improve actions.

Hallucinogen: A class of drugs that distort the perception of reality and cause hallucinations.

Harassment: A type of discrimination that can be based on race, age, disabilities, sex, religion, cultural issues, health, or language barriers.

Initiative: The ability to work without constant supervision and solve problems independently.

Leadership: The ability to set an example for others to follow by exercising authority and responsibility.

Methamphetamine: A highly addictive crystalline drug, derived from amphetamines, that affects the central nervous system.

Opiates: A narcotic painkiller derived from the opium poppy plant or synthetically manufactured. Heroin is the most commonly used opiate.

Professionalism: Integrity and work-appropriate manners.

Self-presentation: The way a person dresses, speaks, acts, and interacts with others.

Sexual harassment: A type of discrimination that results from unwelcome sexual advances, requests for sexual favors, or other verbal or physical behavior with sexual overtones.

Synthetic drugs: A drug with properties and effects similar to known substances but having a slightly altered chemical structure. Such drugs are often not illegal since they are somewhat different than well-defined restricted or illegal substances. The two typical categories are cannabinoids (lab-produced THC or marijuana substitutes) and cathinones, which are designed to mimic the effects of cocaine or methamphetamines.

Tactful: Being aware of the effects of your statements and actions on others.

Work ethic: Work habits that are the foundation of a person's ability to do his or her job.

Zero tolerance: The policy of applying laws or penalties to even minor infringements of a code in order to reinforce its overall importance, typically related to drug and alcohol abuse when applied to the workplace.

A relationship results from the process of interacting with another person, or a group of people. Relationships are affected both by real actions and the perceptions of individuals. Every day, you interact with co-workers, supervisors, and members of the public who see you working. No one wants to work with, hire, or spend time with an unprofessional person.

Craftworkers need to be aware of the appropriate professional conduct for work situations, and follow that conduct at all times. Your actions reflect on your own professional status, that of your colleagues, the company you work for, and the image of your profession as seen by the public.

3.1.0 Personal and Social Skills

Proper self-presentation—the way you dress, speak, act, and interact—is a vital part of any successful work relationship. Self-presentation involves developing good personal and work habits. Personal habits apply to your appearance and general behavior. Work habits, also known as your work ethic, apply to how you do your job. Your personal habits and work ethic make powerful impressions on your colleagues, supervisors, and potential employers. Something as simple as filling out a job application with no errors and good penmanship will give an employer a good first impression.

When you introduce yourself, look the other person in the eyes and stand with confidence. In the United States and a number of other countries, a firm and sincere handshake has long been considered an indicator of a person's character and personality, especially among men. Recent studies indicate that there actually may be a connection. A firm grip (but not so firm as to imply a competition is at hand) and making eye contact at the appropriate moment make a good impression, regardless of gender. Once formed, first impressions are hard to change, so make sure that the first impression you make is a good one.

3.1.1 Personal Habits

Co-workers and supervisors like people who are dependable. Co-workers know that they can trust dependable people to pull their own weight, take their responsibilities seriously, and look after one another's safety. Supervisors know that dependable people will do their best to finish a job correctly and on schedule. When dependable workers say they will do a task, they follow through on that commitment. It also means showing up for work on time every day and not stretching out lunch hours and breaks.

Organizational skills are important as well. Keep your tools and your work areas clean and organized. Know which tools you are responsible for and keep them in good working order. This applies to your personal tools as well as those belonging to your employer. Always plan what you need to do each day before you begin. Although your best plan is likely to change as the day unfolds, a failure to plan is a plan for failure. Approach your work in an organized fashion and follow the schedule that you have been assigned.

The War Against Absenteeism

During World War II, factories across America were humming along, many producing products needed for the war effort. To combat absenteeism in the workplace, employers like Remington Arms related the issue to patriotism and the crucial role that America's workers played in the war effort.

"So you're not coming in tomorrow, Bud"

Come On, Remington --
LET'S STAY ON THE JOB!

00108-15_SA02.EPS

Ensure that you are technically qualified to perform your duties and specific tasks. This means that you know how to use tools, equipment, and machines safely. Take advantage of opportunities to expand your technical knowledge through classes, books and trade periodicals, and mentoring by experienced colleagues. Being technically qualified also means that you will not attempt a task that you are unfamiliar with when any level of safety hazard exists.

Offer to pitch in and help whenever you can. The best workers are willing to take on new tasks and learn new ideas. Supervisors notice employees who are willing, but be careful not to take on more tasks than you can handle; you will not impress anyone if you cannot fulfill what you promised.

Honesty is one of the most important personal habits you can have. Do not abuse your company's system by calling in sick just to take a day off or by leaving early and asking someone else to punch you off the clock. Never steal from the company or from your co-workers; this includes everything from tools and equipment to simple office supplies. If you are struggling with a problem, don't hide it; speak with your supervisor about it truthfully. Be willing to look actively for a solution to problems you are facing.

Professionalism means that you approach your work with integrity and a professional manner. As you learned in *Basic Safety*, there is no place for horseplay or irresponsible behavior on the construction site. Employers want workers who respect the rules, and who understand that rules exist to keep people safe and projects on schedule. A professional employee always respects company confidentiality. In some cases, workers are required to sign documents indicating that they understand the importance of confidentiality and that they agree not to discuss certain information that is shared with them.

Most companies have a dress code, and often they have additional special requirements for specific jobs. Always follow these requirements. Pushing the limits of such policies never impresses anyone of importance. Also, do not forget to attend to the basics of good grooming. People who pay attention to their appearance and develop positive personal habits are more likely to be considered for a job over people who do not have good grooming habits. On the job site, you are a reflection of your trade and employer, as well as yourself.

3.1.2 Work Ethic

Along with good personal habits, a strong work ethic is essential for getting hired and being promoted. Employers look for people who they believe will give them a fair day's work for a fair day's pay. Having a strong work ethic means that you enjoy working and that you always try to do your best on each task. When work is important to you, you believe that you can make a positive contribution to any project you are working on.

Construction work requires people who can work without constant supervision. When there is a problem that you can solve, solve it without waiting for someone to tell you to do it. If you finish your task ahead of time, look for another task that can be finished in the remaining time or ask co-workers if they need help. This type of positive action is called taking **initiative**. Colleagues and supervisors will respect you more when you demonstrate initiative on the job.

An important part of taking initiative, however, is to know when and how to implement it. Suppose a supervisor tells you to perform a task. He demonstrates five steps required to complete the task. Later, as you perform the work, you realize that one of the steps is unnecessary. Leaving out the step could save time—but should you? No; that would not be the right initiative. Instead, tell your supervisor about your discovery, and ask what you should do. Bringing the options to your supervisor is showing initiative. Your supervisor may realize that you are right and allow you to leave out the extra step. After that, every time anyone performs that task, it will take less time and save the company money. Or, your supervisor may explain why the step is important. Then you will have learned something more about the work you are doing. The result of taking the initiative in either case is a positive one.

3.1.3 Tardiness and Absenteeism

The two most common problems supervisors face on the job are tardiness and absenteeism. Tardiness is when a worker habitually shows up late to work. Absenteeism is when a worker consistently fails to show up for work at all, with or without excuses. People with a strong work ethic are rarely late or absent.

To make a profit and stay in business, construction companies operate under tight schedules. These schedules are primarily built around workers. If you are late or do not show up regularly, expensive adjustments become necessary. Your employer may decide that the best way to save money is to stop wasting it on you.

Consider the following suggestions for improving or maintaining your record of punctuality and attendance:

- Think about what would happen if everyone on the job were late or absent frequently.
- Think about how being late or absent affects your co-workers.
- Know and follow your company's policy for reporting legitimate absences or lateness.
- Keep your supervisor informed if you need to be out for more than a day.
- Allow yourself enough time to get to work.
- Explain a late arrival to your supervisor as soon as you get to work.
- Do not abuse lunch and break-time privileges.

3.2.0 Conflict Resolution

Conflict resolution is an important relationship skill, because conflict can happen anywhere, anytime. Conflicts between you and your co-workers can arise because of disagreements over work habits; different attitudes about the job or the company; differences in personality, appearance, culture, or age; or distractions caused by problems at home. Conflicts between you and your supervisor can happen because of a disagreement over workload; lateness or absenteeism; or criticism of mistakes and inefficiencies. Significant disagreements with a supervisor should be reported to the Human Resources (HR) office of your employer.

Most of the time, people are not trying to turn disagreements into conflicts. People are often unaware of the effects of their behavior. Before reacting negatively to a co-worker's behavior, remember to be **tactful**. This means considering how the other person will feel about what you say or do. Do not accuse, embarrass, or threaten the person. This behavior rarely leads to an acceptable solution to anything.

Never let a disagreement or conflict affect job performance, team morale, or—most importantly—site safety. The goal is to keep events from turning into conflicts in the first place. If that is not possible, then the next best thing is to address the conflict quickly and resolve it professionally. If a disagreement with a co-worker is getting out of hand, try one of the following techniques to cool the situation down:

- Think before you react.
- Walk away, explaining that you need some time to think through the situation.
- Try not to take it personally.
- Avoid being drawn into others' disagreements.

Do not let a conflict simmer and then boil over before you take action. Not only is it unprofessional, tempers may flare unexpectedly and make the situation worse. Being proactive when an issue becomes uncomfortable is usually better than being reactive. This is true unless tempers are already out of bounds. In that case, it may best to allow some time to pass. If, despite your best efforts, the conflict escalates, walk away and notify your supervisor immediately.

Keep in mind, however, that there are important differences between the way you resolve conflicts with your co-workers and the way you resolve conflicts with your supervisor. The following sections discuss how to handle these types of conflicts.

3.2.1 Resolving Conflicts with Co-Workers

Remember to have respect for the people you disagree with. After all, they believe they are right, too. Be clear, rational, respectful, and open-minded at all times. Begin by admitting to each other that there is a conflict. Then analyze and discuss the problem. Allow everyone to describe his or her own perception of the conflict. You may realize that the whole problem was simply miscommunication. Begin by asking the following questions:

- How did the conflict start?
- What is keeping the conflict going?
- Is the conflict based on personality issues or a specific event?
- Has this problem been building up for a while, or did it start suddenly?
- Did the conflict start because of a difference in expectations?
- Could the problem have been prevented?
- Do both sides have the same perception of what is happening?

Once you have analyzed the situation, discuss the possible solutions. You will probably have to **compromise** to find a solution that everyone agrees on. When you agree on the solution, act on it, and see if it works. If it does not, then consult your supervisor for help in resolving the conflict. Notice that this process is similar to the problem-solving techniques discussed earlier. If you find that the conflict happened because you were wrong, apologize. An apology is not a sign of weakness; it is a sign of respect.

Once an agreement or compromise has been reached, stand by it in every possible way, and do so with a positive attitude.

3.2.2 Resolving Conflicts with Supervisors

You can usually approach co-workers as equals, because you are working on the same job. However, on the job, your supervisor is in charge of you and your work. This means that you must use a different approach to resolve conflicts with your supervisor.

Before going to your supervisor, take some time to think about the cause of the conflict. Consider writing down your thoughts; organizing the information this way often puts things into perspective. When you approach your supervisor, do so with respect. Wait until your supervisor has a free moment, and then ask if you could arrange a time to talk about something important; or leave a message or note for your supervisor asking to meet. Be willing to meet at a time that is convenient for your supervisor. Remember, supervisors have many responsibilities. They do not have much free time during the regular workday, and you may have to meet before or after work.

When meeting with your supervisor, speak calmly and clearly. Do not be emotional, sarcastic, or accusatory. Do not confront your supervisor in a threatening or angry way. State only the facts as you see them; never say anything that you cannot prove. Do not mention the names of your co-workers unless they are directly involved. If you want to suggest changes or solutions, explain them clearly and discuss how they will benefit the people involved.

Once you have made your case, allow your supervisor to make a decision. You should accept and respect your supervisor's final decision. It may not be the one you wanted, but it will be the one that must be followed.

3.3.0 Giving and Receiving Criticism

As a construction professional, you will always be learning something new about your job. New technologies, materials, and methods appear all the time. For example, talking to an experienced construction worker can help you learn a new way to perform a task. As your skills improve with practice, you will be able to use tools and methods that you were not able to use before. This is a common way of learning on the job (*Figure 8*).

Another way to learn on the job is through **constructive criticism**. Constructive criticism is advice designed to help you correct a mistake or improve an action. Constructive criticism can improve your job performance and relations with co-workers. You have probably heard the word *criticism* used in a negative way to indicate fault or blame; that is not the type of criticism discussed here. Constructive criticism does not mean that colleagues and supervisors think little of you; in fact, it means exactly the opposite. Colleagues and supervisors who offer constructive criticism do so because they believe it is worth their time to help you improve your skills.

As you gain experience on the job, you will be able to give constructive criticism as well as receive it. To make sure that someone does not mistake your constructive criticism for blame, you need to know how to offer it in a positive manner. The following sections offer some general advice on offering and receiving constructive criticism.

00108-15_F08.EPS

Figure 8 A good employee never stops learning on the job.

3.3.1 *Offering Constructive Criticism*

When you are training a less-experienced person or working with co-workers, you might find yourself offering some constructive criticism. How should you offer it? Before you say anything, think about the rules of effective speaking that you learned in *Basic Communication Skills*. Use positive, supportive words and offer facts, not opinions.

Constructive criticism works best when offered occasionally. Do not constantly comment on another person's work or methods; they may block out your criticism or even become angry. Never criticize people in front of their co-workers or supervisors; they may feel embarrassed and will likely resent you. Constructive criticism should include suggested alternatives. Do not criticize how a co-worker does something unless you can suggest another way. Above all, limit your comments to the person's work or methods, without targeting them as a person. There is a great deal of difference in how people react to a statement such as "your pipefitting work is unacceptable" compared to "this threaded joint is unacceptable."

Point out improper behavior or incorrect work techniques the first time they occur. If not addressed, the behavior or technique could become a bad habit. Bad habits, in turn, can lead to accidents. Gently and firmly offer constructive criticism for improper behavior as soon as it is noticed.

Remember to compliment the person you are criticizing. Compliments have a genuinely positive effect on the person receiving them, and they are easy to offer. You do not have to wait to compliment someone until you are prepared to offer constructive criticism. However, if a compliment is to be offered along with constructive criticism, offer the compliment first.

Try to offer compliments on a regular basis. Your appreciative and respectful approach will help keep team spirits high and your supervisor will appreciate your professional attitude.

3.3.2 Receiving Constructive Criticism

Not only can constructive criticism mean the difference between success and failure, but in cases where workplace safety is involved, it can also be the difference between life and death. Think of constructive criticism as a chance to learn and to improve your skills. Take constructive criticism from your supervisor seriously. Treat constructive criticism from experienced co-workers with the same respect; even though they may not be your supervisor, their experience gives them a level of expertise. When someone presents you with constructive criticism in the proper manner, learn from their approach as well.

Always take responsibility for your actions. Never be overly defensive or dispute criticism. If you respond negatively, you might offend a person who is simply trying to help you. As a result, that person will either lose respect for you or will no longer be willing to help. If that happens, the quality of your work, and that of everybody who works with you, will suffer.

When someone offers you constructive criticism, demonstrate your positive personal habits and work ethic. If the criticism is vague, ask the person to suggest specific changes that you should try to make. If you do not understand the criticism, ask for more details. The deepest respect you can show is to improve your performance based on their advice.

You have a right to disagree with criticism that is unacceptable or incorrect. Someone might honestly misunderstand your situation and offer incorrect advice. Or a co-worker might criticize

---

It's All In How You Say It

Constructive criticism should be given in a positive, non-confrontational way. People can be sensitive, and when approached about how they do their jobs, it becomes very personal to them.

When delivering constructive criticism, it is not a good idea to start the sentence off with "You" because it will immediately put the person on the defense. Instead, try starting out with "In the future." This approach will sound less like an attack on their abilities and they will react and listen better. Talk about the work, not the person.

---

you simply because he thinks he performs a task better than you do. In such cases, clearly and respectfully give your reasons for disagreeing, and explain why you believe that you are correct.

3.3.3 Destructive Criticism

Constructive criticism has positive effects. However, you may also encounter destructive criticism. Destructive criticism, as its name suggests, is designed to hurt, not help, the person receiving it. The destructive criticism could actually be about something that needs improvement, but because it is offered negatively, there is no desire to listen.

If you receive destructive criticism, the professional thing to do is to stay calm. Do not get into a fight by replying in a negative tone. Find a way to let the person know how the criticism made you feel. If there is a legitimate criticism beneath a destructive tone, ask the person to offer positive suggestions for ways you can improve. By taking a positive approach to negative criticism, you can set a good example for others.

3.4.0 Social Issues in the Workplace

The modern construction workplace is a cross-section of our society (*Figure 9*). A typical construction project involves men and women from all walks of life, of many ethnic and racial backgrounds, and often from many different countries. Many workers speak more than one language. Workers have grown up in, and currently live in, many different income brackets.

00108-15_F09.EPS

Figure 9 Today's construction workplace is a cross-section of our society.

Construction workers face a wide range of mental and physical demands every day on the job. These demands can sometimes feel overwhelming, and people may seek to escape pressure through illegal drugs or alcohol abuse. As a construction professional, you need to be aware of these issues. You also need to know what to do if someone is behaving inappropriately.

You may encounter the following issues in the workplace:

- Harassment or bullying
- Stress
- Drug and alcohol abuse

3.4.1 Harassment

Harassment is a broad term describing negative social behavior that can be based on race, age, disabilities, gender, religion, cultural issues, health, or language barriers. Harassment often takes the form of ethnic slurs, racial jokes, offensive or derogatory comments, and other verbal or physical conduct. It can create an intimidating, hostile, or offensive working environment. It can also interfere with an individual's work performance.

One type of harassment commonly reported and talked about is sexual harassment. While other forms of harassment may not have been talked about as much or as openly in the past, this issue has made headlines for many years.

When someone makes unwelcome sexual advances, or requests or exhibits verbal or physical behavior with sexual overtones, he or she is guilty of sexual harassment. Contrary to popular belief, sexual harassment can happen between members of the opposite sex or even those of the same gender. The harasser may threaten to fire the victim to keep him or her quiet. Sexual harassment is illegal, and if you experience it, you should report it to your supervisor immediately. Usually, sexual harassment is a pattern of behavior repeated over a period of time, so if you do not report it, you run the risk of experiencing it over and over again.

Bullying may be more common than either sexual harassment or racial discrimination on the job, but it is not reported as often. The term *bullying* is perceived by most people to be related to school-age children. The United States government defines bullying as the unwanted, aggressive behavior among school-aged children that involves a real or perceived power imbalance. That definition covers a lot of ground. When bullying is discussed among adults, it is typically referred to as hazing, stalking, or simply harassment. Most adults are uncomfortable using the term *bullying* when reporting such behavior, as they also perceive it to be related to school-age children. Using the term makes adult victims feel childish to report it as a result. Bullies may target those that appear to be a threat to them and feel the need to control the threat. In some cases, it occurs simply because the individual feels stronger and wishes to exert control to build on their feeling of superiority.

Workers should understand that dealing with an adult bully can be very much like trying to work things out with a school-age bully. Adult bullies are often lifetime bullies, having started the practice at an earlier age. They are not typically interested in compromise or improvement. Bullying makes them feel important and in control. Adult bullies are hard to change through simple interaction and positive discussions.

Workplace bullying can come from an individual or even a group. Bullying, just as any other form of harassment, can cause serious emotional and physical illnesses such as anxiety, depression, and hypertension among victims.

With the use of today's electronic technology, bullies can even cause harm through social media and other public formats; this is referred to as cyberbullying. Once emailed rumors, malicious text messages, embarrassing pictures, videos, or fake profiles are posted on social media, the damage is difficult to undo. While the most common target for cyberbullying is school children, adults are not immune to it.

Once a bullying situation has been reported to your employer, the situation may not change immediately. Your supervisor and/or employer must work within the confines of the law. Careful documentation of events is very important, as it is in all forms of harassment. Clear documentation allows the situation to be handled properly by those in authority. Since there are laws prohibiting such behavior and your co-workers are not school children, legal action will often occur based on careful documentation.

3.4.2 Drug and Alcohol Abuse

Drinking is a common way to avoid or forget about stress. Alcohol is an accepted, legal part of modern social life. The moderate use of alcohol is socially acceptable and is generally believed to cause little or no harm to most people. Alcohol abuse, or habitually drinking to excess, not only is socially unacceptable, but also poses a serious health and safety risk to everyone involved. Although some countries and US states have legalized or decriminalized the recreational and/or medical use of marijuana—and others are contemplating it—it too can be abused and have the same effects as alcohol abuse. Regardless of whether alcohol and marijuana are considered socially acceptable, there is no room in the workplace for either substance. Among employers, this is referred to as **zero tolerance** (*Figure 10*).

Illegal drug use is never acceptable, nor is the misuse of legal prescription drugs. Many companies require a clean drug test before offering a job or allowing an employee to begin work. In fact, many employers in the construction industry are often required to share records related to drug testing as part of their obligations to their clients. Drug testing can also be ordered randomly, to test the current workforce and ensure compliance. Further, an accident in the workplace often leads to immediate drug testing to determine if illegal substances were a factor. There are a number of situations that can lead to a drug test beyond a pre-employment situation. This matter is taken very seriously by the industry as a whole.

00108-15_F10.EPS

Figure 10 Signs like this are common in the construction workplace.

Types of drugs that are sometimes used inappropriately by workers include the following:

- Amphetamines
- Methamphetamines
- Barbiturates
- Hallucinogens
- Opiates
- Synthetic drugs

Amphetamines and methamphetamines are addictive stimulants that affect the central nervous system. Also called uppers, they are used to prolong wakefulness and endurance, and they produce feelings of euphoria, or excitement. They can disturb vision; cause dizziness and an irregular heartbeat; cause the loss of coordination; and can even cause physical collapse or unconsciousness. Cocaine, crack, and crystal meth are examples of illegal amphetamine drugs.

Barbiturates are a sedative, which basically means they cause you to relax. They are also called downers. They can create slurred speech; slow reactions; cause mood swings; and cause a loss of inhibition, which means that they make a person feel less shy or self-conscious. Many abused barbiturates are controlled-substance prescription drugs.

Hallucinogens distort the perception of reality to the point where people see things that are not there (hallucinations). The effects of hallucinogens can range from ecstasy to terror. When someone is hallucinating, they put themselves and others at great risk because they cannot react to situations the right way. Hallucinogens can also cause chills, nausea, trembling, and weakness. Mescaline and LSD are two examples of illegal hallucinogens.

Derived from opium, opiates are used to kill pain. Morphine is the primary psychoactive chemical found in opium, and therefore the basis of most opiates. While opiate use creates a feeling of euphoria and relieves pain, it can cause muscle spasms, cramps, anxiety, nausea, fever, and diarrhea over time. Some prescription-drug forms of opiates are Oxycontin and Vicodin.

Synthetic drugs are relatively new. They are defined as drugs with properties and effects similar to well-defined and familiar restricted or illegal substances, but having a slightly altered chemical structure. Such drugs are often not illegal since they are somewhat different from the illegal substances. However, due to the health risks and other problems associated with them, legislation has begun in some areas to control them. The two typical categories are cannabinoids (lab-produced THC or marijuana substitutes) and cathinones, which are designed to mimic the effects of cocaine or methamphetamines. The cathinones are more commonly known as bath salts (but they are not used for this purpose). Synthetic drugs have quickly become readily available for retail sale. They are often deceptively packaged and do not generally list all the ingredients. Synthetic drugs can be extremely hazardous and even deadly, as users have no idea what their true composition is. Lab testing has shown the potency of some formulas can be as much as 500 times as potent as marijuana.

If you abuse drugs or alcohol, you will probably not be able to get or keep the job you want. If you are discovered using illegal drugs, you can be fired on the spot because of the safety risks related to drug use. Addiction affects your well-being and state of mind, and you cannot do your best on the job or at home. It can also jeopardize your relationships with friends, family, and co-workers. If people you know are addicted to alcohol or drugs, seek help for them or encourage them to get help for themselves. If you are having trouble with alcohol or drugs, talk to your supervisor. Many companies provide substance abuse counseling and will support you with periodic drug testing to help you remain sober and drug-free. The *Appendix* contains a list of organizations that can help people cope with alcohol and drug issues.

3.5.0 Teamwork and Leadership

Every day, you will interact with members of your work crew and your supervisor. Although the ability to get along with people is an important skill, it is not quite enough. The success of any team depends on all of its members filling their roles competently. Everyone must contribute to the effort to ensure that the team achieves its goals. This cooperation is referred to as teamwork.

From the moment your career begins, you will be assigned to different teams for different reasons. Teams can be as small as two people and as large as your entire company. Teams are often made up of people from different trades who work together to complete a task. For example, a team that is assigned the task of burying water and electrical lines might include a backhoe operator to dig the trench, pipefitters to lay and join the pipe, and electricians to run the electrical cable.

Always show respect for the other members of your team. Although you have a specific job to do, always be willing to help a co-worker. Support your co-workers when they need it and they will support you in return.

Good team members are goal-oriented, and those goals are shared among team members. Being goal-oriented means making sure that all activities benefit the team's final objective. Whatever the end result is—fabricating fittings, laying pipe, or framing a house—everyone on the team knows the desired result in advance and concentrates on achieving it.

As a team member, use your skills and strengths to the team's advantage. At the same time, accept your personal limitations and the limitations of others. To be a good team member, strive to do the following:

- Follow your team leader's and/or supervisor's directions.
- Accept that others might be better at some tasks than you.
- Keep a positive attitude when you work with other people.
- Recognize that the work you do is for the benefit of the entire company, not for you personally.
- Learn to work with people who work at different speeds.
- Accept goals that are set by someone else, not by you.
- Trust other members of the team to perform their tasks, just as you perform yours.
- Appreciate the work of others as much as you appreciate your own.

Keep in mind that you are not the only person working hard. Everyone on your team should be focusing on the goal, too. Offer praise and encouragement to your co-workers, and they will do the same for you. This mutual respect will help you feel confident that your team will reach its goal. Share the credit for good work, and be willing to take responsibility for your mistakes and errors. These actions will help you earn the respect of your co-workers.

Did You Know?

Discrimination is Prohibited

Several federal laws prohibit job discrimination:

- **Title VII of the Civil Rights Act of 1964** – Prohibits employment discrimination based on race, color, religion, sex, or national origin.
- **Equal Pay Act of 1963** – Protects men and women who perform substantially equal work in the same establishment from sex-based wage discrimination.
- **Age Discrimination in Employment Act of 1967** – Protects individuals who are 40 years of age or older from workplace discrimination.
- **Title I and Title V of the Americans with Disabilities Act of 1990** – Prohibits employment discrimination against qualified individuals with disabilities in the private sector and in state and local governments.
- **Sections 501 and 505 of the Rehabilitation Act of 1973** – Prohibits discrimination against qualified individuals with disabilities who work in the federal government.
- **Civil Rights Act of 1991** – Provides monetary damages in cases of intentional employment discrimination.

Source: US Equal Employment Opportunity Commission

If goal-oriented teamwork is practiced, the team should be able to meet deadlines and keep projects on schedule. This translates into time and money saved by your company that can be used for pay increases. There will be times when your co-workers or other trades cannot start their tasks until you have completed yours. Out of respect for them and your company's reputation, always finish your work in a timely manner. The success of a team reflects on you, just as your personal behavior and characteristics reflect on the team.

An important part of teamwork is training. As an apprentice, you will receive guidance and advice from more experienced colleagues. As you gain experience, you will be able to help less-experienced colleagues. Training also offers an excellent opportunity to build good relationships.

Being asked to teach someone is an honor. It means that someone believes that you do your job well enough to teach it to others. Such confidence should inspire you to be the best teacher you can be. Be patient with the person you are teaching, and teach by example. Offer encouragement and give constructive criticism, as you have learned in this module. Teach people in the same manner that you would like to be taught. Keep in mind that your teachers in the workplace are not likely to be trained educators. As a result, their approach may be less than perfect. Discount their limitations and focus on the skill or knowledge they are sharing with you.

3.5.1 Leadership Skills

As you gain experience and earn credentials, you will assume positions of greater responsibility. These positions are earned through hard work and dedication. Starting as an apprentice, you can work your way up to team leader, foreman, supervisor, and project manager. Someday, with enough hard work, dedication, and ambition, you may even choose to run your own company.

To progress steadily through your career, you will need to develop leadership skills and learn how to use them. Leaders set an example for others to follow. Because of their skills, leaders are trusted not only with the authority to make decisions, but also with the responsibility to carry them out.

You can become a leader at any stage in your career. As an apprentice, you have the authority to perform the task given to you by your supervisor, and your supervisor expects you to be a responsible worker. By carrying out your task quickly, correctly, and independently, you are setting an example for others to follow. In doing so, you are demonstrating leadership skills.

People with the ability to become leaders often exhibit the following characteristics:

- They lead by example.
- They have a high level of drive, determination, and persistence.
- They are effective communicators.
- They can motivate their team to do its best work.
- They are organized planners.
- They have self-confidence.

The functions of a leader vary with the environment, the group of workers being led, and the tasks to be performed. However, certain functions are common to all situations. Some of these functions include the following:

- Organizing, planning, staffing, directing, and controlling.
- Empowering team members with authority and responsibility.
- Resolving disagreements before they become problems.
- Enforcing company policies and procedures.
- Accepting responsibility for failures as well as for successes.
- Representing the team to different trades, clients, and others.

Leadership styles vary. If leadership styles are classified according to the way a leader makes decisions, the result is three broad categories of leadership: autocratic, democratic, and hands-off. An autocratic leader makes all decisions independently, without seeking recommendations or suggestions from the team. A democratic leader involves the team in the decision-making process; such a leader takes team members' recommendations and suggestions into account before making a decision. A hands-off leader leaves all decision making to the team members themselves.

How Do Your Co-Workers See You?

Do you exhibit any of the following unpopular behaviors on the job? If so, it is likely they are keeping you from being an effective team member. Develop an action plan to deal with them. Your action plan should be a two- or three-step process that will allow you to correct the behavior. Be honest with yourself!

Being a loner	Bad-mouthing the company or your boss
Taking yourself too seriously	Being unable to take a joke
Being uptight	Bragging
Always needing to be the best at everything	Taking credit for others' work
Holding a grudge	Being sarcastic
Arriving late to work	Refusing to listen to other people's ideas
Being inconsiderate	Looking down on other people
Taking breaks that are too long	Being unwilling to pitch in and help
Gossiping about others	Being stingy with a compliment
Sticking your nose into other people's business	Having a chip on your shoulder
Acting as a spy and reporting on the behavior of others to your supervisor	Horsing around when others are trying to work
Saying or doing things to create tension or unhappiness	Thinking you work harder than everyone else
Complaining constantly	Manipulating people

Select a leadership style that is appropriate to the situation. Leading with a single style at all times generally results in some real problems. You will need to consider your authority, experience, expertise, and personality each time a style is chosen. Leaders need to have the respect of people on the team; otherwise, they will be unable to set an example others are willing to follow.

Leaders have to make sure that the decisions they make are ethical. Success in the construction industry demands the highest standards of ethical conduct. The three types of ethics that you will encounter on the job are business or legal ethics, professional ethics, and situational ethics. Business or legal ethics involve adhering to all relevant laws and regulations. Professional ethics involve being fair to everybody, and situational ethics involve appropriate responses to a particular event or situation.

Effective leaders motivate, or inspire, people to do their best. People are motivated by different things at different times. The following are some common ways to motivate people on the job:

- Recognize and praise a job well done.
- Allow people to feel a sense of accomplishment.
- Provide opportunities for advancement.
- Encourage people to feel that their job is important.
- Provide opportunities for change to prevent boredom.
- Reward people for their efforts.

Leaders who can motivate people are more likely to have a team with high morale and a positive work attitude. Morale and attitude are key components of a successful company with satisfied workers, and such a company is more likely to be successful.

Action Plan for Improvement

Example:

Problem: Gossiping about others

Action Plan: 1. Walk away when people start gossiping or bad-mouthing others.

2. Don't repeat what I hear.

3. Focus on my own work, not on that of others.

Problem: _____

Action Plan: _____

_____

Problem: _____

Action Plan: _____

_____

Problem: _____

Action Plan: _____

_____

00108-15_SA03.EPS

Building a Team

In this exercise, you will practice building a team. Consider the following hypothetical situation, and then select an appropriate team. Discuss your selections with your instructor and with the other trainees.

Situation

You are the team leader on a job site where a new single-family home is going to be built. Shortly before construction begins, the architect changes the plans to add a sink to one of the rooms. When you review the plans, you see that an electrical conduit is located where the drain for the new sink should be. In addition, a cabinet needs to be built for the new sink. You may choose up to five of the following construction professionals to be on your team. Whom would you choose and why?

Mason	Painter
Carpenter	Roofer
Plumber	Landscape designer
Welder	Secretary
Electrician	Cabinetmaker

Green Employers

GOING GREEN

Some employers are doing their part to reduce their environmental footprint. Since 2004, the United Parcel Service (UPS) has revised delivery routes of its drivers to minimize left-hand turns. Doing so has made making the deliveries safer (they no longer have to cross traffic) and quicker (due to the ability to make a right on red and green turn arrows). As a result, the company has shaved 30 million miles off its deliveries in 2007 and thus saved the cost of 3 million gallons of gas. It also reduced UPS truck emissions by 32,000 metric tons (equivalent to the emissions of 5,300 passenger cars).

Additional Resources

Bullying and Harassment in the Workplace: Developments in Theory, Research, and Practice, Stale Einarsen and Helge Hoel. 2010. Boca Raton, FL: CRC Press.

The 7 Habits of Highly Effective People: Powerful Lessons in Personal Change, Stephen R. Covey. 2013. New York, NY: Simon & Schuster.

3.0.0 Section Review

1. If you finish a task ahead of time, looking for another task that can be finished in the remaining time is referred to as _____.

   a. stealing someone else's job
   b. being tardy
   c. showing off
   d. taking initiative

2. To find a solution everyone can agree on, you may have to _____.

   a. compromise
   b. discuss it with your supervisor
   c. let someone else solve the problem
   d. come up with more than one solution

3. When does constructive criticism work best?

   a. When offered often
   b. When offered occasionally
   c. When offered by a member of management
   d. When offered by someone of another trade

4. To put an end to personal bullying, one of the most critical things an employee can do is _____.

   a. confront the person one-on-one
   b. stage an intervention
   c. carefully document the events when they happen
   d. avoid the person in spite of work assignments

5. Before displaying their skills as a leader, all craftworkers must wait until they have a great deal of experience in their chosen craft.

   a. True
   b. False

SUMMARY

This module reviewed some of the important non-technical skills that you must learn to be successful as a construction professional. Whatever positions you may achieve throughout your career, these skills are vital.

Whether you are just starting out or moving on to new opportunities and challenges, you should be able to write a top-notch resume, prepare for interviews, and select a job for which you are qualified. Eventually, many construction professionals become entrepreneurs and start their own businesses.

The ability to solve problems using critical thinking skills is important for any employer and employee. Critical thinking involves evaluating and using information to reach conclusions or make decisions.

Good self-presentation skills, including personal habits, work ethic, and honesty, are characteristics of professionalism. Developing your skills in conflict resolution, teamwork, and leadership will help you to advance in your career.

The construction industry demands the best from its people and has great opportunity for personal and career growth. Only you can decide where your career will take you.

1. Someone who can vouch for your skills, experience, and work habits is called a(n) _____.

   a. mission statement
   b. entrepreneur
   c. interviewer
   d. reference

2. One of the most important and productive methods to identify job opportunities is by _____.

   a. networking with family and friends
   b. posting a "Seeking Employment" ad at the grocery store
   c. trying to call Human Resource offices directly
   d. checking the Help Wanted section of the newspaper daily

3. A common barrier to effective problem solving includes _____.

   a. overwork
   b. fear of change
   c. inadequate supervision
   d. procrastination

4. After putting a solution to a problem into effect, it is important to _____.

   a. stop considering other solutions
   b. insist it become a company procedure
   c. monitor the results
   d. develop a business plan for it

5. Depending on the type of project and region of work, one of the most expensive components of a project is often _____.

   a. equipment rental
   b. power tools
   c. hand tools
   d. labor

6. Positive or negative interactions affect _____.

   a. managerial skills
   b. tool use
   c. relationships
   d. work schedules

7. A person who works without constant supervision is showing _____.

   a. initiative
   b. fortitude
   c. respect
   d. self-presentation

8. The process of solving disagreements between co-workers is called _____.

   a. tactfulness
   b. conflict resolution
   c. relationship crafting
   d. addressing conflict

9. Constructive criticism should not be given unless _____.

   a. someone asks your opinion
   b. the supervisor tells you to do so
   c. you feel it is your business
   d. you can also be complimentary

10. A broad term identifying negative social behavior that can be based on race, age, disabilities, sex, religion, cultural issues, health, or language barriers is _____.

    a. absenteeism
    b. racism
    c. cyberbullying
    d. harassment

11. Zero tolerance refers to an employer's policy regarding _____.

    a. being sick
    b. training
    c. alcohol and drug abuse
    d. overtime

12. Barbiturates typically cause a person to _____.

    a. hallucinate
    b. accelerate their reaction times
    c. slow their reaction times
    d. be deprived of oxygen

13. When everyone in a group is focused on the final objective, it is called _____.

    a. being goal-oriented
    b. being team players
    c. taking initiative
    d. being leaders

14. To progress steadily through your career, you will need to develop and learn how to use _____.

    a. leadership skills
    b. classroom skills
    c. studying skills
    d. diversity skills

15. The three categories of leadership are hands-off, democratic, and _____.

    a. automatic
    b. liberal
    c. autocratic
    d. conservative

Trade Terms Quiz

Fill in the blank with the correct term that you learned from your study of this module.

1. A company uses a(n) _____ to state how it does business.

2. Someone who can vouch for your skills, experience, and work habits is a(n) _____.

3. Substances known as _____ can be manufactured by the human body, found in specific plants, or produced in a lab.

4. The way you act, speak, and dress is called _____.

5. Another term for work habit is _____.

6. Although _____ is closely related to self-presentation, it relies more on your integrity and manner.

7. When you break _____, you share with other people information that belongs to the company.

8. You will gain the respect of your supervisor if you regularly take the _____ to complete additional tasks when your work is done.

9. A(n) _____ approach should be used when approaching a co-worker about negative behavior.

10. When a worker has a problem with _____, they are always running late.

11. In order to resolve some situations, there must be a _____.

12. Advice that is given to point out or correct a mistake is called _____.

13. Reinforcing the importance of minor infringements is known as _____.

14. When someone is harassing a person in an offensive or threatening manner, often based on a real or perceived imbalance of power, it can likely be considered _____.

15. A person who sets an example for other people to follow is generally demonstrating good _____ skills.

16. Racial jokes are considered a form of _____.

17. Unwelcome sexual advances or requests are considered _____.

18. Uppers are another word for _____.

19. Drugs that cause you to feel sedated are known as _____.

20. The drug LSD is considered a(n) _____.

21. The crystalline derivative of amphetamines is called _____.

22. Heroin is known to be the most commonly used _____.

23. When a person has a problem with _____, co-workers cannot depend on them.

24. Since _____ are different from specific illegal or restricted substances, they may not be illegal.

Trade Terms

Absenteeism
Amphetamine
Barbiturate
Bullying
Cannabinoids
Compromise
Confidentiality
Constructive criticism

Hallucinogen
Harassment
Initiative
Leadership
Methamphetamine
Mission statement
Opiates
Professionalism

Reference
Self-presentation
Sexual harassment
Synthetic drugs
Tactful
Tardiness
Work ethic
Zero tolerance

Mark Bonda
Indian Land High School
Building Construction Instructor

How did you get started in the construction industry?

My construction career began in high school. I attended Southeastern Regional Vocational Technical High School in Easton, Massachusetts. At Southeastern, I majored in the electrical trade. We learned residential, commercial, and industrial electrical applications. After high school, I moved to South Carolina and earned a Bachelor of Science degree in Industrial Technology Education from Clemson University. I have taken many classes, workshops, and seminars in the construction field over the years.

Who or what inspired you to enter the industry?

My grandfather was a major influence in my life and helped mold my view and understanding of construction. He was the Chief Electrical Inspector for the city of Boston. I remember, while growing up as a kid, learning how to work with wood in his basement shop. That's where my love of construction was awakened.

What do you enjoy most about your career?

The most rewarding and enjoyable thing about my job is exposing young people to different careers in construction. It is exciting working with teens and helping them learn the basic skills required in different trades. A lot of young people are not aware of the opportunities they have before them. Together, we explore these opportunities and help them find a craft they like.

Why do you think training and education are important in construction?

Training and education are extremely important to construction. With an ever-changing environment, there are always revisions to existing practices and new innovations. To keep up with the demand for skilled workers, we need training that meets industry standards. In addition, we need skilled workers who have credentials acceptable to the industry.

Why do you think credentials are important in construction?

Credentials give employers a baseline of what their employees and job applicants can do and what they know. This is also important to employees because it offers them better wages and advancement in a company. Credentials are also valuable to ensure workers are receiving consistent training across the country. With a mobile construction workforce, uniform credentials provide a means of unifying the industry regardless of location.

How has training/construction impacted your life and career?

Training in construction has impacted my life tremendously. I have been able to travel, meet new people, and advance my career within the construction field. It has also helped me learn skills like time management, leadership, and budgeting.

Would you recommend construction as a career to others? Why?

I would recommend a career in construction for several reasons. It is a very rewarding profession. You get to see a project transform from start to finish. You get to go and see places most people would not have the chance to visit. There is lots of room for growth professionally and many opportunities to explore.

What does craftsmanship mean to you?

Craftsmanship is a vital part of a skilled tradesperson's body of work. It means taking pride in your craft and paying attention to detail. Even though times have changed, it is important to produce the best possible result you can. Craftsmanship is an extension of oneself and your capabilities as a leader.

Trade Terms Introduced in This Module

Absenteeism: Consistent failure to show up for work.

Amphetamine: A class of drugs that causes mental stimulation and feelings of euphoria.

Barbiturate: A class of drugs that induces relaxation, slowing the body's ability to react.

Bullying: Unwanted, aggressive behavior that involves a real or perceived power imbalance. This form of harassment may include offensive, persistent, insulting, or physically threatening behavior directed at an individual.

Cannabinoids: A diverse category of chemical substances that repress neurotransmitter releases in the brain. Cannabinoids have a variety of sources; some are created naturally by the human body, while others come from cannabis (marijuana). Still others are synthetic.

Compromise: When people involved in a disagreement make concessions to reach a solution that everyone agrees on.

Confidentiality: Privacy of information.

Constructive criticism: A positive offer of advice intended to help someone correct mistakes or improve actions.

Hallucinogen: A class of drugs that distort the perception of reality and cause hallucinations.

Harassment: A type of discrimination that can be based on race, age, disabilities, sex, religion, cultural issues, health, or language barriers.

Initiative: The ability to work without constant supervision and solve problems independently.

Leadership: The ability to set an example for others to follow by exercising authority and responsibility.

Methamphetamine: A highly addictive crystalline drug, derived from amphetamines, that affects the central nervous system.

Mission statement: A statement of how a company does business.

Opiates: A narcotic painkiller derived from the opium poppy plant or synthetically manufactured. Heroin is the most commonly used opiate.

Professionalism: Integrity and work-appropriate manners.

Reference: A person who can confirm to a potential employer that you have the skills, experience, and work habits that are listed in your resume.

Self-presentation: The way a person dresses, speaks, acts, and interacts with others.

Sexual harassment: A type of discrimination that results from unwelcome sexual advances, requests, or other verbal or physical behavior with sexual overtones.

Synthetic drugs: A drug with properties and effects similar to known substances but having a slightly altered chemical structure. Such drugs are often not illegal since they are somewhat different than well-defined restricted or illegal substances. The two typical categories are cannabinoids (lab-produced THC or marijuana substitutes) and cathinones, which are designed to mimic the effects of cocaine or methamphetamines.

Tactful: Being aware of the effects of your statements and actions on others.

Tardiness: Habitually showing up late for work.

Work ethic: Work habits that are the foundation of a person's ability to do his or her job.

Zero tolerance: The policy of applying laws or penalties to even minor infringements of a code in order to reinforce its overall importance, typically related to drug and alcohol abuse when applied to the workplace.

Additional Resources

This module presents thorough resources for task training. The following resource material is suggested for further study.

Bullying and Harassment in the Workplace: Developments in Theory, Research, and Practice, Stale Einarsen and Helge Hoel. 2010. Boca Raton, FL: CRC Press

Knock 'em Dead Resumes: A Killer Resumé Gets More Job Inteviews!, Martin Yate. 2014. Avon, MA: Adams Media.

Knock 'em Dead: The Ultimate Job Search. Martin Yate. 2014. Avon, MA: Adams Media.

The 7 Habits of Highly Effective People: Powerful Lessons in Personal Change, Stephen R. Covey. 2013. New York, NY: Simon & Schuster.

The Re-Discovery of Common Sense – A Guide to the Lost Art of Critical Thinking, Chuck Clayton. 2007. Lincoln, NE: iUniverse, Inc.

Figure Credits

©iStockphoto.com/NAN104, Module Opener

Mike Casey, National Science Foundation, Figure 1A

Topaz Publications, Inc., Figure 1B

Lincoln Electric Company, Cleveland, OH, USA, Figure 1C

Courtesy of Mark S. Knoke, Figure 1D

Trimble Navigation Limited, Figure 1E

Roanoke Land Surveying, Figure 1F

Ed Gloninger, Figures 3, 6, 9

IPM Global, Figure 7

The Banner Bank Building, SA01

National Archives and Records Administration, SA02

Section Review Answer Key

Answer	Section Reference	Objective
Section One		
1. c	1.1.0	1a
2. b	1.2.0	1b
Section Two		
1. a	2.1.0	2a
2. d	2.2.0	2b
3. a	2.3.1	2c
Section Three		
1. d	3.1.2	3a
2. a	3.2.1	3b
3. b	3.3.1	3c
4. c	3.4.1	3d
5. b	3.5.1	3e

NCCER CURRICULA — USER UPDATE

NCCER makes every effort to keep its textbooks up-to-date and free of technical errors. We appreciate your help in this process. If you find an error, a typographical mistake, or an inaccuracy in NCCER's curricula, please fill out this form (or a photocopy), or complete the online form at **www.nccer.org/olf**. Be sure to include the exact module ID number, page number, a detailed description, and your recommended correction. Your input will be brought to the attention of the Authoring Team. Thank you for your assistance.

Instructors – If you have an idea for improving this textbook, or have found that additional materials were necessary to teach this module effectively, please let us know so that we may present your suggestions to the Authoring Team.

NCCER Product Development and Revision

13614 Progress Blvd., Alachua, FL 32615

Email: curriculum@nccer.org
Online: www.nccer.org/olf

❏ Trainee Guide   ❏ Lesson Plans   ❏ Exam   ❏ PowerPoints   Other _____

Craft / Level: _____ Copyright Date: _____

Module ID Number / Title: _____

Section Number(s): _____

Description: _____

_____

_____

_____

_____

Recommended Correction: _____

_____

_____

_____

_____

Your Name: _____

Address: _____

_____

Email: _____ Phone: _____

00109-15

Introduction to Material Handling

Overview

Lifting, stacking, transporting, and unloading materials such as brick, pipe, and various supplies, are routine tasks on a job site. Whether performing these tasks manually or with the aid of specialized equipment, workers must follow basic safety guidelines to keep themselves and their co-workers safe. This module provides guidelines for using the appropriate PPE for the material being handled and using proper procedures and techniques to carry out the job.

Module Nine

Objectives

When you have completed this module, you will be able to do the following:

1. Identify the basic concepts of material handling and common safety precautions.
   a. Describe the basic concepts of material handling and manual lifting.
   b. Identify common material handling safety precautions.
   c. Identify and describe how to tie knots commonly used in material handling.
2. Identify various types of material handling equipment and describe how they are used.
   a. Identify non-motorized material handling equipment and describe how they are used.
   b. Identify motorized material handling equipment and describe how they are used.

Performance Tasks

Under the supervision of your instructor, you should be able to do the following:

1. Demonstrate safe manual lifting techniques.

2. Demonstrate how to tie two of the following common knots:

   - Square
   - Bowline
   - Half hitch
   - Clove hitch

Trade Terms

Bowline	Hand truck	Spotter
Capsize	Industrial forklift	Square knot
Clove hitch	Material cart	Standing end
Concrete mule	Pallet jack	Standing part
Cylinder cart	Pipe mule	Wheelbarrow
Drum cart	Pipe transport	Work zone
Drum dolly	Powered wheelbarrow	Working end
Freight elevator	Roller skids	
Half hitch	Rough terrain forklift	

Industry Recognized Credentials

If you are training through an NCCER-accredited sponsor, you may be eligible for credentials from NCCER's Registry. The ID number for this module is 00109-15. Note that this module may have been used in other NCCER curricula and may apply to other level completions. Contact NCCER's Registry at 888.622.3720 or go to **www.nccer.org** for more information.

Contents

Topics to be presented in this module include:

Figures ——————————————————————

1.0.0 MATERIAL HANDLING

Objective

Describe the basic concepts of material handling and common safety precautions.

a. Describe the basic concepts of material handling and manual lifting.
b. Identify common material handling safety precautions.
c. Identify and describe how to tie knots commonly used in material handling.

Performance Tasks

1. Demonstrate safe manual lifting techniques.
2. Demonstrate how to tie two of the following common knots:
   - Square
   - Bowline
   - Half hitch
   - Clove hitch

Trade Terms

Bowline: A knot used to form a loop that neither slips nor jams; sometimes referred to as a rescue knot or the king of knots.

Capsize: To change the form and rearrange the parts of a knot, usually by pulling on specific ends of the knot.

Clove hitch: A knot that consists of two half hitches made in opposite directions; used to temporarily secure a rope to an object.

Half hitch: A knot tied by passing the working end of a rope around an object, across the standing part of the rope, and then through the resulting loop; often used as an element in forming other knots or added to make other knots more secure.

Square knot: A knot made of two reverse half-knots and typically used to join the ends of two ropes of similar diameters; also called a reef knot.

Standing end: The end of a rope that is not being knotted.

Standing part: The portion of a rope that is between the standing end and the working end.

Working end: The end of a rope that is being used to tie a knot.

anual material handling is a common task on most construction sites, and it is one of the leading causes of non-fatal injuries in the construction industry. According to the Bureau of Labor Statistics, each year more than one million workers suffer from back injuries, and one of every five workplace injuries or illnesses is associated with back injuries.

Most tasks performed in construction involve the handling of some type of material or load, such as wood, brick, lumber, pipe, or other supplies, on a daily basis. A load is simply the quantity of material that can be carried, transported, or relocated at one time by a machine, vehicle, piece of equipment, or person. Because the risk of injury is present whenever materials are being handled, workers should always be conscious of what they are doing and follow basic material handling safety rules.

1.1.0 Material Handling Basics

To reduce the risk of injury when manually handling material, plan the task before doing it, wear the appropriate personal protective equipment (PPE), and follow proper lifting procedures. Also be aware of hazards when working from heights or working near suspended loads.

1.1.1 Pre-Task Planning

When it comes to material handling, it is just as important to be mentally fit as it is to be physically fit. Most material handling accidents occur when workers are new and relatively inexperienced at a job, or when very experienced workers think that accidents cannot happen to them. Before handling any material, consciously think about what you are doing. Before attempting to lift any material, always assess the situation by doing the following:

- Check to make sure the load is not too big, too heavy, or too hard to grasp.
- Make sure the load does not have protruding nails, wires, or sharp edges.
- Make sure the material is something that you can lift by yourself. If not, ask a co-worker for assistance.
- Inspect the path of travel. Look for slip, trip, or fall hazards. If there are hazards in the path of travel, move them or go around them.
- Always read the warning labels or instructions on materials before they are moved, and be aware of the potential dangers associated with mishandling a particular product.

1.1.2 Personal Protective Equipment

Wear the proper clothing and personal protective equipment (PPE) when moving or handling materials. Remember the following safety guidelines when dressing to perform material handling operations:

- Do not wear loose clothing that can get caught in moving or protruding parts.
- Be sure to button shirt sleeves and tuck in shirt tails.
- Remove all rings and jewelry.
- Tie back and secure long hair underneath your hard hat.
- If wearing a wristwatch, wear one that will easily break away if it gets caught in machinery.
- Wear gloves whenever cuts, splinters, blisters, or other hand injuries are possible. Select gloves that fit properly. Gloves that are too tight may increase hand fatigue. However, the gloves must fit snug and offer reasonable dexterity. Loose gloves reduce grip strength and may get caught on moving objects or machinery.
- Remove gloves when working with rotating machinery and equipment with exposed moving parts.
- Follow the policies of your employer and the job site at large.

1.1.3 Proper Lifting and Lowering Procedures

To reduce the risk of back injuries, use proper lifting techniques. Determine the weight of the load prior to lifting and plan the lift. Know where it is to be unloaded and remove any slipping or tripping hazards from the path of travel.

When lifting objects, avoid unnecessary physical stress and strain. Know your limits and what you are able to handle physically when lifting a load. If the load is too heavy for you to lift on your own, ask a co-worker for help.

As you lift an object, make sure you have firm footing. Bend your knees and get a good grip. Lift with your legs, keep your back straight, and keep your head up. Keep the load close to your body (*Figure 1*). Do not reach out with the load. Never turn or twist until you are standing straight, then pivot your feet and body.

When unloading materials, use your leg muscles to set the load down. Space your feet far enough apart to maintain good balance and control of the load. When placing the load down, move your fingers out of the way to avoid pinching or crushing them. When possible, attach handles to loads to reduce the chance of injuring your fingers.

Ask a co-worker for assistance in the following situations:

- The load is too heavy (weighing 50 pounds or more) for one person to carry alone.
- The objects to be handled are longer than 10 feet, such as lumber, conduit, pipe, and scaffolding poles.
- The objects to be handled, such as plywood and tarps, can be affected by wind gusts.

Sometimes it may not be the weight of a load that is troublesome, but its configuration. Secure long or cumbersome objects. Floppy rods, pipe, or lumber should be tied together in several places before you try to carry them. Never carry more pieces than you can control.

When handling long objects, such as scaffold poles or planks, be aware of objects and persons that may be struck by them as you turn your body. Never swing around quickly or without checking the surroundings. Never pass any material to another worker unless the person is looking directly at you and expecting to receive it.

Most workers have had training about the dangers of trying to lift heavy loads. Likewise, it is important to talk about lowering heavy loads from overhead. Reaching and lowering heavy loads from overhead can be very dangerous. When you are lifting something, you can always stop and put it down if you find that it is too heavy to handle alone. However, if you are lowering something from overhead, it is usually too late by the time you realize it is too heavy. Also, reaching above shoulder height can cause stress to the shoulders and back, and often results in awkward hand and wrist postures.

Before attempting to lower overhead loads, consider the following:

- Size up the load. If it looks too heavy to have been lifted by one person to its current location, then it is probably too heavy for one person to take down.
- Ask yourself, "How did it get up there?" Was it put there by a lift truck or by more than one person? The way it got up is probably the best way to get it back down.
- Lower objects down the same way you would lift them up. Keep your knees bent and your back straight. If you have to place a load to one side or another, move your feet instead of twisting your body.
- Treat the area that the material is being lowered into as a fall zone. Do not allow people to enter the area while the load is being lowered.

Back injuries can easily occur when lifting and lowering materials. Keep in mind that exercise, good posture, and a healthy weight and diet can help prevent such injuries. You can also ask your doctor about muscle-strengthening techniques for your back.

1.2.0 Material Handling Safety

When working on a job site, make sure that all workers, the materials, and the equipment are safe from unexpected movement such as falling, slipping, tipping, rolling, blowing over, or any other uncontrolled motion.

1.2.1 Stacking and Storing Materials

To work efficiently and effectively, workers should properly stack and secure materials. Taking care to stack and secure materials correctly can save space and time, and helps avoid accidents and injuries. Stacking materials keeps them from interfering with other activities and makes them more organized and readily accessible for use. Stacking materials also helps to create safe storage without danger to others and allows more room in the work area or on the job site.

Always properly store materials and equipment when they are not in current use. Learn your company- and site-specific rules, as well

1   2   3

4   5   6

00109-15_F01.EPS

Figure 1 Proper lifting technique.

1.3.0 Knots for Material Handling

Knots and hitches are common ways of fastening ropes or lines either to each other or to another object, such as a beam or container. Most knots are meant to be permanent; they have to be un-woven to be unfastened. However, some knots, which are generally referred to as hitches, are not permanent. They can be undone by simply pulling the ropes in the reverse direction from the direction in which the knots are meant to hold.

Knots and hitches are used for many different material handling tasks, such as securing loads, stabilizing scaffolding, and lifting and lowering tools and supplies. Using the correct type of knot for the task at hand is critical. This section will cover four knots that are commonly used in material handling: the **square knot**, the **bowline**, the **half hitch**, and the **clove hitch**.

First, some knot-tying terminology needs to be reviewed. The **standing end** is the long end of the rope; the end that is not being knotted. The **working end** is the short end; the end of the rope that is being used to tie the knot. The **standing part** is the portion of the rope that lies between the standing end and the working end. Finally, to **capsize** means to change the form and rearrange the parts of a knot, usually by pulling on specific ends of the knot. This may be done intentionally to undo a hitch. On the other hand, if certain knots are used inappropriately they can easily—even spontaneously—capsize, resulting in serious safety hazards.

There are two other basic points to keep in mind when learning how to tie knots. First, many knots can be tied in more than one way. As long as the chosen technique is followed correctly, the final structure of the knot will be the same, no matter which technique was used to tie it. Second, knots can be tied right-handed or left-handed, depending on the dominant hand of the person who is doing the tying. A knot that is tied left-handed will be the mirror image of the same knot tied right-handed. Note, however, that a few knots may have a right-handed element and a left-handed element. Failing to maintain these elements while tying can produce the wrong knot.

1.3.1 The Square Knot

A square knot (*Figure 7*), also called a reef knot, can be used to tie the ends of a rope around an object, such as a parcel or the neck of a sack. It is commonly used to join two lengths of rope together in low-strain applications. The two ropes should be of roughly the same diameter; otherwise the knot is more likely to slip.

00109-15_F07.EPS

Figure 7   Square knot.

> **WARNING!**
> Square knots can capsize easily and should never be used for critical loads. Additional half knots are often added to square knots for increased security.

Follow these steps (see *Figure 8*) to tie a square knot:

Step 1 Bring the two working ends together with one end oriented to the left (red rope) and the other to the right (blue rope).

Step 2 Cross them left over right. (The red rope goes left over the blue.)

Step 3 Pass the left-oriented (red) rope under the right. (The red rope goes under the blue.)

Step 4 This forms a half knot.

Step 5 Repeat the over-and-under process but go right over left this time. (The red rope goes right over the blue rope.)

Step 6 This forms a second half knot. Note that this is a key step in tying a square knot. The working end that starts out on the top when you tie the first half knot must continue to be on top when you switch directions to tie the second half knot.

Step 7 Pull the ends tight to make a neat, symmetrical square knot.

A very common error when attempting to tie a square knot is to tie the second half knot in the wrong direction. This produces a highly insecure knot called the granny knot (*Figure 9*). Compare the granny knot to the properly tied square knot in *Figure 7*.

1.3.2 The Bowline

The bowline (*Figure 10*) is used to form a secure loop in the end of a rope. It is sometimes called a rescue knot, or the king of knots, because it is reliable enough to be used for rescue work (usually backed up with a stopper knot for extra security, or tied twice into a double bowline for extra strength).

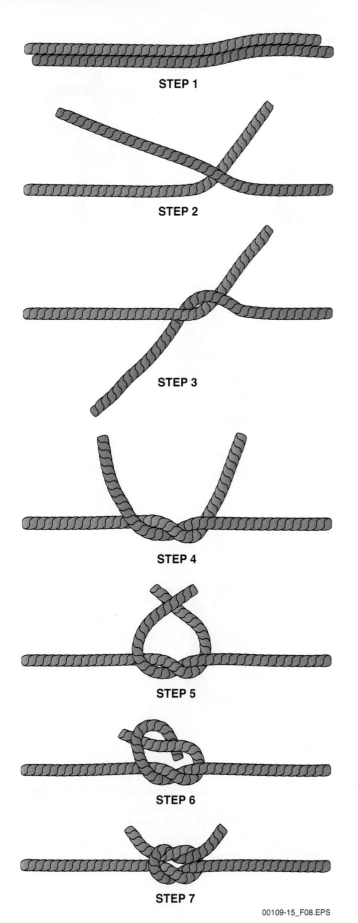

00109-15_F09.EPS

Figure 9 Granny knot.

The bowline is useful for many smaller jobs such as securing a sack of material or a bucket of tools to be hauled up to workers on a platform. Two bowlines can also be linked together to join two ropes when the strain is more than a square knot is capable of bearing.

A bowline does not slip or bind when under load, but it is easy to untie when there is no load. Because a bowline cannot be tied, or untied, when there is a load on the standing end, avoid using this knot if the rope may have to be released while it is under load.

Follow these steps (see *Figure 11*) to tie a bowline:

Step 1 Form a small loop in the rope, leaving a long enough tail to make the desired size of main loop.

Step 2 Pass the tail of the rope through the small loop.

Step 3 Pass the tail under the standing end.

Step 4 Pass the tail back down through the small loop.

Step 5 Adjust the main loop to the required size and tighten the knot, making sure to maintain a sufficiently long tail.

1.3.3 The Half Hitch

A half hitch (*Figure 12*) ties a rope around an object such as a rail, bar, post, or ring. It is commonly used for tasks such as suspending items from overhead beams and carrying light loads that have to be removed easily. The half hitch is also widely used in making other knots, such as the clove hitch. Because the half hitch is not, by itself,

STEP 1

STEP 2

STEP 3

STEP 4

STEP 5

STEP 6

STEP 7

00109-15_F08.EPS

Figure 8 Tying a square knot.

00109-15_F10.EPS

Figure 10 Bowline.

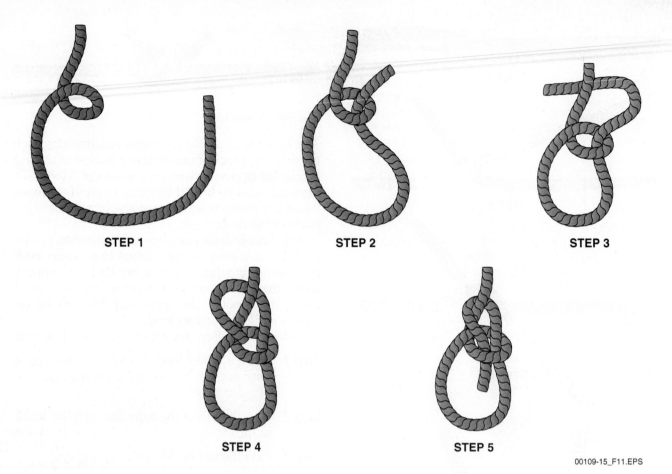

STEP 1 STEP 2 STEP 3

STEP 4 STEP 5

00109-15_F11.EPS

Figure 11 Tying a bowline.

a very stable knot, it is not suitable for heavy loads or tasks in which safety is paramount. It is often used to back up and secure another knot that has already been tied.

Follow these steps (*Figure 13*) to tie a half hitch:

Step 1 Form a loop around the object.

Step 2 Pass the tail of the rope around the standing part and through the loop.

Step 3 Tighten the hitch by pulling on the working end and the standing part of the rope simultaneously.

1.3.4 *The Clove Hitch*

A clove hitch (*Figure 14*), sometimes called a builder's hitch, is one of the most widely used general hitches. It is typically used to make a quick and secure tension knot on a fixed object that serves as an anchor, such as a post, pole, or beam. A clove hitch can also be used as the first knot when lashing items together.

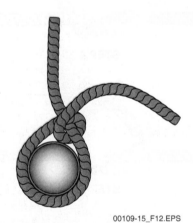

00109-15_F12.EPS

Figure 12 Half hitch.

Because it is a tension knot, a clove hitch comes loose as soon as tension is removed from the rope. On the other hand, it can also bind. For these reasons, a clove hitch should not be used by itself. Additional half hitches can be added to make a clove hitch more secure.

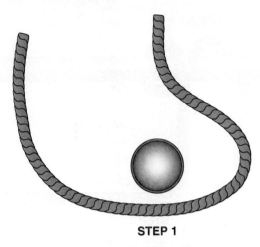

STEP 1

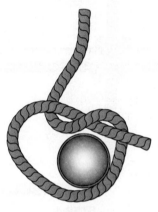

STEP 2

STEP 3

00109-15_F13.EPS

Figure 13 Tying a half hitch.

There are alternate techniques for tying a clove hitch. Regardless of the technique, however, the structure of the resulting knot consists of two half hitches made in opposite directions.

A common technique for tying a clove hitch that will be used to attach a rope to a ring or post is to thread the end, as described in the following steps (see *Figure 15*):

Step 1 Pass the working end of the rope over the object.

Step 2 Pass the working end back over the standing part and then over the object. This forms a half hitch.

Step 3 Thread the working end back under itself and up through the loop to form a second half hitch.

Step 4 Pull both the working and standing ends evenly to tighten the hitch.

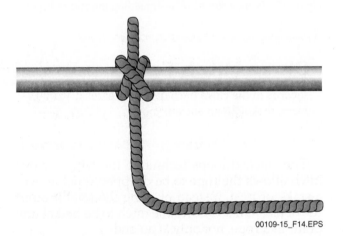

00109-15_F14.EPS

Figure 14 Clove hitch.

The Bowline

The bowline was originally used to secure a ship's square sail forward, closer to the wind. This knot was mentioned in *The Seaman's Dictionary* of 1644.

The bowline reduces the strength of the rope that it is tied in by as much as 40 percent. Bowlines require a long tail for safety. A rarely practiced rule of thumb is that the loose end (tail) of a completed bowline should be as long as 12 times the circumference. This would translate to a tail of more than 18 inches for a rope with a one-half inch diameter.

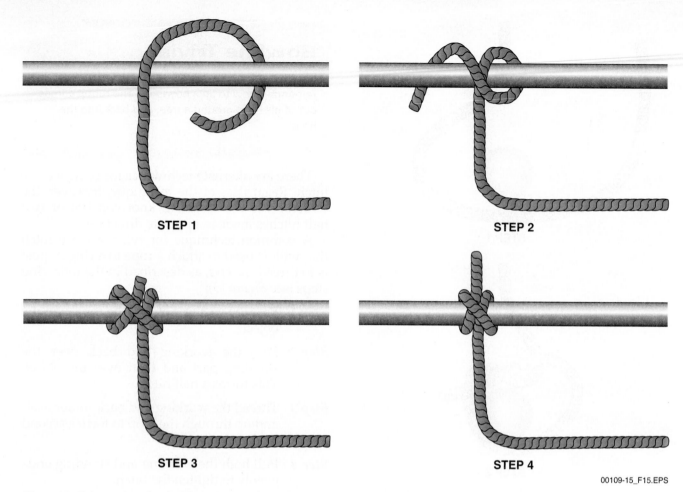

STEP 1 STEP 2

STEP 3 STEP 4

00109-15_F15.EPS

Figure 15 Tying a clove hitch—threading the end technique.

The stacked loops technique for tying a clove hitch allows the rope to be dropped quickly over a stake or post, instead of tying the knot around the object. It also allows the hitch to be tied at any point in a rope, not only at an end.

These are the basic steps of the stacked loops technique (see *Figure 16*):

Step 1 Form two identical loops in a rope, one in the right hand and one in the left.

Step 2 Cross the loops one above the other to form a knot.

Step 3 Place the knot over the stake or post.

Step 4 Tighten the knot by pulling simultaneously on both ends of the rope. The completed clove hitch consists of two half hitches stacked on each other.

Another common technique for tying a clove hitch around a post or timber is to use two half hitches, as described in the following steps (see *Figure 17*):

Step 1 Form a loop in the working end of the rope.

Step 2 Place the loop over the post. (This makes the first half hitch).

Step 3 Form a second loop identical to the first.

Step 4 Place this loop over the post as well (making a second half hitch) and tighten to complete the clove hitch.

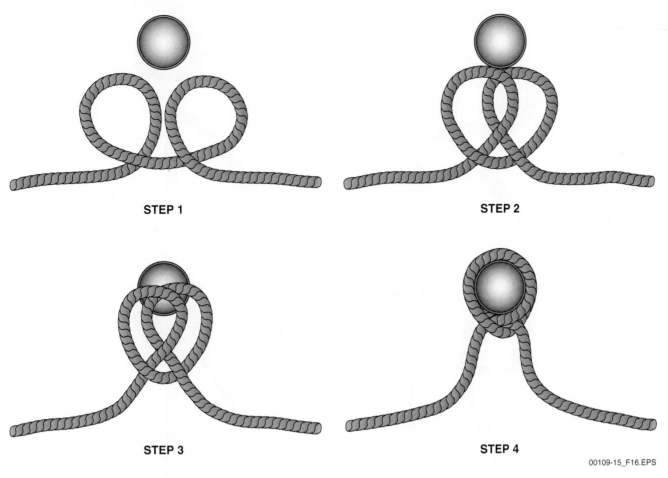

STEP 1

STEP 2

STEP 3

STEP 4

00109-15_F16.EPS

Figure 16 Tying a clove hitch—stacked loops technique.

The Ashley Book of Knots

Although *The Ashley Book of Knots* by Clifford W. Ashley is no longer in print, sailors, climbers, campers, macramé artists—anyone with more than a passing interest in knots—will continue to find this book cited as the definitive work on the subject of knot tying. Originally published by Doubleday & Company in 1944, it contains the histories of and instructions for tying over 3,900 types of knots, accompanied by 7,000 pen-and-ink drawings.

Clifford Ashley was born in 1881, in New Bedford, Massachusetts. By trade he was an author and artist, but while sailing on many types of boats and researching knots he performed a wide variety of jobs. He spent eleven years writing and drawing the illustrations for *The Ashley Book of Knots*. He died three years after the book's publication. Ashley continues to be regarded as a fine marine painter as well as one of the world's leading authorities on knot tying.

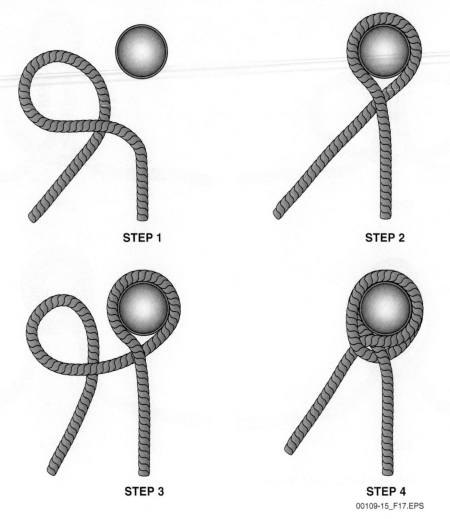

STEP 1

STEP 2

STEP 3

STEP 4

00109-15_F17.EPS

Figure 17 Tying a clove hitch—two half hitches technique.

Additional Resources

Materials Handling Handbook, The American Society of Mechanical Engineers (ASME) and The International Material Management Society (IMMS), Raymond A. Kulwiec, Editor-in-Chief. 1985. New York, NY: Wiley-Interscience.

Manufacturing Facilities Design & Material Handling, Matthew P. Stevens, Fred E. Meyers. 2013. West Lafayette, IN: Purdue University Press.

Knots: The Complete Visual Guide, Des Pawson. 2012. New York, NY: DK Publishing.

Simple Solutions Ergonomics for Construction Workers, US Centers for Disease Control, National Institute for Occupational Safety and Health. Last modified August 2007. **http://www.cdc.gov/niosh/docs/2007-122/**

1.0.0 Section Review

1. For material handling tasks, it is just as important to be mentally fit as it is to be _____.

   a. physically fit
   b. physically aggressive
   c. closely supervised
   d. over 200 pounds

2. When stacking or storing light materials that have a large surface area and could easily be moved by the wind, _____.

   a. keep the stack height less than four feet
   b. weight the top of the stack with clamps
   c. tie down or band the materials
   d. secure the materials in a flameproof cabinet

3. Which of the following is a type of knot that is often used to join the ends of two ropes in non-critical, low-strain applications?

   a. Bowline
   b. Clove hitch
   c. Half hitch
   d. Square knot

2.0.0 MATERIAL HANDLING EQUIPMENT

Objectives

Identify various types of material handling equipment and describe how they are used.

a. Identify non-motorized material handling equipment and describe how they are used.
b. Identify motorized material handling equipment and describe how they are used.

Trade Terms

Concrete mule: A wheeled device used when a concrete pour is in a location that a concrete delivery truck or pump cannot reach; sometimes referred to as a Georgia buggy.

Cylinder cart: A two-wheeled cart that is used to transport cylinders, or bottles, of compressed gases.

Drum cart: A wheeled cart that is used to transport heavier-weight 55-gallon drums/barrels.

Drum dolly: A wheeled circular platform or a caddy with a handle that is used to transport lighter-weight 55-gallon drums/barrels.

Freight elevator: An elevator used to transport materials from floor to floor.

Hand truck: Two-wheeled cart that is used to transport large, heavy loads; also known as a dolly.

Industrial forklift: A vehicle with a power-operated pronged platform that can be raised and lowered for insertion under a load to be lifted and moved.

Material cart: Four-wheeled device used to transport materials around a job site.

Pallet jack: A device used to lift and move heavy or stacked pallets; also known as a pallet truck.

Pipe mule: A two-wheeled device used to transport medium-length pieces of pipe, tubing, or scaffolding; sometimes referred to as a tunnel buggy.

Pipe transport: A wheeled device similar to a pipe mule, but used to move larger pieces of pipe.

Powered wheelbarrow: A vehicle similar to a manual wheelbarrow, but powered by an electric or gas motor; also known as a power buggy.

Roller skids: A device that includes a surface table and two, three, or four roller skids. Materials that are to be moved are placed on the table surface and then pushed on the skids.

Rough terrain forklift: Similar to an industrial forklift, but designed to be used on rough surfaces. Rough terrain forklifts are characterized by large pneumatic tires, usually with deep treads that allow the vehicle to grab onto the roughest of roads or ground cover without sliding or slipping.

Spotter: A person who walks in front of another worker who is carrying or transporting a long load to ensure there is a clear, unobstructed path.

Wheelbarrow: A one- or two-wheeled vehicle with handles at the rear that is used to carry small loads.

Work zone: The area in which a forklift may come in contact with objects or people, either with the rear end or the front forks.

Instead of physically lifting heavy loads at work, use material handling devices when they are available to reduce your risk of back injury. Material handling devices help you to work smarter instead of harder, easing stress on your muscles and joints. These work-saving devices also increase productivity and reduce the chance of dropping and damaging equipment or supplies. There are two types of material handling equipment: non-motorized and motorized.

2.1.0 Non-Motorized Material Handling Equipment

When using non-motorized, or manual, material handling equipment, you use the force and strength of your body to move the equipment. Examples of manual material handling equipment include the following:

- Material carts
- Hand trucks
- Cylinder carts
- Drum dollies
- Drum carts
- Roller skids
- Wheelbarrows
- Pipe mules
- Pipe transports
- Pallet jacks

2.1.1 Material Carts

Material carts, also known as platform trucks, are platforms or boards that are laid horizontally on four caster wheels. These types of carts are typically used to transport materials around a job site (*Figure 18*). When using a material cart, follow these safety guidelines:

- Before using carts with caster wheels, inspect them to ensure that the casters will roll and swivel freely during transport.
- When using a cart with pneumatic tires, check the tires' air pressure before use. Tires that will not hold pressure or those that have flat spots must be replaced.
- Check to make sure that the surface has adequate traction to avoid slips.
- When moving a cart, keep your hands away from the edges of the handles to avoid pinching, crushing, or cutting your hands.
- Make sure your load is centered and secured so it does not roll or fall off.
- If materials are protruding out in front of the cart, have a spotter walk ahead of you to help you avoid contact with objects or other workers.
- Use caution when moving a cart on an inclined or declined surface, and never load a cart past its labeled weight capacity. If the load is too heavy on a decline, it may drag you down with it. If the load is too heavy on an incline, it may back over you. The weight capacity must be marked in plain view on the side of the cart.

2.1.2 Hand Trucks

Hand trucks, also known as dollies, are two-wheeled carts that are used to transport large, heavy loads, such as appliances or stacks of heavy boxes. Hand trucks are vertical material handling

devices that have a metal blade at the bottom that is inserted beneath the load. Many hand trucks are equipped with ratchet straps to secure loads. The entire assembly tilts backward so that it may be easily pulled or pushed (*Figure 19*).

Before using a hand truck, inspect the framework for stress fractures, and check the tires. When loading a hand truck, tilt the object to be moved forward and slide the blade of the hand truck under the object. Hold the object against the hand truck's framework as you tilt back on the wheels to load the object onto the hand truck. Use a ratchet strap to secure the load; rubber or bungee-style cords may not hold the load if the weight shifts. Simply buckle the strap around the load and tighten the belt snugly. Then tilt the hand truck back and move the load.

2.1.3 Cylinder Carts

High-pressure gas cylinders, or bottles, should be transported on cylinder carts, which are designed specifically to transport compressed gases (*Figure 20*). The pressurized gas contained in cylinders can pose explosion risks, even if the gas itself is inert (meaning it is not chemically reactive).

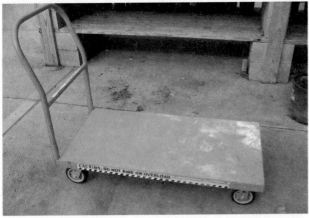

00109-15_F18.EPS

Figure 18 Material cart.

00109-15_F19.EPS

Figure 19 Hand truck.

00109-15_F20.EPS

Figure 20 Cylinder cart.

Follow these guidelines when transporting compressed gas cylinders:

- Check the label on the cylinder to identify the gas that it contains and verify that the cylinder is approved for transporting the gas.
- Never transport a gas cylinder without the safety cap screwed all the way down to the body of the cylinder.
- Never lift a gas cylinder by the safety cap.
- Never transport a gas cylinder with the regulator attached.
- Use chains or straps to secure the cylinder to the cart and keep the cylinder upright.
- Many sites require a partition on the cart to separate the two bottles, as shown in *Figure 20*.
- Wear the appropriate PPE whenever handling gas cylinders.

WARNING!	If the tank of a compressed gas cylinder ruptures or cracks during transport, the tank can become a deadly projectile or it could explode into fragments of shrapnel.

2.1.4 Drum Dollies and Carts

Chemicals, lubricants, and other materials are commonly stored in 55-gallon drums or barrels. Drums that are filled with materials can be very heavy and unwieldy to move. Two methods for safely transporting loaded drums around the workplace are drum dollies and drum carts.

A drum dolly (*Figure 21*) is used to move relatively light drums. The simplest type of drum dolly is a flat wheeled platform that the drum sits on. The dolly may be equipped with a strap to secure it and prevent movement, or a handle to pull and control the load.

A drum cart (*Figure 22*) is typically used to move heavier loads, such as sealed oil drums. One common type is a rotating drum cart. Its main features include prongs, handles, wheels, and a set of heavy duty casters.

To use a rotating drum cart follow these steps:

- With the cart held vertically, slide the two prongs of the cart under the drum.
- Allow the cart to roll back into the horizontal position and sit on the wheels and casters.
- With the drum resting horizontally in the well of the cart, use the cart handles to maneuver the load to its destination.
- Once at the destination, tip the cart into the vertical position to allow the drum contents to be unloaded.

2.1.5 Roller Skids

Roller skids move materials by pushing them on a table surface that is placed on top of two, three, or four roller skids. These devices are available in different capacities, depending on their intended usage. Roller skids used for moving heavy equipment use steel rollers or steel chain rollers, while those used for moving lighter equipment may use polyurethane or nylon rollers. The table surface

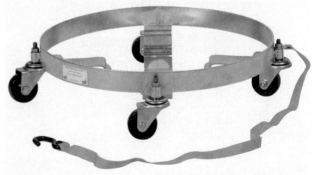

00109-15_F21.EPS

Figure 21 Drum dolly.

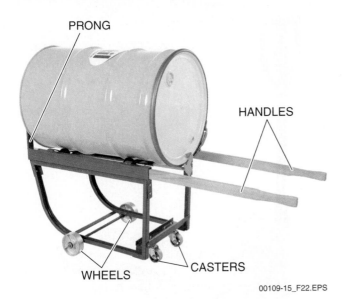

Figure 22 Drum cart.

of roller skids may differ from manufacturer to manufacturer. Some come equipped with a rotating table surface, some have spikes on the table surface for a better grip, and some have a plain surface like any other table. Many roller skids have a pull-push handle attachment for operators to move them easily (*Figure 23*). As with other material handling equipment, inspect the roller skids before use and use roller skids only for their intended purpose and weight capacity.

Roller skids must be used on a relatively smooth surface. The rollers can easily be stopped by small rocks or hardware on the floor surface. Ensure that the route ahead is swept, if necessary, prior to the move.

2.1.6 Wheelbarrows

A wheelbarrow is a one- or two-wheeled vehicle with handles at the rear, used to carry small loads (*Figure 24*). It is recommended that only two-wheeled wheelbarrows be used on job sites, since

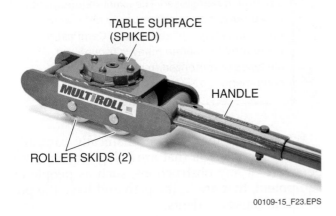

Figure 23 Roller skids.

Figure 24 Wheelbarrow.

they are more stable with heavy loads. Remember to check the air pressure in both tires, and check the body of the wheelbarrow for cracks, breaks, or punctures before using it. Use proper lifting techniques when moving a wheelbarrow. Keep your head up, your back straight, bend your knees, and lift with your legs.

2.1.7 Pipe Mule

A pipe mule is a two-wheeled device used to transport medium-length pieces of pipe, tubing, or scaffolding (*Figure 25*). It is sometimes referred to as a tunnel buggy. It may also be called a V-cart or a grasshopper, which is a brand name. When loading or unloading a pipe mule, use proper lifting techniques. The proper loading method is to have one worker lift one end of the load while another worker rolls the pipe mule under the load. Place the wheels of the pipe mule underneath the center of the load. As with any long load, use a spotter when moving a load with a pipe mule.

Figure 25 Pipe mule.

2.1.8 Pipe Transport

A pipe transport is similar to a pipe mule, but it is used to move larger pieces of pipe. The pipe is slung underneath the frame that is attached to two rubber wheels (*Figure 26*). This device requires two people to operate it. Use caution when moving the pipe into position.

2.1.9 Pallet Jack

A pallet jack, also known as a pallet truck, is a device that typically uses hydraulics to lift and move heavy or stacked pallets (*Figure 27*). Both motorized and non-motorized versions of this device are available. Motorized pallet jacks are sometimes called walkies.

Remember the following safety guidelines when using a pallet jack:

- Plan and review the route before using a pallet jack; be very careful about using it on uneven surfaces or ramps.
- Inspect pallet jacks before use, looking carefully for malfunctions or missing parts.
- Check for oil leaks around fittings or hoses of hydraulic pallet jacks, and repair any leaks before the jack is put into service.
- Never use a pallet jack to lift or transport a load that exceeds the jack's rated capacity. The load rating should be clearly posted on the body of the jack.
- Inspect the wood pallet to ensure it is in good condition to adequately handle the load, and make sure the load is centered between the pallet jack forks.

00109-15_F26.EPS

Figure 26 Pipe transport.

00109-15_F27.EPS

Figure 27 Pallet jack.

2.2.0 Motorized Material Handling Equipment

Motorized material handling equipment is powered by gasoline or electric motors. Operators must be trained, certified, and authorized to operate motorized material handling equipment. Training courses must consist of a combination of formal instruction, hands-on training, and evaluation of the operator's performance in the workplace for each type of material handling equipment being used. All operator training and evaluation must be conducted by a person who has knowledge, training, and experience.

When lifting materials mechanically, carefully secure the materials that are being loaded. Also be aware of any obstructions, such as people or equipment, that are in the path and have the potential to cause accidents.

When handling materials with a machine such as a pallet jack or forklift, always follow these guidelines:

- Know the weight of the object to be handled or moved.
- Know the capacity of the handling device that will be used and never exceed the capacity.
- Ensure that the handling equipment is in good working order and free of damage.

Motorized material handling equipment discussed in this section includes the following:

- **Powered wheelbarrow**
- **Concrete mule**
- **Freight elevator**
- **Industrial forklift**
- **Rough terrain forklift**

2.2.1 Powered Wheelbarrow

A powered wheelbarrow, also known as a power buggy, is similar to a manual wheelbarrow, but is powered by an electric or gas motor (*Figure 28*). Before using a powered wheelbarrow, conduct a thorough inspection of the machine. Check the

00109-15_F28.EPS

Figure 28 Powered wheelbarrow.

brakes and brake linkages for proper operation and adjustment. Inspect tires and wheels for damage and proper tire pressure. Test the directional controls, dumping controls, and speed controls to verify they are all working properly. Be sure to read safety labels and decals for proper safety instructions. Safety labels should be clearly visible and readable on the side of the machine.

Keep the following safety guidelines in mind when operating a powered wheelbarrow:

- Avoid pinch points on dumping mechanisms.
- Be sure the dump bucket is securely down at all times when not dumping.
- Do not move the buggy with the bucket in the dumping position.

Tugs

Some companies are reducing their use of forklifts and pallet jacks and are increasing the use of wheeled carts, called tugs, to transport loads throughout a facility. Material handling tugs are battery-powered devices that can push, as well as pull, loads. One reason the use of tugs has grown is that they can increase productivity by reducing the manpower required to move heavy loads. A single worker operating a tug can safely move loads that weigh hundreds, or even thousands, of pounds. In addition, unlike forklift operators, workers who operate tugs are not required to have special training and certification before they are authorized to operate the device.

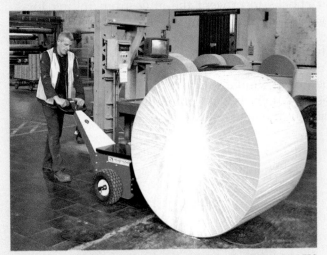

00109-15_SA01.EPS

Tugs are designed to distribute weight ideally and maximize torque, which minimizes the effort required for operators to pull or push loads. The body of the tug acts as a wedge, slightly lifting the load at an angle and transferring all the weight to the tug's drive wheels. Tugs are equipped with safety features to prevent unintended propulsion, including an emergency reverse switch that automatically stops the tug to prevent it from pinning or crushing the operator when a load is being pulled in reverse.

- Do not mount or dismount the buggy while it is moving.
- Shut off the engine and lock the parking brake when you are finished using the machine.
- Do not allow riders.
- Avoid any condition of slope and/or grade which could cause the buggy to tip.
- Do not exceed load limits in weight or height.
- Follow all procedures outlined on the safety labels and decals.

2.2.2 Concrete Mule

A concrete mule, sometimes referred to as a Georgia buggy, is a wheeled device used when a concrete pour is in a location that a concrete delivery truck or pump cannot reach. Concrete mules (*Figure 29*) are designed to carry concrete, sand, and gravel, or materials of that nature, and should not be used to transport undesignated materials.

Before using a concrete mule, you must conduct a thorough inspection of the following:

- Tires
- Fluids
- Mechanical hinges and joints
- Throttle cables and steering mechanisms

When refueling a concrete mule, or any other gas-powered material handling machine, always be sure that the engine is turned off and has cooled down before refueling. Use only fuel and oil types that are specified by the manufacturer.

When discharging concrete mix from a concrete mule near an open pit, use a tire stop board or chocks to prevent the mule from rolling forward into the pit during the emptying process.

When operating a concrete mule, make sure there are no obstacles or other workers in the turning radius of the rear platform. Always look behind the mule before operating in reverse, and never allow anyone other than the operator to ride on the machine.

2.2.3 Freight Elevator

On multi-level sites, workers may use a freight elevator to transport materials from floor to floor. Typically, when freight elevator doors close they travel from the top and bottom simultaneously (*Figure 30*). These doors do not have the safety bumpers that are found on a commercial passenger elevator. Rather, as a person pulls the top door down, by way of a strap or handle, the bottom door rises to meet the top door. When these doors close, there is normally metal-against-metal contact, and this can cause injury. Make sure that all body parts such as arms, hands, and fingers are clear of the door before it closes. Check to make sure that tool belts, body harnesses, and all materials are also clear of the door. Always check the weight capacity shown on the elevator wall before placing material on an elevator.

2.2.4 Industrial Forklift

An industrial forklift is a vehicle with a power-operated pronged platform that can be raised and lowered for insertion under a load to be lifted and moved (*Figure 31*). Forklifts are typically used to lift, lower, and transport large or heavy loads in areas with a smooth terrain, such as warehouses or shops.

00109-15_F29.EPS

Figure 29 Concrete mule.

00109-15_F30.EPS

Figure 30 Freight elevator.

Figure 31 Industrial forklift.

00109-15_F31.EPS

When working in the vicinity of a forklift, remember the following safety guidelines:

- All workers must keep a safe distance from the machine. This perimeter is known as the **work zone**; it is the area in which the machine may come in contact with objects or people either with the rear end or the front forks.

- Workers must also stay clear of the fall zone. This area includes the area directly beneath a load, as well as any portion of the area that suspended materials could possibly fall into.
- Stay in designated walkways and make eye contact with the forklift operator to ensure that he or she sees you. Always avoid the driver's blind spot. Be aware that forklifts pivot on the front wheels and turn with the back wheels, which could cause a potential pinch point.
- Plan a route of escape. Never get caught in a position where you are trapped in a corner or on the side of a truck.

Trained and certified forklift operators must observe certain traffic safety rules and regulations while driving.

Before operating a forklift, put on the proper PPE and fasten the seat belt. If you are not wearing a seat belt and the forklift tips, you can easily be pitched under the forklift, or have some of the load land on you. Falling objects are common when stacking and lifting with a forklift, so wear a hard hat. As with most machinery, when working with a forklift do not wear baggy clothes that can catch on controls, cargo, or stacked objects.

Never play on a forklift or drive recklessly. Accidents often happen because workers forget to respect equipment. A forklift is not a toy and should not be treated like one.

Before operating the forklift each day, inspect the following:

- Tires
- Fluid levels
- Battery/fuel level
- Fire extinguisher
- Brakes
- Deadman control
- Warning lights
- Horn
- Backup alarm

Use a checklist, if one is available. If you find any defects during the inspection, report them to your supervisor.

Forklift operators with the fewest accidents drive with slow deliberation, take their time, and do not make mistakes. Do not attempt to cut corners to speed up your work. Drive slowly and keep your arms, hands, and legs inside the cage of the forklift. Forklifts are top-heavy and, when driven recklessly, can tip over on a curve, drive off the edge of a dock, or fail to stop in time. Drive at a speed that is appropriate and safe for the location and surface condition and use extra caution on hills, corners, and ramps. Keep the forklift to the right on roadways and in wide aisles, and stop at all designated stop signs. Always start and stop smoothly. This helps to preserve the equipment by putting less wear and tear on the forklift. It also keeps loads from shifting, objects from falling, and helps prevent damage to cargo, equipment, and operators.

Remember that pedestrians have the right-of-way. Forklifts, especially electric models, are often so quiet that they are difficult to hear. Pedestrians may not be aware that you are there, so be careful when you are driving near people, especially if they have their back to you. Make sure that the backup alarm is functional. Driving in reverse presents special dangers to pedestrians, especially if you are watching the load in front as you back up.

Workers sometimes attempt to make modifications to forklifts so that they can transport heavier or wider loads than the equipment is designed to lift. This introduces the hazards of broken masts or forks, dropped loads, and tipovers. Never modify a forklift without obtaining the consent from the manufacturer or from a mechanical engineer familiar with the equipment.

Biodiesel

Cranes and other equipment used in rigging operations consume lots of fuel—just like all the other pieces of equipment at a typical job site. Most large trucks and construction equipment run on diesel fuel. These vehicles and machines could go green and use biodiesel instead. Biodiesel is a plant oil based fuel made from soybeans, canola, and other waste vegetable oils. It is even possible to make biodiesel from recycled frying oil from restaurants. Biodiesel is considered a green fuel since it is made using renewable resources and waste products. Biodiesel can be combined with regular diesel at any ratio or be run completely on its own. This means any combination of biodiesel and regular petroleum diesel can be used or switched back and forth as needed.

But what benefits does biodiesel have over traditional fuels?

- It's environmentally friendly. Biodiesel is sustainable and a much more efficient use of our resources than diesel.
- It's non-toxic. Biodiesel reduces health risks such as asthma and water pollution linked with petroleum diesel.
- It produces lower greenhouse gas emissions. Biodiesel is almost carbon-neutral, contributing very little to global warming.
- It can improve engine life. Biodiesel provides excellent lubricity and can significantly reduce wear and tear on your engine.

Think about the environmental impact that would occur if every vehicle and piece of equipment at every job site were converted to biodiesel. The use of biodiesel also continues to increase in Europe, where Germany produces the majority of these fuels. However, even tiny countries such as Malta and Cyprus have some level of production.

2.2.5 Rough Terrain Forklift

Rough terrain forklifts are designed to be used on irregular surfaces. They are typically used for jobs where there are no paved surfaces. Rough terrain forklifts are characterized by large pneumatic tires, usually with deep treads that allow the vehicle to grab onto the roughest of roads or ground cover without sliding or slipping (*Figure 32*). Be sure to apply the same safety guidelines to rough terrain forklifts as you would to industrial forklifts.

One of the most common accidents involving rough terrain forklifts is tipovers. This is because these types of forklifts are often operated on uneven ground. To prevent tipovers with rough terrain forklifts, observe the following safety guidelines:

- Study load charts carefully and only operate the forklift within its stability limits.
- Before placing a load, apply the brakes, shift to neutral, level the frame, and engage the stabilizers, if so equipped.
- Carry the load low. Avoid driving with an elevated load except for short distances.
- Move carefully on slopes and only operate within specific grade limits.
- Retract the boom and lower the forks before moving a forklift.

2.2.6 Hand Signals

The noise generated by motorized lifting equipment, such as forklifts, and the distance between workers when they are working with the equipment can make verbal communication difficult. Standard hand signals (*Figure 33*) allow both the workers and the forklift operator to go through the lifting, loading, moving, and unloading procedures at a more controlled setting, and help to avoid potential misunderstandings and accidents.

Be aware, however, that there is no globally accepted standard for hand signals. The signals shown here are simply some of the most commonly used ones. The differences between other versions are likely to be very minor. Your

00109-15_F32.EPS

Figure 32 Rough terrain forklift.

employer may have a specific set of hand signals that may or may not fully align with the signals presented here. In any case, prior to performing a task the equipment operator and the other workers must agree on the signals that will be used and the exact meaning for each signal.

For instance, it is important for the operator to know when to raise or lower the tines (the L-shaped parts that support the material load) of a fork lift to avoid obstructions that may be in their path. Another very important signal for the operator to know is Dog (or pause) Everything. This signal means to pause, and can be used when potentially risky situations arise, such as when it starts raining, when the load does not fit the space for which it was planned, or when a bystander gets too close.

When providing hand signals, make sure to keep eye contact with the operator at all times. There may be special operations that require adaptations of the basic hand signals, so be sure to review the signals for your work site from time to time. If there is a misunderstanding regarding hand signals, go over the signals with the operator before proceeding.

Forklift Automation

Automated guided vehicles (AGVs) have long been used successfully in manufacturing. Now automated forklifts, which are retrofitted with AGV technology, are poised to generate significant improvements in material handling productivity. Tasks that may be appropriate for forklift automation are similar to the tasks that AGVs have long handled in manufacturing—tasks involving repetitive, horizontal travel along the same route for relatively long distances.

The evolving AGV technology holds particular promise for material handling in the warehouse environment. One application that has emerged as a good fit for forklift automation is high-volume picking where the operator is able to remotely advance the forklift, eliminating the need for the operator to climb on and off the vehicle multiple times while working within an aisle.

Figure 33 Common forklift hand signals.

Additional Resources

Materials Handling Handbook, The American Society of Mechanical Engineers (ASME) and The International Material Management Society (IMMS), Raymond A. Kulwiec, Editor-in-Chief. 1985. New York, NY: Wiley-Interscience.

Manufacturing Facilities Design & Material Handling, Matthew P. Stevens, Fred E. Meyers. 2013. West Lafayette, IN: Purdue University Press.

Simple Solutions Ergonomics for Construction Workers, US Centers for Disease Control, National Institute for Occupational Safety and Health. Last modified August 2007. **http://www.cdc.gov/niosh/docs/2007-122/**

2.0.0 Section Review

1. Which of the following are often used to secure loads on hand trucks?

   a. Dolly spotters
   b. Bungee cords
   c. Wheel chocks
   d. Ratchet straps

2. When refueling a concrete mule, always be sure the _____.

   a. work zone is at least twice the load height
   b. dump bucket is in the dumping position
   c. engine is turned off and has cooled down
   d. boom is in the fully retracted position

SUMMARY

Material handling is one of the most common tasks on a job site, and it is also one that can cause accidents or injuries if not done properly. Workers must follow safety procedures for lifting, carrying, and transporting materials, whether doing so manually or using a piece of material-moving equipment.

This module has presented many of the basic guidelines for ensuring your safety and the safety of your co-workers. These guidelines fall into the following categories:

- Using proper PPE for material handling procedures
- Planning the route before handling materials
- Using proper lifting procedures
- Inspecting material handling equipment before use
- Following proper material handling equipment operating procedures

Review Questions

1. Loose gloves may get caught on moving objects, and they reduce _____.

   a. positive friction
   b. worker mobility
   c. manual torque
   d. grip strength

2. When lifting a load, keep the load _____.

   a. at shoulder height
   b. at arm's length
   c. close to your body
   d. balanced at your hips

3. The area into which a load is being lowered should be treated as a(n) _____.

   a. hot work location
   b. fall zone
   c. primary access route
   d. aerial drop spot

4. How many inches should stacked bricks be tapered for every foot above four feet in the stack?

   a. Two
   b. Four
   c. Six
   d. Twelve

5. The proper way to get tools to a worker on a higher level is to _____.

   a. toss them up carefully
   b. carry them by hand up a ladder
   c. climb up with them in your pocket
   d. hoist them up with a rope and bucket

6. A knot that is *not* meant to be permanent is generally referred to as a(n) _____.

   a. hitch
   b. slip
   c. runner
   d. bypass

00109-15_RQ01.EPS

Figure 1

7. The knot shown in Review Question *Figure 1* is a _____.

   a. square knot
   b. bowline
   c. half hitch
   d. clove hitch

8. To help make sure the path is clear when your view is obstructed while handling materials, use a _____.

   a. mirror
   b. spotter
   c. tow rope
   d. step ladder

9. Before using a hand truck, inspect the framework for _____.

   a. oil residue
   b. popped nails
   c. stress fractures
   d. seal leaks

10. Roller skids for moving heavy equipment use rollers made of _____.

   a. aluminum
   b. brass
   c. polyurethane
   d. steel

11. Which of these devices requires two people to operate it?

    a. Roller skids
    b. Rotating drum cart
    c. Powered wheelbarrow
    d. Pipe transport

12. Before a worker uses a motorized material handling device to move a load, the worker must consider the _____.

    a. moisture content of the load
    b. fall zone for the load
    c. maximum capacity of the device
    d. universal transport hand signals

13. Which of the following is a motorized material handling device?

    a. Concrete mule
    b. Roller skid
    c. Porta-Power®
    d. Pipe mule

14. An industrial forklift pivots on the front wheels and turns with the back wheels, which could cause a _____.

    a. gravity shift
    b. pinch point
    c. rollover
    d. fall zone

15. When the sound of motorized lifting equipment makes verbal communication difficult, workers should use _____.

    a. written notes
    b. shouted commands
    c. hand signals
    d. a relay person

Trade Terms Quiz

Fill in the blank with the correct term that you learned from your study of this module.

1. The area in which a forklift may come in contact with objects or people, either with the rear end or the front forks, is the _____.

2. A knot that is used to form a loop that neither slips nor jams is a(n) _____.

3. A(n) _____ is used to move large pieces of pipe, while a(n) _____ is used to transport medium-length pieces of pipe, tubing, or scaffolding.

4. To ensure there is a clear, unobstructed path, a(n) _____ walks in front of another worker who is carrying or transporting a long load.

5. A(n) _____ is a knot that consists of two half hitches made in opposite directions.

6. Also known as a power buggy, a(n) _____ is a powered by an electric or gas motor.

7. A person may intentionally _____ a hitch to undo it by pulling on specific ends of the knot, which changes the form and rearranges the parts of the knot.

8. A(n) _____ is a two-wheeled cart that is used to transport cylinders, or bottles, of compressed gases.

9. A(n) _____ is a knot made of two reverse half-knots and is typically used to join the ends of two ropes of relatively equal diameter.

10. A(n) _____ is a wheeled circular platform or a caddy with a handle that is used to transport lighter-weight 55-gallon drums or barrels.

11. A(n) _____ is similar to its industrial counterpart, except that it is designed to be used on rough surfaces.

12. A(n) _____ is a machine used to transport materials from floor to floor.

13. The end or part of a rope that is being used to tie a knot is called the _____, while the end or part that is not being knotted is called the _____.

14. A(n) _____ is a device that includes a surface table and two, three, or four skids.

15. A(n) _____ is a knot tied by passing the working end of a rope around an object, across the standing part of the rope, and then through the resulting loop.

16. A(n) _____ is a four-wheeled device used to transport supplies around a job site.

17. Also known as dollies, _____ are two-wheeled carts that are used to transport large, heavy loads.

18. A(n) _____ is a vehicle with a power-operated pronged platform that can be raised and lowered for insertion under a load to be lifted and moved.

19. Also known as a pallet truck, a(n) _____ is a device used to lift and move heavy or stacked pallets.

20. A wheeled cart that is used to transport heavier-weight 55-gallon drums or barrels is a(n) _____.

21. A(n) _____ is a one- or two-wheeled vehicle with handles at the rear that is used to carry small loads.

22. Sometimes referred to as a Georgia buggy, a(n) _____ is a wheeled device that is used when a delivery truck or pump cannot reach the work site.

23. The _____ is the portion of a rope that lies between the end that is being used to tie a knot and the end that is not being knotted.

Trade Terms

Bowline
Capsize
Clove hitch
Concrete mule
Cylinder cart
Drum cart
Drum dolly

Freight elevator
Half hitch
Hand truck
Industrial forklift
Material cart
Pallet jack
Pipe mule

Pipe transport
Powered wheelbarrow
Roller skids
Rough terrain forklift
Spotter
Square knot
Standing end

Standing part
Wheelbarrow
Work zone
Working end

Sid Mitchell
CBI
Trainer

How did you choose a career in the construction industry?
I always enjoyed working with things that required planning and creative thinking, and where environments change frequently.

Who inspired you to enter the industry?
My neighbor, who was a Construction Power Project Manager.

What types of training have you been through?
I have a Business Degree. I have also participated in training related to industrial engineering, decision analysis, process improvement, Kepner Tregoe, time management and motivational techniques, and professional development.

How important is education and training in construction?
Education and training is important enough to provide you with a skill set that will create an opportunity to make an above-average living. The construction industry can often provide you with a paycheck beginning the first week training begins, and without the long-term debt usually associated with a formal education.

How important are NCCER credentials to your career?
NCCER credentials are vital and recognized by most major and many minor construction and industrial employers. The credentials provide you with the ability to safely demonstrate your capabilities and display the confidence necessary to advance in your chosen craft.

How has training/construction impacted your life?
It has provided the skills and confidence necessary to qualify for positions, meet expectations, and add value to the job.

What kinds of work have you done in your career?
I have done many things related to the business side of construction. I also have direct field experience as a supervisor in civil construction activities.

Tell us about your present job.
I am a training instructor providing professional and craft skill development. This training deals primarily with increasing the skill level of new hires and orienting them to the job. This provides newer associates with the fundamental information necessary to safely integrate into our site workforce. I also provide additional training to improve math skills when called upon.

What do you enjoy most about your job?
I truly enjoy the opportunity to transfer some of my experience and insight. I have gained a great deal of both during 40 years of industry experience. I like to see the joy in someone's eyes when they begin to understand some of the principles and knowledge I seek to transfer.

What factors have contributed most to your success?
My willingness to continue learning has helped. I want to transfer my knowledge and experience to others, and have been willing to travel to different project locations to do that. There are few people willing to leave their current environment. It is therefore necessary to travel sometimes and train people, providing them with high-quality, consistent training. They must always do it safely, while protecting the environment as well.

Would you suggest construction as a career to others? Why?
A career in construction will provide an opportunity to be creative and build things that will benefit others for years to come. It will likely offer someone the opportunity to earn above-average pay and, in many cases, without the benefit of a formal education. This is possible while avoiding the burden of debt often associated when taking another path. However, we will certainly continue to need engineers and professional business persons in the future.

What advice would you give to those new to the field?
Pay close attention to what you are doing and learn. Always be mindful of safe working processes. Do not take short cuts, and maintain a high level of respect for equipment and your co-workers.

Interesting career-related fact or accomplishment:
I helped develop project financial modeling and forecasting techniques, working in both domestic and foreign environments.

How do you define craftsmanship?
Craftsmanship is the ability to transfer your skills and knowledge to something that contributes to the overall completion of a project. It is delivering services unmatched in value to your craft, to the benefit of both your team and your clients.

Trade Terms Introduced in This Module

Bowline: A knot used to form a loop that neither slips nor jams; sometimes referred to as a rescue knot or the king of knots.

Capsize: To change the form and rearrange the parts of a knot, usually by pulling on specific ends of the knot.

Clove hitch: A knot that consists of two half hitches made in opposite directions; used to temporarily secure a rope to an object.

Concrete mule: A wheeled device used when a concrete pour is in a location that a concrete delivery truck or pump cannot reach; sometimes referred to as a Georgia buggy.

Cylinder cart: A two-wheeled cart that is used to transport cylinders, or bottles, of compressed gases.

Drum cart: A wheeled cart that is used to transport heavier-weight 55-gallon drums/barrels.

Drum dolly: A wheeled circular platform or a caddy with a handle that is used to transport lighter-weight 55-gallon drums/barrels.

Freight elevator: An elevator used to transport materials from floor to floor.

Half hitch: A knot tied by passing the working end of a rope around an object, across the standing part of the rope, and then through the resulting loop; often used as an element in forming other knots or added to make other knots more secure.

Hand truck: Two-wheeled cart that is used to transport large, heavy loads; also known as a dolly.

Industrial forklift: A vehicle with a power-operated pronged platform that can be raised and lowered for insertion under a load to be lifted and moved.

Material cart: Four-wheeled device used to transport materials around a job site.

Pallet jack: A device used to lift and move heavy or stacked pallets; also known as a pallet truck.

Pipe mule: A two-wheeled device used to transport medium-length pieces of pipe, tubing, or scaffolding; sometimes referred to as a tunnel buggy.

Pipe transport: A wheeled device similar to a pipe mule, but used to move larger pieces of pipe.

Powered wheelbarrow: A vehicle similar to a manual wheelbarrow, but powered by an electric or gas motor; also known as a power buggy.

Roller skids: A device that includes a surface table and two, three, or four roller skids. Materials that are to be moved are placed on the table surface and then pushed on the skids.

Rough terrain forklift: Similar to an industrial forklift, but designed to be used on rough surfaces. Rough terrain forklifts are characterized by large pneumatic tires, usually with deep treads that allow the vehicle to grab onto the roughest of roads or ground cover without sliding or slipping.

Spotter: A person who walks in front of another worker who is carrying or transporting a long load to ensure there is a clear, unobstructed path.

Square knot: A knot made of two reverse half-knots and typically used to join the ends of two ropes of similar diameters; also called a reef knot.

Standing end: The end of a rope that is not being knotted.

Standing part: The portion of a rope that is between the standing end and the working end.

Wheelbarrow: A one- or two-wheeled vehicle with handles at the rear that is used to carry small loads.

Work zone: The area in which a forklift may come in contact with objects or people, either with the rear end or the front forks.

Working end: The end of a rope that is being used to tie a knot.

NCCER – *Core Curriculum* 00109-15

Additional Resources

This module presents a thorough resource for task training. The following resource material is suggested for further study.

Materials Handling Handbook, The American Society of Mechanical Engineers (ASME) and The International Material Management Society (IMMS), Raymond A. Kulwiec, Editor-in-Chief. 1985. New York, NY: Wiley-Interscience.

Manufacturing Facilities Design & Material Handling, Matthew P. Stevens, Fred E. Meyers. 2013. West Lafayette, IN: Purdue University Press.

Knots: The Complete Visual Guide, Des Pawson. 2012. New York, NY: DK Publishing.

Simple Solutions Ergonomics for Construction Workers, US Centers for Disease Control, National Institute for Occupational Safety and Health. Last modified August 2007. **http://www.cdc.gov/niosh/docs/2007-122/**

Figure Credits

Courtesy of Terex Aerial Work Platforms, Module opener

Northern Tool and Equipment, Figures 19, 27

Vestil Manufacturing, Figures 20, 21, 22

Cherry's Industrial Equipment Corp., Figure 23

GAR-BRO Manufacturing Co., Heber Springs, Arkansas, Figure 24

BE & K, Figure 25

Sumner, Figure 26

Power Pusher Company, A division of Nu-Star Inc., SA01

Plug Power, Inc., SA02

Power Barrows, Figure 28

Multiquip Inc., Figure 29

Cianbro Corporation, Figure 30

Taylor Machine Works, Figures 31, 32

Section Review Answer Key

Answer	Section Reference	Objective
Section One		
1. a	1.1.1	1a
2. c	1.2.1	1b
3. d	1.3.1	1c
Section Two		
1. d	2.1.2	2a
2.c	2.2.2	2b

NCCER CURRICULA — USER UPDATE

NCCER makes every effort to keep its textbooks up-to-date and free of technical errors. We appreciate your help in this process. If you find an error, a typographical mistake, or an inaccuracy in NCCER's curricula, please fill out this form (or a photocopy), or complete the online form at **www.nccer.org/olf**. Be sure to include the exact module ID number, page number, a detailed description, and your recommended correction. Your input will be brought to the attention of the Authoring Team. Thank you for your assistance.

Instructors – If you have an idea for improving this textbook, or have found that additional materials were necessary to teach this module effectively, please let us know so that we may present your suggestions to the Authoring Team.

NCCER Product Development and Revision
13614 Progress Blvd., Alachua, FL 32615

Email: curriculum@nccer.org
Online: www.nccer.org/olf

❏ Trainee Guide   ❏ Lesson Plans   ❏ Exam   ❏ PowerPoints   Other _____

Craft / Level: _____   Copyright Date: _____

Module ID Number / Title: _____

Section Number(s): _____

Description: _____

_____

_____

_____

Recommended Correction: _____

_____

_____

_____

Your Name: _____

Address: _____

_____

Email: _____   Phone: _____

Glossary

Abrasive: A substance, such as sandpaper, that is used to wear away material.

Absenteeism: Consistent failure to show up for work.

Accident: Per the US Occupational Safety and Health Administration (OSHA), an unplanned event that results in personal injury or property damage.

Active listening: A process that involves respecting others, listening to what is being said, and understanding what is being said.

Acute angle: Any angle between 0 degrees and 90 degrees.

Adjacent angles: Angles that have the same vertex and one side in common.

Adjustable wrench: A smooth-jawed wrench with an adjustable, moveable jaw used for turning nuts and bolts. Often referred to as a Crescent® wrench due to brand recognition.

Alternating current (AC): The common power supplied to most all wired devices, where the current reverse its direction may times per second. AC power is the type of power generated and distributed throughout settled areas.

Amphetamine: A class of drugs that causes mental stimulation and feelings of euphoria.

Angle: The shape made by two straight lines coming together at a point. The space between those two lines is measured in degrees.

Appendix: A source of detailed or specific information placed at the end of a section, a chapter, or a book.

Arbor: The end of a circular saw shaft where the blade is mounted.

Arc welding: The joining of metal parts by fusion, in which the necessary heat is produced by means of an electric arc.

Architect: A qualified, licensed person who creates and designs drawings for a construction project.

Architect's scale: A specialized ruler used in making or measuring reduced scale drawings. The ruler is marked with a range of calibrated ratios for laying out distances, with scales indicating feet, inches, and fractions of inches. Used on drawings other than site plans.

Architectural plans: Drawings that show the design of the project. Also called architectural drawings.

Area: The surface or amount of space occupied by a two-dimensional object such as a rectangle, circle, or square.

Auger bit: A drill bit with a spiral cutting edge for boring holes in wood and other materials.

Ball-peen hammer: A hammer with a flat face that is used to strike cold chisels and punches. The rounded end—the peen—is used to bend and shape soft metal.

Barbiturate: A class of drugs that induces relaxation, slowing the body's ability to react.

Base: As it relates to triangles, the base is the line forming the bottom of the triangle.

Beam: A large, horizontal structural member made of concrete, steel, stone, wood, or other structural material to provide support above a large opening.

Bell-faced hammer: A claw hammer with a slightly rounded, or convex, face.

Bevel: To cut on a slant at an angle that is not a right angle (90-degree). The angle or inclination of a line or surface that meets another at any angle but 90-degree.

Bisect: To divide something into two parts that are often equal. When an angle is bisected for example, the two resulting angles are equal.

Block and tackle: A simple rope-and-pulley system used to lift loads.

Blueprints: The traditional name used to describe construction drawings.

Body language: A person's facial expression, physical posture, gestures, and use of space, which communicate feelings and ideas.

Bowline: A knot used to form a loop that neither slips nor jams; sometimes referred to as a rescue knot or the king of knots.

Box-end wrench: A wrench, usually double-ended, that has a closed socket that fits over the head of a bolt.

Brazing: A process using heat in excess of 800°F (427°C) to melt a filler metal that is drawn into a connection. Brazing is commonly used to join copper pipe.

Bridle: A configuration using two or more slings to connect a load to a single hoist hook.

Bull ring: A single ring used to attach multiple slings to a hoist hook.

Bullets: Large, vertically aligned dots that highlight items in a list.

Bullying: Unwanted, aggressive behavior that involves a real or perceived power imbalance. This form of harassment may include offensive, persistent, insulting, or physically threatening behavior directed at an individual.

Cannabinoids: A diverse category of chemical substances that repress neurotransmitter releases in the brain. Cannabinoids have a variety of sources; some are created naturally by the human body, while others come from cannabis (marijuana). Still others are synthetic.

Capsize: To change the form and rearrange the parts of a knot, usually by pulling on specific ends of the knot.

Carbide: A very hard material made of carbon and one or more heavy metals. Commonly used in one type of saw blade.

Carpenter's square: A flat, steel square commonly used in carpentry.

Cat's paw: A straight steel rod with a curved claw at one end that is used to pull nails that have been driven flush with the surface of the wood or slightly below it.

Change order: A written order by the owner of a project for the contractor to make a change in time, amount, or specifications.

Chisel: A metal tool with a sharpened, beveled edge used to cut and shape wood, stone, or metal.

Chisel bar: A tool with a claw at each end, commonly used to pull nails.

Chuck: A clamping device that holds an attachment; for example, the chuck of the drill holds the drill bit.

Chuck key: A small, T-shaped steel piece used to open and close the chuck on power drills.

Circle: A closed curved line around a central point. A circle measures 360 degrees.

Circumference: The distance around the curved line that forms a circle.

Civil plans: Drawings that show the location of the building on the site from an aerial view, including contours, trees, construction features, and dimensions.

Claw hammer: A hammer with a flat striking face. The other end of the head is curved and divided into two claws to remove nails.

Clove hitch: A knot that consists of two half hitches made in opposite directions; used to temporarily secure a rope to an object.

Combination square: An adjustable carpenter's tool consisting of a steel rule that slides through an adjustable head.

Combination wrench: A wrench with an open end and a closed end.

Combustible: Capable of easily igniting and rapidly burning; used to describe a fuel with a flash point at or above 100°F (38°C).

Competent person: An individual capable of identifying existing and predictable hazards in the surroundings and working conditions which are unsanitary, hazardous, or dangerous to employees; an individual who is authorized to take prompt corrective measures to eliminate these issues.

Compromise: When people involved in a disagreement make concessions to reach a solution that everyone agrees on.

Computer-aided drafting (CAD): The making of a set of construction drawings with the aid of a computer.

Concrete mule: A wheeled device used when a concrete pour is in a location that a concrete delivery truck or pump cannot reach; sometimes referred to as a Georgia buggy.

Confidentiality: Privacy of information.

Confined space: A work area large enough for a person to work in, but with limited means of entry and exit and not designed for continuous occupancy. Tanks, vessels, silos, pits, vaults, and hoppers are examples of confined spaces. Also see *permit-required confined space*.

Constructive criticism: A positive offer of advice intended to help someone correct mistakes or improve actions.

Contour lines: Solid or dashed lines showing the elevation of the earth on a civil drawing.

Core: Center support member of a wire rope around which the strands are laid.

Countersink: A bit or drill used to set the head of a screw at or below the surface of the material.

Cross-bracing: Braces (metal or wood) placed diagonally from the bottom of one rail to the top of another rail to add support to a structure.

Cube: A three-dimensional square, with the measurements in all three dimensions being equal.

Cylinder cart: A two-wheeled cart that is used to transport cylinders, or bottles, of compressed gases.

Decimal: A part of a number represented by digits to the right of a point, called a decimal point. For example, in the number 1.25, .25 is the decimal portion of the number. In this case, it represents 25% of the whole number 1.

Degree: A unit of measurement for angles. For example, a right angle is 90 degrees, an acute angle is between 0 and 90 degrees, and an obtuse angle is between 90 and 180 degrees.

Denominator: The part of a fraction below the dividing line. For example, the 2 in ½ is the denominator. It is equivalent to the divisor in a long division problem.

Detail drawings: Enlarged views of part of a drawing used to show an area more clearly.

Diagonal: Line drawn from one corner of a rectangle or square to the farthest opposite corner.

Diameter: The length of a straight line that crosses from one side of a circle, through the center point, to a point on the opposite side. The diameter is the longest straight line you can draw inside a circle.

Difference: The result of subtracting one number from another. For example, in the problem 8 − 3 = 5, 5 is the difference between the two numbers.

Digit: Any of the numerical symbols 0 to 9.

Dimension line: A line on a drawing with a measurement indicating length.

Direct current (DC): An electric power supply where the current flows in one direction only. DC power is supplied by batteries and by transformer-rectifiers that change AC power to DC.

Dividend: In a division problem, the number being divided is the dividend.

Divisor: In a division problem, the number that is divided into another number is called the divisor.

Dowel: A pin, usually round, that fits into a corresponding hole to fasten or align two pieces.

Drum cart: A wheeled cart that is used to transport heavier-weight 55-gallon drums/barrels.

Drum dolly: A wheeled circular platform or a caddy with a handle that is used to transport lighter-weight 55-gallon drums/barrels.

Electrical plans: Engineered drawings that show all electrical supply and distribution.

Electronic signature: A signature that is used to sign electronic documents by capturing handwritten signatures through computer technology and attaching them to the document or file.

Elevation (EL): Height above sea level, or other defined surface, usually expressed in feet or meters.

Elevation drawing: Side view of a building or object, showing height and width.

Engineer: A person who applies scientific principles in design and construction.

Engineer's scale: A straightedge measuring device divided uniformly into multiples of 10 divisions per inch so that drawings can be made with decimal values. Used mainly for land measurements on site plans.

Equation: A mathematical statement that indicates that the value of two mathematical expressions, such as 2×2 and 1×4, are equal. An equation is written using the equal sign in this manner: $2 \times 2 = 1 \times 4$.

Equilateral triangle: A triangle that has three equal sides and three equal angles.

Equivalent fractions: Fractions having different numerators and denominators but still have equal values, such as the two fractions ½ and ¼.

Excavation: Any man-made cut, cavity, trench, or depression in an earth surface, formed by removing earth. It can be made for anything from basements to highways. Also see *trench*.

Fastener: A device such as a bolt, clasp, hook, or lock used to attach or secure one material to another.

Fire protection plan: A drawing that shows the details of the building's sprinkler system.

Flammable: Capable of easily igniting and rapidly burning; used to describe a fuel with a flash point below 100°F (38°C).

Flash burn: The damage that can be done to eyes after even brief exposure to ultraviolet light from arc welding. A flash burn requires medical attention.

Flash point: The temperature at which fuel gives off enough gases (vapors) to burn.

Flats: The straight sides or jaws of a wrench opening. Also, the sides on a nut or bolt head.

Floor plan: A drawing that provides an aerial view of the layout of each room.

Font: The type style used for letters and numbers.

Foot-pounds: Unit of measure used to describe the amount of pressure exerted (torque) to tighten a large object.

Force: A push or pull on a surface. In this module, force is considered to be the weight of an object or fluid. This is a common approximation.

Formula: A mathematical process used to solve a problem. For example, the formula for finding the area of a rectangle is side A times side B = Area, or $A \times B = Area$.

Forstner bit: A bit designed for use in wood or similar soft material. The design allows it to drill a flat-bottom blind hole in material.

Foundation plan: A drawing that shows the layout and elevation of the building foundation.

Fraction: A portion of a whole number represented by two numbers. The upper number of a fraction is known as the numerator and the bottom number is known as the denominator.

Freight elevator: An elevator used to transport materials from floor to floor.

Glossary: An alphabetical list of terms and definitions.

Graph: Information shown as a picture or chart. Graphs may be represented in various forms, including line graphs and bar charts.

Grit: A granular, sand-like material used to make sandpaper and similar materials abrasive. Grit is graded according to its texture. The grit number indicates the number of abrasive granules in a standard size (per inch or per cm). The higher the grit number, the more particles in a given area, indicating a finer abrasive material.

Ground: The conducting connection between electrical equipment or an electrical circuit and the Earth.

Ground fault: An unintentional, electrically conducting connection between an ungrounded conductor of an electrical circuit and the normally noncurrent-carrying conductors, metal objects, or the Earth.

Ground fault circuit interrupter (GFCI): A circuit breaker designed to protect people from electric shock and to protect equipment from damage by interrupting the flow of electricity if a circuit fault occurs; a device that interrupts and de-energizes an electrical circuit to protect a person from electrocution.

Ground fault protection: Protection against short circuits; a safety device cuts power off as soon as it senses any imbalance between incoming and outgoing current.

Guarded: Enclosed, fenced, covered, or otherwise protected by barriers, rails, covers, or platforms to prevent dangerous contact.

Half hitch: A knot tied by passing the working end of a rope around an object, across the standing part of the rope, and then through the resulting loop; often used as an element in forming other knots or added to make other knots more secure.

Hallucinogen: A class of drugs that distort the perception of reality and cause hallucinations.

Hand line: A line attached to a tool or object so a worker can pull it up after climbing a ladder or scaffold.

Hand truck: Two-wheeled cart that is used to transport large, heavy loads; also known as a dolly.

Harassment: A type of discrimination that can be based on race, age, disabilities, sex, religion, cultural issues, health, or language barriers.

Hazard Communication Standard (HAZCOM): The standard that requires contractors to educate employees about hazardous chemicals on the job site and how to work with them safely.

Hex key wrench: A hexagonal steel bar that is bent to form a right angle. Often referred to as an Allen® wrench.

Hidden line: A dashed line showing an object obstructed from view by another object.

Hitch: The rigging configuration by which a sling connects the load to the hoist hook. The three basic types of hitches are vertical, choker, and basket.

Hoist: A device that applies a mechanical force for lifting or lowering a load.

HVAC: Heating, ventilating, and air conditioning.

Hydraulic: Powered by fluid under pressure.

Improper fraction: A fraction whose numerator is larger than its denominator. For example, ¾ and ⅔ are improper fractions.

Inch-pounds: Unit of measure used to describe the amount of pressure exerted (torque) to tighten a small object.

Incident: Per the US Occupational Safety and Health Administration (OSHA), an unplanned event that does not result in personal injury but may result in property damage or is worthy of recording.

Index: An alphabetical list of topics, along with the page numbers where each topic appears.

Industrial forklift: A vehicle with a power-operated pronged platform that can be raised and lowered for insertion under a load to be lifted and moved.

Initiative: The ability to work without constant supervision and solve problems independently.

Invert: To reverse the order or position of numbers. In fractions, inverting means to reverse the positions of the numerator and denominator, such that ¾ becomes ⁴⁄₃. When you are dividing by fractions, one fraction is inverted.

Isosceles triangle: A triangle that has two equal sides and two equal angles.

Italics: Letters and numbers that lean to the right rather than stand straight up.

Jargon: Specialized terms used in a specific industry.

Joint: The point where members or the edges of members are joined. The types of welding joints are butt joint, corner joint, and T-joint.

Joist: Lengths of wood or steel that usually support floors, ceiling, or a roof. Roof joists will be at the same angle as the roof itself, while floor and ceiling joists are usually horizontal.

Kerf: A cut or channel made by a saw; the channel created by a saw blade passing through the material, which is equal to the width of the blade teeth.

Lanyard: A short section of rope or strap, one end of which is attached to a worker's safety harness and the other to a strong anchor point above the work area.

Leader: In drafting, the line on which an arrowhead is placed and used to identify a component.

Leadership: The ability to set an example for others to follow by exercising authority and responsibility.

Legend: A description of the symbols and abbreviations used in a set of drawings.

Level: Perfectly horizontal; completely flat; also, a tool used to determine if an object is level.

Lifting clamp: A device used to move loads such as steel plates or concrete panels without the use of slings.

Load: The total amount of what is being lifted, including all slings, hitches, and hardware.

Load control: The safe and efficient practice of load manipulation, using proper communication and handling techniques.

Load stress: The strain or tension applied on the rigging by the weight of the suspended load.

Loadbearing: Carrying a significant amount of weight and/or providing necessary structural support. A loadbearing wall typically carries some portion of the roof weight and cannot be removed without risking structural failure or collapse.

Lockout/tagout: A formal procedure for taking equipment out of service and ensuring that it cannot be operated until a qualified person has removed the lock and/or warning tag.

Management system: The organization of a company's management, including reporting procedures, supervisory responsibility, and administration.

Masonry bit: A drill bit with a carbide tip designed to penetrate materials such as stone, brick, or concrete.

Mass: The quantity of matter present.

Master link: The main connection fitting for chain slings.

Material cart: Four-wheeled device used to transport materials around a job site.

Maximum intended load: The total weight of all people, equipment, tools, materials, and loads that a ladder can hold at one time.

Mechanical plans: Engineered drawings that show the mechanical systems, such as motors and piping.

Memo: Informal written correspondence. Another term for memorandum (plural: *memoranda*).

Methamphetamine: A highly addictive crystalline drug, derived from amphetamines, that affects the central nervous system.

Metric scale: A straightedge measuring device divided into centimeters, with each centimeter divided into 10 millimeters. Usually used for architectural drawings and sometimes referred to as a metric architect's scale.

Midrail: Mid-level, horizontal board required on all open sides of scaffolds and platforms that are more than 14 inches (35 cm) from the face of the structure and more than 10 feet (3.05 m) above the ground. It is placed halfway between the toeboard and the top rail.

Mission statement: A statement of how a company does business.

Miter joint: A joint made by fastening together usually perpendicular parts with the ends cut at an angle.

Mixed number: A combination of a whole number with a fraction or decimal. Examples of mixed numbers are 3⁷⁄₁₆, 5.75, and 1¼.

Nail puller: A tool used to remove nails.

Negative numbers: Numbers less than zero. For example, –1, –2, and –3 are negative numbers.

Newton-meter: A measure of torque or moment equal to the force of one Newton applied to a lever one meter long.

Nonverbal communication: All communication that does not use words. This includes tone of voice, appearance, personal environment, use of time, and body language.

Not to scale (NTS): Describes drawings that show relative positions and sizes only, without scale.

Numerator: The part of a fraction above the dividing line. For example, the 1 in ½ is the numerator. It is the equivalent of the dividend in a long division problem.

Obtuse angle: Any angle between 90 degrees and 180 degrees.

Occupational Safety and Health Administration (OSHA): An agency of the US Department of Labor. Also refers to the Occupational Safety and Health Act of 1970, a law that applies to more than more than 111 million workers and 7 million job sites in the country.

One-rope lay: The lengthwise distance it takes for one strand of a wire rope to make one complete turn around the core.

Open-end wrench: A non-adjustable wrench with a fixed opening at each end that is typically different, allowing it to be used to fit two different nut or bolts sizes.

Opiates: A narcotic painkiller derived from the opium poppy plant or synthetically manufactured. Heroin is the most commonly used opioid.

Opposite angles: Two angles that are formed by two straight lines crossing. They are always equal.

Pallet jack: A device used to lift and move heavy or stacked pallets; also known as a pallet truck.

Paraphrase: Express something heard or read using different words.

Peening: The process of bending, shaping, or cutting material by striking it with a tool.

Perimeter: The distance around the outside of a closed shape, such as a rectangle, circle, square, or any irregular shape.

Permit: A legal document that allows a task to be undertaken.

Permit-required confined space: A confined space that has been evaluated and found to have actual or potential hazards, such as a toxic atmosphere or other serious safety or health hazard. Workers need written authorization to enter a permit-required confined space. Also see *confined space*.

Personal protective equipment (PPE): Equipment or clothing designed to prevent or reduce injuries.

Pi: A mathematical value of approximately 3.14 (or ²²⁄₇) used to determine the area and circumference of circles. It is sometimes symbolized by π.

Pipe mule: A two-wheeled device used to transport medium-length pieces of pipe, tubing, or scaffolding; sometimes referred to as a tunnel buggy.

Pipe transport: A wheeled device similar to a pipe mule, but used to move larger pieces of pipe.

Pipe wrench: A wrench for gripping and turning a pipe or pipe-shaped object; it tightens when turned in one direction.

Piping and instrumentation drawings (P&IDs): Schematic diagrams of a complete piping system.

Place value: The exact value a digit represents in a whole number, determined by its place within the whole number or by its position relative to the decimal point. In the number 124, the number 2 represents 20, since it is in the tens position.

Plane: A surface in which a straight line joining two points lies wholly within that surface.

Plane geometry: The mathematical study of two-dimensional (flat) shapes.

Planed: Describing a surface made smooth by using a tool called a plane.

Planked: Having pieces of material 2 inches (5 cm) thick or greater and 6 inches (15 cm) wide or greater used as flooring, decking, or scaffold decks.

Pliers: A scissor-shaped type of adjustable wrench equipped with jaws and teeth to grip objects.

Plumb: Perfectly vertical; the surface is at a right angle (90 degrees) to the horizon or floor and does not bow out at the top or bottom.

Plumbing isometric drawing: A type of three-dimensional drawing that depicts a plumbing system.

Plumbing plans: Engineered drawings that show the layout for the plumbing system.

Pneumatic: Powered by air pressure, such as a pneumatic tool.

Points: Teeth on the gripping part of a wrench. Also refers to the number of teeth per inch on a handsaw.

Positive numbers: Numbers greater than zero. For example, 1, 2, and 3 are positive numbers. Any number without a negative (−) sign in front of it is considered to be a positive number.

Powered wheelbarrow: A vehicle similar to a manual wheelbarrow, but powered by an electric or gas motor; also known as a power buggy.

Product: The answer to a multiplication problem. For example, the product of 6 × 6 is 36.

Professionalism: Integrity and work-appropriate manners.

Proximity work: Work done near a hazard but not actually in contact with it.

Punch: A steel tool used to indent metal.

Punch list: A written list that identifies deficiencies requiring correction at completion.

Qualified person: A person who, through the possession of a recognized degree, certificate, or professional standing, or one who has gained extensive knowledge, training, and experience, has successfully demonstrated his or her ability to solve problems relating to the subject matter, the work, or the project.

Quotient: The result of a division problem. For example, when dividing 6 by 2, the quotient is 3.

Radius: The distance from a center point of a circle to any point on the curved line, or half the width (diameter) of a circle.

Rafter angle square: A type of carpenter's square made of cast aluminum that combines a protractor, try square, and framing square.

Rated capacity: The maximum load weight a sling or piece of hardware or equipment can hold or lift; also referred to as the working load limit (WLL).

Reciprocating: Moving backward and forward on a straight line.

Rectangle: A four-sided shape with four 90-degree angles. Opposite sides of a rectangle are always parallel and the same length. Adjacent sides are perpendicular and are not equal in length.

Reference: A person who can confirm to a potential employer that you have the skills, experience, and work habits that are listed in your resume.

Rejection criteria: Standards, rules, or tests on which a decision can be based to remove an object or device from service because it is no longer safe.

Remainder: The amount left over in a division problem. For example, in the problem 34 ÷ 8, 8 goes into 34 four times (8 × 4 = 32) with 2 left over; in other words, 2 is the remainder.

Respirator: A device that provides clean, filtered air for breathing, no matter what is in the surrounding air.

Revolutions per minute (rpm): The rotational speed of a motor or shaft, based on the number of times it rotates each minute.

Rigging hook: An item of rigging hardware used to attach a sling to a load.

Right angle: An angle that measures 90 degrees. The two lines that form a right angle are perpendicular to each other. This is the angle used most in the trades.

Right triangle: A triangle that includes one 90-degree angle.

Ring test: A method of testing the condition of a grinding wheel. The wheel is mounted on a rod and tapped. A clear ring means the wheel is in good condition; a dull thud means the wheel is in poor condition and should be disposed of.

Ripping bar: A tool used for heavy-duty dismantling of woodwork, such as tearing apart building frames or concrete forms.

Roller skids: A device that includes a surface table and two, three, or four roller skids. Materials that are to be moved are placed on the table surface and then pushed on the skids.

Roof plan: A drawing of the view of the roof from above the building.

Rough terrain forklift: Similar to an industrial forklift, but designed to be used on rough surfaces. Rough terrain forklifts are characterized by large pneumatic tires, usually with deep treads that allow the vehicle to grab onto the roughest of roads or ground cover without sliding or slipping.

Round off: To smooth out threads or edges on a screw or nut.

Safety culture: The culture created when the whole company sees the value of a safe work environment.

Safety data sheet (SDS): A document that must accompany any hazardous substance. The SDS identifies the substance and gives the exposure limits, the physical and chemical characteristics, the kind of hazard it presents, precautions for safe handling and use, and specific control measures.

Scaffold: An elevated platform for workers and materials.

Scale: The ratio between the size of a drawing of an object and the size of the actual object.

Scalene triangle: A triangle with sides of unequal lengths.

Schematic: A one-line drawing showing the flow path for electrical circuitry or the relationship of all parts of a system.

Section drawing: A cross-sectional view of a specific location, showing the inside of an object or building.

Self-presentation: The way a person dresses, speaks, acts, and interacts with others.

Sexual harassment: A type of discrimination that results from unwelcome sexual advances, requests, or other verbal or physical behavior with sexual overtones.

Shackle: Coupling device used in an appropriate lifting apparatus to connect the rope to eye fittings, hooks, or other connectors.

Shank: The smooth part of a drill bit that fits into the chuck.

Sheave: A grooved pulley-wheel for changing the direction of a rope's pull; often found on a crane.

Shielding: A structure used to protect workers in trenches but lacking the ability to prevent cave-ins.

Shoring: Using pieces of timber, usually in a diagonal position, to hold a wall in place temporarily.

Signaler: A person who is responsible for directing a vehicle when the driver's vision is blocked in any way.

Six-foot rule: A rule stating that platforms or work surfaces with unprotected sides or edges that are 6 feet (1.83 m) or higher than the ground or level below it require fall protection.

Sling: Wire rope, alloy steel chain, metal mesh fabric, synthetic rope, synthetic webbing, or jacketed synthetic continuous loop fibers made into forms, with or without end fittings, used to handle loads.

Sling legs: The parts of the sling that reach from the attachment device around the object being lifted.

Sling reach: A measure taken from the master link of the sling, where it bears weight, to either the end fitting of the sling or the lowest point on the basket.

Solid geometry: The mathematical study of three-dimensional shapes.

Specifications: Precise written presentation of the details of a plan.

Splice: To join together.

Spoil: Material such as earth removed while digging a trench or excavation.

Spotter: A person who walks in front of another worker who is carrying or transporting a long load to ensure there is a clear, unobstructed path.

Square: Exactly adjusted; any piece of material sawed or cut to be rectangular with equal dimensions on all sides; a tool used to check angles.

Square: (1) A special type of rectangle with four equal sides and four 90-degree angles. (2) The product of a number multiplied by itself. For example, 25 is the square of 5; 16 is the square of 4.

Square knot: A knot made of two reverse half-knots and typically used to join the ends of two ropes of similar diameters; also called a reef knot.

Standing end: The end of a rope that is not being knotted.

Standing part: The portion of a rope that is between the standing end and the working end.

Straight angle: A 180-degree angle or flat line.

Strand: A group of wires wound, or laid, around a center wire, or core. Strands are laid around a supporting core to form a rope.

Striking (or slugging) wrench: A non-adjustable wrench with an enclosed, circular opening designed to lock on to the fastener when the wrench is struck.

Strip: To damage the head or threads on a screw, nut, or bolt.

Structural plans: A set of engineered drawings used to support the architectural design.

Stud: A vertical support inside the wall of a structure to which the wall finish material is attached. The base of a stud rests on a horizontal baseplate, and a horizontal cap plate rests on top of a series of studs.

Sum: The resulting total in an addition problem. For example, in the problem $7 + 8 = 15$, 15 is the sum.

Symbol: A drawing that represents a material or component on a plan.

Synthetic drugs: A drug with properties and effects similar to known substances but having a slightly altered chemical structure. Such drugs are often not illegal since they are somewhat different than well-defined restricted or illegal substances. The two typical categories are cannabinoids (lab-produced THC or marijuana substitutes) and cathinones, which are designed to mimic the effects of cocaine or methamphetamines.

Table: A way to present important text and numbers so they can be read and understood at a glance.

Table of contents: A list of book chapters or sections, usually located at the front of the book.

Tactful: Being aware of the effects of your statements and actions on others.

Tag line: Rope that runs from the load to the ground. Riggers hold on to tag lines to keep a load from swinging or spinning during the lift.

Tang: Metal handle-end of a file. The tang fits into a wooden or plastic file handle.

Tardiness: Habitually showing up late for work.

Tattle-tail: Cord attached to the strands of an endless loop sling. It protrudes from the jacket. A tattle-tail is used to determine if an endless sling has been stretched or overloaded.

Tempered: Treated with heat to create or restore hardness in steel.

Tenon: A piece that projects out of wood or another material for the purpose of being placed into a hole or groove to form a joint.

Threaded shank: A connecting end of a fastener, such as a bolt, with a series of spiral grooves cut into it. The grooves are designed to mate with grooves cut into another object in order to join them together.

Title block: A part of a drawing sheet that includes some general information about the project.

Toeboard: A vertical barrier at floor level attached along exposed edges of a platform, runway, or ramp to prevent materials and people from falling.

Top rail: A top-level, horizontal board required on all open sides of scaffolds and platforms that are more than 14 inches (36 cm) from the face of the structure and more than 10 feet (3 m) above the ground.

Torque: A rotating or twisting force applied to an object such as a nut, bolt, or screw, using a socket wrench or screwdriver. Torque wrenches allow a specific torque value to be set and applied.

Trench: A narrow excavation made below the surface of the ground that is generally deeper than it is wide, with a maximum width of 15 feet (4.6 m). Also see *excavation*.

Triangle: A closed shape that has three sides and three angles.

Trigger lock: A small lever, switch, or part that can be used to activate a locking catch or spring to hold a power tool trigger in the operating mode without finger pressure.

Try square: A square whose legs are fixed at a right angle.

Unit: A definite standard of measure.

Unstranding: Describes wire rope strands that have become untwisted. This weakens the rope and makes it easier to break.

Vertex: A point at which two or more lines or curves come together.

Volume: The amount of space contained in a given three-dimensional shape.

Warning yarn: A component of the sling that shows the rigger whether the sling has suffered too much damage to be used.

Weld: To heat or fuse two or more pieces of metal so that the finished piece is as strong as the original; a welded joint.

Welding curtain: A protective screen set up around a welding operation designed to safeguard workers not directly involved in that operation.

Wheelbarrow: A one- or two-wheeled vehicle with handles at the rear that is used to carry small loads.

Whole numbers: Complete number units without fractions or decimals.

Wind sock: A cloth cone open at both ends mounted in a high place to show which direction the wind is blowing.

Wire rope: A rope made from steel wires that are formed into strands and then laid around a supporting core to form a complete rope; sometimes called cable.

Work ethic: Work habits that are the foundation of a person's ability to do his or her job.

Work zone: The area in which a forklift may come in contact with objects or people, either with the rear end or the front forks.

Working end: The end of a rope that is being used to tie a knot.

Zero tolerance: The policy of applying laws or penalties to even minor infringements of a code in order to reinforce its overall importance, typically related to drug and alcohol abuse when applied to the workplace.

Index

A

Attitudes affecting safety, (00101):6
Auger bit, (00104):1, 2, 43
Automated guided vehicles (AGVs), (00109):23–24

B

Back injuries, (00109):1, 2
Ball-peen hammer, (00103):1, 49
Band clamp, (00103):40
Banner Bank Building, (00108):13
Barbiturate, (00108):15, 23, 35
Bar clamp, (00103):40
Barricades, (00101):17, 56
Barriers, (00101):17
Base, (00102):48, 51, 71
Base-twelve systems, (00102):21
Basket hitches, (00106):27, 29
Batteries, recycling, (00104):9
Beam grabs, (00101):26, 27
Beams, (00105):1, 9, 50
Bell-faced hammer, (00103):1, 49
Bench grinders, (00104):24–25, 26, 27–28
Benching systems, trench protection, (00101):44, 45
Bevel, (00103):1, 49
Beveled, (00103):6
BIM. *See* Building Information Modeling (BIM)
Biodiesel, (00109):22
Biological hazards, (00101):70–71
Biometric identification, (00108):6
Birdcaging, (00106):15
Bisect, (00102):48, 51, 71
Block and tackle, (00106):1, 20, 22, 35
Bloodborne pathogen exposure, (00101):68
Blue collar, (00108):1
Blueprints, (00105):1, 50
Body language, (00107):1, 7, 29
Bonda, Mark, (00108):34
Boredom, effect on communication, (00107):8
Bowline, (00109):1, 6, 32
Box-end wrench, (00103):1, 10–11, 49
Brazil. *See* Green Building Council (GBC), Brazil
Brazing, (00101):67, 75, 98
Break lines, (00105):31
Bricklayers, (00103):22
Bridle, (00106):1, 8, 35
Bridle hitches, (00106):24
Building Information Modeling (BIM), (00105):3
Built-up scaffolds, (00101):35, 37
Bullets, (00107):12, 29
Bull ring, (00106):1, 24, 25, 35
Bullying, (00108):15, 21–22, 35
Burr bits, (00104):29

C

CAD. *See* Computer-aided drafting (CAD)
Calculators, using, (00102):21
Cannabinoids, (00108):15, 24, 35
Capacity ratings, ladders, (00101):27, 29
Capsize, (00109):1, 6, 32
Carabiners, PFAS, (00101):23–24
Carbide, (00104):1, 2, 43
Carpenter's square, (00103):19, 24, 49
Catastrophe inspections, OSHA, (00101):95
Cathinones, (00108):23
Cat's paw, (00103):1, 4, 49
Caught-in/caught-between hazards
   cranes, (00101):47–48
   death from, (00101):3, 4, 10, 43–48
   defined, (00101):3

excavations, (00101):43–46
   guards, tool and machine, (00101):46–47
   heavy equipment, (00101):47–48
   trenches, (00101):43–46
Caution signs, (00101):6
Cave-ins, (00101):43–44
C-clamp, (00103):39
Cell phone guidelines, (00107):9–10, 22–23
Cell phone recycling, (00107):11
Celsius to Fahrenheit conversions, (00102):45–46, 66, 68
Center lines, (00105):31
Center punch, (00103):7–8
CFR. *See Code of Federal Regulations (CFR)*
Chain falls, (00103):37–38
Chain hoists, (00106):20, 22–23
Chain shackles, (00106):16
Chain slings, (00106):8–9
Change order, (00107):12, 13, 29
Chemical splashes, (00101):68
Chest straps, PFAS, (00101):21–22
Chisel, (00103):1, 3, 49
Chisel bar, (00103):1, 4, 49
Choker hitches, (00106):24–26
Chuck, (00104):1, 3, 43
Chuck key, (00104):1, 3, 43
Circle
   area of, (00102):68
   circumference
      calculating, (00102):68
      defined, (00102):48, 51, 71
   defined, (00102):48, 49, 71
Circular saws, (00104):15–18
Circumference
   calculating, (00102):68
   defined, (00102):48, 51, 71
Civil plans, (00105):1, 4, 37, 50
Civil Rights Act of 1964, (00108):24
Civil Rights Act of 1991, (00108):24
Clamps, (00103):39–40
Claw hammer, (00103):1, 49
Click-type torque wrenches, (00103):14
Closed-mindedness, problem solving and, (00108):8
Clove hitch, (00109):1, 6, 32
Clutch-drive screwdriver, (00103):8
Coast Guard, US, (00108):6
Coating thickness, (00102):22
Codebooks, (00107):13
Code of Federal Regulations (CFR), (00101):94
Cold chisels, (00103):6–7
Cold stress, (00101):72–73
Combination square, (00103):19, 24, 49
Combination wrench, (00103):1, 10, 11, 49
Combustibles, (00101):1, 9, 77–78, 98
Come-alongs, (00103):38, (00106):22–23
Communication
   boredom and, (00107):8
   cell phone guidelines, (00107):9–10
   distractions to, (00107):8
   ego and, (00107):8
   emotion and, (00107):8, 22
   jargon, (00107):1, 2, 22, 29
   labels and signs
      danger, (00101):6
      fire extinguishers, (00101):80, 81
      hazardous material containers, (00101):69–70
      informational, (00101):5–6
      pictograms, (00101):69–70, 81
      radioactivity warnings, (00101):71

Welders, PPE for, (00101):60, 74–75
Welding curtain, (00101):67, 75, 99
WGBC. *See* World Green Building Council (WGBC)
Wheelbarrow
  defined, (00109):32
  non-motorized, (00109):17
  powered, (00109):14, 19–20, 32
White collar, (00108):1
Whole numbers
  adding, (00102):3
  defined, (00102):, 1, 72
  multiplying, (00102):4–5
  negative, (00102):1, 2, 72
  place values of, (00102):2
  positive, (00102):1, 2, 72
  reading, (00102):2
  subtracting, (00102):3
  zero (0), (00102):2, 21
Wide-body shackles, (00106):16–17
Williams, Chris, (00103):47–48
Wind sock, (00101):67, 99
Wire brushes, grinders, (00104):29
Wire rope, (00106):2, 35
Wire rope slings, (00106):3, 9–10, 13–15
Wood chisels, (00103):6–7
Wooden folding rule, (00103):19–20
Wooden ladders, (00101):29, 34
Work cages, (00101):35, 38
Worker complaint and referral inspections, OSHA, (00101):95–96
Work ethic, (00108):15, 16, 17, 35
Work habits affecting safety, (00101):6
Working end, (00109):1, 6, 32
Work zone, (00109):14, 21, 32
World Green Building Council (WGBC), (00105):14
Worm-drive saw, (00104):17
Wrecking bar, (00103):49
Wrenches
  adjustable, (00103):1, 10, 11–13, 49
  box-end, (00103):1, 10–11, 49
  click-type torque, (00103):14
  combination, (00103):1, 10, 11, 49
  Crescent®, (00103):11
  digital electronic torque, (00103):14
  hex-key, (00103):1, 8, 9, 49
  impact, (00103):(00104)11–13
  no-hub torque, (00103):14
  non-adjustable, (00103):10–13
  open-end, (00103):2, 10, 49
  pipe, (00103):2, 12, 49
  socket, (00103):13
  spud, (00103):12
  striking (slugging), (00103):2, 10, 11, 50
  torque, (00103):14–15

Writing on the job
  electronic signatures, (00107):12, 20, 29
  email, (00107):20–21
  good, characteristics of
    clarity and concision, (00107):19, 22–23
    considerate of the reader, (00107):20
    correctness and completeness, (00107):19
  hot work permit (example), (00107):17–18
  the journalist's questions, (00107):19
  texting, (00107):22–23
  text presentation, (00107):16
Writing process
  prewriting, (00107):16
  proofreading, (00107):19
  revisions, (00107):19
  rough draft, (00107):16
  taking breaks, (00107):16
Writing resumes, (00108):3
Writing skills, (00102):1, (00107):12–13

Y

Y-configured shock absorbing lanyard, (00101):25

Z

Zero (0), (00102):2, 21
Zero tolerance, (00108):15, 22, 35